SECOND EDITI

ethics

for the INFORMATION AGE

SECOND EDITION

ethics

for the INFORMATION AGE

Michael J. Quinn

Oregon State University

Boston San Francisco New York

London Toronto Sydney Tokyo Singapore Madrid

Mexico City Munich Paris Cape Town Hong Kong Montreal

Publisher Greg Tobin
Senior Acquisitions Editor Michael Hirsch
Production Supervisor Marilyn Lloyd
Editorial Assistant Lindsey Triebel
Cover Designer Joyce Cosentino Wells
Cover Image © 2005 Digital Vision / Pete Gardner
Marketing Manager Michelle Brown
Marketing Assistant Dana Lopreato
Project Management Windfall Software
Composition Windfall Software, using ZzTEX
Technical Illustration George Nichols
Copyeditor Richard Camp
Proofreaders Jennifer McClain & MaryEllen N. Oliver
Indexer Ted Laux
Prepress and Manufacturing Caroline Fell
Printer Courier Stoughton, Inc.

Access the latest information about Addison-Wesley titles from our World Wide Web site: http://www.aw-bc.com/computing

ISBN 0-321-37334-0
2 3 4 5 6 7 8 9 10—CRS—08 07 06

Brief Contents

Preface xix

1 Catalysts for Change 1

An Interview with Douglas Engelbart 51

2 Introduction to Ethics 53

An Interview with James Moor 103

3 Networking 107

An Interview with Jennifer Preece 153

4 Intellectual Property 155

An Interview with Wendy Seltzer 209

5 Privacy 211

An Interview with Ann Cavoukian 275

6 Computer and Network Security 279

An Interview with Matt Bishop 321

7 Computer Reliability 325

8 Work and Wealth 369

An Interview with Jerry Berman 411

9 Professional Ethics 415

An Interview with Paul Axtell 455

Appendix A: Plagiarism 459

Index 463

Contents

Preface xix

1 Catalysts for Change 1

1.1 Introduction 2

1.2 Milestones in Computing 6

1.2.1 Mechanical Adding Machines 6
1.2.2 The Analytical Engine 9
1.2.3 Boolean Algebra 10
1.2.4 Zuse's Z Series 11
1.2.5 Harvard Mark 1 12
1.2.6 Colossus 13
1.2.7 The Atanasoff-Berry Computer 13
1.2.8 ENIAC (Electronic Numerical Integrator and Computer) 13
1.2.9 Small-Scale Experimental Machine 14
1.2.10 First Commercial Computers 15
1.2.11 Transistor 17
1.2.12 Integrated Circuit 18
1.2.13 IBM System/360 19
1.2.14 Microprocessor 20

1.3 Milestones in Networking 22

1.3.1 Electricity and Electromagnetism 22
1.3.2 Telegraph 23
1.3.3 Telephone 24
1.3.4 Typewriter and Teletype 25
1.3.5 Radio 26
1.3.6 Television 27
1.3.7 Remote Computing 28
1.3.8 ARPANET 28
1.3.9 Email 29
1.3.10 Internet 29
1.3.11 NSFNET 31
1.3.12 Broadband 31

1.4 Milestones in Information Storage and Retrieval 31
1.4.1 Codex 31
1.4.2 Gutenberg's Printing Press 32
1.4.3 Newspapers 32
1.4.4 Hypertext 33
1.4.5 Personal Computers 34
1.4.6 Single-Computer Hypertext Systems 36
1.4.7 Networked Hypertext: World Wide Web 36
1.4.8 Search Engines 38

1.5 Information Technology Issues 39
Summary 40
Review Questions 42
Discussion Questions 46
In-Class Exercises 46
Further Reading 47
References 48

An Interview with **Douglas Engelbart** 51

2 Introduction to Ethics 53

2.1 Introduction 53
2.1.1 Defining Terms 54
2.1.2 Four Scenarios 55
2.1.3 Overview of Ethical Theories 59

2.2 Subjective Relativism 60
2.2.1 The Case for Subjective Relativism 60
2.2.2 The Case against Subjective Relativism 61

2.3 Cultural Relativism 62
2.3.1 The Case for Cultural Relativism 63
2.3.2 The Case against Cultural Relativism 64

2.4 Divine Command Theory 66
2.4.1 The Case for the Divine Command Theory 66
2.4.2 The Case against the Divine Command Theory 67

2.5 Kantianism 69
2.5.1 Good Will and the Categorical Imperative 69
2.5.2 Evaluating a Scenario Using Kantianism 71
2.5.3 The Case for Kantianism 72
2.5.4 The Case against Kantianism 73

2.6 **Act Utilitarianism** 74
2.6.1 Principle of Utility 74
2.6.2 Evaluating a Scenario Using Act Utilitarianism 75
2.6.3 The Case for Act Utilitarianism 76
2.6.4 The Case against Act Utilitarianism 77

2.7 **Rule Utilitarianism** 79
2.7.1 Basis of Rule Utilitarianism 79
2.7.2 Evaluating a Scenario Using Rule Utilitarianism 79
2.7.3 The Case for Rule Utilitarianism 80
2.7.4 The Case against Utilitarianism in General 81

2.8 **Social Contract Theory** 82
2.8.1 The Social Contract 82
2.8.2 Rawls's Theory of Justice 84
2.8.3 Evaluating a Scenario Using Social Contract Theory 87
2.8.4 The Case for Social Contract Theory 87
2.8.5 The Case against Social Contract Theory 89

2.9 **Comparing Workable Ethical Theories** 90

2.10 **Morality of Breaking the Law** 91
2.10.1 Social Contract Theory Perspective 91
2.10.2 Kantian Perspective 91
2.10.3 Rule Utilitarian Perspective 92
2.10.4 Act Utilitarian Perspective 93
2.10.5 Conclusion 93

Summary 94
Review Questions 95
Discussion Questions 97
In-Class Exercises 98
Further Reading 99
References 99

An Interview with **James Moor** 103

3 Networking 107

3.1 **Introduction** 107

3.2 **Email and Spam** 109
3.2.1 How Email Works 109
3.2.2 The Spam Epidemic 110
3.2.3 Ethical Evaluations of Spamming 112

3.3 Fighting Spam 115

3.3.1 Mail Abuse Prevention System 115
3.3.2 Ethical Evaluations of Blacklisting by MAPS 116
3.3.3 Proposed Solutions to the Spam Epidemic 117
3.3.4 CAN SPAM Act of 2003 118
3.3.5 Emergence of "Spim" 120

3.4 The World Wide Web 120

3.4.1 Attributes of the Web 120
3.4.2 How We Use the Web 121
3.4.3 Too Much Control or Too Little? 123

3.5 Ethical Perspectives on Pornography 124

3.5.1 Analyses Concluding Pornography Is Immoral 124
3.5.2 Analyses Concluding Adult Pornography Is Moral 125
3.5.3 Commentary 125
3.5.4 Summary 125

3.6 Censorship 126

3.6.1 Direct Censorship 126
3.6.2 Self-Censorship 127
3.6.3 Challenges Posed by the Internet 127
3.6.4 Ethical Perspectives on Censorship 128

3.7 Freedom of Expression 129

3.7.1 History 129
3.7.2 Freedom of Expression Not an Absolute Right 130
3.7.3 *FCC v. Pacifica Foundation et al.* 131

3.8 Children and the Web 132

3.8.1 Web Filters 132
3.8.2 Child Internet Protection Act 133
3.8.3 Ethical Evaluations of CIPA 133

3.9 Breaking Trust on the Internet 135

3.9.1 Identity Theft 135
3.9.2 Chat-Room Predators 137
3.9.3 Ethical Evaluations of Police "Sting" Operations 138
3.9.4 False Information 139

3.10 Internet Addiction 140

3.10.1 Is Internet Addiction Real? 140
3.10.2 Contributing Factors 142
3.10.3 Ethical Evaluation of Internet Addiction 143

Summary 144
Review Questions 145
Discussion Questions 146

In-Class Exercises 147
Further Reading 148
References 148

An Interview with **Jennifer Preece** 153

4 Intellectual Property 155

4.1 Introduction 155

4.2 Intellectual Property Rights 157

4.2.1 What Is Intellectual Property? 157
4.2.2 Property Rights 157
4.2.3 Extending the Argument to Intellectual Property 159
4.2.4 Benefits of Intellectual Property Protection 161
4.2.5 Limits to Intellectual Property Protection 161

4.3 Protecting Intellectual Property 163

4.3.1 Trade Secrets 163
4.3.2 Trademarks and Service Marks 164
4.3.3 Patents 165
4.3.4 Copyrights 165

4.4 Fair Use 169

4.4.1 *Sony v. Universal City Studios* 170
4.4.2 *RIAA v. Diamond Multimedia Systems Inc.* 171
4.4.3 Digital Technology and Fair Use 173

4.5 New Restrictions on Use 174

4.5.1 Digital Millennium Copyright Act 174
4.5.2 Digital Rights Management 175
4.5.3 Secure Digital Music Initiative 175
4.5.4 Encrypting DVDs 176
4.5.5 Making CDs Copyproof 177
4.5.6 Criticisms of Digital Rights Management 177

4.6 Peer-to-Peer Networks 178

4.6.1 Napster 179
4.6.2 FastTrack 179
4.6.3 BitTorrent 180
4.6.4 RIAA Lawsuits 180
4.6.5 MP3 Spoofing 181
4.6.6 Universities Caught in the Middle 182
4.6.7 *MGM v. Grokster* 183
4.6.8 Legal Music Services on the Internet 184

4.7 **Protections for Software** 184
4.7.1 Software Copyrights 185
4.7.2 Violations of Software Copyrights 185
4.7.3 Software Patents 186
4.7.4 Safe Software Development 187

4.8 **Open-Source Software** 188
4.8.1 Consequences of Proprietary Software 188
4.8.2 Open-Source Definition 189
4.8.3 Beneficial Consequences of Open-Source Software 189
4.8.4 Examples of Open-Source Software 190
4.8.5 The GNU Project and Linux 191
4.8.6 Impact of Open-Source Software 192
4.8.7 Critique of the Open-Source Software Movement 192

4.9 **Legitimacy of Intellectual Property Protection for Software** 193
4.9.1 Rights-Based Analysis 193
4.9.2 Utilitarian Analysis 194
4.9.3 Conclusion 196

4.10 **Creative Commons** 196
Summary 198
Review Questions 202
Discussion Questions 203
In-Class Exercises 204
Further Reading 204
References 205

An Interview with **Wendy Seltzer** 209

5 Privacy 211

5.1 **Introduction** 211

5.2 **Perspectives on Privacy** 213
5.2.1 Defining Privacy 213
5.2.2 Harms and Benefits of Privacy 214
5.2.3 Is There a Natural Right to Privacy? 217
5.2.4 Privacy and Trust 220

5.3 **Disclosing Information** 221

5.4 **Public Information** 223
5.4.1 Rewards or Loyalty Programs 223
5.4.2 Body Scanners 224
5.4.3 Digital Video Recorders 224

5.4.4 Automobile "Black Boxes" 225
5.4.5 Enhanced 911 Service 225
5.4.6 RFIDs 226
5.4.7 Implanted Chips 227
5.4.8 Cookies 227
5.4.9 Spyware 228

5.5 U.S. Legislation 228

5.5.1 Fair Credit Reporting Act 228
5.5.2 The Family Education Rights and Privacy Act 229
5.5.3 Video Privacy Protection Act 229
5.5.4 Financial Services Modernization Act 229
5.5.5 Children's Online Privacy Protection Act 229
5.5.6 Health Insurance Portability and Accountability Act 230

5.6 Public Records 230

5.6.1 Census Records 230
5.6.2 Internal Revenue Service Records 231
5.6.3 FBI National Crime Information Center 2000 231
5.6.4 Privacy Act of 1974 232

5.7 Covert Government Surveillance 234

5.7.1 Wiretaps and Bugs 235
5.7.2 Operation Shamrock 237

5.8 U.S. Legislation Authorizing Wiretapping 238

5.8.1 Title III 239
5.8.2 Electronic Communications Privacy Act 239
5.8.3 Communications Assistance for Law Enforcement Act 239
5.8.4 USA PATRIOT Act 240

5.9 Data Mining 244

5.9.1 Marketplace: Households 246
5.9.2 IRS Audits 246
5.9.3 Syndromic Surveillance System 246
5.9.4 Total Information Awareness 247
5.9.5 Criticisms of the TIA Program 247
5.9.6 Who Should Own Information about a Transaction? 248
5.9.7 Opt-in Versus Opt-out 248
5.9.8 Platform for Privacy Preferences (P3P) 249

5.10 Identity Theft 249

5.10.1 Background 249
5.10.2 History and Role of the Social Security Number 251
5.10.3 Debate over a National ID Card 252
5.10.4 The REAL ID Act 253

5.11 **Encryption** 254
5.11.1 Symmetric Encryption 254
5.11.2 Public-Key Cryptography 255
5.11.3 Pretty Good Privacy 255
5.11.4 Clipper Chip 258
5.11.5 Effects of U.S. Export Restrictions 258
5.11.6 Digital Cash 259
Summary 262
Review Questions 264
Discussion Questions 266
In-class Exercises 268
Further Reading 269
References 269

An Interview with **Ann Cavoukian** 275

6 Computer and Network Security 279

6.1 **Introduction** 279
6.2 **Viruses, Worms, and Trojan Horses** 280
6.2.1 Viruses 280
6.2.2 Worms 283
6.2.3 The Internet Worm 286
6.2.4 Trojan Horses 290
6.2.5 Defensive Measures 291
6.3 **Phreaks and Hackers** 292
6.3.1 Hackers 292
6.3.2 Phone Phreaking 296
6.3.3 The Cuckoo's Egg 297
6.3.4 Legion of Doom 297
6.3.5 Fry Guy 298
6.3.6 *U.S. v. Riggs* 299
6.3.7 Steve Jackson Games 300
6.3.8 Retrospective 301
6.3.9 Penalties for Hacking 302
6.3.10 Recent Incidents 303
6.4 **Denial-of-Service Attacks** 304
6.4.1 Attacks that Consume Scarce Resources 304
6.4.2 Defensive Measures 307
6.4.3 Distributed Denial-of-Service Attacks 307
6.4.4 SATAN 308

6.5 Online Voting 308
6.5.1 Motivation for Online Voting 308
6.5.2 Proposals 309
6.5.3 Ethical Evaluation 310
Summary 313
Review Questions 314
Discussion Questions 315
In-Class Exercises 316
Further Reading 317
References 318

An Interview with **Matt Bishop** 321

7 Computer Reliability 325
7.1 Introduction 325
7.2 Data-Entry or Data-Retrieval Errors 326
7.2.1 Disfranchised Voters 326
7.2.2 False Arrests 327
7.2.3 Analysis: Accuracy of NCIC Records 327
7.3 Software and Billing Errors 328
7.3.1 Errors Leading to System Malfunctions 328
7.3.2 Errors Leading to System Failures 329
7.3.3 Analysis: E-Retailer Posts Wrong Price, Refuses to Deliver 329
7.4 Notable Software System Failures 330
7.4.1 Patriot Missile 331
7.4.2 Ariane 5 332
7.4.3 AT&T Long-Distance Network 333
7.4.4 Robot Missions to Mars 333
7.4.5 Denver International Airport 335
7.5 Therac-25 336
7.5.1 Genesis of the Therac-25 337
7.5.2 Chronology of Accidents and AECL Responses 337
7.5.3 Software Errors 340
7.5.4 Post Mortem 342
7.5.5 Moral Responsibility of the Therac-25 Team 343
7.6 Computer Simulations 344
7.6.1 Uses of Simulation 344
7.6.2 Validating Simulations 345

7.7 Software Engineering 347
7.7.1 Specification 347
7.7.2 Development 348
7.7.3 Validation 349
7.7.4 Software Quality Is Improving 350
7.8 Software Warranties 350
7.8.1 Shrinkwrap Warranties 351
7.8.2 Are Software Warranties Enforceable? 352
7.8.3 Uniform Computer Information Transaction Act 355
7.8.4 Moral Responsibility of Software Manufacturers 357
Summary 359
Review Questions 361
Discussion Questions 363
In-class Exercises 364
Further Reading 365
References 365

8 Work and Wealth 369
8.1 Introduction 369
8.2 Automation and Unemployment 370
8.2.1 Automation and Job Destruction 371
8.2.2 Automation and Job Creation 372
8.2.3 Effects of Increase in Productivity 375
8.2.4 Rise of the Robots? 376
8.3 Workplace Changes 379
8.3.1 Organizational Changes 379
8.3.2 Telework 381
8.3.3 Temporary Work 383
8.3.4 Monitoring 383
8.3.5 Multinational Teams 384
8.4 Globalization 385
8.4.1 Arguments for Globalization 385
8.4.2 Arguments against Globalization 387
8.4.3 Dot-Com Bust Increases IT Sector Unemployment 387
8.4.4 Foreign Workers in the American IT Industry 388
8.4.5 Foreign Competition 389
8.5 The Digital Divide 389
8.5.1 Evidence of the Digital Divide 390
8.5.2 Models of Technological Diffusion 391

8.5.3 Critiques of the Digital Divide 392

8.6 The "Winner-Take-All Society" 394

8.6.1 The Winner-Take-All Phenomenon 394

8.6.2 Harmful Effects of Winner-Take-All 395

8.6.3 Reducing Winner-Take-All Effects 397

8.7 Access to Public Colleges 397

8.7.1 Effects of Tuition Increases 397

8.7.2 Ethical Analysis 398

Summary 401

Review Questions 403

Discussion Questions 403

In-class Exercises 404

Further Reading 406

References 406

An Interview with **Jerry Berman** 411

9 Professional Ethics 415

9.1 Introduction 415

9.2 Is Software Engineering a Profession? 416

9.2.1 Characteristics of a Profession 416

9.2.2 Certified Public Accountants 417

9.2.3 Software Engineers 418

9.3 Software Engineering Code of Ethics 419

9.3.1 Preamble 420

9.3.2 Principles 421

9.4 Analysis of the Code 428

9.4.1 Preamble 428

9.4.2 Virtue Ethics 429

9.4.3 Alternative List of Fundamental Principles 432

9.5 Case Studies 433

9.5.1 Software Recommendation 434

9.5.2 Child Pornography 435

9.5.3 Anti-Worm 436

9.6 Whistleblowing 438

9.6.1 Morton Thiokol/NASA 439

9.6.2 Hughes Aircraft 441

9.6.3 Morality of Whistleblowing 443

Summary 446
Review Questions 448
Discussion Questions 449
In-class Exercises 450
Further Reading 451
References 451

An Interview with **Paul Axtell** 455

Appendix A: Plagiarism 459

Consequences of Plagiarism 459
Types of Plagiarism 459
Guidelines for Citing Sources 460
How to Avoid Plagiarism 460
Misuse of Sources 460
Additional Information 461
References 461

Index 463

Preface

Computers and high-speed communication networks are transforming our world. These technologies have brought us many benefits, but they have also raised many social and ethical concerns. My view is that we ought to approach every new technology in a thoughtful manner, considering not just its short-term benefits, but also how its long term use will affect our lives. A thoughtful response to information technology requires a basic understanding of its history, an awareness of current information-technology-related issues, and a familiarity with ethics. I have written *Ethics for the Information Age* with these ends in mind.

Ethics for the Information Age is suitable for college students at all levels. The only prerequisite is some experience using computers. The book is appropriate for a stand-alone "computers and society" or "computer ethics" course offered by a computer science, business, or philosophy department. It can also be used as a supplemental textbook in a technical course that devotes some time to social and ethical issues related to computing.

As students discuss controversial issues related to information technology, they learn from each other and improve their critical thinking skills. The provocative discussion questions raised in every chapter, combined with dozens of in-class exercises, provide many opportunities for students to express their viewpoints. They will learn how to evaluate complex issues and logically defend their conclusions.

WHAT'S NEW IN THE SECOND EDITION

Rapid changes in the field of information technology make the study of ethics in this area exciting and challenging. Nearly every day the media reports on a new invention, controversy, or court ruling. The second edition of *Ethics for the Information Age* has been updated to include many important developments; among them are:

- The emergence of the BitTorrent network and how some universities are responding to the problem to illegal file sharing
- The U.S. Supreme Court decision in the entertainment industry lawsuit against peer-to-peer network operators Grokster and StreamCast
- Ramifications of the USA PATRIOT Act and the debate surrounding the renewal of its most controversial provisions

- Passage of the REAL ID Act, which may result in a de facto national identification card for the United States
- The creation of autonomous robots controlled by artificial intelligence
- The emergence of China and India as legitimate competitors in the global information technology industry

Eight end-of-chapter interviews with leaders from industry and academia have been added to provide important new insights and perspectives to the book. Besides being informative, these interviews can serve as catalysts for in-class discussions.

Other sections have been added or enhanced in response to requests from readers. A new appendix describes what plagiarism is and how to avoid it. The history of the Internet is now told in greater detail. A new section discusses the problem of Internet addiction. An extended example illustrates how public key encryption works. Numerous ethical analyses have been sharpened. Throughout the book, new references to the latest news stories and analyses ensure that facts and figures are as up-to-date as possible.

ORGANIZATION OF THE BOOK

The book is divided into nine chapters. Chapter 1 has three objectives: to get the reader thinking about the process of technological change; to present a brief history of computing, networking, and information storage and retrieval; and to provide examples of moral problems brought about by the introduction of information technology.

Chapter 2 is an introduction to ethics. It presents seven different theories of ethical decision-making, weighing the pros and cons of each one. Four of these theories—Kantianism, act utilitarianism, rule utilitarianism, and social contract theory—are the most appropriate "tools" for analyzing moral problems in the remaining chapters.

Chapters 3–8 discuss a wide variety of issues related to the introduction of information technology into society. I think of these chapters as forming concentric rings around a particular computer user.

Chapter 3 is the innermost ring, dealing with what can happen when people communicate over the Internet using the Web, email, and chat rooms. Issues such as the increase in spam, easy access to pornography, and Internet addiction raise important questions related to quality of life, free speech, and censorship.

The next ring, Chapter 4, deals with the creation and exchange of intellectual property. It discusses intellectual property rights, legal safeguards for intellectual property, the definition of fair use, abuses of peer-to-peer networks, the rise of the open-source movement, and the legitimacy of intellectual property protection for software.

Chapter 5 focuses on privacy. What is privacy exactly? Is there a natural right to privacy? How do others learn so much about us? The chapter describes the electronic trail that people leave behind when they use cell phones, make credit card purchases, open bank accounts, or apply for loans. Other topics in this chapter include the difference between public information and public records, covert governmental surveillance, the USA PATRIOT Act, data mining, identity theft, encryption, and attempts to create anonymous digital cash.

Chapter 6 focuses on the vulnerabilities of networked computers. Students will learn the difference between a virus, a worm, and a Trojan horse. The chapter chronicles the transformation of hacker culture, the emergence of phone phreaks, and the hacker crackdown of 1990. The chapter also discusses denial-of-service attacks, the reliability of proposed on-line voting systems, and the important role system administrators play in keeping computers and networks secure.

Computerized system failures have led to lost business, the destruction of property, human suffering, and even death. Chapter 7 describes some notable software system failures, including the story of the Therac-25 radiation therapy system. It also discusses the reliability of computer simulations, the emergence of software engineering as a distinct discipline, the validity of software warranties, and the controversial Uniform Computer Information Transaction Act.

Chapter 8 raises a wide variety of issues related to how information technology has impacted work and wealth. Topics include workplace monitoring, telecommuting, and globalization. Does automation increase unemployment? Is there a "digital divide" separating society into "haves" and "have nots?" Is information technology widening the gap between rich and poor? Are we systematically excluding the poor from opportunities to succeed in our high-tech economy? These are just a few of the important questions the chapter addresses.

Chapter 9 is particularly relevant for those readers who plan to become software engineers. The chapter presents the Software Engineering Code of Ethics and Professional Practice, followed by an analysis of the code and a list of fundamental principles underlying it. Several case studies illustrate how to use the code as a tool for the evaluation of moral problems related to software engineering. The chapter concludes with an ethical evaluation of whistle blowing, an extreme example of organizational dissent.

NOTE TO INSTRUCTORS

In December 2001 a joint task force of the IEEE Computer Society and the Association for Computing Machinery released the final draft of *Computing Curricula 2001* (www.computer.org/education/cc2001/final). The report recommends that every undergraduate computer science degree program incorporate 40 hours of instruction related to social and professional issues related to computing. For those departments that choose to dedicate an entire course to these issues, the report provides a model syllabus for CS 280T, Social and Professional Issues. *Ethics for the Information Age* covers all of the major topics listed in the syllabus. Table 1 shows the mapping between the 10 units of CS 280T and the chapters of this book.

The organization of the book makes it easy to adapt to your particular needs. If your syllabus does not include the history of information technology, you can easily skip the middle three sections of Chapter 1 and still expose your students to examples motivating

TABLE 1 Mapping between the units of the Social and Professional Issues course in Computing Curricula 2001 and the chapters in this book.

Unit	*Name*	*Chapter(s)*
SP1	History of computing	1
SP2	Social context of computing	1, 3, 8
SP3	Methods and tools of analysis	2–9
SP4	Professional and ethical responsibilities	9
SP5	Risks and liabilities of computer-based systems	7
SP6	Intellectual property	4
SP7	Privacy and civil liberties	5
SP8	Computer crime	3, 5, 6
SP9	Economic issues in computing	8
SP10	Philosophical frameworks	2

the formal study of ethics in Chapter 2. After Chapter 2, you may cover the remaining chapters in any order you choose, because Chapters 3–9 do not depend on each other.

Many departments choose to incorporate discussions of social and ethical issues throughout the undergraduate curriculum. The independence of Chapters 3–9 makes it convenient to use *Ethics for the Information Age* as a supplementary textbook. You can simply assign readings from the chapters most closely related to the course topic.

SUPPLEMENTS

The following supplements are available to qualified instructors on Addison-Wesley's Instructor Resource Center. Please contact your local Addison-Wesley Sales Representative, or visit www.aw.com/irc to access this material.

- An instructor's manual provides tips for teaching a course in computer ethics. It also contains answers to all of the review questions.
- A test bank contains more than 250 multiple-choice, fill-in-the-blank, and essay questions that you can use for quizzes, midterms, and final examinations.
- A set of PowerPoint lecture slides outlines the material covered in every chapter.

FEEDBACK

Ethics for the Information Age cites hundreds of sources and includes dozens of ethical analyses. Despite the best efforts of myself and many others, the book is bound to contain errors. I appreciate getting comments (both positive and negative), corrections, and suggestions from readers. Please send them to `InformationAge@cs.orst.edu` or Michael J. Quinn, Oregon State University, School of Electrical Engineering and Computer Science, Corvallis, OR 97331.

ACKNOWLEDGMENTS

An outstanding team of dedicated professionals at Addison-Wesley and Windfall Software made the creation of the second edition a pleasure. My editor, Michael Hirsch, enthusiastically backed the project and played a large role in shaping the final product by suggesting the addition of interviews between the chapters. Editorial Assistant Lindsey Triebel edited the interviews and took care of dozens of important details for me. Maite Suarez-Rivas and Christopher Hu saved me a lot of time by taking over the tasks of finding images and gathering the necessary permissions. Marilyn Lloyd ensured the production process went smoothly. Michelle Brown led an effective marketing effort and provided me with regular updates on new adoptions. Copyeditor Richard Camp polished my prose and provided many excellent suggestions that made the book easier to read. Proofreaders Jennifer McClain and MaryEllen N. Oliver corrected numerous discrepancies. Paul C. Anagnostopoulos and Jacqui Scarlott produced beautiful camera-ready copy in a short amount of time. I thank them and everyone else who helped produce this edition.

I appreciate the contributions of all who participated in the creation of the first edition or provided useful suggestions for the second edition: Valerie Anctil, Beth Anderson, Bob Baddeley, George Beekman, Brian Breck, Sherry Clark, Thomas Dietterich, Beverly Fusfield, Peter Harris, Michael Johnson, Pat McCutcheon, Beth Paquin, Shauna Quinn, Stuart Quinn, Charley Renn, and Susan Hartman.

I thank the faculty members who have reviewed the book, supplying many insightful reactions and suggestions for improvements: John Clark, University of Colorado at Denver; Timothy Colburn, University of Minnesota-Duluth; Lorrie Faith Cranor, Carnegie Mellon University; Lee D. Cornell, Minnesota State University, Mankato; David Goodall, State University of New York at Albany; Fritz H. Grupe, University of Nevada, Reno; Tamara A. Maddox, George Mason University; Richard D. Manning, Nova Southeastern University; John G. Messerly, University of Texas at Austin; Joe Oldham, Centre College; Mimi Opkins, California State University, Long Beach; Holly Patterson-McNeill, Lewis-Clark State College; Michael Scanlan, Oregon State University; Matthew Stockton, Portland Community College; Leon Tabak, Cornell College; Renée Turban, Arizona State University; John Wright, Juniata College; and Matthew Zullo, Wake Technical Community College. I am particularly grateful to my colleague Michael Scanlan, who helped me refine the philosophical analyses.

Finally, I am indebted to my wife Victoria for her support and encouragement. You are a wonderful helpmate. Thanks for everything.

Michael J. Quinn
Corvallis, Oregon

We never know how high we are
Till we are called to rise;
And then, if we are true to plan,
Our statures touch the skies.

The heroism we recite
Would be a daily thing,
Did not ourselves the cubits warp
For fear to be a king.

—Emily Dickinson, *Aspiration*

I dedicate this book to my children: *Shauna*, *Brandon*, and *Courtney*.

Know that my love goes with you, wherever your aspirations may lead you.

CHAPTER

1 Catalysts for Change

A tourist came in from Orbitville,
parked in the air, and said:

The creatures of this star
are made of metal and glass.

Through the transparent parts
you can see their guts.

Their feet are round and roll
on diagrams of long

measuring tapes, dark
with white lines.

They have four eyes.
The two in back are red.

Sometimes you can see a five-eyed
one, with a red eye turning

on the top of his head.
He must be special–

the others respect him
and go slow

when he passes, winding
among them from behind.

They all hiss as they glide,
like inches, down the marked

tapes. Those soft shapes,
shadowy inside

the hard bodies–are they
their guts or their brains?

–MAY SWENSON, *Southbound on the Freeway*[1]

1.1 Introduction

MOST OF US WHO LIVE IN WESTERN DEMOCRATIC NATIONS take technological change for granted. In the past decade alone, we have witnessed the emergence of exciting new technologies, including cell phones, email, and the World Wide Web. There is good reason to say we are living in the Information Age. Never before have so many people had such easy access to information. The two principal catalysts for the Information Age have been low-cost computers and high-speed communication networks (Figure 1.1). Even in a society accustomed to change, the rate at which computers and communication networks have transformed our lives is breathtaking.

In 1950 there were no more than a handful of electronic digital computers in the world. Today we are surrounded by devices containing embedded computers. We rely upon microprocessors to control our heating and cooling systems, microwaves, stereos, elevators, and a multitude of other devices we use every day. Thanks to microprocessors, our automobiles get better gas mileage and produce less pollution. On the other hand, the days of the do-it-yourself tune-up are gone. It takes a mechanic with computerized diagnostic equipment to work on a modern engine.

In 1990 few people other than college professors used email. Today more than 600 million people around the world have email accounts. We consider people without access to email as deprived, even though most of us also complain about the amount of spam we receive.

The World Wide Web was still being designed in 1990; today it contains billions of pages and is an extraordinarily valuable information retrieval system. Teachers expect grade school children to use the Web when writing their reports. However, many parents worry that their Web-surfing children may be exposed to pornographic images or other inappropriate material.

May Swenson has vividly described our ambivalent feelings toward technology. In her poem "Southbound on the Freeway," an alien hovers above an expressway and watches the cars move along [1]. The alien notes "soft shapes" inside the automobiles

1. "Southbound on the Freeway" by May Swenson used with permission of the Literary Estate of May Swenson.

FIGURE 1.1 Low-cost computers and high-speed communication networks make possible the products of the Information Age, such as the Nokia 6800 messaging device. It functions as a phone, email client, multimedia messaging device, calculator, clock, stopwatch, and FM radio. (Courtesy of Nokia)

and wonders, "are they their guts or their brains?" It's fair to ask: Do we drive technology, or does technology drive us?

Our relationship with technology is complicated. We create technology and choose to adopt it. However, once we have adopted a technological device, it can change us and how we relate to other people and our environment.

The choice to use a new technology can affect you physically. For example, anecdotal evidence from physicians and physical therapists reveals that the growing popularity of laptop computers is increasing the number of people suffering from wrist, neck, shoulder, and back pain. That's not surprising, given the awkward places many people use laptop computers, such as traditional college lecture halls with cramped seating and tiny writing surfaces. A chiropractor remarks, "Have you seen pictures of kids using computers? They lie on their stomachs on the floor and work on their elbows. That's a prescription for a lifetime of neck pain, back pain, and lower back pain" [2].

Use of a technology can also change your outlook. For example, more than 90 percent of cell phone users report that having a cell phone makes them feel safer. On the other hand, once people get used to carrying a cell phone, losing the phone may make them feel more vulnerable than they ever did before they began carrying one. A Rutgers University professor asked his students to go without their cell phones for 48 hours. A female student reported to the student newspaper, "I felt like I was going to get raped if I didn't have my cell phone in my hand." Some parents purchase cell phones for their children so that a child may call a family member in an emergency. However,

FIGURE 1.2 The Amish carefully evaluate new technologies, choosing those that enhance family and community solidarity. (AP/Wideworld Photos)

parents who provide a cell phone "lifeline" may be implicitly communicating to their children the idea that people in trouble cannot expect help from strangers [3].

The Amish understand that the adoption of a new technology can affect the way people relate to each other (Figure 1.2). Amish bishops meet twice a year to discuss matters of importance to the church, including whether any new technologies should be allowed. Their discussion about a new technology is driven by the question, "Does it bring us together, or draw us apart?" You can visit an "Old Order" Amish home and find a gas barbecue on the front porch, but no telephone inside, because they believe gas barbecues bring people together, while telephones interfere with face-to-face conversations [4].

New technologies are adopted to solve problems, but they often create problems, too. The automobile has given people the ability to travel where they want, when they want to. On the other hand, millions of people spend an hour or more each day stuck in traffic commuting between home and work. Refrigerators make it possible for us to keep food fresh for long periods of time. We save time because we don't have to go grocery shopping every day. Unfortunately, freon leaking from refrigerators has contributed to the depletion of the ozone layer that protects us from harmful ultraviolet rays. New communication technologies have made it possible for us to get access to news and entertainment from around the world. However, the same technologies have enabled major software companies to move thousands of jobs to India, China, and Vietnam, putting downward pressure on the salaries of computer programmers in the United States [5].

We may not be able to prevent a new technology from being invented, but we do have control over whether to adopt it. Nuclear power is a case in point. Nuclear power plants create electricity without producing the carbon dioxide emissions that lead to global warming, but they also produce radioactive waste products that must be safely stored for 100,000 years. Although nuclear power technology is available, no new nuclear power plants have been ordered in the United States since the accident at Three Mile Island in 1979 [6].

Finally, we *can* influence the rate at which new technologies are developed. Some societies, such as the United States, have a history of nurturing and exploiting new inventions. Congress has passed intellectual property laws that allow people to make money from their creative work, and the federal income tax structure allows individuals to accumulate great wealth. At the other extreme, the Aboriginal peoples of Australia have developed a culture in which new technologies are developed at a very slow pace.

Of course, most of us do not want to live as the Aborigines do. We appreciate the many beneficial changes that technology has brought into our lives. In health care alone: computed tomography (CT) and magnetic resonance imaging (MRI) scanners have greatly improved our ability to diagnose major illnesses; new vaccines and pharmaceuticals have eradicated some deadly diseases and brought others under control; and pacemakers, hearing aids, and artificial joints have improved the physical well-being of millions.

The point is that we should be making informed decisions about how to use technology to maximize its benefits and minimize its harms. To that end, this book will help you gain a better understanding of contemporary issues related to the use of information technology.

This chapter sets the stage for the remainder of the book. Electronic digital computers and high-performance communication networks are central to contemporary information technology. While the impact of these inventions has been dramatic in the past few decades, their roots go back hundreds of years. Section 1.2 tells the story of the computer, showing how it evolved from early efforts to construct machines that could perform arithmetic into personal computers and microprocessors embedded in everyday devices. In Section 1.3 we describe two centuries of progress in networking technology, starting with the semaphore telegraph and culminating in the creation of an email system connecting hundreds of millions of users. Section 1.4 shows how information storage and retrieval evolved from the creation of the codex (paginated book) to the invention of the World Wide Web. In these three sections, you will see that revolutionary changes are rare. Instead, most progress occurs when one person or team makes an incremental improvement over the prior state of the art. You will also see that it is not unusual for more than one person to get the same idea at about the same time. Finally, Section 1.5 discusses some of the moral issues that have arisen from the deployment of information technology.

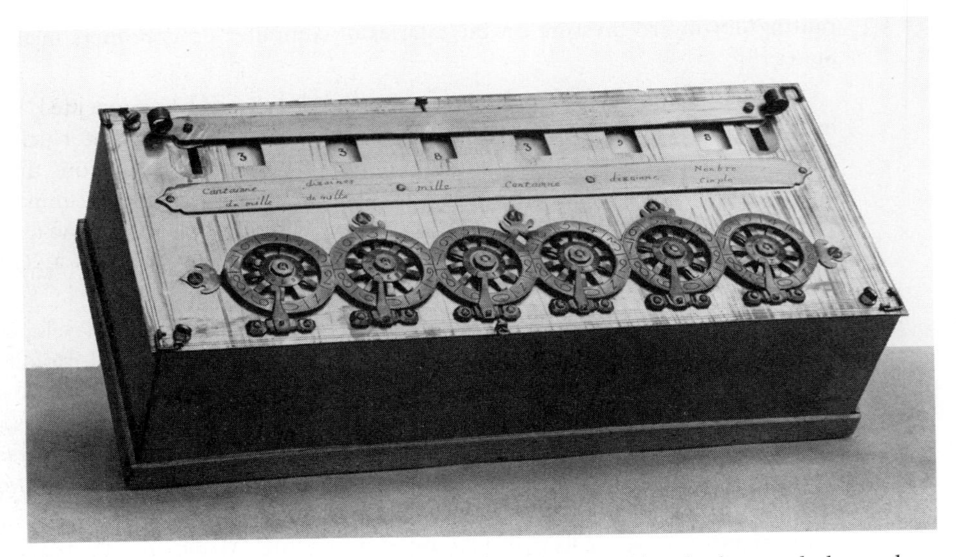

FIGURE 1.3 The Arithmetic Machine of Blaise Pascal could add and subtract whole numbers. (©Bettmann/CORBIS)

1.2 Milestones in Computing

1.2.1 Mechanical Adding Machines

Blaise Pascal (1623–1662) had a weak physique but a powerful mind. Although he was not introduced to mathematics until he was twelve, he published an important new result before his sixteenth birthday. Pascal's father was a French tax collector and frequently needed to compute sums. In 1640 Pascal constructed a mechanical calculator to assist his father in his work (Figure 1.3). The calculator was capable of adding and subtracting whole numbers containing up to six digits. The arithmetic machine contained six wheels, each with the digits from 0 to 9. Linkages between adjacent wheels enabled them to implement carry operations. Small windows enabled the user to read the sum, similar to the way you can read a car's odometer.

Inspired by Pascal's invention, the German Baron Gottfried von Leibniz (1646–1716) constructed a more sophisticated calculator that could add, subtract, multiply, and divide whole numbers (Figure 1.4). The hand-cranked machine, which he called the Step Reckoner, performed multiplications and divisions through repeated additions and subtractions, respectively. Leibniz also advocated the use of the binary number system in calculating machines.

These calculators were not reliable, and for more than a century after Pascal and Leibniz died, most arithmetic computations continued to be done by hand. In the days of pencil-and-paper calculations, lookup tables for logarithmic and trigonometric functions were essential. Suppose I asked you to figure out—by hand—the length of the shadow cast by a 40-foot-high tree when the sun is at a 20 degree angle in the sky

Figure 1.4 The Step Reckoner of Gottfried von Leibniz used repeated addition or subtraction to perform multiplication or division of whole numbers. (Science & Society Picture Library, London)

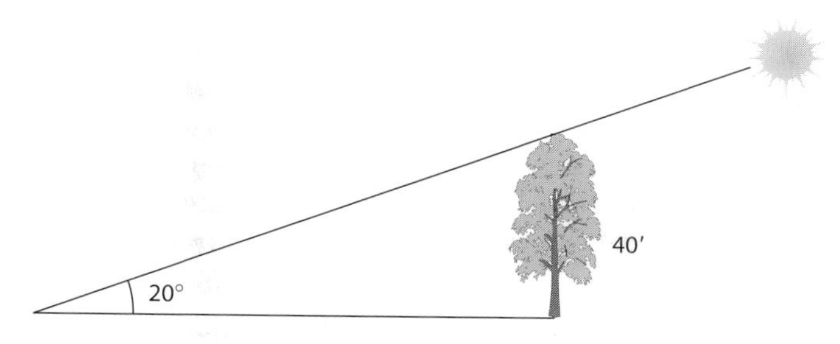

Figure 1.5 Computing the length of the shadow is much, much simpler if you are given the value of the tangent of 20 degrees (0.34907 radians).

(Figure 1.5). Using simple geometry, you could determine that the shadow's length is 40 feet divided by the tangent of 20 degrees. If you had access to a lookup table, you could quickly find that the tangent of 20 degrees is 0.36397, meaning the shadow is 109.90 feet long. Without the table, you would have to perform many more calculations to find the tangent. That begs the question of where the lookup table comes from, but once a group of people computes the values in the table and publishes it, everyone can benefit from it.

Unfortunately, people who solve mathematical problems by hand make mistakes. An even greater source of error is the printer who typesets the tables. As a result, all published mathematical tables in the early nineteenth century contained errors. Navigators relied upon tables of logarithms to guide ships. For a maritime nation such as England, navigation errors could have had significant economic consequences [7].

Charles Babbage (1791–1871) was born into a wealthy London family and educated at Cambridge, where he studied mathematics. Around 1812 he first began to think of calculating mathematical tables mechanically in order to save time and avoid the introduction of human error. He knew that logarithmic and trigonometric functions can be computed to arbitrary precision using polynomials (Figure 1.6). Babbage also knew that polynomials could be computed through repeated addition using the method of differences (Figure 1.7).

In 1819 Babbage began constructing a small machine capable of computing polynomials using the method of differences. He completed the machine in 1822 and presented a paper describing his invention to the Royal Astronomical Society in June of that year. Babbage said his machine could compute successive terms of the sequence $n^2 + n + 41$ at the rate of about one every five seconds.

$$\tan x \approx x + \frac{x^3}{3} + \frac{2x^5}{15} + \frac{17x^7}{315}$$

$$\tan 0.34907 \approx 0.34907 + \frac{0.34907^3}{3} + \frac{2 \times 0.34907^5}{15} + \frac{17 \times 0.34907^7}{315} \approx 0.36397$$

FIGURE 1.6 Polynomials can approximate logarithmic and trigonometric functions to arbitrary precision. This example illustrates how to compute the tangent of x radians.

n	n^3	*First Difference*	*Second Difference*	*Third Difference*
1	1	1	0	6
2	8	7	6	6
3	27	19	12	6
4	64	37	18	6
5	125	61	24	6
6	216			

FIGURE 1.7 A machine can compute polynomial functions through repeated addition. This table illustrates how to compute cubes in this manner. After values are inserted into the first row, we can compute each subsequent row from the row above it. The value in the second column is the sum of the values in columns 2–5 of the prior row: $8 = 1 + 1 + 0 + 6$, $27 = 8 + 7 + 6 + 6$, and so on. Try to figure out how to compute the first and second differences from values in prior rows.

Figure 1.8 In 1991 the Science Museum of London constructed a replica of the computational portion of the Difference Engine. It has 4,000 parts and weighs 3 tons. (Science Museum, London)

After the Royal Astronomical Society gave Babbage a gold medal for this invention in 1823, it supported his request to the British government to fund the construction of a larger system, which he called the Difference Engine. The Difference Engine would compute the values of sixth-degree polynomials to 20 digits of precision and print the answers in tables, eliminating all chance of human error (Figure 1.8). The government granted him £1,500 for the three-year project. Eleven years later, after the government had invested £17,000 and Babbage had put in £6,000 of his own money, work stopped on the unfinished Difference Engine. However, in the 1850s Georg and Edvard Scheutz independently constructed a working Difference Engine, validating Babbage's design.

1.2.2 The Analytical Engine

After the disappointing suspension of his Difference Engine project, Babbage began work on a follow-up machine he called the Analytical Engine. The Analytical Engine would have been much more versatile than the Difference Engine, capable of computing arbitrary mathematical formulas. The system would have been able to multiply, divide, and find square roots of numbers through repeated addition and subtraction. In this sense the Analytical Engine can be seen as a precursor to the modern computer. However, key concepts such as an instruction set, conditional execution of instructions, and memory addressing are missing from the plans of the Analytical Engine [8]. Hence even if it had been completed, it would not have been a true mechanical computer.

Shortly after Augusta Ada Byron (1815–1852) was born, her mother asked for a legal separation from her father, the English poet Lord Byron. Ada's mother did not want her daughter to turn out to be a wild romantic like her husband, so she ensured that Ada received tutoring in mathematics. After Ada Byron heard Charles Babbage lecture on the Difference Engine when she was 18, she engaged in an extensive correspondence with him. In this correspondence she predicted that computing machines would be able to compose music and produce graphics. She translated from French to English a paper on the Analytical Engine written by Menebrea. In the process she added extensive notes, including a description of how to compute Bernoulli numbers[2] on the Analytical Engine.

Ada Byron assumed the title Countess of Lovelace when she married the Earl of Lovelace. For this reason she is sometimes mistakenly called Ada Lovelace. Because of her description of a method to compute Bernoulli numbers on the Analytical Engine, Ada Byron is often generously given the title "the first programmer" [9].

1.2.3 Boolean Algebra

The background of George Boole (1815–1864) could hardly have been more different from that of fellow Englishman Charles Babbage. While Babbage was born into the upper class and attended Cambridge, Boole was born into the lower classes and received a mediocre formal education. Despite the handicap of his low birth, Boole's talent and perseverance enabled him to make a far more significant contribution to the development of the modern computer.

Boole overcame his inadequate education by teaching himself. By the time he was sixteen he was supporting his parents and himself as a schoolteacher. A few years later he opened his own school. In order to instruct his students in mathematics, he began a thorough study of the subject. Boole excelled in this field, and eventually he became a professor of mathematics at Queen's College in Cork, Ireland. There, he wrote a book demonstrating how logical reasoning could be formalized by manipulating symbols using simple algebraic rules [10].

The branch of mathematics now known as Boolean algebra focuses on the manipulation of just two values—*true* and *false*—with the operators AND, OR, and NOT. Boole's work lay dormant for about 80 years after his death, until Massachusetts Institute of Technology Ph.D. student Claude Shannon demonstrated how switching circuits could be used to perform logical functions (Figure 1.9). Today, Boolean algebra is ingrained in the design of modern hardware and software systems. For example, if you type

2. Bernoulli numbers play a role in the polynomial expansion of some trigonometric functions.

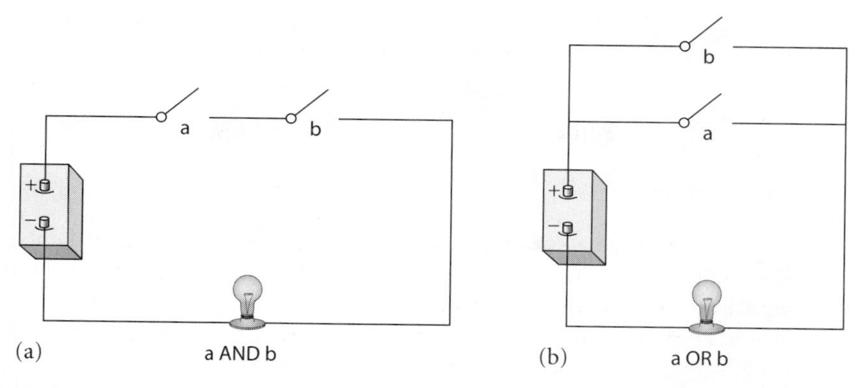

Figure 1.9 Switches can be combined to compute logical functions. In these examples, a closed switch represents the value *true*; an open switch represents the value *false*. If the result of the logical function is *true*, the light is illuminated; the light is dark if the result is *false*. (a) The expression a AND b is *true* only if both a and b are *true*. (b) The expression a OR b is *true* if at least one of a and b is *true*.

the query LINCOLN GETTYSBURG into the Web search engine Google, it assumes you are interested in Web pages containing information about both Lincoln and Gettysburg. In other words, your query is a Boolean expression.

1.2.4 Zuse's Z Series

In the late 1930s the conditions were ripe for the creation of the world's first fully electronic, general-purpose, stored-program computer. Which country would win the race: Germany, the United Kingdom, or the United States? (The answer may surprise you.)

Konrad Zuse (1910–1995) studied construction engineering in Berlin and went to work for the airplane industry. Tired of doing repetitive calculations, he quit his job, moved in with his parents, and began building a mechanical computer in their apartment. Two years later he and his friends had completed the system, which had 64 words of binary memory and could multiply two floating-point numbers in 5 seconds (Figure 1.10).

The mechanical memory of the Z1 was unreliable, and Zuse set out to construct two follow-on computers, the Z2 and the Z3, which used relays for memory. The Z3, completed in 1941, was a completely electronic, programmable device, and many people call Konrad Zuse "the inventor of the computer." However, two features of the Z3 differentiate it from a modern computer. The program was stored outside the Z3 on a punched tape, and (perhaps as a consequence) the system had no branch or jump instruction, meaning it could not loop through a series of instructions multiple times.

FIGURE 1.10 Reconstruction of Konrad Zuse's Z1 computer. The original Z1 was destroyed during the Allied bombing of Berlin in World War II. (Courtesy of Dr. Horst Zuse)

1.2.5 Harvard Mark I

As a graduate student in theoretical physics at Harvard in the 1930s, Howard Aiken (1900–1973) worked on solving nonlinear equations related to transmitting television signals through the ionosphere. Like some of his contemporaries who had tired of pencil-and-paper calculations, Aiken conceived of building a large-scale electromechanical calculating machine to help him solve these equations. In late 1937 he met with representatives of IBM to discuss his ideas for such a machine, and in March 1939 IBM's Board of Directors approved construction of the calculator. Aiken and IBM engineers worked together on the detailed design of the system, which was completed in 1943 and delivered to Harvard University in 1944, where it became known as the Harvard Mark I.

Much of the Harvard Mark I was constructed out of existing IBM products, including punched card readers and writers, electronic typewriters, and 73 IBM Automatic Accounting Machines. The completed system was 51 feet long and 8 feet high. It contained more than 750,000 moving parts and more than 500 *miles* of wiring. The rather noisy system was controlled by programs stored externally on paper tape. During World War II it computed gunnery tables for the U.S. Navy. In July 1944 Navy Lieutenant Grace Murray Hopper (1906–1992) arrived at Harvard and became the third programmer for the Harvard Mark I.

1.2.6 Colossus

During World War II the Allies set up a codebreaking team at Bletchley Park in England. The team, which included Alan Turing (1912–1954), set out to decrypt German messages transmitted using the Enigma, an electromechanical cipher machine. By 1943 a special-purpose electronic computer known as Colossus was up and running. Colossus was capable of translating encrypted messages into plain text as quickly as the message could be read into the machine via a paper tape reader. Because all work on Colossus was done under a veil of secrecy, it did not have a significant impact on the design of future computing systems.

1.2.7 The Atanasoff–Berry Computer

Between 1939 and 1941 Iowa State College professor John Atanasoff (1903–1995) and graduate student Clifford Berry (1918–1963) constructed an electronic device for solving systems of linear equations. The Atanasoff-Berry Computer had two important innovations. It was the first computer built with vacuum tubes, and it used a rotating drum to serve as a dynamic random access memory.

For a long time the work of Atanasoff and Berry was overshadowed by the accomplishments of fellow Americans J. Presper Eckert and John Mauchly. An antitrust case in the early 1970s resulted in a trial that revealed the extent to which Eckert and Mauchly had relied upon concepts developed in the Atanasoff-Berry Computer. In his ruling, the judge stated that Atanasoff was the inventor of the computer. Today, many cite this ruling when they call John Atanasoff the "father of the computer."

However, while it was a fully electronic computing device, the Atanasoff-Berry Computer was especially designed to solve systems of linear equations. It could not be programmed to perform other functions.

1.2.8 ENIAC (Electronic Numerical Integrator and Computer)

The Ballistic Research Laboratory (BRL) at the U.S. Army's Aberdeen Proving Ground in Maryland was responsible for testing all new ordnance. One of the laboratory's responsibilities was creating artillery tables used by gun crews to aim their weapons. In 1935 the BRL installed a Bush differential analyzer to help perform the calculations needed to create these tables. The differential analyzer was an analog computer that used rotating wheels and gears to calculate its results.

With the outbreak of World War II, the BRL was unable to keep up with the rapidly increasing workload. The Moore School at the University of Pennsylvania had a faster Bush differential analyzer. The Moore School signed a contract with the U.S. Army in which it agreed to devote its analyzer exclusively to solving problems of interest to the Army. However, even with two differential analyzers working, the Army could not perform all of the necessary computations in a timely manner.

Dr. John W. Mauchly (1907–1980), a physics professor at the University of Pennsylvania, had visited John Atanasoff at Iowa State College in 1941 to learn more about the Atanasoff-Berry Computer. After he returned to Penn, Mauchly worked with J. Presper

Eckert (1919–1995) to create a design for an electronic computer, and in October 1942 he wrote a memorandum to the U.S. Army outlining the design of the ENIAC (Electronic Numerical Integrator and Computer). The U.S. Army signed a contract with the Moore School in June 1943, and the completed system was dedicated in February 1946. The speed of the ENIAC was truly impressive. A person with a desk calculator could compute a 60-second trajectory in 20 hours. The Bush differential analyzer required 15 minutes to compute the same result. The ENIAC performed the computation in 30 seconds. In other words, the ENIAC was 2,400 times faster than a person with a desk calculator.

While the ENIAC represented numbers internally in base 10, it had many features of a modern computer. All of its internal components were electronic, and it could be programmed to perform a variety of computations. Many people today call Eckert and Mauchly the inventors of the computer. However, the ENIAC falls short of our definition of a modern computer. Most importantly, its program was not stored inside memory. Instead, it was "wired in" from the outside. Reprogramming the computer meant removing and reattaching many wires. This process could take several days.

Even before the ENIAC was completed, work began on a follow-on system called the EDVAC (Electronic Discrete Variable Automatic Computer). The design of the EDVAC incorporated many improvements over the ENIAC. The most important improvement was that the EDVAC would store the program in primary memory, along with the data manipulated by the program. Not only does this simplify reprogramming, it also enables the computer to modify its own programs, making the system much more powerful from a theoretical point of view. John von Neumann (1903–1957) produced a document called "First Draft of a Report on the EDVAC" that elaborated the design of the computer, including the key concept of storing programs in primary memory. Others on the design team were dismayed to see that von Neumann had made himself sole author of the paper. The report was widely circulated and extremely influential. As a result, von Neumann has received most of the credit for the stored program concept, even though Eckert had written a paper describing the idea months before von Neumann joined the project [11].

In July and August 1946 Eckert, Mauchly, and several other computer pioneers gave a series of 48 lectures at the Moore School. While some of the lectures discussed lessons learned from the ENIAC, others focused on the design of its successor, the EDVAC. These lectures influenced the design of future machines built in the United States and the United Kingdom.

1.2.9 Small-Scale Experimental Machine

During World War II the Allies launched a major effort to improve and deploy radar systems. The scientists and engineers discovered that as they increased the power of the radar, the amount of background clutter displayed on the screen, a cathode ray tube (CRT), also increased. The clutter made it difficult for radar operators to locate a moving target. To solve this problem, somebody suggested that if the system could store the background clutter, then it could use this information to cancel out the clutter on the next sweep of the radar. That way, only moving objects would be displayed on the screen. This idea was successfully implemented. Hence the first use of electrostatic memory was in radar systems [12].

British engineer F.C. Williams (1911–1977) was actively involved in the development of radar. After the war, he decided to put his knowledge to use by figuring out how to use a CRT as a storage device for digital information. In November 1946 he successfully stored a bit on a CRT. A month later he joined the University of Manchester, where he continued to work on CRT storage with associate Tom Kilburn (1921–2001). In 1947 Kilburn discovered a better way to store bits (1s and 0s) on a CRT, and by the end of the year the group could store 2048 bits on a 6-inch-diameter screen, which became known as a Williams Tube.

Incredibly, the University of Manchester team decided to build a computer in order to more fully test the capabilities of the Williams Tube. In their initial experiments with the Williams Tube, they had to reset bits by hand, which means they could not test its ability to set and read bits at electronic speeds. In early 1948 they set out to build a small computer that would more fully exercise the capabilities of the Williams Tube.

They called their system the Small-Scale Experimental Machine (Figure 1.11). Inside the computer one Williams Tube held 1024 bits of random access memory, divided into 32 words of 32 bits each. Two more Williams Tubes held the contents of three registers: the accumulator, the current instruction, and the address of the current instruction. On June 21, 1948, the computer successfully executed its first program.

The Small-Scale Experimental Machine was the first system that had all of the essential features of a modern computer: fully electronic operation, program and data stored in random access memory, and a jump instruction that enables the program to execute loops. For this reason, F.C. Williams, Tom Kilburn, and Geoff Tootill are properly recognized as the creators of the first true modern computer. However, it should be clear by now that their achievement relied upon the prior accomplishments of a great many scientists and engineers.

The Manchester team continued work on its computer in anticipation of commercialization by Ferranti, Ltd. They called their improved system the Mark 1.

1.2.10 First Commercial Computers

In February 1951 British corporation Ferranti, Ltd. introduced the Ferranti Mark 1, the world's first commercial computer. As suggested by its name, the computer was the direct descendent of the research computers constructed at the University of Manchester.

After completing their work on the ENIAC, Americans J. Presper Eckert and John Mauchly formed the Eckert-Mauchly Computer Corporation. They signed a preliminary agreement with the National Bureau of Standards (representing the Census Bureau) in 1946 to develop a commercial computer, which they called the UNIVAC, for UNIVersal Automatic Computer. Work on the UNIVAC went badly, and the final contract was not signed until 1948. The project experienced huge cost overruns, and by

FIGURE 1.11 A portion of the University of Manchester's Small-Scale Experimental Machine, the first true modern computer. (Reproduced with permission of the Department of Computer Science, University of Manchester, United Kingdom)

1950 the Eckert-Mauchly Computer Corporation was on the brink of bankruptcy. Remington Rand (the electric razor company) bailed them out, forming the Univac Division of Remington Rand. Remington Rand delivered the UNIVAC I to the U.S. Bureau of the Census at the end of March 1951 [13].

In a public relations coup, Remington Rand cooperated with CBS to use a UNIVAC computer to predict the outcome of the 1952 Presidential election (Figure 1.12). Adlai Stevenson had led Dwight Eisenhower in polls taken before the election, but less than an hour after the polls closed, with just 7 percent of the votes tabulated, the UNIVAC was giving 100-1 odds that Dwight Eisenhower would win the election in a landslide. Sig Mickelson, director of news and public affairs for CBS, was disappointed, saying, "We had been convinced that the UNIVAC would have the right answer." Rather than announce the prediction, Mickelson decided to keep the prediction under wraps and let the UNIVAC try again when more returns were in. Even with more data, the UNIVAC persisted in its prediction that Eisenhower would win in a landslide. As it turns out, the computer was right and the pundits were wrong. UNIVAC predicted Eisenhower would

Figure 1.12 Walter Cronkite (right) visits the UNIVAC computer at a Philadelphia factory to prepare for the 1952 Presidential election night coverage. The man standing to his right is J. Presper Eckert. The console operator is Harold Sweeney. (Courtesy of Unisys Corporation)

win 438 electoral votes to 93 for Stevenson. The official result was a 442-89 victory for Eisenhower [14].

1.2.11 Transistor

American inventor Lee De Forest invented the vacuum tube in 1906 and demonstrated how it could amplify signals. AT&T bought the patent from De Forest and made significant improvements in its design. Using vacuum tubes, AT&T constructed amplifiers that made long-distance telephone calls possible. However, vacuum tubes required a lot of power, generated a lot of heat, and burned out like lightbulbs. AT&T was on the lookout for a better technology.

The development of the radar in World War II led to advances in semiconductors. AT&T put together a team of Bell Labs scientists, led by Bill Shockley, to develop a semiconductor replacement for the vacuum tube. On June 30, 1948, Bell Labs announced the invention of such a device, which they called the **transistor.** Shockley shares the title of "inventor of the transistor" with colleagues John Bardeen and Walter Brattain, who actually created the first working prototype [15].

Figure 1.13 The eight founders of Fairchild Semiconductor on the factory floor. Gordon Moore is second from the left and Robert Noyce is on the right. (Magnum Photos, Inc. ©1960 Wayne Miller)

1.2.12 Integrated Circuit

While industry largely ignored the invention of the transistor, Bill Shockley understood its potential. He left Bell Labs and moved to Palo Alto, California, where he founded Shockley Semiconductor in 1956. He hired an exceptional team of engineers and physicists, but many disliked his heavy-handed management style [15].

In September 1957 eight of Shockley's most talented employees, including Gordon Moore and Robert Noyce, walked out. The group, soon to be known as the "traitorous eight," founded Fairchild Semiconductor (Figure 1.13). By this time transistors were being used in a wide variety of devices, from transistor radios to computers. While transistors were far superior to vacuum tubes, they were still too big for some applications. Fairchild Semiconductor set out to produce a single semiconductor device containing transistors, capacitors, and resistors; in other words, an **integrated circuit.** Another firm, Texas Instruments, was on the same mission. Today, Robert Noyce of Fairchild Semiconductor and Jack Kilby of Texas Instruments are credited for independently inventing the integrated circuit [16].

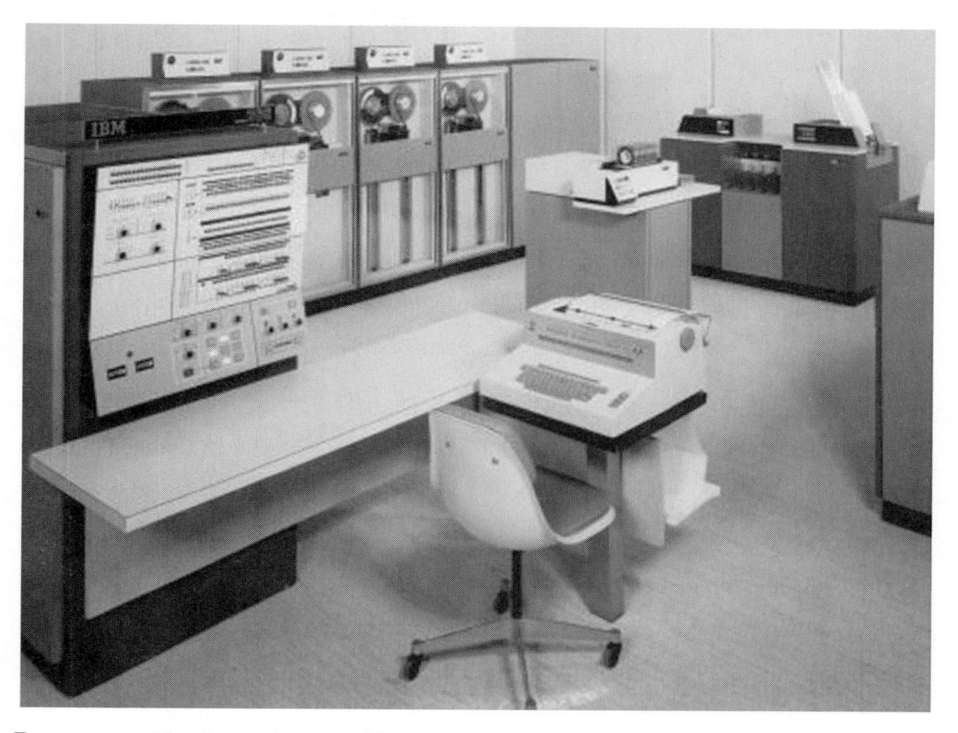

Figure 1.14 The System/360 Model 40, a medium-sized IBM mainframe introduced in 1964. (Courtesy of IBM Corporate Archives)

1.2.13 IBM System/360

The integrated circuit made possible the construction of much more powerful and much more reliable computers. The 1960s was the era of mainframe computers, large computers designed to serve the data processing needs of large businesses. Mainframe computers enabled enterprises to centralize all of their data processing applications in a single system. In the United States IBM dominated the mainframe market.

In 1964 IBM announced the System/360, a series of 19 compatible computers with varying levels of computing speed and memory capacity (Figure 1.14). The low-end system could perform 33,000 additions per second, while the high-end processor was capable of 750,000 additions per second. The smallest system had a mere 8 kilobytes of random access memory, but the largest system could be expanded to an astonishing (for the time) 8 megabytes of memory.[3] Because the systems were compatible, a business could upgrade its computer without having to rewrite its application programs.

3. A kilobyte is 1024 bytes; a megabyte is 1024 kilobytes, or slightly more than one million bytes.

FIGURE 1.15 The Intel 4004, the world's first microprocessor. (Copyright © Intel Corporation. All rights reserved)

1.2.14 Microprocessor

In 1968 Robert Noyce and Gordon Moore left Fairchild Semiconductor to found another semiconductor manufacturing company, which they named Intel. A year later Japanese calculator manufacturer Busicom approached Intel and asked it to design 12 custom chips for use in a new scientific calculator. Intel agreed to provide the chips and assigned responsibility for the project to Marcian "Ted" Hoff (1937–). After reviewing the project, Hoff suggested that it was not in Intel's best interests to manufacture a custom chip for every customer. As an alternative, he suggested that Intel create a general-purpose chip that could be programmed to perform a wide variety of tasks. Each customer could program the chip to meet its particular needs. Intel and Busicom agreed to the plan, which reduced the required number of chips for Busicom's calculator from 12 to 4. A year of development by Ted Hoff, Stanley Mazor (1941–), and Federico Faggin (1941–) led to the release of the Intel 4004, the world's first **microprocessor** (Figure 1.15). Inside the 1/8″ × 1/6″ chip were 2,300 transistors, giving the Intel 4004 the same computing power as the ENIAC, which had occupied 3,000 cubic feet.

Microprocessors made it possible for hobbyists to construct personal computers. After Intel released the 8008, an 8-bit microprocessor, Jon Titus designed a computer called the Mark-8 and published an article about its construction in the July 1974 *Radio Electronics* magazine. A hobbyist could purchase the plans for the computer, the Intel 8008, and all the other needed parts for less than $500.

In January 1975 *Popular Electronics* magazine introduced another computer for hobbyists, the Altair 8800, based on Intel's 8080 microprocessor. A company called MTS, in Albuquerque, New Mexico, sold an Altair 8800 kit for $395 or assembled for $495. More than 4,000 people bought the system in the first three months.

Paul Allen and his longtime friend, Harvard sophomore Bill Gates, saw the potential in the Altair and determined to be the first to produce a workable programming

Figure 1.16 Tandy sold its TRS-80 in 1977 for $599.95 through Radio Shack stores. The computer had a 1.77 MHz Zilog Z80 microprocessor and 4 kilobytes of random access memory. A tape cassette player served as the external storage device. Tandy expected to sell 600 to 1,000 the first year. Instead, it sold 10,000 in the first month! In all, more than 200,000 of these computers were sold. (Courtesy of Radio Shack Corporation)

language for it. In about a month of marathon programming sessions they produced a BASIC interpreter for the Altair and licensed it to MTS. Soon both were doing software development in Albuquerque. In a letter to Allen, Gates referred to their partnership as "Micro-Soft."

By the end of the 1970s many companies, including Apple and Tandy, were producing personal computers (Figure 1.16). While hundreds of thousands of people bought personal computers for home use, businesses were reluctant to move to the new computer platform. Two significant developments made personal computers more attractive to businesses.

The first development was the computer spreadsheet program. For decades firms have used spreadsheets to perform financial predictions. Manually computing spreadsheets is monotonous and error prone, since changing a value in a single cell can require updating many other cells. In the fall of 1979 Bob Frankston and Harvard MBA student Dan Bricklin released their program, called VisiCalc. VisiCalc's labor-saving potential was obvious to businesses. After a slow start, it quickly became one of the most popular application programs for personal computers.

The second development was the release of the IBM PC in 1981. The IBM name exuded safety and respectability, making it easier for companies to make the move to desktop systems for their employees. As the saying went, "Nobody ever got fired for buying from IBM." In contrast to the approach taken by Apple Computer, IBM decided to make its PC an open architecture, meaning the system was built from off-the-shelf parts and other companies could manufacture "clones" with the same functionality. This decision helped to make the IBM PC the dominant personal computer architecture.

The success of IBM-compatible PCs fueled the growth of Microsoft. In 1980 IBM contracted with Microsoft to provide the DOS operating system for the IBM PC. Microsoft let IBM have DOS for practically nothing, but in return IBM gave Microsoft the right to collect royalties from other companies manufacturing PC-compatible computers. Microsoft profited handsomely from this arrangement when PC-compatibles manufactured by other companies gained more than 80 percent of the PC market [17].

Microprocessors also made it possible to integrate computers into everyday devices. Today we're surrounded by devices containing microprocessors: ATM machines, automobiles, clock radios, microwave ovens, thermostats, traffic lights, and much more.

1.3 Milestones in Networking

In the early nineteenth century the United States fell far behind Europe in networking technology. The French had begun constructing a network of telegraph towers in the 1790s, and forty years later there were towers all over the European continent (Figure 1.17). At the top of each tower was a pair of semaphores. Operators raised and lowered the semaphores; each pattern corresponded to a letter or symbol. A message initiated at one tower would be seen by another tower within viewing distance. The receiving tower would then repeat the message for the next tower in the network, and so on. This optical telegraph system could transmit messages at the impressive rate of about 350 miles per hour when the skies were clear.

In 1837 Congress asked for proposals to create a telegraph system between New York and New Orleans. John Parker and Samuel Reid, harbormaster of the Port of New York, submitted a proposal based on proven European technology. Samuel Morse submitted a radically different proposal. He suggested constructing a telegraph system that used electricity to communicate the signals. Let's step back and review some of the key discoveries and inventions that enabled Morse to make his dramatic proposal.

1.3.1 Electricity and Electromagnetism

Amber is a hard, translucent, yellowish brown fossil resin often used to make beads and other ornamental items. About 2,600 years ago the Greeks discovered that if you rub amber, it becomes charged with a force enabling it to attract light objects such as feathers and dried leaves. The Greek word for amber is ηλεκτρων (electron). Our word "electric" literally means "to be like amber."

For more than two thousand years amber's ability to attract other materials was seen as a curiosity with no practical value, but in the seventeenth and eighteenth centuries scientists began to study electricity in earnest. Alessandro Volta (1745–1827), a professor of physics at the University of Pavia, made a key breakthrough when he discovered that electricity could be generated chemically. He produced an electric current by submerging two different metals close to each other in an acid. In 1799 Volta used this principle to create the world's first battery. Volta's battery produced an electric charge more than 1,000 times as powerful as that produced by rubbing amber. Scientists soon put this power to practical use.

FIGURE 1.17 A semaphore telegraph tower on the first line from Paris to Lille (1794). (Coll. Musee de la Poste, Paris)

In 1820 Danish physicist Christian Oersted (1777–1851) discovered that an electric current creates a magnetic field. Five years later, British electrician William Sturgeon (1783–1850) constructed an electromagnet by coiling wire around a horseshoe-shaped piece of iron. When he ran an electric current through the coil, the iron became magnetized. Sturgeon showed how a single battery was capable of producing a charge strong enough to pick up a nine-pound metal object.

In 1830 American professor Joseph Henry (1797–1878) rigged up an experiment that showed how a telegraph machine could work. He strung a mile of wire around the walls of his classroom at the Albany Academy. At one end he placed a battery; at the other end he connected an electromagnet, a pivoting metal bar, and a bell. When Henry connected the battery, the electromagnet attracted the metal bar, causing it to ring the bell. Disconnecting the battery allowed the bar to return to its original position. In this way he could produce a series of rings.

1.3.2 Telegraph

Samuel Morse (1791–1872), a professor of arts and design at New York University, worked on the idea of a telegraph during most of the 1830s, and in 1838 he patented his design of a telegraph machine. The U.S. Congress did not approve Morse's proposal to construct a New York–to–New Orleans telegraph system in 1837, but it did not fund any of the other proposals, either. Morse persisted with his lobbying, and in 1843 Congress appropriated $30,000 to Morse for the construction of a 40-mile telegraph line between Washington, D.C., and Baltimore, Maryland.

On May 1, 1844, the Whig party convention in Baltimore nominated Henry Clay for President. The telegraph line was only complete to Annapolis Junction at that time. A courier hand-carried a message about Clay's nomination from Baltimore to Annapolis Junction, where it was telegraphed to Washington. This was the first news reported via telegraph. The line officially opened on May 24. Morse, seated in the old Supreme Court chamber inside the U.S. Capitol, sent his partner in Baltimore a verse from the Bible: "What hath God wrought?"

The value of the telegraph was immediately apparent, and the number of telegraph lines quickly increased. By 1846 telegraph lines connected Washington, D.C., Baltimore, Philadelphia, New York, Buffalo, and Boston. In 1850 20 different companies operated 12,000 miles of telegraph lines. The first transcontinental telegraph line was completed in 1861, putting the Pony Express out of business. The telegraph was the sole method of rapid long-distance communication until 1877. By this time the United States was networked by more than 200,000 miles of telegraph wire [18].

The telegraph was a versatile tool, and people kept finding new applications for it. For example, by 1870 fire alarm telegraphs were in use in 75 major cities in the United States. New York City alone had 600 fire alarm telegraphs. When a person pulled the lever of the alarm box, it automatically transmitted a message identifying its location to a fire station. These devices greatly improved the ability of fire departments to dispatch equipment quickly to the correct location [18].

1.3.3 Telephone

Alexander Graham Bell (1847–1922) was born in Edinburgh, Scotland into a family focused on impairments of speech and hearing. His father and grandfather were experts in elocution and the correction of speech. His mother was almost completely deaf. Bell was educated to follow in the same career path as his father and grandfather, and he became a teacher of deaf students. Later, he married a deaf woman.

Bell pursued inventing as a means of achieving financial independence. At first he focused on making improvements to the telegraph. A significant problem with early telegraph systems was that a single wire could transmit only one message at a time. If multiple messages could be sent simultaneously along the same wire, communication delays would be reduced, and the value of the entire system would increase.

Bell's solution to this problem is called a harmonic or musical telegraph. If you imagine hearing Morse code, it's obvious that all of the dots and dashes are the same note played for a shorter or longer period of time. The harmonic telegraph assigned a different note (different sound frequency) to each message. At the receiving end different receivers could be tuned to respond to different notes, as you can tune your radio to hear only what is broadcast by a particular station.

Bell knew that the human voice is made up of sounds at many different frequencies. From his work on the harmonic telegraph, he speculated that it should be possible to capture and transmit human voice over a wire. He and Thomas A. Watson succeeded in transmitting speech electronically in 1876. Soon after, they commercialized their invention.

Nearly all early telephones were installed in businesses. Leasing a telephone was expensive, and most people focused on its commercial value rather than its social value. However, the number of phones placed in homes increased rapidly in the 1890s, after Bell's first patent expired.

Once telephones were placed in the home, the traditional boundaries between private, family life and public, business life became blurred. People enjoyed being able to conduct business transactions from the privacy of their home, but they also found that a ringing telephone could be an unwelcome interruption [19].

Another consequence of the telephone was that it eroded traditional social hierarchies. An 1897 issue of *Western Electrician* reports that Governor Chauncey Depew of New York was receiving unwanted phone calls from ordinary citizens: "Every time they see anything about him in the newspapers, they call and tell him what a 'fine letter he wrote' or 'what a lovely speech he made,' or ask if this or that report is true; and all this from people who, if they came to his office, would probably never say more than 'Good morning' " [20].

People also worried about the loss of privacy brought about by the telephone. In 1877 *The New York Times* reported that telephone men responsible for operating an early system in Providence, Rhode Island, overheard many confidential conversations. The writer fretted that telephone eavesdropping would make it dangerous for anyone in Providence to accept a nomination for public office [19].

The telephone enabled the creation of the first "online" communities. In rural areas the most common form of phone service was the party line: a single circuit connecting multiple phones to the telephone exchange. Party lines enabled farmers to gather by their phones every evening to talk about the weather and exchange gossip [21].

The power of this new medium was demonstrated in the Bryan/McKinley Presidential election of 1896. For the first time, Presidential election returns were transmitted directly into people's homes. "Thousands sat with their ear glued to the receiver the whole night long, hypnotized by the possibilities unfolding to them for the first time" [22].

1.3.4 Typewriter and Teletype

For hundreds of years people dreamed of a device that would allow an individual to produce a document that looked as if it had been typeset, but the dream was not realized until 1867, when Americans Christopher Sholes, Carlos Glidden, and Samual Soule patented the first typewriter. In late 1873 Remington & Sons Company, famous for guns and sewing machines, produced the first commercial typewriter. It was difficult to use and was not well received; Remington & Co. sold only 5,000 machines in the first five years. However, the typewriter did get the attention of Mark Twain, who used it to produce *Tom Sawyer*, which may have been the world's first typewritten manuscript. By 1890, more reliable typewriters were being produced, and the typewriter became a common piece of office equipment [7].

In 1908 Charles and Howard Krum succeeded in testing an experimental machine that allowed a modified Oliver typewriter to print a message transmitted over a telegraph line. They called their invention the teletype. During the 1920s news organizations began using teletype machines to transmit stories between distant offices, and Wall Street firms began sending records of stock transactions over teletypes.

1.3.5 Radio

Earlier we described how the experiments of Oersted, Sturgeon, and Henry led to the development of the electromagnet and the telegraph. The connection between electricity and magnetism remained mysterious, however, until Scottish physicist James Clerk Maxwell (1831–1879) published a mathematical theory demonstrating their relationship. This theory predicted the existence of an electromagnetic wave spreading with the velocity of light. It also predicted that light itself was an electromagnetic phenomenon. In 1885 Heinrich Hertz (1857–1894) successfully generated electromagnetic waves, proving the correctness of Maxwell's theory.

Guglielmo Marconi (1874–1937) put Hertz's discovery to practical use by successfully transmitting radio signals in the hills outside Bologna, Italy, in 1895. Unable to attract the attention of the Italian government, he took his invention to England, where he founded the Marconi Wireless Telegraph Company. The name of the company reflects Marconi's concept of how his invention would be used. To Marconi, radio, or "wireless," was a superior way to transmit telegraph messages.

David Sarnoff (1891–1971) emigrated from Russia to the United States with his family when he was nine. When he had completed school, he landed a position with the Marconi Wireless Telegraph Company. In 1912 Sarnoff made a name for himself when he intercepted the first distress signal from the *Titanic* and spent the next three days relaying information about the rescue effort to the rest of the world. Four years later, Sarnoff suggested the use of radio as an entertainment device, writing: "I have in mind a plan of development which would make radio a household utility in the same sense as the piano or phonograph . . . The receiver can be designed in the form of a simple 'Radio Music Box' . . . (which) can be placed in the parlor or living room" [23]. In two decades, Sarnoff's vision had become a reality.

The power of radio as a medium of mass communication was demonstrated on the evening of October 30, 1938 (the night before Halloween). From CBS Radio Studio One in New York, Orson Welles and the Mercury Theater put on a one-hour dramatization of H. G. Wells's *War of the Worlds*. To increase suspense, the play was performed as a series of news bulletins interrupting a concert of dance music. These bulletins described events occurring on a farm near Grovers Mill, New Jersey. Many listeners panicked. "People packed the roads, hid in cellars, loaded guns, even wrapped their heads in wet towels as

Figure 1.18 On July 20, 1969, television images of Neil Armstrong walking on the Moon were broadcast to hundreds of millions of viewers around the world. (Courtesy of NASA)

protection from Martian poison gas, oblivious to the fact that they were acting out the role of the panic-stricken public that actually belonged in a radio play" [24].

1.3.6 Television

Broadcasting video over a wire began in 1884 with the invention of an electromechanical television by Paul Nipkow (1860–1940), but the first completely electronic television transmission was made in 1927 by Philo Farnsworth (1906–1971). Millions of Americans were formally introduced to the television at the 1939 World's Fair held in New York City, which had as its theme, "The World of Tomorrow." Since a television set cost about as much as an automobile, televisions remained a rarity in American households until the 1950s, when prices fell dramatically.

Television's ability to send a message around the world was demonstrated in July 1969. Hundreds of millions of people watched on live TV as astronaut Neil Armstrong stepped from the lunar module onto the surface of the Moon (Figure 1.18).

Television has created many opportunities for "news junkies" to get their fixes. The major commercial broadcast television networks have been supplemented by CNN and other cable news organizations and a myriad of Web sites. The various organizations compete with each other to be the first to break news stories. Increasingly, the media have turned to computer technology to help them provide information to the public. Sometimes this has led to embarrassing mistakes, as in the 2000 U.S. Presidential election.

About 7:50 p.m. on the evening of Tuesday, November 7, 2000, before the polls had even closed in the Florida panhandle, the major networks began announcing that Al Gore would be the winner in Florida. Based on the expected result of the Florida election, the networks went on to predict—while people were still voting in the Western states—that Al Gore would be the next President of the United States.

You might be wondering how it is possible to predict the outcome of an election before everyone has voted. In a practice known as exit polling, a company called Voter News Service questions people leaving polling places. It combines the information it collects with early returns to predict the outcome of elections. Since 1988 the television networks have relied upon the Voter News Service to provide them with exit polling results.

As it turns out, Voter News Service's prediction was wrong. More than a month after the election, after a series of recounts and court decisions, George W. Bush was declared the victor in Florida. With Florida's electoral votes in hand, Bush won the Presidency.

1.3.7 Remote Computing

About the same time Claude Shannon figured out that electronic circuits could compute Boolean functions, Bell Labs researcher George Stibitz (1904–1995) got the same idea. Working at his kitchen table in 1937, he built a binary adder out of telephone relays, batteries, flashlight bulbs, tin strips, and wire. He took his invention back to Bell Labs and enlisted the help of Samuel Williams. Over the next two years they built the Complex Number Calculator, an electromechanical system that would add, subtract, multiply, and divide complex numbers.

Stibitz's next action is what sets him apart from other computer pioneers. He made a teletype machine the input/output device for the Complex Number Calculator. With this innovation, he did not have to be in the same room as the calculator to use it; he could operate it remotely.

In 1940 Stibitz demonstrated remote computing to members of the American Mathematical Society who were meeting in New Hampshire. He typed numbers into the teletype, which transmitted the data 250 miles to the calculator in New York City. After the calculator had computed the answer, it transmitted the data back to the teletype, which printed the result.

1.3.8 ARPANET

In reaction to the launch of Sputnik by the Soviet Union in 1957, the Department of Defense created the Advanced Research Projects Agency (ARPA). ARPA funded research and development at prominent universities. The agency's first director, J.C.R. Licklider, imagined a "Galactic Network"—a global computer network that would facilitate the exchange of programs and data.[4] This view of the computer as a device to improve communication was in stark contrast to the mindset of computer manufacturers, which continued to think of computers as number-crunching machines.

4. The primary source document for this description of the evolution of the Internet is "A Brief History of the Internet" by Barry M. Leiner et al. [25].

Conventional, circuit-switched telephone networks were not a good foundation upon which to build a global computer network (Figure 1.19a). Between 1961 and 1967 three research teams independently came up with an alternative to circuit-switched networks. These teams were led by Donald Davies and Roger Scantlebury at NPL in England, Paul Baran at RAND, and Leonard Kleinrock at MIT. Eventually the new design came to be called a packet-switched network (Figure 1.19b).

In 1967 ARPA initiated the design and construction of the ARPANET. Fear of a nuclear attack led to the crucially important design decision that the network should be decentralized. In other words, the loss of any single computer or communication link would not prevent the rest of the network from working. Every computer on the network would have the ability to make decisions about how message traffic should be routed. Packet-switched networks met this condition; circuit-switched networks did not.

BBN in Boston was responsible for the Interface Message Processor (IMP) that connected a computer to the telephone network. In 1969 BBN delivered its first four IMPs to UCLA, the Stanford Research Institute, the University of California at Santa Barbara, and the University of Utah.

1.3.9 Email

During the earliest years of ARPANET, the networked computers could transfer programs and data only. ARPANET users still relied upon the telephone for personal communications. In March 1972 Ray Tomlinson at BBN wrote the first software enabling email messages to be sent and received by ARPANET computers. A few months later, Lawrence Roberts created the first "killer app" for the network: an email utility that gave individuals the ability to list their email messages, selectively read them, reply to them, forward them to others, and save them. Email quickly became the most popular network application.

Today, email is one of the most important communication technologies on the planet. More than two billion email messages a day are sent in the United States alone.

1.3.10 Internet

ARPA researchers anticipated the need to connect the ARPANET with other networks based on different designs. Robert Kahn developed the concept of open architecture networking, which means individual networks could be quite different as long as they shared a common "internetworking architecture." Vinton Cerf and Robert Kahn designed the TCP/IP protocol that would support open architecture networking [26]. TCP (Transmission Control Protocol) is responsible for dividing a message into packets at the sending computer and reassembling the packets at the receiving computer. IP (Internet Protocol) is the set of rules used to route data from computer to computer. The **Internet** is the network of networks that communicate using TCP/IP. You could call January 1, 1983, the birth date of the Internet, because that was the date on which all ARPANET hosts converted to TCP/IP.

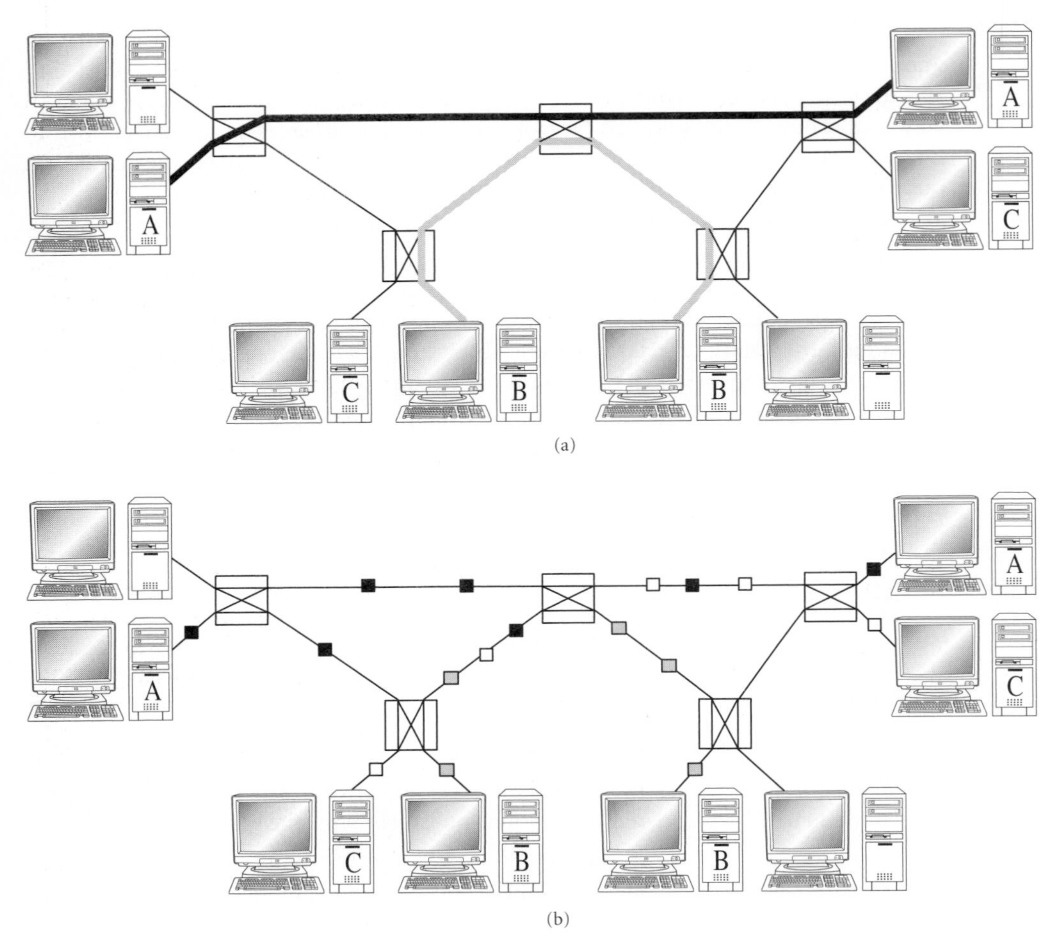

FIGURE 1.19 Comparison of circuit-switched networking and packet-switched networking. (a) In a **circuit-switched network**, a single physical connection is established between the two ends. The physical connection cannot be shared. In this illustration, one circuit links the two computers labeled A, and another circuit links the two computers labeled B. The computers labeled C may not communicate at this time, because no circuit can be established. (b) In a **packet-switched network**, a message is divided up into small bundles of data called packets. Every packet has the address of the computer where it should be routed. If there is more than one path from the message source to the message destination, different message packets may take different routes. Packets from different messages may share the same wire. In this illustration, three pairs of computers (labeled A, B, and C) are communicating simultaneously over a packet-switched network.

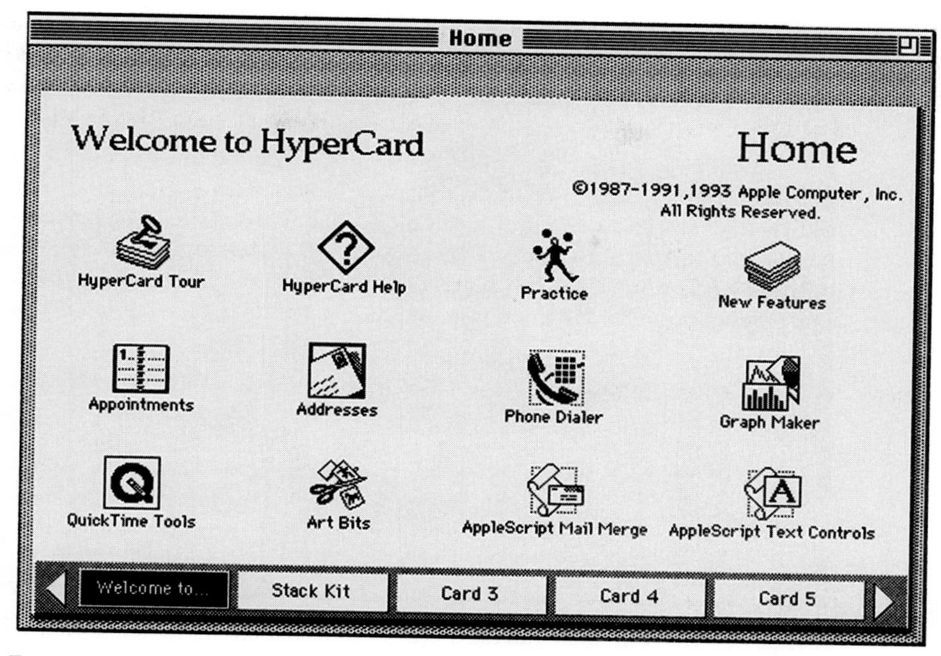

Figure 1.22 Apple's HyperCard is a notable example of a single-computer hypertext system.

operating system and programming environment. His boss okayed the purchase, then puckishly suggested that maybe Berners-Lee ought to try implementing his proposed hypertext system on it [36].

Unlike earlier commercial hypertext systems, Berners-Lee's system allowed links between information stored on different computers connected by a network. Because it is built on top of the TCP/IP protocol, links can connect any two computers on the Internet, even if they have different hardware or are running different operating systems.

A **Web browser** is a program that allows a user to view Web pages and traverse hyperlinks between pages. Berners-Lee completed the first Web browser on the NeXT Computer on Christmas Day, 1990. He called his browser WorldWideWeb. In March 1991 he released the browser to some computer users at CERN.

The first widely used Web browser was Mosaic, developed at the National Center for Supercomputer Applications at the University of Illinois, Urbana-Champaign. Other well-known Web browsers are Netscape Navigator, Mozilla Firefox, and Microsoft Internet Explorer. Since 1999 Internet Explorer has been the most widely used Web browser.

Today, you can use the Web to retrieve text, still images, movies, songs, computer programs—in theory, anything that can be digitized. The Web has also become a convenient way for organizations to provide people access to news updates and dynamically changing information (Figure 1.23).

FIGURE 1.23 Using the Web, you can access up-to-date traffic information from many major cities. (Digital Stock)

1.4.8 Search Engines

A **search engine** is a program that accepts a list of keywords from a user, searches a database of documents, and returns those documents most closely matching the specified keywords. Today, the term search engine is most frequently used to describe programs that search databases of Web pages. Web search engines are the most powerful information retrieval devices ever invented. The most popular Web search engine, Google, has indexed more than 8 billion Web pages.

There are two types of Web search engines. Crawler-based search engines, such as Google and AltaVista, automatically create the database of information about Web pages. In a process similar to Web surfing, programs called spiders follow hyperlinks, eventually visiting millions of different Web pages. Summary information about these pages is collected into massive databases. When you perform a query, the search engine consults its database to find the closest matches.

The second type of Web search engine relies upon humans to build the database of information about various Web pages. People who develop a Web site can submit a summary of their site to the keepers of the search engine. Alternatively, those responsible for the search engine may create their own reviews of Web sites. The advantage of this kind of search engine is that humans can create more accurate summaries of a Web page than a spider program. The disadvantage of this approach is that only a small fraction of the Web can be catalogued. Open Directory is an example of this kind of search engine.

Today, some search engines are taking a hybrid approach. For example, MSN Search uses LookSmart to present results from a human-created database of Web pages, but it supplements these results with pages found by the crawler-based search engine Inktomi.

1.5 Information Technology Issues

Information technology (IT) refers to devices used in the creation, storage, manipulation, exchange, and dissemination of data, sound, and/or images. Computers, telephones, and video cameras are examples of IT. The cost of IT devices continues to fall, while their capabilities continue to increase. As a result, people are making greater use of IT in their everyday lives. Some of these uses create new issues that need to be resolved. In this section we describe a few of the issues raised by the growth of IT.

The great power of email is that it allows (at least in principle) anyone to send email to anyone else with an email address. Now that just about everyone has an email account, it is easier than ever to contact friends and family. Parents who used to complain because they never got letters from their children at college found out that it was much easier to keep in touch via email. On the other hand, about half of all email traffic is spam: unsolicited, bulk, commercial email. Is spam destroying the value of email?

Thanks to the Web, it is easier than ever to share information with people all over the world. Imagine I live in Canada and post some files on my Web site. Some Americans visit my Web site and download the files, an action that violates U.S. laws. Should I be prevented from posting material that is legal in Canada but illegal in the United States?

For many items of value, making the original copy is expensive, but making copies of the original is inexpensive. For example, a record company may spend hundreds of thousands of dollars to produce a CD, but copies can be burned for just a few cents each. Once CDs have been ripped into MP3 files, the Internet provides a fast and efficient way to distribute them. As a result, unauthorized copies of songs, movies, and computer programs are proliferating. Should we continue to give ownership rights to creators of intellectual property, or is it hopeless? If we no longer give ownership rights to creators of intellectual property, will creativity be stifled?

If I use a credit card to purchase an item, the credit card company now has information about my spending habits. Who has a right to that information? For example, if I buy a pair of water skis with my credit card, does the credit card company have a right to sell my name, address, and phone number to other companies that may want to sell me some of their products?

The use of IT has changed the way that banks process loan applications. Rather than make a local decision regarding my creditworthiness, the bank will check a national credit bureau and ask for my credit rating. What are the advantage and disadvantages of this alternative approach to lending money?

Computers are now embedded in many devices on which we depend, from traffic signals to pacemakers. Software errors have resulted in injury and even death. When bugs result in harm to humans, what should the liability be for the people or corporations that produced the software?

When employees use IT devices in their work, companies can monitor their actions closely. For example, a company can track the number of calls per minute each of its telephone operators is handling. It can document the number of keystrokes per minute of its data entry operators. It can log all of the Web sites its employees visit, and it can read the email they send and receive at work. How does such monitoring affect the workplace? Does it create an unacceptable level of stress among employees?

IT is allowing more people than ever to telecommute; in other words, work from home. What are the advantages and disadvantages of telecommuting?

IT capabilities are leading to changes in the IT industry itself. Silicon Valley used to be the epicenter of the IT industry, but improvements in the speed and reliability of communication networks have led to a more decentralized landscape. New hot spots of innovation include Redmond, Washington (Microsoft), Austin, Texas (Dell Computer), Armonk, New York (IBM), Walldorf, Germany (SAP), and Bangalore, India (Wipro Technologies). U.S.-based software companies are doing more development in countries where salaries are much lower, such as India, China, and Vietnam [5]. Will this trend continue? How many software jobs in the United States will be lost to countries where labor is significantly cheaper?

Finally, the World Wide Web has provided an unprecedented opportunity for individuals and nongovernmental organizations to have their points of view made available to millions. This could bring about new levels of citizen involvement and democratic reform. On the other hand, some countries are making large portions of the Web unavailable to its citizens. Will the Web prove to be a tool for democracy, or will it be muzzled by repressive regimes?

Summary

We are living in the Information Age, an era characterized by ubiquitous computing and communication devices that have made information much easier to collect, transmit, store, and retrieve. These devices are the culmination of centuries of technological progress.

In the seventeenth century Blaise Pascal and Gottfried von Leibniz constructed slow, unreliable mechanical adding machines. The mechanical era of computing continued into the nineteenth century, when Charles Babbage designed an analytical engine that many cite as a precursor to the modern computer. However, the theoretical work of nineteenth-century mathematician George Boole had a much greater impact on the development of the electronic computer. Boole invented an algebra that manipulated only two symbols: false and true (or 0 and 1). In the 1930s Claude Shannon and George Stibitz independently discovered that Boole's logical functions could be performed by electronic switching circuits. Their discovery opened the way for the development of electronic computing devices.

Several groups invented computing machines that had many of the attributes of the modern electronic, programmable computer. However, the first system to have all of the essential features of a modern computer—fully electronic operation, program and data stored in random access memory, and a jump instruction—was the Small-Scale Experimental Machine developed at the University of Manchester in England. Soon after, computers were commercially available.

The invention of the transistor and the integrated circuit paved the way for the creation of the microprocessor, or computer on a chip. The rapid decline in the cost of manufacturing microprocessors has made possible the use of computers in items as mundane as electric mixers and door locks, as well as solar-powered calculators selling for less than $5, a far cry from the pioneering efforts of Pascal and Leibniz.

Advances in computing technology have been accompanied by equally dramatic improvements in communications networks. Two centuries ago mechanical semaphore telegraph systems could transmit messages at about 350 miles per hour. The discovery of electromagnetism led to the invention of the modern telegraph that could transmit messages at the speed of electricity. Alexander Graham Bell worked on improving the telegraph. He ended up inventing the telephone, which enabled the creation of the first "online" communities. Radio and television made it possible for millions of people around the planet to receive the same message simultaneously. Today's Internet binds hundreds of millions of computers across the globe, creating new opportunities for sending as well as receiving messages.

The codex represented a significant improvement over the scroll as a way of storing information. The codex was more durable than the papyrus scroll, and it was much easier for readers to find a particular passage they were looking for. The availability of paper and printing presses based on movable metal type made it possible for ordinary people to afford codices. Even today, books and magazines are typically produced using this technology.

Vannevar Bush, Ted Nelson, and Douglas Engelbart all envisioned more powerful ways of storing and retrieving information. Bush suggested that a machine be used to mimic the associative memory of the brain. Nelson invented the word "hypertext," meaning a linked network of nodes containing information. Engelbart conceived of a computing system built around a graphical display device. His investigations led to many innovations, including the computer mouse and a video display divided into windows. Engelbart's system became practical with the invention of the PC at Xerox PARC and the subsequent availability of low-cost systems with graphical user interfaces, such as the Apple Macintosh. Single-user hypertext systems were available for the Macintosh a couple of years after its initial release, but the true power of hypertext was revealed when Tim Berners-Lee created the World Wide Web, which allowed links between information stored on different computers. The Web has become a popular and powerful information storage and retrieval mechanism.

What conclusions can we draw from our study of the development of computers, communication networks, and information storage and retrieval devices? First, revolutionary discoveries are rare. Rather, most innovations represent simply the next step in a

long staircase of evolutionary changes. Each inventor, or team of inventors, relies upon prior work. In many cases different inventors come up with the same "original" idea at the same time.

A second conclusion we can draw from these stories is that information technology did not begin with the personal computer and the World Wide Web. Many other inventions, including the telegraph, the telephone, the radio, and the television, led to significant changes in society when they were adopted.

Nevertheless, in the past two decades the rate of technological change has accelerated, thanks to low-cost computers and high-speed communication networks. The pace of technological progress is so rapid that we may feel hard-pressed to keep up. It's good to reflect on Seymour Papert's observation:

> So we are entering this computer future; but what will it be like? What sort of a world will it be? There is no shortage of experts, futurists, and prophets who are ready to tell us, but they don't agree. The Utopians promise us a new millennium, a wonderful world in which the computer will solve all our problems. The computer critics warn us of the dehumanizing effect of too much exposure to machinery, and of disruption of employment in the workplace and the economy.
>
> Who is right? Well, both are wrong—because they are asking the wrong question. The question is not "What will the computer do to us?" The question is "What will we make of the computer?" The point is not to predict the computer future. The point is to make it. [37]

Review Questions

1. Why do some people call this the Information Age?
2. What can the Amish teach us about our relationship with technology?
3. Using a pencil and paper (no calculators allowed!), calculate the tangent of 0.5 radians to four digits of precision. (The formula you need is in Figure 1.6.)
4. What were Charles Babbage's motivations for constructing the Difference Engine?

5. Use the method of differences to compute the squares of the integers 2, 3, 4, 5, and 6.
6. Using the diagram in Figure 1.9 as your guide, draw circuits for the following Boolean expressions:
   a. (a AND b) OR (c AND d)
   b. (a AND b) OR (b AND c)
7. What was the principal innovation of the IBM System/360?
8. Draw a large table to illustrate the capabilities of various experimental computers. Put each of these computers in its own row:
   - Analytical Engine
   - Arithmetic Machine (Pascal)
   - Atanasoff-Berry Computer
   - ENIAC
   - Harvard Mark 1
   - Small-Scale Experimental Machine
   - Step Reckoner (Leibniz)
   - Z1
   - Z3

   Put each of these attributes in its own column:
   - Stores whole numbers
   - Addition and subtraction
   - Multiplication and division
   - Stores fractional numbers
   - Automatic multiple-step calculation
   - Programmable for different functions
   - Program stored internally
   - Jump instruction implemented
   - Electronic calculation (no moving parts)

   Fill in the table with Xs to indicate which of these machines had each of these attributes.
9. Whom does the author recognize as the creators of the first true computer? What was their nationality?
10. How did the invention of the vacuum tube help AT&T introduce long-distance telephone service?
11. Name three ways the development of radar in World War II stimulated advances in computing.
12. Can you think of a practical reason why the semaphore telegraph was adopted more rapidly on the continent of Europe than in the British Isles?
13. Briefly describe three ways in which society changed by adopting the telephone.
14. What is the difference between a circuit-switched network and a packet-switched network?
15. Why does the Internet have a decentralized structure?
16. How did the National Science Foundation stimulate the creation of commercial, long-distance data networks in the United States?

17. Describe two ways in which the codex represented an improvement over the scroll.
18. What is hypertext?
19. Who invented the computer mouse?
20. Who invented the personal computer?
21. The Apple Macintosh succeeded in the marketplace, while the Apple Lisa failed. Give two reasons why this happened.
22. How is a hypertext link similar to a citation in a book? How is it different?
23. In what fundamental way is an Apple HyperCard stack different from the World Wide Web?
24. Berners-Lee decided to build the World Wide Web on top of the TCP/IP protocol. Why did this decision help ensure the success of the Web?
25. What was the first widely used Web browser? What is the most popular browser in use today?
26. Use four different search engines (www.altavista.com, www.google.com, www.msn.com, www.yahoo.com) to perform a search on the phrase "Information Technology." Create a table that compares the top 10 Web pages returned by each search engine. Which engines were the most similar?
27. Link each person in the left column with the appropriate invention or discovery in the right column. A single person may link to more than one invention. More than one person may link to the same invention.

Howard Aiken	Analytical Engine
Paul Allen	Apple Computer
John Atanasoff	Arithmetic Machine
Charles Babbage	Atanasoff-Berry Computer
Alexander Graham Bell	BASIC for Altair 8800, then Microsoft
Tim Berners-Lee	Battery
Clifford Berry	Boolean algebra
George Boole	Colossus
Peter Brown	Difference Engine
Vannevar Bush	EDVAC
Augusta Ada Byron	Electromagnet
Vinton Cerf	Electromagnetic waves
Lee De Forest	Electromagnetism
J. Presper Eckert	ENIAC
Douglas Engelbart	Harvard Mark 1
Federico Faggin	Hypertext
Philo Farnsworth	Integrated circuit
Bill Gates	Internet
Carlos Glidden	Mark-8 microcomputer
Johannes Gutenberg	Microprocessor

Joseph Henry
Heinrich Hertz
Tedd Hoff
Grace Murray Hopper
Steve Jobs
Robert Kahn
Alan Kay
Tom Kilburn
Jack Kilby
Charles Krum
Howard Krum
Gottfried von Leibniz
Guglielmo Marconi
John Mauchly
James Clerk Maxwell
Stanley Mazor
Samuel Morse
Ted Nelson
John von Neumann
Robert Noyce
Christian Oersted
Blaise Pascal
Georg and Edvard Scheutz
Christopher Sholes
Bill Shockley
Samual Soule
George Stibitz
William Sturgeon
Jon Titus
Geoff Tootill
Alan Turing
Alessandro Volta
F. C. Williams
Steve Wozniak
Konrad Zuse

Mouse
Printing press
Radio
Remote computing
Small-Scale Experimental Machine
Step Reckoner
Telegraph
Telephone
Teletype
Television
Transistor
Typewriter
UNIVAC
Vacuum tube
World Wide Web
Xerox Alto
Z1, Z2, Z3

28. What is a search engine? Describe the two types of search engines.

29. What is information technology?

30. Name three inventions described in this chapter that were created for a military application.

31. This chapter names the two most popular applications of the Internet. What are they?

Discussion Questions

32. Think about the last piece of consumer electronics you purchased. How did you first learn about it? What factors (features, price, ease of use, etc.) did you weigh before you purchased it? Which of these factors were most influential in your purchase decision? Are you still happy with your purchase?
33. Do you tend to acquire new technological devices before or after the majority of your friends? What are the pros and cons of being an early adopter of a new technology? What are the pros and cons of being a late adopter of a new technology?
34. Have you ever gone camping or had another experience where you went for at least a few days without access to a phone, radio, television, or computer? (In other words, there was no communication between you and the outside world.) What did you learn from your experience?
35. Are there any technologies that you wish had never been adopted? If so, which ones?
36. Some say that no technology is inherently good or evil; rather, any technology can be used for either good or evil purposes. Do you share this view? Explain.
37. More than 90 percent of personal computers run a version of the Microsoft Windows operating system. In what ways is this situation beneficial to computer users? In what ways does this situation harm computer users?
38. Martin Carnoy writes, "Thanks to a communications and software revolution, we are more 'connected' than ever before—by cell phone, email, and video conferencing—yet more disconnected than in the past from social interaction" [38]. Do you agree?

In-Class Exercises

39. Many cell phones now come equipped with cameras. Managers of health clubs are concerned that people in locker rooms may be secretly photographed by other members carrying small cell phones. Debate the following proposition: "Health clubs should ban all cell phone use within their premises."
40. In the 1984 Presidential election, all the major television networks used computers to predict that Republican Ronald Reagan would defeat Democrat Walter Mondale, even before the polls closed on the West Coast. When they heard this news, some Mondale supporters who had been waiting in line to vote simply went home without voting. This may have influenced the results of some statewide and local elections. Debate the following proposition: "In Presidential elections polls should close at the same time everywhere in the United States."
41. Honda Motor Company has begun offering a collision mitigation brake system on its new Inspire sedan sold in Japan. The system uses a radar hidden behind the H logo in the grill to detect vehicles within 100 meters. When a vehicle is detected, the system warns the driver by sounding a buzzer, flashing a light, tightening the seat belt, and braking slightly. If the driver fails to respond, the system brakes more and tightens the seat belt further to reduce the impact of the collision.

Discuss possible benefits and risks associated with Honda's collision detection system.

42. IBM's T. J. Watson Laboratory is working on software called the Artificial Passenger. Suppose you're driving alone to a faraway destination. The Artificial Passenger will engage you in conversation and analyze your responses. For example, the program may ask you, "Who was the first person you ever dated?" If your response is mumbled, incorrect, or even just too slow in coming, the program may sound an alarm, roll down a window, or start music playing. It may even suggest you pull over or find a hotel at which to spend the night.

    Discuss possible benefits and risks associated with IBM's Artificial Passenger.

Further Reading

George Beekman and Michael J. Quinn. *Computer Confluence: Exploring Tomorrow's Technology*. 7th ed. Prentice Hall, Upper Saddle River, NJ, 2006.

Tim Berners-Lee with Mark Fischetti. *Weaving the Web: The Original Design and Ultimate Destiny of the World Wide Web by Its Inventor*. HarperCollins, New York, NY, 1999.

Alice R. Burks and Arthur W. Burks. *The First Electronic Computer: The Atanasoff Story*. The University of Michigan Press, Ann Arbor, MI, 1988.

Martin Campbell-Kelly and William Aspray. *Computer: A History of the Information Machine*. BasicBooks, New York, NY, 1996.

Paul Carroll. *Big Blues: The Unmaking of IBM*. Crown Publishers, New York, NY, 1993.

William H. Dutton, editor, with Malcom Peltu. *Information and Communication Technologies: Visions and Realities*. Oxford University Press, Oxford, England, 1996.

Paul Freiberger and Michael Swaine. *Fire in the Valley: The Making of the Personal Computer*. Osborne/McGraw-Hill, Berkeley, CA, 1984.

Barry M. Leiner, Vinton G. Cerf, David D. Clark, Robert E. Kahn, Leonard Kleinrock, Daniel C. Lynch, Jon Postel, Larry G. Roberts, and Stephen Wolff. "A Brief History of the Internet." Internet Society Web site, www.isoc.org/internet/history/brief/shtml, December 10, 2003.

Stephen Manes and Paul Andrews. *Gates: How Microsoft's Mogul Reinvented an Industry—And Made Himself the Richest Man in America*. Doubleday, New York, NY, 1993.

Carolyn Marvin. *When Old Technologies Were New: Thinking About Electric Communication in the Late Nineteenth Century*. Oxford University Press, New York, NY, 1988.

John Naughton. *A Brief History of the Future: From Radio Days to Internet Years in a Lifetime*. The Overlook Press, Woodstock, NY, 1999.

Ithiel de Sola Pool, editor. *The Social Impact of the Telephone*. The MIT Press, Cambridge, MA, 1977.

Michael Riordan and Lillian Hoddeson. *Crystal Fire: The Birth of the Information Age*. W. W. Norton & Company, New York, NY, 1997.

David Ritchie. *The Computer Pioneers: The Making of the Modern Computer*. Simon & Schuster, New York, NY, 1986.

Frederick Seitz and Norman G. Einspruch. *Electronic Genie: The Tangled History of Silicon*. University of Illinois Press, Urbana, IL, 1998.

Joel Shurkin. *Engines of the Mind: The Evolution of the Computer from Mainframes to Microprocessors*. W. W. Norton & Company, New York, NY, 1996.

Nancy Stern. *From ENIAC to UNIVAC: An Appraisal of the Eckert-Mauchly Computers*. Digital Press, Bedford, MA, 1981.

Transparency (Web site). www.transparencynow.com.

James Wallace and Jim Erickson. *Hard Drive: Bill Gates and the Making of the Microsoft Empire*. John Wiley & Sons, New York, NY, 1992.

What You Need to Know About (Web site). Inventors section, www.inventors.about.com.

Michael R. Williams. *A History of Computing Technology*. 2nd ed. IEEE Computer Society Press, Washington, DC, 1997.

References

[1] May Swenson. "Southbound on the Freeway." *The New Yorker*, February 16, 1963.

[2] Steve Friess. "Laptop Design Can Be a Pain in the Posture." *USA Today*, April 12, 2005.

[3] Christine Rosen. "Our Cell Phones, Ourselves." *The New Atlantis: A Journal of Technology & Society*, Summer 2004.

[4] Howard Rheingold. "Look Who's Talking." *Wired*, (7.01), January 1999.

[5] "The New Geography of the IT Industry." *The Economist*, pages 47–49, July 19, 2003.

[6] D'Arcy Jenish and Catherine Roberts. "Heating Up Nuclear Power." *Maclean's*, 113(24): 19, June 11, 2001.

[7] Martin Campbell-Kelly and William Aspray. *Computer: A History of the Information Machine*. BasicBooks, New York, NY, 1996.

[8] Maurice V. Wilkes. "Babbage's Expectations for His Engines." *IEEE Annals of the History of Computing*, 13(2):141–145, April–June 1991.

[9] George Beekman and Michael J. Quinn. *Computer Confluence: Exploring Tomorrow's Technology*. 7th ed. Prentice Hall, Upper Saddle River, NJ, 2006.

[10] George Boole. *An Investigation of the Laws of Thought: On Which Are Founded the Mathematical Theories of Logic and Probabilities*. Dover Publications, New York, NY, 1951. Originally published in 1854 by Walton and Maberly of London.

[11] Nancy Stern. *From ENIAC to UNIVAC: An Appraisal of the Eckert-Mauchly Computers*. Digital Press, Bedford, MA, 1981.

[12] Tony Sale. The "Williams Tube Revisited." *Resurrection, The Bulletin of the Computer Conservation Society*, 1(5), Spring 1993.

[13] Joel Shurkin. *Engines of the Mind: The Evolution of the Computer from Mainframes to Microprocessors*. W.W. Norton & Company, New York, NY, 1996.

[14] Leslie Goff. "Univac Predicts Winner of 1952 Election." *CNN.com*, April 30, 1999.

[15] Michael Riordan and Lillian Hoddeson. *Crystal Fire: The Birth of the Information Age*. W. W. Norton & Company, New York, NY, 1997.

[16] Frederick Seitz and Norman G. Einspruch. *Electronic Genie: The Tangled History of Silicon*. University of Illinois Press, Urbana, IL, 1998.

[17] Paul Carroll. *Big Blues: The Unmaking of IBM*. Crown Publishers, New York, NY, 1993.

[18] Sidney H. Aronson. "Bell's Electrical Toy." In *The Social Impact of the Telephone*, edited by Ithiel de Sola Pool. The MIT Press, Cambridge, MA, 1977.

[19] Carolyn Marvin. *When Old Technologies Were New: Thinking about Electric Communications in the Late Nineteenth Century*. Oxford University Press, New York, NY, 1988.

[20] "Telephone Cranks." *Western Electrician (Chicago)*, page 37, July 17, 1897.

[21] Ithiel de Sola Pool. Introduction. In *The Social Impact of the Telephone*, edited by Ithiel de Sola Pool. The MIT Press, Cambridge, MA, 1977.

[22] Asa Briggs. "The Pleasure Telephone." In *The Social Impact of the Telephone*, edited by Ithiel de Sola Pool. The MIT Press, Cambridge, MA, 1977.

[23] "Radio: The Roots of Broadcasting." *Technical Press*. www.tvhandbook.com/History.

[24] "War of the Worlds, Orson Welles, and the Invasion from Mars." *Transparency*. www.transparencynow.com.

[25] Barry M. Leiner, Vinton G. Cerf, David D. Clark, Robert E. Kahn, Leonard Kleinrock, Daniel C. Lynch, Jon Postel, Larry G. Roberts, and Stephen Wolff. *A Brief History of the Internet, Version 3.32*, December 10, 2003. www.isoc.org/internet/history/brief.shtml.

[26] Vinton G. Cerf and Robert E. Kahn. "A Protocol for Packet Network Intercommunication." *IEEE Transactions on Communications*, COM-22(5), May 1974.

[27] Frank Rose. "Seoul Machine." *Wired*, pages 126–131, May 2005.

[28] "Prophet of American Technodoom." *The Economist*, page 66, April 23, 2005.

[29] Thomas Cahill. *How the Irish Saved Civilization: The Untold Story of Ireland's Role from the Fall of Rome to the Rise of Medieval Europe*. Anchor Books, New York, NY, 1995.

[30] Elizabeth L. Eisenstein. *The Printing Press as an Agent of Change*. Volume 1. Cambridge University Press, Cambridge, England, 1979.

[31] Vannevar Bush. "As We May Think." *The Atlantic Monthly*, 176(1):101–108, August 1945.

[32] Owen Edwards. "Ted Nelson." *Forbes*, August 25, 1997.

[33] Lauren Wedeles. "Prof. Nelson Talk Analyzes P.R.I.D.E." *Vassar Miscellany News*, February 3, 1965.

[34] T. O'Brien. "The Mouse." *SiliconValley.com*, 2000.

[35] "Internet Pioneers: Doug Englebart." *ibiblio (Web site)*. www.ibiblio.org/pioneers/englebart.html.

[36] Tim Berners-Lee. *Weaving the Web*. HarperCollins Publishers, New York, NY, 1999.

[37] Seymour Papert. "A Critique of Technocentrism in Thinking About the School of the Future." Technical report, MIT Media Lab, September 1990. Epistemology and Learning Memo No. 2.

[38] Martin Carnoy. *Sustaining the New Economy: Work, Family, and the Community in the Information Age*. Russel Sage Foundation/Harvard University Press, New York, NY/Cambridge, MA, 2000.

AN INTERVIEW WITH

Douglas Engelbart

Raised on a small farm near Portland, Oregon, Douglas C. Engelbart earned a B.S. in electrical engineering from Oregon State College in 1948. After working three years at NASA's Ames Research Lab, he went to the University of California, Berkeley and earned a Ph.D. in electrical engineering in 1955. A few years later he began the Augmentation Research Center at the Stanford Research Institute. He and his research team are credited with prototyping the computer mouse, multi-window displays, shared-screen teleconferencing, a working hypermedia system, and dozens of other inventions.

Dr. Engelbart has won many prizes and awards, including the National Medal of Technology and the world's largest prize for invention and innovation, the Lemelson-MIT Prize of $500,000. In 1998 he was inducted into the National Inventors Hall of Fame.

He is currently the Director of his own company, Bootstrap Institute, which uses speaking engagements, consulting, development projects, management seminars, and strategic alliances to help organizations achieve peak performance by improving their processes of improvement.

What thinkers influenced you the most?

When I was stationed in The Philippines after World War II, I ran across an issue of *Life* magazine that reprinted Vannevar Bush's *Atlantic Monthly* article "As We May Think." At the time, I didn't think about it that much, but 13 years later I remembered his description of links that would take you from one microfiche frame to another.

Another person who influenced my thinking was S. I. Hayakawa, who wrote a book called *Language in Thought and Action.* It talks about how perception converts sensory stimuli into recognition in particular concept terms. When you are learning how to read, you first learn how to convert letters into a word. After a while, you just perceive the word, not the individual letters. As you mature, you don't even say the word, the meaning just flows. It got me thinking about how the human brain, which was created for survival, was adapted for reading. The current paradigm says the way you are meant to read is on a book page. But wait—that's just the current technology. A computer can present new options, and our perception can adapt. I'm still waiting for the paradigms to come around.

Your laboratory, the Augmentation Research Center, is credited with dozens of innovations. In your opinion, which of them have had the most impact on society?

Boy, a lot of them were just rejected! But not all. We were the first ones to have a system that was using hyperlinks. We had email by 1970 and spreadsheets by the mid-1970s. We also had distance collaboration with shared screens. Two widely separated people could be talking on the phone, looking at exactly the same working display and passing its control back and forth between them.

I started using the word "augmentation" because I didn't think the phrase "office automation" was accurate. An electronic typewriter doesn't really automate writing. It augments the human by neatly printing characters on the page. The word "augmentation" suggests that a significant new artifact can create a whole bunch of new capabilities. You can't predict all the changes that a new technology will bring. The best thing you can do is facilitate these changes.

How can organizational improvements "boost mankind's collective IQ," as it says on your Web site?

The world is getting so much more complex, and there is a mindlessness in our approach to many issues. If we don't raise our collective IQ, things will just collapse. For example, the United States is collectively very stupid about global warming. The everyday populace doesn't have a realization of what's going on. How can we boost the collective IQ of American citizens? If we improve the collective ability of people to digest information, they can make better decisions. We need to create a capability infrastructure that gives people the best understanding of the current situation, the best understanding of the possible solutions, and the best understanding of the resources available. A higher collective IQ means being alert to a situation, understanding it, and having a plan.

What do you mean by bootstrapping collective IQ?

If we'd like to get collectively smarter, how fast and in what corners can we start making progress toward it? When we make headway, let's fix it so that we are that much more effective in making more headway. Strategically, we must consider how to invest resources to boost collective IQ on broad scale.

For the past five decades, your visions for the future of computing have been often ignored and frequently ridiculed. Did you ever lose heart?

No, but I certainly got dented. Many of the things for which I got prizes and awards thirty years later were ridiculed when I thought them up. I was fortunate to have reflected on this paradigm thing enough to understand that when someone criticized me, it's because in their paradigm I was talking nonsense. When I got my big idea in the 1950s, there were only about three working computers in the country. When I talked about people interacting with a computer, it was equivalent to saying someday people will have their own private helicopters. I got terribly frustrated waiting for the paradigms to change.

Do you have a prediction for the next big information-technology breakthrough?

There are going to be computer displays incorporated into glasses or contact lenses. Our interactions with computers will become more subtle. They will respond to our twitches or facial expressions. We'll have cell phones in our ears. Cranial implants will provide amazing extensions to our senses. These implants will speed our perceptions. On the other hand, you won't have to get out of bed to interact with a computer. If you don't move, imagine what kind of a deplorable physical life you will have.

We will have a dynamic knowledge repository. That's a prime thing to pursue. It's one thing to Google something, to find instances of a fact or concept—but, what is the current understanding? What is the upshot? What does it all mean? What does it add up to be? It's a little understood process by which all of this activity evolves into a common picture. Accountants can look at each person's entry and come up with an overall financial picture, but we don't know how to do this "integration" effectively for general knowledge.

Imagine what would happen if artificial intelligence gave machines more intelligence than humans. Why keep us around? We need to improve our collective IQ.

CHAPTER

2 Introduction to Ethics

No man is an island, entire of itself; every man is a piece of the continent, a part of the main. If a clod be washed away by the sea, Europe is the less, as well as if a promontory were, as well as if a manor of thy friend's or of thine own were. Any man's death diminishes me, because I am involved in mankind; and therefore never send to know for whom the bell tolls; it tolls for thee.

–JOHN DONNE, *Meditation XVII*

2.1 Introduction

IMAGINE HOVERING ABOVE THE EARTH IN A SPACECRAFT on a cloudless night. Looking down upon our planet, you see beautiful constellations of artificial light (Figure 2.1). The stars in these incandescent galaxies are our communities.

Forming communities allows us to enjoy better lives than if we lived in isolation. Communities facilitate the exchange of goods and services. Instead of each family assuming responsibility for all of its needs, such as food, housing, clothing, education, and health care, individuals can focus on particular activities. Specialization results in higher productivity that increases everyone's quality of life. Communities also make people more secure against external dangers.

There is also a price associated with being part of a community. Communities prohibit certain actions and make other actions obligatory. Those who do not conform with these prohibitions and obligations can be punished. Still, the fact that people *do*

Figure 2.1 Looking down on London, England, at night from space. (Courtesy of NASA)

live in communities is strong evidence that the advantages of community life outweigh the disadvantages.

Virtually everybody values life, happiness, and the ability to accomplish goals. These are examples of what philosopher James Moor calls "core values" [1]. You have done nothing virtuous if you act in accord with your core values. Even evil people seek these things for themselves. To move from a selfish point of view to the "ethical point of view," you must decide that other people and their core values are worthy of your respect as well [1].

People who take the ethical point of view may still disagree over what is the proper course of action to take in a particular situation. Some disagreements are caused because people cannot agree on the facts of the matter. At other times, different value judgements lead people to opposite conclusions. Often, the source of different value judgements is the use of different ethical theories to evaluate the problem. For this reason, it is worthwhile to have a basic understanding of some of the most popular ethical theories. In this chapter we will describe the difference between morality and ethics, discuss a variety of ethical theories, evaluate their pros and cons, and show how to use the more viable ethical theories to solve moral problems.

2.1.1 Defining Terms

A **society** is an association of people organized under a system of rules designed to advance the good of its members over time [2]. Cooperation among individuals helps promote the common good. However, people in a society also compete with each other; for example, when deciding how to divide limited benefits among themselves. Sometimes the competition is relatively trivial, such as when many people vie for tickets to a movie premiere. At other times the competition is much more significant, such as when two start-up companies seek control of an emerging market. Every society has rules of conduct describing what people ought and ought not to do in various situations. We call these rules **morality**.

Ethics is the philosophical study of morality, a rational examination into people's moral beliefs and behavior. Consider the following analogy (Figure 2.2). Society is like a town full of people driving cars. Morality is the road network within the town. People ought to keep their cars on the roads. Those who choose to "do ethics" are in balloons floating above the town. From this perspective, an observer can evaluate individual roads (particular moral guidelines) as well as the quality of the entire road network (moral system). The observer can also judge whether individual drivers are staying on the roads (acting morally) or taking shortcuts (acting immorally). Finally, the observer can propose and evaluate various ways of constructing road networks (alternative moral systems). While there may in fact be a definite answer regarding the best way to construct and operate a road network, it may be difficult for the observers to identify and agree upon this answer, because each observer has a different viewpoint.

The study of ethics is particularly important right now. Our society is changing rapidly as it incorporates the latest advances in information technology. Just think about how cell phones, portable CD players, laptop computers, and the World Wide Web have changed how we spend our time and interact with others! These inventions have brought us many benefits. However, some people selfishly exploit new technologies for personal gain, even if that reduces their overall benefit for the rest of us. Here are two examples. While most of us are happy to have the ability to send email to people all over the world, we are dismayed at the amount of spam—unsolicited bulk email—we receive. Access to the World Wide Web provides libraries with an important new information resource for its patrons, but should children be exposed to pop-up advertisements for pornographic Web sites?

When we encounter new problems such as spam or pornographic Web sites, we need to decide which activities are "good," which are "neutral," and which are "bad." Unfortunately, existing moral guidelines sometimes seem old-fashioned or unclear. If we can't always count on "common wisdom" to help us answer these questions, we need to learn how to work through these problems ourselves.

2.1.2 Four Scenarios

As an initiation into the study of ethics, carefully read each of the following scenarios. After reflection, come up with your answer to each question.

SCENARIO 1

Alexis, a gifted high school student, wants to become a doctor. Because she comes from a poor family, she will need a scholarship in order to attend college.

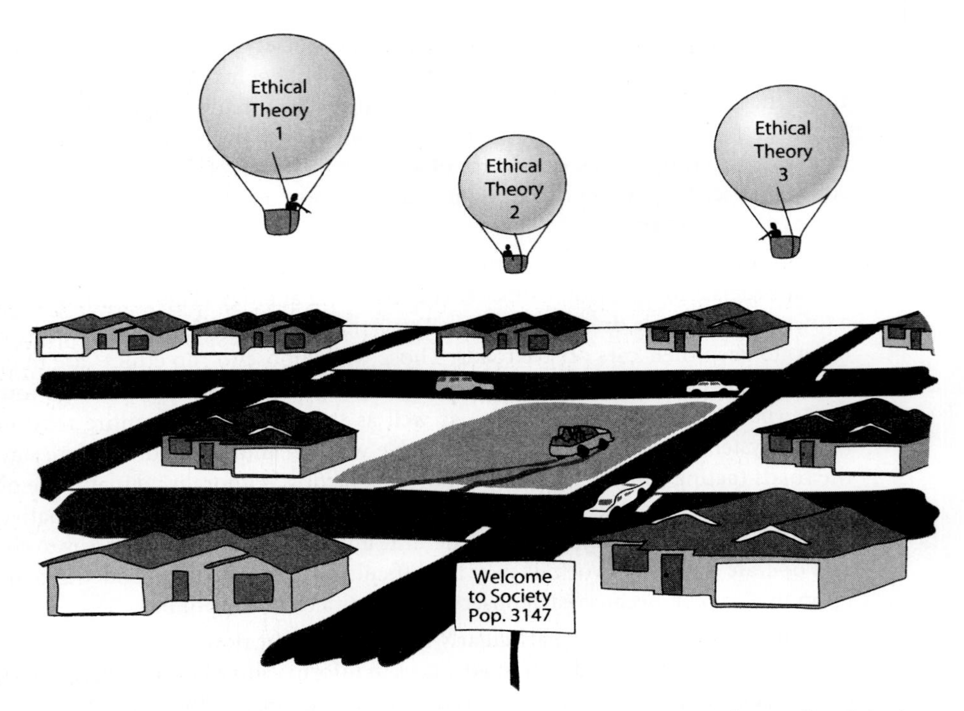

FIGURE 2.2 An analogy explaining the difference between ethics and morality. Imagine society as a town. Morality is the road network within the town. People doing ethics are in balloons floating above the town.

Some of her classes require extra research projects in order to get an A. Her high school has a few, older PCs, but there are always long lines of students waiting to use them during the school day. After school, she usually works at a part-time job to help support her family.

On some evenings Alexis goes to the library of a private college a few miles from her family's apartment, where she always finds plenty of unused PCs connected to the Internet. On the few occasions when a librarian asks her if she is a student at the college, she says "Yes," and the librarian leaves her alone. Using the resources of this library, Alexis efficiently completes the extra research projects, graduates from high school with straight As, and gets a full-ride scholarship to attend a prestigious university.

Questions

1. Did Alexis do anything wrong?
2. Who benefited from Alexis's course of action?
3. Who was hurt by Alexis's course of action?

4. Did Alexis have an unfair advantage over her high school classmates?
5. Would any of your answers change if it turns out Alexis did not win a college scholarship after all?
6. In what other ways could Alexis have accomplished her objective?

SCENARIO 2

An organization dedicated to reducing spam tries to get Internet service providers (ISPs) in an East Asian country to stop the spammers by protecting their mail servers. When this effort is unsuccessful, the anti-spam organization puts the addresses of these ISPs on its "black list." Many ISPs in the United States consult the black list and refuse to accept email from the blacklisted ISPs. This action has two results. First, the amount of spam received by the typical email user in the United States drops by 25 percent. Second, tens of thousands of innocent computer users in the East Asian country are unable to send email to friends and business associates in the United States.

Questions

1. Did the anti-spam organization do anything wrong?
2. Did the ISPs that refused to accept email from the blacklisted ISPs do anything wrong?
3. Who benefited from the organization's action?
4. Who was hurt by the organization's action?
5. Could the organization have achieved its goals through a better course of action?

SCENARIO 3

In an attempt to deter speeders, the East Dakota State Police (EDSP) connects video cameras to all of its freeway overpasses. The cameras are connected to computers that can reliably detect cars traveling more than five miles per hour above the speed limit. These computers have sophisticated image recognition software that enables them to read license plate numbers and capture high-resolution pictures of vehicle drivers. If the picture of the driver matches the driver's license photo of one of the registered owners of the car, the system issues a speeding ticket to the driver, complete with photo evidence. Six months after the system is put into operation, the number of people speeding on East Dakota freeways is reduced by 90 percent.

The FBI asks the EDSP for real-time access to the information collected by the video cameras. The EDSP complies with this request. Three months later, the FBI uses this information to arrest five members of a terrorist organization.

Questions

1. Did the East Dakota State Police do anything wrong?
2. Who benefited from the actions of the EDSP?
3. Who was harmed by the actions of the EDSP?
4. What other courses of action could the EDSP have taken to achieve its objectives? Examine the advantages and disadvantages of these alternative courses of action.

SCENARIO 4

You are the senior software engineer at a start-up company developing an exciting new product for handheld computers that will revolutionize the way nurses keep track of their hospitalized patients. Your company's sales force has led hospital administrators to believe your product will be available next week. Unfortunately, at this point the package still contains quite a few bugs. The leader of the testing group has reported that all of the known bugs appear to be minor, but it will take another month of testing for his team to be confident the product contains no catastrophic errors.

Because of the fierce competition in the medical software industry, it is critical that your company be the "first to market." To the best of your knowledge, a well-established company will release a similar product in a few weeks. If its product appears first, your start-up company will probably go out of business.

Questions

1. Should you recommend release of the product next week?
2. Who will benefit if the company follows your recommendation?
3. Who will be harmed if the company follows your recommendation?
4. Do you have an obligation to any group of people that may be affected by your decision?

Reflect on the process you used in each scenario to come up with your answers. How did you decide if particular actions or decisions were right or wrong? Were your reasons consistent from one case to the next? Did you use the same methodology in more than one scenario? If someone disagreed with you on the answer to one of these questions, how would you try to convince that person that your position makes more sense?

Ethics is the rational, systematic analysis of conduct that can cause benefit or harm to other people. Because ethics is based in reason, people are required to explain *why* they hold the opinions they do. This gives us the opportunity to compare ethical evaluations. When two people reach different conclusions, we can weigh the facts and the reasoning process behind their conclusions to determine the stronger line of thinking.

It's important to note that ethics is focused on the *voluntary, moral* choices people make because they have decided they ought to take one course of action rather than an alternative. Ethics is not concerned about involuntary choices or choices outside the moral realm.

For example, if I am ordering a new car, I may get to choose whether it is red, white, green, or blue. This choice is not in the moral realm.

Now, suppose I'm driving my new, red car down a city street. A pedestrian, obscured from my view by a parked car, runs out into traffic. In an attempt to miss the pedestrian, I swerve, lose control of my car, and kill another pedestrian walking along the sidewalk. While my action caused harm to another person, this is not an example of ethical decision-making, because my decision was a reflex action rather than a reasoned choice.

However, suppose I did not have full control of the car because I had been driving while intoxicated. In that case the consequences of my voluntary choice to drink affected another moral being (the innocent pedestrian). Now the problem has entered the realm of ethics.

2.1.3 Overview of Ethical Theories

The formal study of ethics goes back at least 2,400 years, to the Greek philosopher Socrates. Socrates did not put any of his philosophy in writing, but his student Plato did. In Plato's dialogue called the *Crito*, imprisoned Socrates uses ethical reasoning to explain why he ought to face an unjust death penalty rather than take advantage of an opportunity to flee into exile with his family [3].

In the past two millennia, philosophers have proposed many ethical theories. In this chapter we review some of them. How do we decide if a particular theory is useful? A useful theory allows its proponents to examine moral problems, reach conclusions, and defend these conclusions in front of a skeptical, yet open-minded audience (Figure 2.3).

Figure 2.3 A good ethical theory should enable you to make a persuasive, logical argument to a diverse audience.

Suppose you and I are debating a moral problem in front of a nonpartisan crowd. You have concluded that a particular course of action is right, while I believe it is wrong. It is only natural for me to ask you, "Why do you think doing such-and-such is right?" If you are unable to give any logical reasons why your position is correct, you are unlikely to persuade anyone. On the other hand, if you can explain the chain of reasoning that led you to your conclusion, you will be more likely to convince the audience that your position is correct. At the very least you will help reveal where there are disputed facts or values. Hence we will reject proposed ethical theories that are not based on reasoning from facts or commonly accepted values.

In the following sections we will consider seven ethical theories—seven frameworks for moral decision-making. We will present the motivation or insight underlying each theory, explain how it can be used to determine whether an action is right or wrong, and give the "case for" and the "case against" the theory. The workable theories will be those that make it possible for a person to present a persuasive, logical argument to a diverse audience of skeptical, yet open-minded people.

The principal sources for these brief introductions to ethical theories are *Ethical Insights: A Brief Introduction, Second Edition* by Douglas Birsch [4] and *The Elements of Moral Philosophy, Fourth Edition* by James Rachels [5]. Consult one or both of these books if you'd like to explore any of these theories in greater depth.

2.2 Subjective Relativism

Relativism is the theory that there are no universal moral norms of right and wrong. Different individuals or groups of people can have completely opposite views of a moral problem, and both can be right. Two particular kinds of relativism we'll discuss are subjective relativism and cultural relativism.

Subjective relativism holds that each person decides right and wrong for himself or herself. This notion is captured in the popular expression "What's right for you may not be right for me."

2.2.1 The Case for Subjective Relativism

1. *Well-meaning and intelligent people can have totally opposite opinions about moral issues.*

   For example, consider the issue of legalized abortion in the United States. There are a significant number of rational people on each side of the issue. The reason people cannot reach the same conclusion is that morality is not like gravity; it is not something "out there" that rational people can discover and try to understand. Instead, each of us creates his or her own morality.

2. *Ethical debates are disagreeable and pointless.*

   Going back to the example of abortion, the debate in the United States has been going on for more than 30 years. An agreement about whether abortion is right or wrong may never be reached. Nobody is all-knowing. When faced with a difficult moral problem, who is to say which side is correct? If morality is relative, we do not have to try to reconcile opposing views. Both sides are right.

2.2.2 The Case against Subjective Relativism

1. *With subjective relativism the line between doing what you think is right and doing what you want to do is not sharply drawn.*

   People are good at rationalizing their bad behavior. Subjective relativism provides an ideal last line of defense for someone whose motives or behavior are being challenged. When pressed to explain a decision or action, a subjective relativist can reply, "Who are *you* to tell *me* what I should and should not do?" If morality means doing whatever you want to do, it doesn't mean much, if it means anything at all.

2. *By allowing each person to decide right and wrong for himself or herself, subjective relativism makes no moral distinction between the actions of different people.*

   The fact is that some people have caused millions to suffer, while others have led lives of great service to humanity. Suppose both Adolf Hitler and Mother Teresa spent their entire lives doing what they thought was the right thing to do. Do you want to give both of them credit for living good lives?

   A modification of the original formulation of subjective relativism might be: "I can decide what's right for me, as long as my actions don't hurt anybody else." That solves the problem of Adolf Hitler versus Mother Teresa. However, as soon as you introduce the idea that you shouldn't harm others, you must come to an agreement with others about what it means to harm someone. At this point the process is no longer subjective or completely up to the individual. In other words, a statement of the form "I can decide what's right for me, as long as my actions don't hurt anyone else" is inconsistent with subjective relativism.

3. *Subjective relativism and tolerance are two different things.*

   Some people may be attracted to relativism because they believe in tolerance. There is a lot to be said for tolerance. It allows individuals in a diverse society like the United States to live in harmony. However, tolerance is not the same thing as subjective relativism. Subjective relativism holds that individuals decide for themselves what is right and what is wrong. If you are a tolerant person, is it okay with you if some people decide they want to be intolerant? What if a person decides that he will only interact with people of his own racial group? Note that any statement of the form "People ought to be tolerant" is an example of a universal moral **norm,** or rule. Relativism is based on the idea that there are no universal moral norms, so a blanket statement about the need for tolerance is incompatible with subjective relativism.

4. *We should not give legitimacy to an ethical theory that allows people to make decisions based on something other than reason.*

   If individuals decide for themselves what is right and what is wrong, they can reach their conclusions by any means they see fit. They may choose to base their decisions on something other than logic and reason, such as the rolling of dice or the turning of Tarot cards. This path is contrary to using logic and reason.

If your goal is to persuade others that your solutions to actual moral problems are correct, subjective relativism is self-defeating. It is based on the idea that each person decides for himself or herself what is right and what is wrong. According to ethical relativism, nobody's conclusions are any more valid than anyone else's, no matter how these conclusions are drawn. Because of its self-defeating nature, we reject subjective relativism as a workable ethical theory.

2.3 Cultural Relativism

If subjective relativism is unworkable, what about different views of right and wrong held by different societies at the same point in time, or those held by the same society at different points in time?

In the modern era, anthropologists have collected evidence of societies with moral codes markedly different from those of the societies of Europe and North America. William Graham Sumner described the evolution of *folkways*, which he argues eventually become institutionalized into the moral guidelines of a society:

> The first task of life is to live . . . The struggle to maintain existence was not carried on individually but in groups. Each profited by the other's experience; hence there was concurrence towards that which proved to be the most expedient. All at last adopted the same way for the same purpose; hence the ways turned into customs and became mass phenomena. Instincts were learned in connection with them. In this way folkways arise. The young learn by tradition, imitation, and authority. The folkways, at a time, provide for all the needs of life then and there. They are uniform, universal in the group, imperative, and invariable. As time goes on, the folkways become more and more arbitrary, positive, and imperative. If asked why they act in a certain way in certain cases, primitive people always answer that it is because they and their ancestors always have done so . . . The morality of a group at a time is the sum of the taboos and prescriptions in the folkways by which right conduct is defined . . . 'Good' mores are those which are well adapted to the situation. 'Bad' mores are those which are not so well adapted [6].

Cultural relativism is the ethical theory that the meaning of "right" and "wrong" rests with a society's actual moral guidelines. These guidelines vary from place to place and from time to time.

Charles Hampden-Turner and Fons Trompenaars conducted a modern study that reveals how notions of right and wrong vary widely from one society to another. Here is a dilemma they posed to people from 46 different countries:

> You are riding in a car driven by a close friend. He hits a pedestrian. You know he was going at least 35 miles per hour in an area of the city where the maximum allowed speed is 20 miles per hour. There are no witnesses other than you. His

lawyer says that if you testify under oath that he was driving only 20 miles per hour, you will save him from serious consequences.

What right has your friend to expect you to protect him?

- My friend has a definite right as a friend to expect me to testify to the lower speed.
- He has some right as a friend to expect me to testify to the lower speed.
- He has no right as a friend to expect me to testify to the lower speed.

What do you think you would do in view of the obligations of a sworn witness and the obligation to your friend?

- Testify that he was going 20 miles per hour.
- Not testify that he was going 20 miles per hour [7].

About 90 percent of Norwegians would not testify to the lower speed and do not believe that the person's friend has a definite right to expect help. In contrast, only about 10 percent of Yugoslavians feel the same way. About three-quarters of Americans and Canadians agree with the dominant Norwegian view, but Mexicans are fairly evenly divided [7]. Cultural relativists say we ought to pay attention to these differences.

2.3.1 The Case for Cultural Relativism

1. *Different social contexts demand different moral guidelines.*

   It's unrealistic to assume that the same set of moral guidelines can be expected to work for all human societies in every part of the world for all ages. Just think about how our relationship with our environment has changed. For most of the past 10,000 years, human beings have spent most of their time trying to produce enough food to survive. Thanks to technology, the human population of the Earth has increased exponentially in the past century. The struggle for survival has shifted away from people to the rest of Nature. Overpopulation has created a host of environmental problems, such as the extinction of many species, the destruction of fisheries in the world's oceans, and the accumulation of greenhouse gases. People must change their ideas about what is acceptable conduct and what is not, or they will destroy the planet.

2. *It is arrogant for one society to judge another.*

   Anthropologists have documented many important differences among societies with respect to what they consider proper and improper moral conduct. We may have more technology than people in other societies, but we are no more intelligent than they are. It is arrogant for a person living in twenty-first-century America to judge the actions of another person who lived in Peru in the fifteenth century.

3. *Morality is reflected in actual behavior.*

   We often find people saying that certain actions are wrong, but then they do them anyway. Some parents tell their children, "Do as I say, not as I do." Looking at the actual behavior of people (their *de facto* values) gives a truer picture of what a society believes is right and wrong than listening to their hypothetical discussions about how they ought to behave.

2.3.2 The Case against Cultural Relativism

1. *Just because two societies do have different views about right and wrong doesn't imply that they ought to have different views.*

   Perhaps one society has good guidelines and another has bad guidelines. Perhaps neither society has good guidelines.

   Suppose two societies are suffering from a severe drought. The first society constructs an aqueduct to carry water to the affected cities. The second society makes human sacrifices to appease the rain god. Are both "solutions" equally acceptable? No, they are not. Yet, if we accept cultural relativism, we cannot speak out against this wrongdoing, because no person in one society can make any statements about the morality of another society.

2. *Cultural relativism does not explain how an individual determines the moral guidelines of a particular society.*

   Suppose I am new to a society and I understand I am supposed to abide by its moral guidelines. How do I determine what those guidelines are?

   One approach would be to poll other people, but this begs the question. Here's why. Suppose I ask other people whether the society considers a particular action to be morally acceptable. I'm not interested in knowing whether they personally feel the action is right or wrong. I want them to tell me whether the society as a whole thinks the action is moral. That puts the people I poll in the same position I'm in—trying to determine the moral guidelines of a society. I still do not know how these guidelines are discovered.

   Perhaps the guidelines are summarized in the society's laws, but laws take time to enact. Hence the legal code reflects at best the moral guidelines of the same society at some point in the past, but that's not the same society I am living in today, because the morals of any society change over time. That leads us to our next objection.

3. *Cultural relativism does not do a good job of explaining how moral guidelines evolve.*

   Until the 1960s many southern American states had segregated universities. Today these universities are integrated. This change in attitudes was accelerated by the actions of a few brave people of color who challenged the status quo and enrolled in universities that had been the exclusive preserve of white students. At the time these students were doing what they "ought not" to have done; they were doing something wrong according to the moral guidelines of the time. By today's standards, they did nothing wrong, and many people view them as heroic figures. Doesn't it make more sense to believe that their actions were the right thing to do all along?

4. *Cultural relativism provides no framework for reconciliation between cultures in conflict.*

   Think about the culture of the poverty-stricken Palestinians who have been crowded into refugee camps in the Gaza Strip for the past 50 years. Many of these people are completely committed to an armed struggle against Israel. Meanwhile, many people in Israel believe the Jewish state ought to be larger and are completely committed to the expansion of settlements into the Gaza Strip. The values of each society lead to actions that harm the other, yet cultural relativism says each society's moral guidelines are right. Cultural relativism provides no way out—no way for the two sides to find common ground.

5. *The existence of many acceptable cultural practices does not imply that any cultural practice would be acceptable.*

   Judging *many* options to be acceptable and then reaching the conclusion that *any* option is acceptable is called the **many/any fallacy**. To illustrate this fallacy, consider documentation styles for computer programs. There are many good ways to add comments to a program; that does not mean that any commenting style is good.

   It is false that all possible cultural practices have equal legitimacy. Certain practices must be forbidden and others must be mandated if a society is to survive [1]. This observation leads us directly to our next point.

6. *Societies do, in fact, share certain core values.*

   While a superficial observation of the cultural practices of different societies may lead you to believe they are quite different, a closer examination often reveals similar values underlying these practices. James Rachels argues that all societies, in order to maintain their existence, must have a set of core values [5]. For example, newborn babies are helpless. A society must care for its infants if it wishes to continue on. Hence a core value of every society is that babies must be cared for. Communities rely upon people being able to believe each other. Hence telling the truth is another core value. Finally, in order to live together, people must not constantly be on guard against attack from their community members. For this reason a prohibition against murder is a core value of any society.

   Because societies do share certain core values, there is reason to believe we could use these values as a starting point in the creation of a universal ethical theory that would not have the deficiencies of cultural relativism.

7. *Cultural relativism is only indirectly based on reason.*

   As Sumner observed, many moral guidelines are a result of tradition. You behave in a certain way because it's what you're supposed to do, not because it makes sense.

Cultural relativism has significant weaknesses as a tool for ethical persuasion. According to cultural relativism, the ethical evaluation of a moral problem made by a person in one society may be meaningless when applied to the same moral problem in another society. Cultural relativism suggests there are no universal moral guidelines. It gives tradition more weight in ethical evaluations than facts and reason. For these reasons cultural relativism is not a powerful tool for constructing ethical evaluations persuasive to a diverse audience, and we consider it no further.

2.4 Divine Command Theory

The three great religious traditions that arose in the Middle East—Judaism, Christianity, and Islam—teach that a single God is the creator of the universe and that human beings are part of God's creation. Each of these religions has sacred writings containing God's revelation. If you are a religious person, living your life aligned with the will of God may be very important to you.

Jews, Christians, and Muslims all believe that God inspired the Torah. Here is a selection of verses from Chapter 19 of the third book of the Torah, called Leviticus:

> You shall each revere his mother and his father, and keep My sabbaths. When you reap the harvest of your land, you shall not reap all the way to the edges of your field, or gather the gleanings of your harvest. You shall not pick your vineyard bare, or gather the fallen fruit of your vineyard; you shall leave them for the poor and the stranger. You shall not steal; you shall not deal deceitfully or falsely with one another. You shall not swear falsely by My name. You shall not defraud your neighbor. You shall not commit robbery. The wages of a laborer shall not remain with you until morning. You shall not insult the deaf, or place a stumbling block before the blind. You shall not take vengeance or bear a grudge against your kinsfolk. Love your neighbor as yourself [8].

The **divine command theory** is based on the idea that good actions are those aligned with the will of God and bad actions are those contrary to the will of God. Since the holy books contain God's directions, we can use the holy books as moral decision-making guides. God says we should revere our mothers and fathers, so revering our parents is good. God says do not lie or steal, so lying and stealing are bad (Figure 2.4).

2.4.1 The Case for the Divine Command Theory

1. *We owe obedience to our Creator.*

   God is the creator of the universe. God created each one of us. We are dependent upon God for our lives. Hence we are obligated to follow God's rules.

2. *God is all-good and all-knowing.*

   God loves us and wants the best for us. God is omniscient; we are not. Hence God knows better than we do what we must do to be happy. For this reason we should align ourselves with the will of God.

3. *God is the ultimate authority.*

   Since most people are religious, they are more likely to submit to God's law than to a law made by people. Our goal is to create a society where everyone obeys the moral laws. Hence our moral laws should be based on God's directions to us.

Figure 2.4 The divine command theory of ethics is based on two premises: good actions are those actions aligned with the will of God, and God's will has been revealed to us.

2.4.2 The Case against the Divine Command Theory

1. *There are many holy books, and some of their teachings disagree with each other.*

   There is no single holy book that is recognized by people of all faiths, and it is unrealistic to assume everyone in a society will adopt the same religion. Even among Christians there are different versions of the Bible. The Catholic Bible has six books not found in the Protestant Bible. Some Protestant denominations rely upon the King James version, but others use more modern translations. Every translation has significant differences. Even when people read the same translation, they often interpret the same verse in different ways.

2. *It is unrealistic to assume a multicultural society will adopt a religion-based morality.*

   An obvious example is the United States. In the past two centuries, immigrants representing virtually every race, creed, and culture have made America their home. Some Americans are atheists. When a society is made up of people with different religious beliefs, the society's moral guidelines should emerge from a secular authority, not a religious authority.

3. *Some moral problems are not addressed directly in scripture.*

   For example, there are no verses in the Bible mentioning the Internet. When we discuss moral problems arising from information technology, a proponent of the divine command theory must resort to analogy. At this point the conclusion is based not simply on what appears in the sacred text but also on the insight of the person who invented the analogy. The holy book alone is not sufficient to solve the moral problem.

4. *It is fallacious to equate "the good" with "God."*

   Religious people are likely to agree with the statement "God is good." That does not mean, however, that God and "the good" are exactly the same thing. Trying to equate two things that are similar is called the **equivalence fallacy.** Instead, the statement "God is good" means there is something outside of God that is good.

   Here's another way to put the question. Is an action good because God commands it, or does God command it because it's good? This is an ancient question: Plato raised it about 2,400 years ago in the Socratic dialogue *Euthyphro*. In this dialogue Socrates concludes, "The gods love piety because it is pious, and it is not pious because they love it" [9]. In other words, good is something that exists outside of God.

   We can reason our way to the same conclusion. If good means "commanded by God," then good is arbitrary. Why should we praise God for being good if good is whatever God wills? According to this view of the good, it doesn't matter whether God commanded, "Thou shalt not commit adultery" or "Thou shalt commit adultery." Either way, the command would have been good by definition. If you object that there is no way God would command us to commit adultery, because marital fidelity is good and adultery is bad, then you are admitting that there is a standard of right and wrong separate from God. In that case, we can talk about the good without talking about God. That opens the door to a rational discussion of the good, which we will pursue in the next section.

5. *The divine command theory is based on obedience, not reason.*

   If good means "willed by God," and if religious texts contain everything we need to know about what God wills, then there is no room left for collecting and analyzing facts. Hence the divine command theory is not based on reaching sound conclusions from premises through logical reasoning. There is no need for a person to question a commandment. The instruction is right because it's commanded by God, period.

   Consider the story of Abraham in the book of Genesis. God commands Abraham to take his only son, Isaac, up on a mountain, kill him, and make of him a burnt offering. Abraham obeys God's command and is ready to kill Isaac with his knife when an angel calls down and tells him not to harm the boy. Because he does not withhold his only son from God, God blesses Abraham [10]. Earlier in Genesis God condemns Cain for killing Abel [11]. How, then, can Abraham's sacrifice of Isaac be considered good? To devout readers, the logic of God's command is irrelevant to this story. Abraham is a good person, a heroic model of faith, because he demonstrated his obedience to the will of God.

The fact that moral guidelines are not the result of a logical progression from a set of underlying principles is a significant obstacle. While you may choose to align your personal actions with the divine will, the divine command theory often fails to produce arguments that can persuade skeptical listeners whose religious beliefs are different. Hence we conclude the divine command theory is not a powerful weapon for ethical debate in a secular society, and we reject it as a workable theory for the purposes of this book.

2.5 Kantianism

Kantianism is the name given to the ethical theory of the German philosopher Immanuel Kant (1724–1804). Kant spent his entire life in or near Königsberg in East Prussia, where he was a professor at the university. Kant believed that people's actions ought to be guided by moral laws, and that these moral laws were universal. He held that in order to apply to all rational beings, any supreme principle of morality must itself be based on reason. Hence while many of the moral laws Kant describes can also be found in the Bible, Kant's methodology allows these laws to be derived through a reasoning process. A Kantian is able to go beyond simply stating *that* an action is right or wrong by citing chapter and verse; a Kantian can explain *why* it is right or wrong.

2.5.1 Good Will and the Categorical Imperative

Kant begins his inquiry by asking, "What is always good without qualification?" Many things, such as intelligence and courage, can be good, but they can also be used in a way that is harmful. For example, a group of gangsters may use intelligence and courage to rob a bank. Kant's conclusion is that the only thing in the world that can be called good without qualification is *good will*. People with good will often accomplish good deeds, but producing beneficial outcomes is not what makes good will good. Good will is good in and of itself. Even if a person's best efforts at doing good should fall short and cause harm, the good will behind the efforts is still good. Since good will is the only thing that is universally good, the proper function of reason is to cultivate a will that is good in itself.

Most of us have probably had many experiences when we've been torn between *what we want to do* and *what we ought to do*. According to Kant, what we want to do is of no importance. Our focus should be on what we ought to do. Our sense of "ought to" is called **dutifulness** [12]. A dutiful person feels compelled to act in a certain way out of respect for some moral rule. Our will, then, should be grounded in a conception of moral rules. The moral value of an action depends upon the underlying moral rule. It is critical, therefore, that we be able to determine if our actions are grounded in an appropriate moral rule.

What makes a moral rule appropriate? To enable us to answer this question, Kant proposes the Categorical Imperative.

CATEGORICAL IMPERATIVE (FIRST FORMULATION)

Act only from moral rules that you can at the same time will to be universal moral laws.

To illustrate the Categorical Imperative, Kant poses the problem of an individual in a difficult situation who must decide if he will make a promise with the intention of later breaking it. The translation of this into a moral rule could be: "I may make promises with the intention of later breaking them."

To evaluate this moral rule, we universalize it. What would happen if everybody in extreme circumstances made false promises? If that were the case, promises would be meaningless. There would cease to be such a thing as a promise. Hence our moral rule self-destructs when we try to make it a universal law. Therefore, it is wrong for me to make a promise with the intention of breaking it.

It is important to see that Kant is *not* arguing that the consequences of everybody breaking promises would be to undermine interpersonal relationships, increase violence, and make people miserable, and that is why we cannot imagine turning our hypothetical moral rule into a universal law. Rather, Kant is saying that simply willing that our moral rule become a universal law produces a logical contradiction.

Let's see how. On the one hand, it is my will that I be able to make a promise that is believed. After all, that's what promises are for. If my promise isn't believed, I won't be able to get out of the difficult situation I am in. But when I universalize the moral rule, I am willing that everybody be able to break promises. If that were a reality, then promises would not be believable, which means there would be no such thing as a promise [13]. If there were no such thing as a promise, I would not be able to make a promise. Trying to universalize our proposed moral rule leads to a contradiction.

Kant also presents a second formulation of the Categorical Imperative, which many find more useful.

CATEGORICAL IMPERATIVE (SECOND FORMULATION)

Act so that you always treat both yourself and other people as ends in themselves, and never only as a means to an end.

To use popular terminology, the second formulation of the Categorical Imperative says it is wrong for one person to "use" another (Figure 2.5). Instead, every interaction with other people must respect them as rational beings.

Here is an example that illustrates how we can apply the second formulation. Suppose I manage a semiconductor fabrication plant for a large corporation. The plant manufactures integrated circuits on eight-inch wafers. I know that in one year the cor-

FIGURE 2.5 The second formulation of the Categorical Imperative states that it is wrong for one person to use himself or another person solely as a means to an end.

poration is going to shut down the plant and move all of its production to other sites capable of producing twelve-inch wafers. In the meantime, I need new employees to work in the clean room. Many of the best applicants are from out of state. I am afraid that if they knew the plant was going to shut down next year, they would not want to go through the hassle and expense of moving to this area. If that happens, I'll have to hire less-qualified local workers. Should I disclose this information to the job applicants?

According to the second formulation of the Categorical Imperative, I have an obligation to inform the applicants, since I know this information is likely to influence their decision. If I deny them this information, I am treating them as a means to an end (a way to get wafers produced), not as ends in themselves (rational beings).

2.5.2 Evaluating a Scenario Using Kantianism

SCENARIO

Carla is a single mother who is working hard to complete her college education while taking care of her daughter. Carla has a full-time job and is taking two evening courses per semester. If she can pass both courses this semester, she will graduate. She knows her child will benefit if she can spend more time at home.

One of her required classes is modern European history. In addition to the midterm and final examinations, the professor assigns four lengthy reports, which is far more than the usual amount of work required for a single class. Students must submit all four reports in order to pass the class.

Carla earns an "A" on each of her first three reports. At the end of the term, she is required to put in a lot of overtime where she works. She simply does not have time to research and write the final report. Carla uses the Web to identify a company that sells term papers. She purchases a report from the company and submits it as her own work.

Was Carla's action morally justifiable?

Analysis

Many times it is easier to use the second formulation of the Categorical Imperative to analyze a moral problem from a Kantian point of view, so that's where we begin. By submitting another person's work as her own, Carla treated her professor as a means to an end. She deceived her professor with the goal of getting credit for someone else's work. It was wrong for Carla to treat the professor as a grade-generating machine rather than a rational agent with whom she could have communicated her unusual circumstances.

We can also look at this problem using the first formulation of the Categorical Imperative. Carla wants to be able to get credit for turning in a report she has purchased. A proposed moral rule might be: "I may claim credit for a report written by someone else." However, if everyone followed this rule, reports would cease to be credible indicators of the students' knowledge, and professors would not give academic credit for reports. Her proposed moral rule is self-defeating. Therefore, it is wrong for Carla to purchase a report and turn it in as her own work.

Commentary

Note that the Kantian analysis of the moral problem focuses on the will behind the action. It asks the question: "What was Carla trying to do when she submitted under her own name a term paper written by someone else?" The analysis ignores circumstances that some may find to excuse her behavior.

2.5.3 The Case for Kantianism

1. *Kantianism is rational.*

   Unlike the moral theories we have already described, Kantianism is based on the premise that rational beings can use logic to explain the "why" behind their solutions to ethical problems.

2. *Kantianism produces universal moral guidelines.*

   Kantianism aligns with the intuition of many people that the same morality ought to apply to all people for all of history. These guidelines allow us to make clear moral judgements. For example, one such judgment might be, "Sacrificing living human beings to appease the gods is wrong." It is wrong in North America in the twenty-first century, and it was wrong in South America in the fifteenth century.

3. *All persons are treated as moral equals.*

   A popular belief is that "all people are created equal." Because it holds that people in similar situations should be treated in similar ways, Kantianism provides an ethical framework to combat discrimination.

2.5.4 The Case against Kantianism

1. *Sometimes no single rule fully characterizes an action.*

   Kant holds that every action is motivated from a rule. The appropriate rule depends upon how we characterize the action. Once we know the rule, we can test its value using the Categorical Imperative. What happens when no single rule fully explains the situation? Douglas Birsch gives this example: Suppose I'm considering stealing food from a grocery store to feed my starving children [4]. How should I characterize this action? Am I stealing? Am I caring for my children? Am I trying to save the lives of innocent people? Until I characterize my action, I cannot determine the rule and test it against the Categorical Imperative. Yet no single one of these ways of characterizing the action seems to capture the ethical problem in its fullness.

2. *There is no way to resolve a conflict between rules.*

   We may try to address the previous problem by allowing multiple rules to be relevant to a particular action. In the previous example, we might say that the relevant rules are (1) You should not steal, and (2) You should try to protect the lives of innocent persons. Unfortunately, Kantianism does not provide us a way to put moral laws in order of importance. Even if we could rank moral laws in order of importance, how would we compare a minor infraction of a more important law against a major infraction of a less important law? One conclusion is that Kantianism does not provide a practical way to solve ethical problems when there is a conflict between moral rules.

3. *Kantianism allows no exceptions to moral laws.*

   Common sense tells us that sometimes we ought to "bend" the rules a bit if we want to get along with other people. For example, suppose your mother asks you if you like her new haircut, and you think it is the ugliest haircut you have ever seen. What should you say? Common sense dictates that there is no point in criticizing your mother's hair. She certainly isn't going to get her hair un-cut, no matter what you say. If you compliment her, she will be happy, and if you criticize her looks, she will be angry and hurt. She expects you to say something complimentary, even if you don't mean it. There just seems to be no downside to lying. Yet a Kantian would argue that lying is wrong because it goes against the moral law. Many people hold that any ethical theory so unbending is not going to be useful for solving "real world" problems.

While these objections point out weaknesses with Kantianism, the theory does support moral decision-making based on logical reasoning from facts and commonly held values. It is culture neutral and treats all humans as equals. Hence it meets our criteria for a workable ethical theory, and we will use it as a way of evaluating moral problems in the rest of the book.

2.6 Act Utilitarianism

2.6.1 Principle of Utility

The English philosophers Jeremy Bentham (1748–1832) and John Stuart Mill (1806–1873) proposed a theory that is in sharp contrast to Kantianism. According to Bentham and Mill, an action is good if it benefits someone; an action is bad if it harms someone. Their ethical theory, called **utilitarianism,** is based upon the Principle of Utility, also called the Greatest Happiness Principle.

PRINCIPLE OF UTILITY (GREATEST HAPPINESS PRINCIPLE)

An action is right (or wrong) to the extent that it increases (or decreases) the total happiness of the affected parties.

Utility is the tendency of an object to produce happiness or prevent unhappiness for an individual or a community. Depending on the circumstances, you may think of "happiness" as advantage, benefit, good, or pleasure, and "unhappiness" as disadvantage, cost, evil, or pain.

We can use the Principle of Utility as a yardstick to judge all actions in the moral realm. To evaluate the morality of an action, we must determine, for each affected person, the increase or decrease in that person's happiness, and then add up all of these values to reach a grand total. If the total is positive (meaning the total increase in happiness is greater than the total decrease in happiness), the action is moral; if the total is negative (meaning the total decrease in happiness is greater than the total increase in happiness), the action is immoral. The Principle of Utility is illustrated in Figure 2.6.

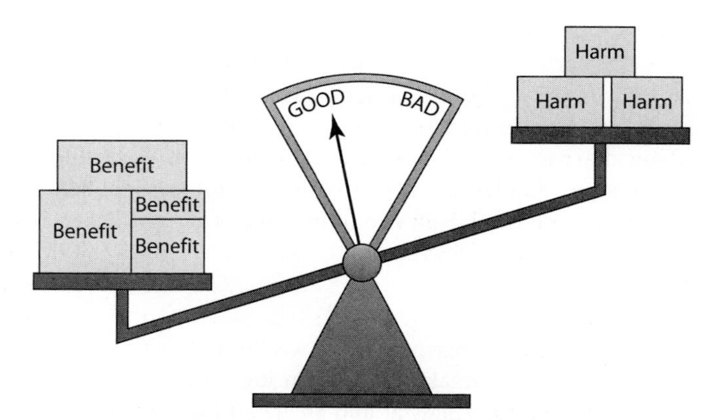

FIGURE 2.6 Utilitarianism is based on the Principle of Utility, which states that an action is good (or bad) to the extent that it increases (or decreases) the total happiness of the affected parties.

Note that the morality of an action has nothing to do with the attitude behind the action. Bentham writes: "There is no such thing as any sort of motive that is in itself a bad one. If [motives] are good or bad, it is only on account of their effects" [21]. We call utilitarianism a **consequentialist** theory, because the focus is on the consequences of an action.

Act utilitarianism is the ethical theory that an action is good if its net effect (over all affected beings) is to produce more happiness than unhappiness. Suppose we measure pleasure as a positive number and pain as a negative number. To make a moral evaluation of an action, we simply add up, over all affected beings, the change in their happiness. If the sum is positive, the action is good. If the sum is negative, the action is bad.

Did you notice that I used the word "beings" rather than "persons" in the previous paragraph? An important decision an act utilitarian must make is determining which beings are considered to be morally significant. Bentham noted that at one time only adult white males were considered morally significant beings. Bentham felt that any being that can experience pain and pleasure ought to be seen as morally significant. Certainly women and people of color are morally significant beings by this definition, but in addition all mammals (and perhaps other animals) are morally significant beings, because they, too, can experience pain and pleasure. Of course, as the number of morally significant beings increases, the difficulty of evaluating the consequences of an action also increases. It means, for example, that the environmental impacts of decisions must often be included when performing the utilitarian calculus.

2.6.2 Evaluating a Scenario Using Act Utilitarianism

Scenario

A state is considering replacing a curvy stretch of highway that passes along the outskirts of a large city. Would building the highway be a good action?

Analysis

To perform the analysis of this problem, we must determine who is affected and the effects of the highway construction on them. Our analysis is in terms of dollars and cents. For this reason we'll use the terms "benefit" and "cost" instead of "happiness" and "unhappiness."

About 150 houses lie on or very near the proposed path of the new, straighter section of highway. Using its power of eminent domain, the state can condemn these properties. It would cost the state $20 million to provide fair compensation to the homeowners. Constructing the new highway, which is three miles long, would cost the taxpayers of the state another $10 million. Suppose the environmental impact of the new highway in terms of lost habitat for morally significant animal species is valued at $1 million.

Every weekday 15,000 cars are expected to travel on this section of highway, which is one mile shorter than the curvy highway it replaces. Assuming it costs 40 cents per mile to operate a motor vehicle, construction of the new highway will save drivers $6,000 per weekday in operating costs. The highway has an expected operating lifetime of 25 years. Over a 25-year period, the expected total savings to drivers will be $39 million.

We'll assume the highway project will have no positive or negative effects on any other people. Since the overall cost of the new highway is $31 million and the benefit of the new highway is $39 million, building the highway would be a good action.

Commentary

Performing the benefit/cost (or happiness/unhappiness) calculations is crucial to the utilitarian approach, yet it can be controversial. In our example, we translated everything into dollars and cents. Was that reasonable? Neighborhoods are the site of many important relationships. We did not assign a value to the harm the proposed highway would do to these neighborhoods. There is a good chance that many of the homeowners will be angry about being forced out of their houses, even if they are paid a fair price for their properties. How do we put a dollar value on their emotional distress? On the other hand, we can't add apples and oranges. Translating everything into dollars and cents is the only way we can do the calculation.

Bentham acknowledged that a complete analysis must look beyond simple benefits and harms. Not all benefits have equal weight. To measure them, he proposed seven attributes that can be used to increase or decrease the weight of a particular pleasure or pain:

- *intensity:* magnitude of the experience
- *duration:* how long the experience lasts
- *certainty:* probability it will actually happen
- *propinquity:* how close the experience is in space and time
- *fecundity:* its ability to produce more experiences of the same kind
- *purity:* extent to which pleasure is not diluted by pain, or vice versa
- *extent:* number of people affected

As you can see, performing a complete calculation for a particular moral problem can be a daunting prospect!

2.6.3 The Case for Act Utilitarianism

1. *It focuses on happiness.*

   By relying upon the Greatest Happiness Principle as the yardstick for measuring moral behavior, utilitarianism fits the intuition of many people that the purpose of life is to be happy.

2. *It is down-to-earth.*

   The utilitarian calculus provides a straightforward way to determine whether a particular action is good or bad. By grounding everything in terms of happiness and unhappiness resulting from an action, it seems more practical than Kantian ethics, which is focused on the Categorical Imperative. For this reason it is a good way for a diverse group of people to come to a collective decision about a controversial topic.

For example, suppose your state needs to build a new prison because the number of prisoners is growing. Everybody understands the prison must be built somewhere in the state, but nobody wants the prison in their neighborhood. A panel of trusted citizens considers a variety of siting options and, after a series of public hearings to gather evidence, weighs the pluses and minuses of each location. At the end of the process the panel recommends the site with the highest total net good. While some will be unhappy at the prospect of a prison being built near their homes, an open and impartial process can speed their acceptance of the decision.

3. *It is comprehensive.*

   Act utilitarianism allows the moral agent to take into account all the elements of a particular situation. Recall the problem of having to decide what to say about your mother's haircut? Since telling the truth would cause more pain to all parties involved than lying, deciding what the right thing to do would be a "no brainer" using the utilitarian calculus.

2.6.4 The Case against Act Utilitarianism

1. *When performing the utilitarian calculus, it is not clear where to draw the line, yet where we draw the line can change the outcome of our evaluation.*

   In order to perform our calculation of total net happiness produced by an action, we must determine whom to include in our calculation and how far into the future to consider the consequences. In our highway example, we counted the people who lost their homes and the people who would travel the new highway in the next 25 years. The proposed highway may cut neighborhoods in two, making it more difficult for some children to get to school, but we did not factor in consequences for neighbors. The highway may cause people to change their commutes, increasing traffic congestion in other parts of town, but we did not count those people either. The highway may be in existence more than 25 years, but we didn't look beyond that date. We cannot include all morally relevant beings for all time into the future. We must draw the line somewhere. Deciding where to draw the line can be a difficult problem.

2. *It is not practical to put so much energy into every moral decision.*

   Correctly performing the utilitarian calculus requires a great deal of time and effort. It seems unrealistic that everyone would go to so much trouble every time they were faced with a moral problem.

   A response to this criticism is that act utilitarians are free to come up with moral "rules of thumb." For example, a moral rule of thumb might be "It is wrong to lie." In most situations it will be obvious this is the right thing to do, even without performing the complete utilitarian calculus. However, an act utilitarian always reserves the right to go against the rule of thumb if particular circumstances should warrant it. In these cases, the act utilitarian will perform a detailed analysis of the consequences to determine the best course of action.

3. *Act utilitarianism ignores our innate sense of duty.*

   Utilitarianism seems to be at odds with how ordinary people make moral decisions. People often act out of a sense of duty or obligation, yet the act utilitarian theory gives no weight to these notions. Instead, all that matters are the consequences of the action.

   W.D. Ross gives the following example [22]. Suppose I've made a promise to A. If I keep my word, I will perform an action that produces 1,000 units of good for him. If I break my promise, I will be able to perform an action that produces 1,001 units of good for B. According to act utilitarianism, I ought to break my promise to A and produce 1,001 units of good for B. Yet most people would say the right thing for me to do is keep my word.

   Note that it does no good for an act utilitarian to come back and say that the hard feelings caused by breaking my word to A will have a negative impact on total happiness of −N units, because all I have to do is change the scenario so that breaking my promise to A enables me to produce 1,001 + N units of good for B. We've arrived at the same result: breaking my promise results in 1 more unit of good than keeping my word. The real issue is that utilitarianism forces us to reduce all consequences to a positive or negative number. "Doing the right thing" has a value that is difficult to measure.

4. *Act utilitarianism is susceptible to the problem of moral luck.*

   Sometimes actions do not have the intended consequences. Is it right for the moral worth of an action to depend solely on its consequences when these consequences are not fully under the control of the moral agent? This is called the **problem of moral luck.**

   Suppose I hear that one of my aunts is in the hospital, and I send her a bouquet of flowers. After the bouquet is delivered, she suffers a violent allergic reaction to one of the exotic flowers in the floral arrangement, extending her stay in the hospital. My gift gave my aunt a bad case of hives and a much larger hospital bill. Since my action had far more negative consequences than positive consequences, an act utilitarian would say my action was bad. Yet many people would say I did something good. For this reason, some philosophers prefer a theory in which the moral agent has complete control over the factors determining the moral worth of an action.

Two additional arguments have been raised against utilitarianism in general. We'll save these arguments for the end of the section on rule utilitarianism.

While it is not perfect, act utilitarianism is an objective, rational ethical theory that allows a person to explain why a particular action is right or wrong. It joins Kantianism on our list of workable ethical theories we can use to evaluate moral problems.

2.7 Rule Utilitarianism

2.7.1 Basis of Rule Utilitarianism

The weaknesses of act utilitarianism have led some philosophers to develop another ethical theory based on the Principle of Utility. This theory is called rule utilitarianism. Some philosophers have concluded that John Stuart Mill was actually a rule utilitarian, but others disagree.

Rule utilitarianism is the ethical theory that holds we ought to adopt those moral rules which, if followed by everyone, will lead to the greatest increase in total happiness. Hence a rule utilitarian applies the Principle of Utility to moral rules, while an act utilitarian applies the Principle of Utility to individual moral actions.

Both rule utilitarianism and Kantianism are focused on rules, and the rules these two ethical theories derive may have significant overlap. Both theories hold that rules should be followed without exception. However, the two ethical theories derive moral rules in completely different ways. A rule utilitarian chooses to follow a moral rule because its universal adoption would result in the greatest happiness. A Kantian follows a moral rule because it is in accord with the Categorical Imperative: all human beings are to be treated as ends in themselves, not merely as means to an end. In other words, the rule utilitarian is looking at the consequences of the action, while the Kantian is looking at the will motivating the action.

2.7.2 Evaluating a Scenario Using Rule Utilitarianism

Scenario

A worm is a self-contained program that spreads through a computer network by taking advantage of security holes in the computers connected to the network. In August 2003 the Blaster worm infected many computers running the Windows 2000, Windows NT, and Windows XP operating systems. The Blaster worm caused computers it infected to reboot every few minutes.

Soon another worm was exploiting the same security hole in Windows to spread through the Internet. However, the purpose of the new worm, named Nachi, was benevolent. Since Nachi took advantage of the same security hole as Blaster, it could not infect computers that were immune to the Blaster worm. Once Nachi gained access to a computer with the security hole, it located and destroyed copies of the Blaster worm. It also automatically downloaded from Microsoft a patch to the operating system software that would fix the security problem. Finally, it used the computer as a launching pad to seek out other Windows PCs with the security hole.

Was the action of the person who released the Nachi worm morally right or wrong?

Analysis

To analyze this moral problem from a rule utilitarian point of view, we must think of an appropriate moral rule and determine if its universal adoption would increase the happiness of the affected parties. In this case, an appropriate moral rule might be: "If a harmful computer worm is infecting the Internet, and I can write a helpful worm that automatically removes the harmful worm from infected computers and shields them from future attacks, then I should write and release the helpful worm."

What would be the benefits if everyone followed the proposed moral rule? Many people do not keep their computers up to date with the latest patches to the operating system. They would benefit from a worm that automatically removed their network vulnerabilities.

What harm would be caused by the universal adoption of the rule? If everyone followed this rule, the appearance of every new harmful worm would be followed by the release of many other worms designed to eradicate the harmful worm. Worms make networks less usable by creating a lot of extra network traffic. For example, the Nachi worm disabled networks of Diebold ATM machines at two financial institutions [23]. The universal adoption of the moral rule would reduce the usefulness of the Internet while the various worms were circulating.

Another negative consequence would be potential harm done to computers by the supposedly helpful worms. Even worms designed to be benevolent may contain bugs. If many people are releasing worms, there is a good chance some of the worms may accidentally harm data or programs on the computers they infect.

A third harmful consequence would be the extra work placed on system administrators. When system administrators detect a new worm, it is not immediately obvious whether the worm is harmful or beneficial. Hence the prudent response of system administrators is to combat every new worm that attacks their computers. If the proposed moral rule is adopted, more worms will be released, forcing system administrators to spend more of their time fighting worms [24].

In conclusion, the harms caused by the universal adoption of this moral rule appear to outweigh the benefits. Therefore, the action of the person who released the Nachi worm is morally wrong.

2.7.3 The Case for Rule Utilitarianism

1. *Performing the utilitarian calculus is simpler.*

   When calculating the expected total happiness resulting from an action, act utilitarians struggle with determining whom to include in the calculation and how far into the future to look. It's easier for a rule utilitarian to think in general terms about the long-term consequences on society of the universal adoption of a particular moral rule.

2. *Not every moral decision requires performing the utilitarian calculus.*

   A person that relies on rules of behavior does not have to spend a lot of time and effort analyzing every particular moral action in order to determine if it is right or wrong.

3. *Exceptional situations do not overthrow moral rules.*

   Remember the problem of choosing between keeping a promise to A and producing 1,000 units of good for A, or breaking the promise to A and producing 1,001 units of good for B? A rule utilitarian would not be trapped on the horns of this dilemma. A rule utilitarian would reason that the long-term consequences of everyone keeping their promises produce more good than giving everyone the liberty to break their promises. Hence in this situation a rule utilitarian would conclude the right thing to do is keep the promise to A.

4. *Rule utilitarianism solves the problem of moral luck.*

   Since it is interested in the typical result of an action, the occasional atypical result does not affect the goodness of an action. A rule utilitarian would conclude that sending flowers to people in the hospital is a good action.

5. *It appeals to a wide cross section of society.*

   Bernard Gert points out that utilitarianism is "paradoxically, the kind of moral theory usually held by people who claim that they have no moral theory. Their view is often expressed in phrases like the following: 'It is all right to do anything as long as no one gets hurt,' 'It is the actual consequences that count, not some silly rules,' or 'What is important is that things turn out for the best, not how one goes about making that happen.' On the moral system, it is not the consequences of the particular violation that are decisive in determining its justifiability, but rather the consequences of such a violation being publicly allowed" [25]. In other words, an action is justifiable if allowing that action would, as a rule, bring about greater net happiness than forbidding that action.

2.7.4 The Case against Utilitarianism in General

As we have just seen, rule utilitarianism seems to solve several problems associated with act utilitarianism. However, two criticisms have been leveled at utilitarian theories in general. These problems are shared by both act utilitarianism and rule utilitarianism.

1. *Utilitarianism forces us to use a single scale or measure to evaluate completely different kinds of consequences.*

   In order to perform the utilitarian calculus, all consequences must be put into the same units. Otherwise, we cannot add them up. For example, if we are going to determine the total amount of happiness resulting from the construction of a new highway, many of the costs and benefits (such as construction costs and the gas expenses of car drivers) are easily expressed in dollars. Other costs and benefits are intangible, but we must express them in terms of dollars in order to find the total amount of happiness created or destroyed as a result of the project. Suppose a sociologist informs the state that if it condemns 150 homes, it is likely to cause 15 divorces among the families being displaced. How do we assign a dollar value to that unhappy consequence?

2. *Utilitarianism ignores the problem of an unjust distribution of good consequences.*

   The second, and far more significant, criticism of utilitarianism is that the utilitarian calculus is solely interested in the total amount of happiness produced. Suppose one course of action results in every member of a society receiving 100 units of good, while another course of action results in half the members of society receiving 201 units of good each, with the other half receiving nothing. According to the calculus of utility, the second course of action is superior because the total amount of good is higher. That doesn't seem right to many people.

   A possible response to this criticism is that our goal should be to promote the greatest good of the greatest number. In fact, that is how utilitarianism is often described. A person subscribing to this philosophy might say that we ought to use two principles to guide our conduct: (1) we should act so that the greatest amount of good is produced, and (2) we should distribute the good as widely as possible. The first of these principles is the Principle of Utility, but the second is a principle of justice. In other words, "act so as to promote the greatest good of the greatest number" is not pure utilitarianism. The proposed philosophy is not internally consistent, because there are times when the two principles will conflict. In order to be useful, the theory also needs a procedure to resolve conflicts between the two principles. We'll talk more about the principle of justice in the next section.

The criticisms leveled at utilitarianism point out circumstances in which it seems to produce the "wrong" answer to a moral problem. However, rule utilitarianism treats all persons as equals and provides its adherents with the ability to give the reasons why a particular action is right or wrong. Hence we consider it a third workable theory for evaluating moral problems, joining Kantianism and act utilitarianism.

2.8 Social Contract Theory

In the spring of 2003 a coalition of military forces led by the United States invaded Iraq and removed the government of Saddam Hussein. When the police disappeared, thousands of Baghdad residents looted government ministries [26]. Sidewalk arms merchants did a thriving business selling AK-47 assault rifles to homeowners needing protection against thieves. Are Iraqis much different from residents of other countries, or should we view the events in Baghdad as the typical response of people to a lack of governmental authority and control?

2.8.1 The Social Contract

Philosopher Thomas Hobbes (1603–1679) lived during the English civil war and saw firsthand the terrible consequences of social anarchy. In his book *Leviathan* he argues that without rules and a means of enforcing them, people would not bother to create anything of value, because nobody could be sure of keeping what they created. Instead, people would be consumed with taking what they needed and defending themselves against the attacks of others. They would live in "continuall feare, and danger of violent death," and the life of man would be "solitary, poore, nasty, brutish, and short" [27].

To avoid this miserable condition, which Hobbes calls the *state of nature*, rational people understand that cooperation is essential. However, cooperation is possible only when people mutually agree to follow certain guidelines. Hence moral rules are "simply the rules that are necessary if we are to gain the benefits of social living" [5]. Hobbes argues that everybody living in a civilized society has implicitly agreed to two things: (1) the establishment of such a set of moral rules to govern relations among citizens, and (2) a government capable of enforcing these rules. He calls this arrangement the **social contract.**

The Franco-Swiss philosopher Jean-Jacques Rousseau (1712–1778) continued the evolution of social contract theory. In his book *The Social Contract* he writes, "Since no man has any natural authority over his fellows, and since force alone bestows no right, all legitimate authority among men must be based on covenants" [28]. Rousseau states that the critical problem facing society is finding a form of association that guarantees everybody their safety and property, yet enables each person to remain free. The answer, according to Rousseau, is for everybody to give themselves and their rights to the whole community. The community will determine the rules for its members, and each of its members will be obliged to obey the rules. What prevents the community from enacting bad rules is that no one is above the rules. Since everyone is in the same situation, no one will want to put unfair burdens on others.

While everyone might agree to this in theory, it's easy for a single person to rationalize selfish behavior. How do we prevent individuals from shirking their duties to the group? Suppose Bill owes the government $10,000 in taxes, but he discovers a way to cheat on his taxes so that he only has to pay $8,000. Bill thinks to himself, "The government gets billions of dollars a year in taxes. So to the government another $2,000 is just a drop in the bucket. But to me, $2,000 is a lot of money." What restrains Bill from acting selfishly is the knowledge that if he is caught, he will be punished. In order for the social contract to function, society must provide not only a system of laws, but a system of enforcing the laws as well.

According to Rousseau, living in a civil society gives a person's actions a moral quality they would not have if that person lived in a state of nature. "It is only then, when the voice of duty has taken the place of physical impulse, and right that of desire, that man, who has hitherto thought only of himself, finds himself compelled to act on other principles, and to consult his reason rather than study his inclinations" [28].

James Rachels summarizes these ideas in an elegant definition of social contract theory:

Social Contract Theory

"Morality consists in the set of rules, governing how people are to treat one another, that rational people will agree to accept, for their mutual benefit, on the condition that others follow those rules as well" [5].

Both social contract theory and Kantianism are based on the idea that there are universal moral rules that can be derived through a rational process. However, there is a subtle, but important difference in how we decide what makes a moral rule ethical. Kantianism has the notion that it is right for me to act according to a moral rule if the rule can be universalized. Social contract theory holds that it is right for me to act according to a moral rule if rational people would collectively accept it as binding because of its benefits to the community.

Hobbes, Locke, and many other philosophers of the seventeenth and eighteenth centuries held that all morally significant beings have certain rights, such as the right to life, liberty, and property. Some modern philosophers would add other rights to this list, such as the right to privacy.

There is a close correspondence between rights and duties. If you have the right to life, then others have the duty or obligation not to kill you. If you have a right to free health care when you are ill, then others have the duty to make sure you receive it. Rights can be classified according to the duties they put on others. A **negative right** is a right that another can guarantee by leaving you alone to exercise your right. For example, the right of free expression is a negative right. In order for you to have that right, all others have to do is not interfere with you when you express yourself. A **positive right** is a right that obligates others to do something on your behalf. The right to a free education is a positive right. In order for you to have that right, the rest of society must allocate resources so that you may attend school.

Another way to view rights is to consider whether they are absolute or limited. An **absolute right** is a right that is guaranteed without exception. Negative rights are usually considered absolute rights. For example, there is no situation in which it would be reasonable for another person to interfere with your right to life. A **limited right** is a right that may be restricted based on the circumstances. Typically, positive rights are considered to be limited rights. For example, states guarantee their citizens the right to an education. However, because states do not have unlimited budgets, they typically provide a free education for everyone up through the 12th grade but require people to pay for at least some of the costs of their higher education.

Proponents of social contract theory evaluate moral problems from the point of view of moral rights. Kant argued that rights follow from duties. Hence Kantians evaluate moral problems from duties or obligations.

2.8.2 Rawls's Theory of Justice

One of the criticisms of utilitarianism is that the utilitarian calculus is solely interested in the total amount of happiness produced. From a purely utilitarian standpoint, an unequal distribution of a certain amount of utility is better than an equal distribution of a lesser amount of utility.

Social contract theory recognizes the harm that a concentration of wealth and power can cause. According to Rousseau, "the social state is advantageous to men only when all possess something and none has too much" [28]. John Rawls (1921–2002), who did much to revive interest in social contract theory in the twentieth century, proposed two principles of justice that extend the definition of the social contract to include a principle dealing with unequal distributions of wealth and power.

John Rawls's Principles of Justice

1. Each person may claim a "fully adequate" number of basic rights and liberties, such as freedom of thought and speech, freedom of association, the right to be safe from harm, and the right to own property, so long as these claims are consistent with everyone else having a claim to the same rights and liberties.
2. Any social and economic inequalities must satisfy two conditions: first, they are associated with positions in society that everyone has a fair and equal opportunity to assume; and second, they are "to be to the greatest benefit of the least-advantaged members of society (the **difference principle**)" [29].

Rawls's first principle of justice, illustrated in Figure 2.7, is quite close to our original definition of social contract theory, except that it is stated from the point of view of rights and liberties rather than moral rules. The second principle of justice, however, focuses on the question of social and economic inequalities. It is hard to imagine a society in which every person has equal standing. For example, it is unrealistic to expect every person to be involved in every civic decision. Instead, we elect representatives who vote in our place and officials who act on our behalf. Likewise, it is hard to imagine everybody

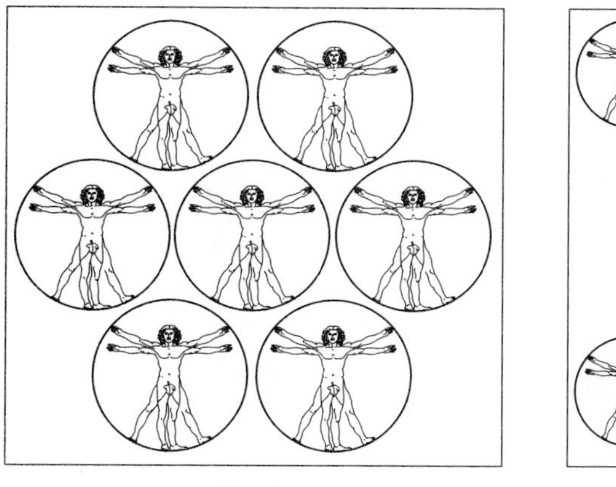

Figure 2.7 Rawls's first principle of justice states that each person may have a "fully adequate" number of rights and liberties as long as they are consistent with everyone else having the same rights and liberties.

in a society having equal wealth. If we allow people to hold private property, we should expect that some people will acquire more than others. According to Rawls, social and economic inequalities are acceptable if they meet two conditions.

First, every person in the society should have an equal chance to assume a position of higher social or economic standing. That means that two people born with equal intelligence, equal talents, and equal motivation to use them wisely should have the same probability of reaching an advantaged position, regardless of the social or economic class to which they were born. For example, the fact that someone's last name is Bush or Kennedy should not give that person a greater probability of being elected President of the United States than any other American born with equal intelligence, talent, and determination.

The second condition, called the difference principle, states that social and economic inequalities must be justified. The only way to justify a social or economic inequality is to show that its overall effect is to provide the most benefit to the least advantaged. The purpose of this principle, illustrated in Figure 2.8, is to help maintain a society composed of free *and equal* citizens. An example of the difference principle in action is a graduated income tax system in which people with higher incomes pay a higher percentage of their income in taxes. An example of a violation of the difference principle would be a military draft system in which poor people had a higher probability of being drafted than wealthy people.

FIGURE 2.8 Rawls's difference principle states that social and economic inequalities must be arranged so that they are of the greatest benefit to the least-advantaged members of society.

2.8.3 Evaluating a Scenario Using Social Contract Theory

Scenario

Bill, the owner of a chain of DVD rental stores in a major metropolitan area, uses a computer to keep track of the DVDs rented by each customer. Using this information, he is able to construct profiles of the customers. For example, a customer that rents a large number of Disney titles is likely to have children. Bill sells these profiles to mail order companies. The customers begin receiving many unsolicited mail order catalogs. Some of the customers are happy to receive these catalogs and make use of them to order products. Others are unhappy at the increase in the amount of "junk mail" they are receiving.

Analysis

To analyze this scenario using social contract theory, we think about the rights of the rational agents involved. In this case, the rational agents are Bill, his customers, and the mail order companies. The morality of Bill's actions revolve around the question of whether he violated the privacy rights of his customers. If someone rents a DVD from one of Bill's stores, both the customer and Bill have information about the transaction. Are their rights to this information equal? If both the customer and Bill have equal rights to this information, then you may conclude there is nothing wrong with him selling this information to a mail order company. On the other hand, if customers have the right to expect transactions to be confidential, you may conclude that Bill was wrong to sell this information without gaining the permission of the customer.

2.8.4 The Case for Social Contract Theory

1. *It is framed in the language of rights.*

   The cultures of many modern countries, particularly Western-style democracies, promote individualism. For people raised in these cultures, the concept of individual rights is powerful and attractive.

2. *It explains why rational people act out of self-interest in the absence of a common agreement.*

   Suppose we are living in a city experiencing a gasoline shortage. If every car owner uses public transportation two days a week, there will be enough gasoline to go around. I need to decide if I will take the bus two days a week.

   Suppose no other car owners ride the bus two days a week. If I decide to ride the bus, I will have to put up with the inconvenience and the city will still run out of gas. Alternatively, I can do what everybody else is doing and continue driving my car until the gasoline supply is exhausted. Since the city will run out of gas either way, I experience less inconvenience by continuing to drive my car every day.

   On the other hand, suppose all the other car owners decide to ride the bus two days a week. If I decide to ride the bus, I will have plenty of company, which is good, but I will still have to adjust my work schedule to fit the bus schedule, waste time waiting at the bus stop, and so on. Alternatively, I can continue to drive my car. That will be more convenient for me. The amount of gasoline my car consumes is insignificant compared to the needs of the city, and the city will not run out of gasoline. Since the city will not run out of gas either way, I experience less inconvenience by continuing to drive my car every day.

To summarize, if no one else rides the bus, it's better for me if I drive my car. If everyone else rides the bus, it's better for me if I drive my car. I have used logic to conclude that I should continue to drive my car. *Unfortunately, everyone else in the town logically reaches the same conclusion!* As a result, the city runs out of gasoline.

The reason we all decided to act selfishly was because we did not have a common agreement. If all of us agreed that everyone should ride the bus two days a week, and those who did not would be punished, then logic would have led people to choose to use public transportation.

Social contract theory is based on the idea that morality is the result of an implicit agreement among rational beings who understand that there is a tension between self-interest and the common good. The common good is best realized when everyone cooperates. Cooperation occurs when those acting selfishly suffer negative consequences.

3. *It provides a clear ethical analysis of some important moral issues regarding the relationship between people and government.*

For example, social contract theory provides a logical explanation of why it is morally acceptable to punish someone for a crime. You might ask, "If everyone has a right to liberty, how can we put in prison someone who has committed a crime?" The social contract is based on the notion that everyone benefits when everyone bears the burden of following certain rules. Knowledge that those who do not follow the rules will be punished restrains individuals from selfishly flouting their obligations. People will have this knowledge only if society punishes those who commit crimes.

Another example is the problem of civil disobedience. While civil disobedience is difficult to justify under Kantianism and utilitarianism, social contract theory provides a straightforward explanation of why civil disobedience can be the morally right decision.

Consider the lunch counter sit-ins of the 1960s. On February 1, 1960, four African-American students from North Carolina A&T walked into the Woolworth's store on South Elm Street in Greensboro, sat down at a whites-only lunch counter, and asked for service. When they were denied service, they refused to leave. Two days later, 85 students participated in the "sit-in" at Woolworth's. All of these students were breaking segregation laws, but according to social contract theory their actions could be considered morally justified. As we have said, the social contract is based on the idea that everyone receives certain benefits in return for bearing certain burdens. The segregation laws were designed to give people of color greater burdens and fewer benefits than white people. Hence they were unjust.

2.8.5 The Case against Social Contract Theory

1. *None of us signed the social contract.*

   The social contract is not a real contract. Since none of us have actually agreed to the obligations of the citizens of our society, why should we be bound to them?

   Defenders of social contract theory point out that the social contract is a theoretical notion that is supposed to explain the rational process through which communities adopt moral guidelines. As John Rawls puts it, social contract agreements are *hypothetical* and *nonhistorical*. They are hypothetical in the sense that they are what reasonable people "could, or would, agree to, not what they have agreed to" [29]. They are nonhistorical because they "do not suppose the agreement has ever, or indeed ever could actually be entered into" [29]. Furthermore, even if it could be entered into, that would make no difference. The reason it would make no difference is because the moral guidelines are supposed to be the result of analysis (facts and values plus logical reasoning), not history. Social contract theory is *not* cultural relativism in disguise.

2. *Some actions can be characterized multiple ways.*

   This is a problem social contract theory shares with Kantianism. Some situations are complicated and can be described in more than one way. Our characterization of a situation can affect the rules or rights we determine to be relevant to our analysis.

3. *Social contract theory does not explain how to solve a moral problem when the analysis reveals conflicting rights.*

   This is another problem social contract theory shares with Kantianism. Consider the knotty moral problem of abortion, in which the mother's right to privacy is pitted against the fetus's right to life. As long as each of these rights is embraced by one side in the controversy, the issue cannot be resolved. What typically happens in debates is that advocates on one side of the issue "solve" the problem by discounting or denying the right invoked by their adversaries.

4. *Social contract theory may be unjust to those people who are incapable of upholding their side of the contract.*

   Social contract theory provides every person with certain rights in return for that person bearing certain burdens. When a person does not follow the moral rules, he or she is punished. What about human beings who, through no fault of their own, are unable to follow the moral rules?

   A response to this objection is that there is a difference between someone who deliberately chooses to break a moral rule and someone who is incapable of understanding a rule. Society must distinguish between these two groups of people. People who deliberately break moral rules should be punished, but people who cannot understand a rule must be cared for.

   However, this response overlooks the fact that distinguishing between these two groups of people can be difficult. For example, how should we treat drug addicts who steal to feed their addiction? Some countries treat them as criminals and put them in a prison. Other countries treat them as mentally ill people and put them in a hospital.

These criticisms demonstrate some of the weaknesses of social contract theory. Nevertheless, social contract theory is logical and analytical. It allows people to explain why a particular action is moral or immoral. According to our criteria, it is a workable ethical theory, joining Kantianism, act utilitarianism, and rule utilitarianism.

2.9 Comparing Workable Ethical Theories

The divine command theory, Kantianism, act utilitarianism, rule utilitarianism, and social contract theory share the viewpoint that moral good and moral precepts are objective. In other words, morality has an existence outside the human mind. For this reason we say these theories are examples of **objectivism.**

What distinguishes Kantianism, utilitarianism, and social contract theory from the divine command theory is the assumption that ethical decision-making is a rational process by which people can discover objective moral principles with the use of logical reasoning based on facts and commonly held values. While each of these four theories has weaknesses, all of them are workable in the sense that they pass this test.

We can make several important distinctions among the four workable theories.

1. *Faced with a moral problem, what is the motivation for taking a particular action?*

   Do we think about rights, responsibilities, and duties, or do we consider the consequences of the action? Kantianism and social contract theory are clearly oriented toward the notion that people should "do the right thing." Kantianism starts more from the viewpoint of duty, while social contract theory begins by considering the rights of the persons involved. Utilitarian theories are oriented toward the consequences of actions, the notion that people should "do good." Note, however, once a complete analysis has been done, rule utilitarians adopt rules that people are obliged to follow without exception. Hence rule utilitarianism ends up with a mixed motivation.

2. *What criteria are used to determine if an action is ethical or unethical?*

   Kantianism, rule utilitarianism, and social contract theory use universal moral rules as their metric. An act utilitarian computes the total change in utility to determine if an action is right or wrong.

3. *Is the focus on the individual or the group?*

   Kantianism and social contract theory focus on the individual decision-maker. In contrast, act and rule utilitarianism must consider all affected parties when evaluating the consequences of an action.

Theory	*Motivation*	*Criteria*	*Focus*
Kantianism	Dutifulness	Rules	Individual
Act Utilitarianism	Consequence	Actions	Group
Rule Utilitarianism	Consequence/Duty	Rules	Group
Social Contract	Rights	Rules	Individual

Table 2.1 Comparison of four workable ethical theories. All of these theories are based on objectivism and reasoning from facts or commonly held values.

Table 2.1 provides a summary of these differences among Kantianism, act utilitarianism, rule utilitarianism, and social contract theory.

2.10 Morality of Breaking the Law

What is moral and what is legal are not identical. Certain actions may be wrong, even if there are no laws forbidding these actions. Is it possible that an illegal action may be the right action?

Let's analyze this question from the point of view of our four workable ethical theories. To ground our analysis, we will consider a particular illegal action: violating a licensing agreement by copying a CD containing copyrighted music and giving it to a friend.

2.10.1 Social Contract Theory Perspective

Social contract theory is based on the assumption that everyone in society ought to bear certain burdens in order to receive certain benefits. The legal system is instituted to guarantee that people's rights are protected. It guarantees people will not choose their selfish interests over the common good. For this reason we have a *prima facie* obligation to obey the law (Figure 2.9). That means, everything else being equal, we should be law-abiding. In return, our own legal rights will be respected. Our obligation to obey the law should only be broken if we are compelled to follow a higher-order obligation.

From the point of view of social contract theory, then, it is wrong to copy a CD containing copyrighted music, because that action violates the legal rights of the person or organization owning the copyright.

2.10.2 Kantian Perspective

According to the Categorical Imperative, we should act only from moral rules that we can at the same time will to be universal moral laws. Suppose I think the current copyright laws are unjust because they unfairly favor the producers of intellectual property

FIGURE 2.9 According to social contract theory, we have a *prima facie* obligation to obey the law. (Beth Anderson)

rather than the consumers. I could propose the following rule: "I may ignore a law that I believe to be unjust."

What happens when we universalize this rule? If everyone acted according to this rule—ignoring laws they felt to be unjust—then the authority of Congress to legislate laws would be fatally undermined. Yet the goal of Congress is to create laws that ensure we live in a just society. Hence there is a logical contradiction, because I cannot both will there be justice (by ignoring an unjust law) and will there be no justice (by denying Congress the authority it needs to create a just society).

Another line of Kantian reasoning leads us to the same conclusion. If I copy a CD containing copyrighted material, I am violating the legal rights of the person who owns the copyright. No matter how good my intended use of the CD, I am using the copyright owner if I make a copy without their permission. This violates the second formulation of the Categorical Imperative. Hence it is wrong to copy the CD.

2.10.3 Rule Utilitarian Perspective

What would be the consequences of people ignoring laws they felt to be unjust? A beneficial consequence is the happiness of the people who are doing what they please rather than obeying the law. There are, however, far more harmful consequences. First, the people directly affected by lawless actions will be harmed. Second, people in general would have less respect for the law. Third, assuming increased lawlessness puts an additional burden on the criminal justice system, society as a whole would have to pay for having additional police officers, prosecutors, judges, and prisons. Hence, from a rule utilitarian viewpoint, breaking the law is wrong.

2.10.4 Act Utilitarian Perspective

We will do an act utilitarian analysis to show there can be situations where the benefits of breaking a law are greater than the harms. Suppose I purchase a music CD. I play it, and I think it is great. A friend of mine is in a terrible automobile accident. While he recovers, he will need to stay quiet for a month. I know he has no money to spend on music. In fact, people are doing fundraisers simply to help his family pay the medical bills. I don't have money to contribute to a fundraiser, but I think of another way I could help him out. I give my friend a copy of the CD. He is grateful for having a diversion during his time of bed rest.

What are the consequences of my action? As far as I can tell, there is no lost sale, because even if I had not bought my friend the CD, he would not have bought it. In fact, giving a copy of the CD to my friend may actually increase the sales of the CD if my friend likes it and recommends it to other people who do have money to spend on CDs. I am not likely to be prosecuted for what I did. Therefore, there will be no impact on the legal system. No extra police detectives, prosecutors, or judges will need to be hired as a result of my action. The principal harm I have done is to have violated the legal rights of the owner of the copyright. The benefits are that my friend is thrilled to have something to do during his recovery and I am happy to have been able to do something to help him out during his time of need. Overall, the benefits appear to outweigh the harms.

2.10.5 Conclusion

There is nothing intrinsically immoral about copying a CD. However, our society has chosen to enact laws that grant intellectual property rights to people who do creative work and distribute it on CDs. From the viewpoint of Kantianism, rule utilitarianism, and social contract theory, breaking the law is wrong unless there is a strong overriding moral obligation. Copying a disc to save a few dollars or help a friend does not fall into that category. Copying a CD containing copyrighted music is immoral *because* it is illegal.

From an act utilitarian viewpoint, it is not hard to devise particular instances where making a copy of a copyrighted CD is the right action. Put another way, a blanket prohibition against copying cannot be morally justified from an act utilitarian point of view.

Summary

We live together in communities for our mutual benefit. Every society has guidelines indicating what people are supposed to do in various circumstances. We call these guidelines morality. Ethics, also called moral philosophy, is a rational examination into people's moral beliefs and behaviors. In this chapter we have considered a variety of ethical theories with the purpose of identifying those that will be of most use to us as we consider the effects of information technology on society.

Relativistic theories are based on the idea that people *invent* morality. A relativist claims there are no universal moral principles. Subjective relativism is the theory that morality is an individual creation. Cultural relativism is the idea that each society determines its own morality. If morality is invented, and no set of moral guidelines is any better than another, then there are no objective criteria that can be used to determine if one set of guidelines is better than another. Under these circumstances, the study of ethics is extremely difficult, if not impossible. For this reason we shall not make use of relativistic theories.

In contrast, objectivism is based on the idea that morality has an existence outside the human mind. It is the responsibility of people to *discover* morality. An objectivist claims there are certain universal moral principles that are true for all people, regardless of their historical or cultural situation.

The first objectivist theory we considered was the divine command theory. The divine command theory is based on the idea that God has provided us with moral guidelines designed to promote our well-being. These guidelines are to be followed because they reflect the will of God, not because we understand them. Because this theory does not rationally derive moral guidelines from facts and commonly held values, we reject it as a useful ethical theory.

The second objectivist theory we considered was Kantianism, named after the German philosopher Immanuel Kant. Kantianism is focused on dutifulness. If we are dutiful, we will feel compelled to act in certain ways out of respect for moral rules. A moral rule is appropriate if it is consistent with the Categorical Imperative. Kant provides two formulations of the Categorical Imperative. The first is: "Act only from moral rules that you can at the same time will to be universal laws." The second is: "Act so that you always treat both yourself and other people as ends in themselves, and never solely as a means to an end." While both Kantianism and the divine command theory hold that actions should be motivated by the desire to obey universal moral rules, Kantianism holds that rational beings can discover these rules without relying upon divine inspiration. Kantianism is considered a non-consequentialist theory because the morality of an action is determined by evaluating the moral rule upon which the will to act is grounded rather than the action's consequences.

Utilitarianism, developed by Jeremy Bentham and John Stuart Mill, is based upon the Principle of Utility, also called the Greatest Happiness Principle. According to this principle, an action is right (or wrong) to the extent that it increases (or decreases) the total happiness of the affected parties. Utilitarianism is called a consequentialist theory, because its focus is on the consequences of an action. Act utilitarianism is the theory that an action is good if its net effect (over all affected beings) is to produce more happiness than unhappiness. An action is bad if its net effect is to produce more unhappiness than happiness. Rule utilitarianism is the ethical theory that holds we ought to adopt those moral rules which, if followed by everyone, will lead to the greatest increase in total happiness. In other words, rule utilitarianism applies the Principle of Utility to moral rules, while act utilitarianism applies the Principle of Utility to individual moral actions. Both of these theories hold that rational beings can perform the analysis needed to determine if a moral action or moral rule is good or evil.

The final ethical theory we considered was social contract theory, identified with Thomas Hobbes, Jean-Jacques Rousseau, and John Rawls. Social contract theory holds that "morality consists in the set of rules, governing how people are to treat one another, that rational people will agree to accept, for their mutual benefit, on the condition that others follow those rules as well" [5]. Rawls proposed two principles of justice that are designed to maintain society over time as an association of free and equal citizens. Like Kantianism and both forms of utilitarianism, social contract theory is based on the premise that there are universal, objective moral rules that can be discovered through rational analysis.

Our survey identified four practical ethical theories: Kantianism, act utilitarianism, rule utilitarianism, and social contract theory. We used these theories to analyze the question, "Is it morally acceptable to break the law?" According to social contract theory, Kantianism, and rule utilitarianism, the answer to this question is "No." It is wrong to break the law unless there is an overriding moral concern. From an act utilitarian perspective, however, it is possible to devise a situation in which the benefits of breaking the law outweigh the harms.

Our discussion of the strengths and weaknesses of Kantianism, act utilitarianism, rule utilitarianism, and social contract theory revealed that none of these theories is perfect. Considering any one of the theories, we will find some moral problems that it is able to solve easily. We will find other moral problems that it is unable to solve. While it is disappointing that no one ethical theory is clearly superior to the others, these four theories together have a lot of power.

Consider the analogy between ethical theories and tools in a toolbox. A toolbox that contains only a hammer is not very useful, but a well-equipped toolbox enables a handy person to fix a wide range of household problems. In the chapters that follow, we'll use our "toolbox" of Kantianism, act utilitarianism, rule utilitarianism, and social contract theory to propose solutions to many problems arising from the introduction of information technology into society.

Review Questions

1. Define in your own words what "the ethical point of view" means.
2. Define morality and ethics in your own words.

3. What is the difference between morality and ethics?
4. What is the difference between relativism and objectivism?
5. What are the advantages of using an ethical theory in which all humans are treated equally and guidelines are developed through a process of logical reasoning?
6. What do we mean when we say an ethical theory is rational?
7. What is the many/any fallacy? Invent your own example of this fallacy.
8. What is the equivalence fallacy? Invent your own example of this fallacy.
9. Come up with your own example of a moral rule that would violate the Categorical Imperative.
10. What is plagiarism? Describe four different ways that a person can commit plagiarism.
11. What is the difference between plagiarism and misuse of sources?
12. What is the difference between a consequentialist theory and a non-consequentialist theory?
13. Give three examples of a situation in which your action would be primarily motivated by a sense of duty or obligation. Give three examples of a situation in which your action would be primarily motivated by its expected consequences.
14. What is the problem of moral luck?
15. Why do businesses and governments often use utilitarian thinking to determine the proper course of action?
16. What is the difference principle?
17. Is social contract theory as first presented a consequentialist theory or a non-consequentialist theory? Is social contract theory as articulated in Rawls's two principles of justice a consequentialist theory or a non-consequentialist theory?
18. Describe similarities and differences between divine command theory and Kantianism.
19. Describe similarities and differences between subjective relativism and act utilitarianism.
20. Describe similarities and differences between Kantianism and rule utilitarianism.
21. Describe similarities and differences between act utilitarianism and rule utilitarianism.
22. Describe similarities and differences between cultural relativism and social contract theory.
23. Describe similarities and differences between Kantianism and social contract theory.
24. Evaluate the four scenarios presented in Section 2.1 from a Kantian perspective.
25. Evaluate the four scenarios presented in Section 2.1 from an act utilitarian perspective.
26. Evaluate the four scenarios presented in Section 2.1 from a rule utilitarian perspective.
27. Evaluate the four scenarios presented in Section 2.1 from the perspective of social contract theory.
28. A college student attached a webcam to his laptop computer and left the computer running in his dormitory room in order to broadcast video images of his roommate

and his roommate's girlfriend engaged in sexual intercourse. They were unaware of his actions. The student's Web site accumulated thousands of hits for the two weeks it was up. Copies of some images were posted on at least one other Web site [30]. Using each of the four workable ethical theories presented in this chapter, evaluate the actions of the college student.

Discussion Questions

29. If everyone agreed to take the ethical point of view by respecting others and their core values, would there be any need for a rigorous study of ethics?
30. If you had to choose only one of the ethical theories presented in this chapter and use it for all of your personal ethical decision-making, which theory would you choose? Why? How would you respond to the arguments raised against the theory you have chosen?
31. Most ethical theories agree on a large number of moral guidelines. For example, it is nearly universally held that it is wrong to steal. What difference, then, does it make whether someone subscribes to the divine command theory, Kantianism, utilitarianism, or one of the other ethical theories? (Hint: Think about which theories are more persuasive when they lead to different conclusions about the right thing to do.)
32. Suppose a spaceship lands in your neighborhood. Friendly aliens emerge and invite humans to enter the galactic community. You learn that this race of aliens has colonized virtually the entire galaxy; Earth is one of the few inhabitable planets to host a different intelligent species. The aliens seem to be remarkably open-minded. They ask you to outline the ethical theory that should guide the interactions between our two species. Which ethical theory would you describe? Why?
33. According to the Golden Rule, you should do unto others as you would want them to do unto you. Is the Categorical Imperative simply the Golden Rule in disguise?
34. Are there any ethical theories described in this chapter that would allow someone to use the argument "Everybody is doing it" to show that an activity is not wrong?
35. What are some examples of contemporary information technology issues for which our society's moral guidelines seem to be nonexistent or unclear? (Hint: Think about issues that are generating a lot of media coverage or lawsuits.)
36. People give a variety of reasons for copying a music CD from a friend instead of buying it [31]. Refute each of the reasons given below using one of the viable theories described in this chapter. (You don't have to use the same theory each time.)
    a. I don't have enough money to buy it.
    b. The retail price is too high. The company is gouging customers.
    c. Since I wouldn't have bought it anyway, the company didn't lose a sale.
    d. I'm giving my friend the opportunity to do a good deed.
    e. Everyone else is doing it. Why should I be the only person to buy it when everyone else is getting it for free?
    f. This is a drop in the bucket compared to Chinese pirates who sell billions of dollars worth of copied music.

g. This is insignificant compared to the billions of dollars worth of music being exchanged over the Internet.

37. Suppose a society holds that it is wrong for one individual to eavesdrop on the telephone conversations of another citizen. Should that society also prohibit the government from listening in on its citizens' telephone conversations?

38. Should moral guidelines for individuals apply to nation-states as well? Are the interactions of nation-states analogous to the interactions of individuals? Should there be a different kind of morality to guide the actions of nation-states, or are the actions of nation-states with each other outside the moral realm?

39. Are the citizens of a representative democracy morally responsible for the actions of their government?

In-Class Exercises

40. Students in a history class are asked to take a quiz posted on the course Web site. The instructor has explained the following rules to the students: First, they are supposed to do their own work. Second, they are free to consult their lecture notes and the textbook while taking the quiz. Third, in order to get credit for the quiz, they must correctly answer at least 80 percent of the questions. If they do not get a score of 80 percent, they may retake the quiz as many times as they wish.

    Mary and John are both taking the quiz. They are sitting next to each other in the computer room. John asks Mary for help in answering one of the questions. He says, "What's the difference if you tell me the answer, I look it up in the book, or I find out from the computer that my answer is wrong and retake the quiz? In any case, I'll end up getting credit for the right answer." Mary tells John the correct answer to the question.

    Discuss the morality of Mary's decision.

41. In Plato's dialogue *The Republic*, Glaucon argues that people do not voluntarily do what is right [32]. According to Glaucon, anyone who has the means to do something unjust and get away with it will do so. Glaucon illustrates his point by telling the story of Gyges.

    Gyges, a shepherd, discovers a magic ring. He accidentally discovers that wearing this ring renders him invisible. He uses the power of the ring to seduce the queen, kill the king, and take over the kingdom.

    Glaucon believes that whenever people have the opportunity to act unjustly without any fear of getting caught or anyone thinking the worse of them, they do so. If they do not act to their own advantage when given the opportunity, others will think they are fools. Do you agree with Glaucon?

42. Is the right to life a negative right or a positive right? In other words, when we say someone has the right to life, are we simply saying we have an obligation not to harm that person, or are we saying we have an obligation to provide that person what he or she needs in order to live, such as food and shelter?

43. Which of the following rights should be considered legitimate positive rights by our society?

a. The right to a K–12 education
b. The right to a higher education
c. The right to housing
d. The right to health care
e. The right of a Presidential candidate to receive time on television

Further Reading

Douglas Birsch. *Ethical Insights: A Brief Introduction.* 2nd ed. McGraw-Hill, New York, NY, 2002.

Brian Hansen. "Combating Plagiarism: Is the Internet Causing More Students to Copy?" *The CQ Researcher*, September 19, 2003 (entire issue).

Oliver A. Johnson. *Ethics: Selections from Classical and Contemporary Writers.* 8th ed. Harcourt Brace, Fort Worth, TX, 1999.

Immanuel Kant. *Foundations of the Metaphysics of Morals and What Is Enlightenment?* Translated, with an introduction, by Lewis White Beck. Prentice Hall, Upper Saddle River, NJ, 1997.

John Stuart Mill, *On Liberty and Utilitarianism.* with an introduction by Alan M. Dershowitz. Bantam Books, New York, NY, 1993.

Plato. *Gorgias.* Translated, with an introduction, by Walter Hamilton. Penguin Books, Harmondsworth, England, 1960.

James Rachels. *The Elements of Moral Philosophy.* 4th ed. McGraw-Hill, New York, NY, 2003.

Jean-Jacques Rousseau. *The Social Contract.* Penguin Books, London, England, 1968.

References

[1] James H. Moor. "Reason, Relativity, and Responsibility in Computer Ethics. In *Readings in CyberEthics.* 2nd ed. Edited by Richard A. Spinello and Herman T. Tavani. Jones and Bartlett, Sudbury, MA, 2004.

[2] John Rawls. *A Theory of Justice, Revised Edition.* The Belknap Press of Harvard University Press, Cambridge, MA, 1999.

[3] Plato. *Portrait of Socrates: Being the Apology, Crito and Phaedo of Plato in an English Translation.* Translated by Sir R. W. Livingstone. Clarendon Press, Oxford, England, 1961.

[4] Douglas Birsch. *Ethical Insights: A Brief Introduction.* 2nd ed. McGraw-Hill, Boston, MA, 2002.

[5] James Rachels. *The Elements of Moral Philosophy.* 4th ed. McGraw-Hill, Boston, MA, 2003.

[6] William Graham Sumner. *Folkways: A Study of the Sociological Importance of Usages, Manners, Customs, Mores, and Morals.* Ginn and Company, Boston, MA, 1934.

[7] Charles M. Hampden-Turner and Fons Trompenaars. *Building Cross-Cultural Competence: How to Create Wealth from Conflicting Values*. Yale University Press, New Haven, CT, 2000.

[8] *The Torah: A Modern Commentary*. Union of American Hebrew Congregations, New York, NY, 1981.

[9] Plato. *Plato's Euthyphro: with Introduction and Notes and Pseudo-Platonica*. Arno Press, New York, NY, 1976.

[10] *The Holy Bible, New Revised Standard Version*. Genesis, Chapter 22. Oxford University Press, Oxford, England, 1995.

[11] *The Holy Bible, New Revised Standard Version*. Genesis, Chapter 4. Oxford University Press, Oxford, England, 1995.

[12] Lewis White Beck. "Translator's Introduction." In *Foundations of the Metaphysics of Morals*. 2nd ed. Library of Liberal Arts / Prentice Hall, Upper Saddle River, NJ, 1997.

[13] William K. Frankena. *Ethics*. 2nd ed. Prentice Hall, Englewood Cliffs, NJ, 1973.

[14] Council of Writing Program Administrators. "Defining and Avoiding Plagiarism: The WPA Statement on Best Practices." January, 2003. www.wpacouncil.org.

[15] Corrections. *The New York Times*, May 2, 2003.

[16] Scott Smallwood. "Arts Professor at New School U. Resigns after Admitting Plagiarism." *The Chronicle of Higher Education*, September 20, 2004.

[17] Brian Hansen. Combating Plagiarism: Is the Internet Causing More Students to Copy? *The CQ Researcher*, 13(32), 2003.

[18] Katie Hafner. "Lessons in Internet Plagiarism." *The New York Times*, June 28, 2001.

[19] Jay Vegso. "Interest in CS as a Major Drops among Incoming Freshmen." *Computing Research News*, 17(3), May 2005.

[20] Cass Sunstein. *republic.com*. Princeton University Press, Princeton, NJ, 2001.

[21] Jeremy Bentham. *An Introduction to the Principles of Morals and Legislation*. Oxford, 1823.

[22] W. D. Ross. *The Right and the Good*. 2nd ed. Oxford University Press, Oxford, England, 2003.

[23] Kevin Poulsen. "Nachi Worm Infected Diebold ATMs." *The Register*, November 25, 2003. www.theregister.co.uk.

[24] Florence Olsen. "Attacks Threaten Computer Networks as Students Arrive for the Fall Semester." *The Chronicle of Higher Education*, September 5, 2003.

[25] Bernard Gert. "Common Morality and Computing." In *Readings in CyberEthics*. 2nd ed. Edited by Richard A. Spinello and Herman T. Tavani. Jones and Bartlett, Sudbury, MA, 2004.

[26] John Daniszewski and Tony Perry. "War with Iraq; U.S. in Control; Baghdad in U.S. Hands; Symbols of Regime Fall As Troops Take Control." *The Los Angeles Times*, April 10, 2003.

[27] Thomas Hobbes. *Leviathan*. Penguin Books, London, England, 1985.

[28] Jean-Jacques Rousseau. *The Social Contract*. Translated by Maurice Cranston. Penguin Books, London, England, 1968.

[29] John Rawls. *Justice as Fairness: A Restatement*. The Belknap Press of Harvard University Press, Cambridge, MA, 2001.

[30] Schellene Clendenin. "Student Punished for Webcam Misuse." *The Daily Barometer (Oregon State University)*, November 26, 2002.

[31] Sara Baase. *A Gift of Fire*. 2nd ed. Prentice Hall, Upper Saddle River, NJ, 2003.

[32] Plato. *The Republic of Plato*. Translated by F. M. Cornford. The Oxford University Press, London, England, 1941.

AN INTERVIEW WITH

James Moor

James Moor is a professor of philosophy at Dartmouth College. He is currently President of the International Society for Ethics and Information Technology, as well as Editor-in-chief of the philosophical journal *Minds and Machines.*

Professor Moor has written extensively on computer ethics, the philosophy of artificial intelligence, the philosophy of mind, the philosophy of science, and logic. His publications include "The Future of Computer Ethics: You Ain't Seen Nothin' Yet," *Ethics and Information Technology*, Vol. 3, No. 2 (2001). He and Terrell Bynum co-edited *The Digital Phoenix: How Computers Are Changing Philosophy* (Oxford: Basil Blackwell Publishers, 1998 and revised edition 2000) and *Cyberphilosophy: The Intersection of Computing and Philosophy* (Oxford: Basic Blackwell Publishers, 2002).

In 2003 Dr. Moor received the Making a Difference Award from the Association for Computing Machinery's Special Interest Group on Computers and Society. He holds a Ph.D. from Indiana University.

What stimulated your interest in studying the philosophy of technology?

My interest developed initially through a fascination with computing. The philosophy of computing is a combination of logic, epistemology, metaphysics, and value theory—the complete philosophical package wrapped up in a very practical and influential technological form. Who wouldn't be interested in that? Many standard philosophical issues are brought to life in a computer setting. Consider a simple example: In the *Republic* Plato tells a story about the Ring of Gyges, in which a shepherd finds a ring that, when he wears it and turns it, makes him invisible. Being a clever but rather unethical shepherd, he uses the power of the ring to take over the kingdom, including killing the king and marrying the queen. Through this story Plato raises a deep and important philosophical question: Why be just if one can get away with being unjust? Today the Internet offers each of us our own ring of Gyges. Agents on the Internet can be largely invisible. The question for us, echoing Plato, is why be just while using the Internet if one can get away with being unjust?

What distinguishes ethical problems in computing from ethical problems in other fields?

Some have argued that the ethical problems in the field are unique. This is difficult to show, because the problems involving computing usually connect with our ordinary ethical problems in some way. Nevertheless, what makes the field of computer ethics special and important, though probably not unique, is the technology itself–the computer. Computers are logically malleable machines in that they can be shaped to do any task that one can design, train, or evolve them to do. Computers are universal tools, and this explains why they are so commonplace and culturally transforming. Because they are used in so many ways, new situations continually arise for which we do not have clear policies to guide actions. The use of computing creates policy vacuums. For instance, when wireless technology first appeared, there were questions about whether one should be allowed to access someone else's wireless system, e.g., when driving down the street. Should such access be considered trespassing? Ethical rights and duties of novel situations are not always clear. Because computers are universal tools and can be applied in so many diverse ways, they tend to create many more policy vacuums than other technologies. This is one respect in which the ethical problems in computing are different from other fields at least in degree if not in kind. This makes computer ethics an extraordinarily important discipline for all of us.

How has information technology affected the field of ethics in the past two decades?

Twenty years ago I had to search newspapers and magazines to find stories on computer/information ethics. Such stories were uncommon. Now many such stories appear daily. They are so common that the fact that computing is involved is unremarkable. Stories about body parts being sold on eBay or identity theft over the Internet or spam legislation all presuppose computing, but computing has so permeated our culture that it is not something uncommon, but something almost everybody uses. In a sense, much of ethics has become computer ethics!

Why do you believe it is helpful to view computer ethics issues in terms of policies?

When we act ethically, we are acting such that anyone in a similar situation would be allowed to do the same kind of action. I am not allowed to have my own set of ethical policies that allow me to do things that others in a relevantly similar situation cannot do. Ethical policies are public policies. An act utilitarian, by contrast, would consider each situation individually. On this view, cheating would not only be justified but required if the individual doing the cheating benefited and others were not harmed because they did not know about it. This seems to me to be a paradigm of unethical behavior, and hence I advocate a public policy approach. If cheating is allowed for some, then everyone should be allowed to cheat in similar situations.

Rather than using "policies" I could use "rules." But ethical rules are sometimes regarded as binding without exceptions. A system of exceptionless rules will never work as an ethical theory, for rules can conflict and sometimes exceptions must be made because of extraordinary consequences. One might be justified in lying to save a life, for example. I prefer using the word "policy" because I want to suggest modification may be necessary in cases of conflict or in extraordinary circumstance. Notice that the policies involving exceptions must themselves be treated as public policy. If it is justifiable for someone to lie to save a life, it will be justified for others to lie to save a life in similar circumstances.

Please explain the process of resolving an ethical issue using your theory of "just consequentialism."

The view is somewhat like rule utiliatrianism and somewhat like Kantian ethics, but differs crucially from both of them. Rule utilitarians wish to maximize the good, but typically without concern for justice. Just consequentialism does not require maximization of the good, which is in general unknowable, and does not sanction unjust policies simply because they have good consequences. Kant's theory requires us to act only on those maxims that we can will to be a universal law. But Kant's theory does not allow for exceptions. Kant thought one ought never lie. Moreover, the typical Kantian test question of what would happen if everyone did a certain kind of action is not the right question, for this test rules out far too much, e.g., becoming a computer programmer (what if everyone were to become a computer programmer?). For just consequentialism, the test question is what would happen if everyone were *allowed* to do a certain kind of action. We need to consider both the consequences and the justice of our public policies.

In ethics we are concerned about rights and duties, and consequences of actions. Just consequentialism is a mixed system in that it is part deontological and part consequential. Rights and duties can be challenged if they are unfair or cause significant harm, but usually are properly taken as normative guides. One's rights as a citizen and one's duties as a parent are examples. In evaluating consequences we need to consider values that all people share, because we want to develop a policy that we can impartially publicly advocate. Everyone in similar circumstances should be allowed to follow it. At least some of these universal values to be considered will be happiness, life, ability, security, knowledge, freedom, opportunity, and resources. Notice that these are core goods that any sane human wants regardless of which society the human is in.

In the ethical decision process, step one is to consider a set of policies for acting in the kind of situation under consideration. Step two is to consider the relevant duties, rights, and consequences involved with each policy. Step three is to decide whether the policy can be impartially advocated as a public policy, i.e., anyone should be allowed to act in a similar way in similar circumstances. Many policies may be readily acceptable. Many may be easily rejected. And some may be in dispute, as people may weigh the relevant values differently or disagree about the factual outcomes.

In general, rights and duties will carry prima facie weight in ethical decision making, and in general cannot be overridden lightly. But if the consequences of following certain rights and duties are bad enough, then overriding them may be acceptable as long as this kind of exception can be an acceptable public policy. In controversial cases there will be rational disagreements. Just consequentialism does not require complete agreement on every issue. Note that we have disagreements in ordinary non-ethical decision making as well. But just consequentialism does guide us in determining where and why the disagreements occur so that further discussion and resolution may be possible.

You have also studied the field of artificial intelligence from a philosophical point of view. Do you believe it is possible to create a truly intelligent machine capable of ethical decision making? If so, how far are we from making such a machine a reality?

Nobody has shown that it is impossible, but I think we are very far away from such a possibility. The problem may have less to do with ethics than with epistemology. Computers (expert systems) sometimes possess considerable knowledge about special topics, but they lack common-sense knowledge. Without even the ability to understand simple things that any normal child can grasp, computers will not be able to make considered ethical decisions in any robust sense.

Can an inanimate object have intrinsic moral worth, or is the value of an object strictly determined by its utility to one or more humans?

I take values or moral worth to be a judgment based on standards. The standards that count for us are human. We judge other objects using our standards. This may go beyond utility, however, as we might judge a non-useful object to be aesthetically pleasing. Our human standards might be challenged sometime in the future if robots developed consciousness or if we become cyborgs with a different set of standards. Stay tuned.

CHAPTER

3 Networking

Lo, soul, seest thou not God's purpose from the first?
The earth to be spann'd, connected by network,
The races, neighbors, to marry and be given in marriage,
The oceans to be cross'd, the distant brought near,
The lands to be welded together.

–WALT WHITMAN, *Passage to India*

3.1 Introduction

YOU CAN PUT AN ISOLATED COMPUTER TO A LOT OF GOOD USES—such as word processing, touching up digital photographs, constructing spreadsheets, and playing games—but a computer's utility increases tremendously when it is connected to a network. Networked computers can share resources such as printers or extra storage. Networks also support the exchange of email and files.

The Internet has greater value still, because it connects millions of computers. If your computer is connected to the Internet, you can send email to anyone else in the world that also has an email account. You can surf the World Wide Web's billion-plus pages for information, products, and services, or you can use the Web to promote your own company.

Taken in total, the Internet has enormous raw computational power. Some groups have harnessed thousands of computers to tackle extraordinarily complicated scientific problems. The best-known example of this capability is SETI@home, hosted by the University of California at Berkeley. SETI stands for "search for extraterrestrial intelligence." Participants in the project download a special kind of screen saver. Unlike other screen savers, which simply waste CPU cycles when they are running, the SETI@home screen saver analyzes data collected by the world's largest radio telescope. More recently, United Devices has begun distributing a screen saver with the goal of harnessing the power of up to two million PCs to find a cure for smallpox [1, 2, 3]. Another group used 200,000 computers and 25,000 years of computer time to find the largest known prime number [4, 5]. (In case you're curious, the prime is $2^{20,996,011} - 1$. That's a number 6,320,430 digits long.)

As impressive as the total computational power of the Internet is, it has had even more impact as a communication medium. Suppose you have a device connected to a communication network. The usefulness of this device increases as the network grows. For example, if you share a phone network with one other person, your phone isn't very useful. However, a 10-phone network allows you to call 9 other people, a 100-phone network lets you call 99 other people, and so on. The same concept holds for the Internet. Suppose you have a computer connected to the Internet. As more servers supporting additional email accounts and Web pages are added, the opportunities open to you increase. Since the Internet connects millions of computers, its potential power as a communications device is truly staggering.

In order for the utility per person to increase with the number of users, however, two conditions must be met. First, the network infrastructure must be able to support all the data exchanges people are trying to perform. If the network is saturated, adding more users can have the negative consequence of slowing everything down. Second, people sharing the network need to act responsibly. A few people acting in an antisocial manner can make life miserable for everyone else.

In this chapter we explore moral issues associated with our use of the Internet. We begin by focusing on email, the most popular Internet application. After describing how email is routed, we discuss how the increase in unsolicited bulk email, or spam, has degraded the quality of email service. Evaluating the actions of spammers and those who combat spam provide two good opportunities for us to use the four practical ethical theories described in Chapter 2.

The World Wide Web has proven to be the most popular way of organizing information on the Internet. Many governments and individuals find much of the information available on the Web to be subversive, dangerous, or immoral. When people discuss inappropriate Web content, they often mention pornography as a prime example. We use our workable ethical theories to evaluate the morality of producing and using pornography. That discussion leads us into a discussion of the different kinds of censorship, the challenges posed to censorship by the Internet, and the morality of censorship.

Next we turn to the issue of freedom of expression. We explore its history in England and the United States, and examine how it became enshrined as the First Amendment to the United States Constitution. While the First Amendment protects freedom of expression, it is not an absolute right. The U.S. Supreme Court has ruled that personal freedom of expression must be balanced against the public good.

We focus on the issue of children and the Web. We discuss how Web filters work, and we summarize the Child Internet Protection Act, which requires Web filters to be installed in public libraries receiving federal funds. We conclude the section by evaluating the morality of this law.

The Internet provides new ways to commit fraud and deceive people. Email has become the most common way in which identity thieves capture credit card numbers and other personal information. Pedophiles have used chat rooms to arrange meetings with children. Police have responded to the pedophile threat with "sting" operations, which are themselves morally questionable. Web surfers must be aware that the Web contains a great deal of low-quality information. We describe one way in which search engines attempt to direct Web surfers to higher-quality sites.

The widespread availability of the Internet has increased the number of people who spend 40 or more hours a week online. Some psychologists claim there are a vast number of Internet addicts. Others say these fears are overblown. In the last section of this chapter we discuss this issue and evaluate the problem of excessive Internet use from an ethical point of view.

3.2 Email and Spam

3.2.1 How Email Works

Email refers to messages embedded in files transferred from one computer to another via a telecommunications system. An **email address** uniquely indicates a virtual mailbox in cyberspace. Every email address has two parts. The first part (before the `@` sign) identifies the individual user. The second part (after the `@` sign) identifies the domain name. If you are a college student, your college may provide you with an email account, in which case some or all of your domain name is the domain name of your college. Another way to get an email account is through an Internet service provider (ISP). Each ISP has its own domain name.

Suppose you want to send an email to your friend Alyssa Allbright (login name `AA`) at East Dakota State University (domain name `edsu.edu`). You compose the message, indicate the recipient is `AA@edsu.edu`, and send the message. Your mail server uses the domain name system (DNS) to look up `edsu.edu` and find its Internet Protocol (IP) address. This address uniquely identifies a mail server at East Dakota State University. Next, if your email message to Alyssa is more than a few lines long, it is broken up into two or more pieces, called **packets.** At the front of each packet is the IP address of East Dakota State University.

There is a good chance that your mail server is not directly connected to Alyssa's mail server. The Internet contains thousands of interconnected routers that cooperate to get IP packets to their destination (Figure 3.1). Your server sends the packets to a router that is on the path to East Dakota State. It forwards the packets to the next router on the path, and so on, until the packets arrive at Alyssa's mail server. Her mail server reassembles the packets into an email message and puts it in her mailbox.

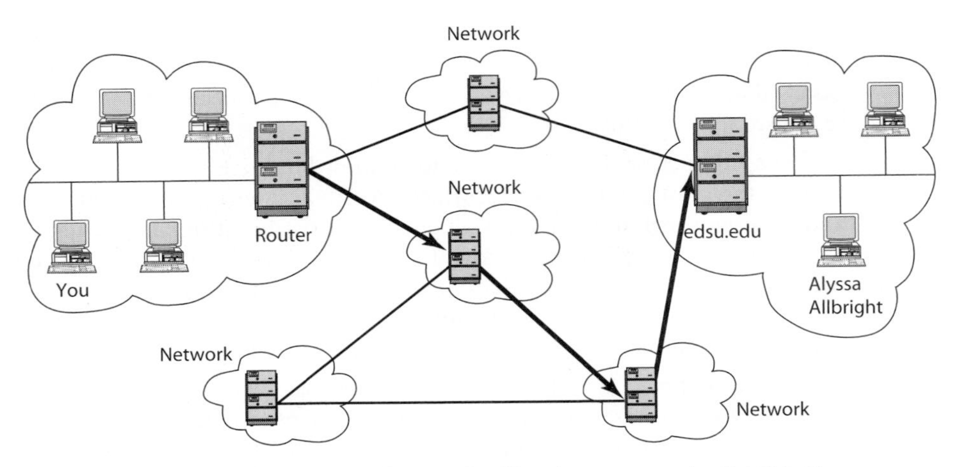

FIGURE 3.1 The Internet connects thousands of local area networks (LANs). Routers pass email and other messages from one LAN to another. Usually there are multiple possible routes.

3.2.2 The Spam Epidemic

The growth of email has been phenomenal. Surveys reveal that in September 2002 there were about 180 million people with access to email in the United States and Canada, and about 600 million worldwide [7, 8]. Every day billions of email messages are exchanged. Unfortunately, a significant percentage of this traffic consists of unsolicited bulk email, or **spam.** Table 3.1 lists the 10 most frequently used subject lines for spam in 2004, as determined by America Online. Do any of them look familiar?

- We carry the most popular medications
- You've been sent an Insta-Kiss!
- You Have 17 New Pictures
- STEAMY HOT LESBIAN ACTION LIVE ON CAMERA!
- All orders are shipped from authorized locations
- 2005 Digital Cable Filters
- F R E E* 30 Second Pre-Qualification MORTGAGE Application
- HURRY HURRY Hot Stock on the RISE
- Sale PRICES ARE BEST ONLINE!
- Breaking news on the TOP Pick stock

TABLE 3.1 According to American Online, these were the 10 most common subject lines for spam in 2004 [6].

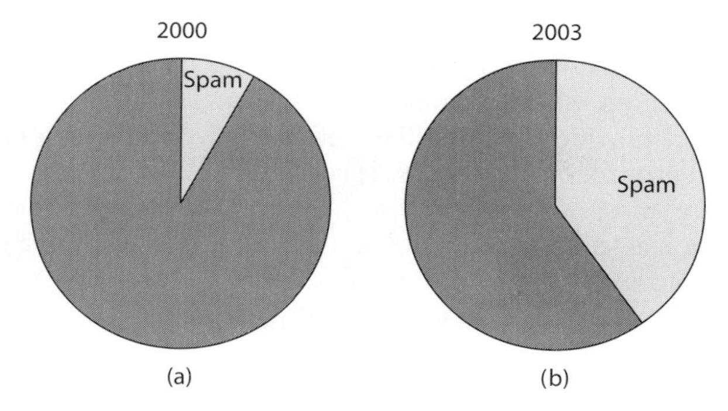

Figure 3.2 The increase in spam between 2000 and 2003. (a) In 2000 spam accounted for 8 percent of all email. (b) By 2003, about 40 percent of email messages were spam.

Why is spam called spam? Brad Templeton, Chairman of the Board of the Electronic Frontier Foundation, traces the term back to the SPAM sketch from *The Final Rip Off* by Monty Python's Flying Circus, in which a group of Vikings drown out a cafe conversation by loudly and obnoxiously repeating the word "spam" [9].

Dealing with spam has become one of the Internet's biggest problems. As recently as 2000, spam accounted for only about 8 percent of all email (Figure 3.2a). Back in those days, it was viewed as a problem for individuals managing their mailboxes. By 2003, about 40 percent of all emails were spam (Figure 3.2b) [10]. Currently, spam consumes a large percentage of the Internet's bandwidth and huge amounts of storage space on mail servers and individual computers. The cost to businesses is estimated at billions of dollars per year in wasted productivity.

The volume of spam is increasing because companies have found it to be effective. The principal advantage of spam is its low cost compared to other forms of advertising. For between $500 and $2,000, a company can hire an Internet marketing firm to send an advertisement to a million different email addresses. Sending the same advertisement to a million addresses using the U.S. Postal Service costs at least $40,000 for the mailing list and $190,000 for bulk-rate postage. And that doesn't include the cost of the brochures! In other words, an email advertisement is more than 100 times cheaper than a traditional flyer sent out in the mail. The cost is so low that a company can make money even if only one in 100,000 recipients of the spam actually buys the product or service [11].

How do direct marketing firms build email lists with millions of addresses? One way a spammer can get your email address is through an opt-in list. Have you ever entered a contest on the Web? There is a good chance the fine print on the entry form said you agree to receive "occasional offers of products you might find valuable" from the company's marketing partners; in other words, spam [11]. Sign-ups for email lists often contain this fine print, too.

Another way spammers get your email address is through so-called dictionary attacks. The term comes from programs that try to guess passwords by trying every entry in an online dictionary. In this case, it means spammers bombard Internet service providers with millions of emails with made-up email addresses, such as AdamA@isprovider.com, AdamB@isprovider.com, AdamC@isprovider.com, and so on. Of course, most of these emails will bounce back, because the addresses are no good. However, if an email *doesn't* bounce, the spammer knows there is a user with that email address and adds it to its mailing list.

To keep networks from being flooded with spam, ISPs have installed spam filters to block spam from reaching users' mailboxes. America Online's spam filters block more than *one billion* email messages a day [12]. These filters look for a large number of messages coming from the same email address, messages with suspicious subject lines, or messages with spam-like content.

Despite these measures, a tremendous amount of spam is still being delivered to email users. Spammers are changing their tactics to confound the efforts of spam blockers.

Spammers disguise themselves by changing email addresses and IP addresses to disguise the sending machine. (See Figure 3.3.) Some innocent victims of "spam spoofing" have received tens of thousands of bounced-back emails. Others have lost their email privileges for violating their ISP's anti-spam policy. Unfortunately, spam spoofing is "easy to do, difficult to trace, and impossible to prevent" [13].

Another trick used by spammers is to divert attention to an innocent bystander by using that person's system as the launch pad for a spam attack. At Colgate University someone discovered an insecure Unix server and used it to send spam [14]. Since the emails originated from a Colgate University computer, Colgate had to handle the complaints, not the spammer. A student at Tufts University received $20 a month from a spammer in exchange for the use of his computer to send spam [15].

Spammers buy spam-screening software and check to see if their messages are blocked. If so, they modify their messages until they pass through the screens. Spammers have changed to more complicated (or more misleading) subject lines, such as "How are you?"; "Thanks for requesting more information"; and "Error in your favor."

Once screeners started checking the bodies of messages for key words, some spammers began switching to images containing the message. The introduction of explicit pornographic images in spam has infuriated many email users. It has also raised concerns that the appearance of pornographic images in people's mailboxes will lead to legal charges against companies for tolerating a work environment that is hostile to women.

3.2.3 Ethical Evaluations of Spamming

If you're like most people, you hate spam. Just because you hate getting spam, does that mean spammers are doing anything wrong? No. As we learned in Chapter 2, an emotional response isn't the same as an ethical evaluation. Let's look at the facts and use logical reasoning to reach our conclusion.

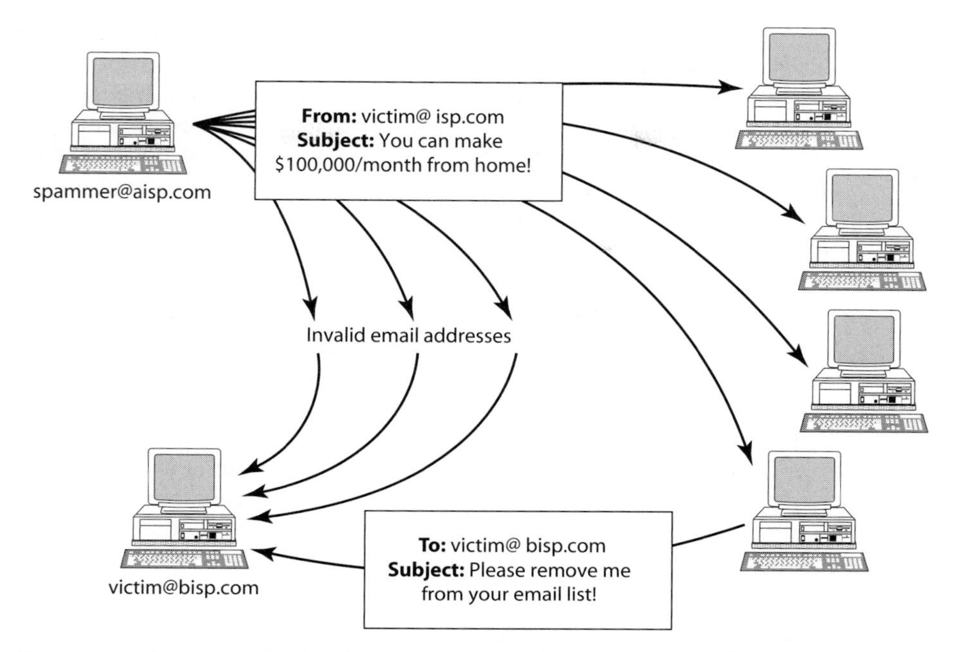

Figure 3.3 Spam spoofing is when a spammer falsifies outgoing emails to make it appear that they originated with someone else. The innocent victim receives the emails that bounce back due to an invalid address, as well as complaints from irate recipients of the spam.

KANTIAN EVALUATION

First, let's evaluate spamming from a Kantian perspective. Suppose I have a great idea for a product or a service that I wish to sell. I choose to send an unsolicited email to a large group of people, knowing that only a tiny fraction will be interested in what I am trying to sell. Some of the recipients of my email message will pay money for it to be held in their mailboxes. If the recipients are accessing their email via a cell phone or another network that charges by the minute, they will pay money to read my email message. Nearly all of the recipients will not be interested in what I have to sell. My email message has cost them time and money. I am treating them as a means to an end (my profit) rather than respecting them as ends in themselves. Hence it is wrong for me to send a spam message.

ACT UTILITARIAN EVALUATION

Now let's look at spamming from an act utilitarian perspective. Suppose I choose to send a spam message to 100 million people. According to *The New York Times*, I can make money if only 1 in 100,000 recipients reply and purchase my product. Let's suppose I have a wildly successful product, and 1 in 10,000 actually buy it. What are the consequences of this scenario? One entrepreneur has made lots of money and is very happy. Suppose 90 percent of the customers are happy with their purchase and 10 percent feel ripped off. That means we have 9,000 happy consumers and 1,000 unhappy ones. We also have 99,990,000 unhappy people. They spent time deleting the unwanted email. They may have paid for the disk space occupied by the email, and they may have paid money to retrieve it. They may be offended by the content of the email message. In addition, everyone on the paths of the email messages lived with slightly slower connections while these 100 million emails were being transmitted. Since 99.991 percent of the people are unhappy, it is safe to say that the action is wrong.

RULE UTILITARIAN EVALUATION

Now let's consider spamming from a rule utilitarian perspective. The general case is likely to be similar to the specific case we just considered from the act utilitarian point of view. If by some chance everyone were interested in a particular product or service, there is no way that many customers could be accommodated. By design, we know that only a tiny fraction of recipients will respond. In addition, as the amount of spam increases, users are more inclined to drop their email accounts. If the number of people using email decreases, the usefulness of the email system will fall for everyone. It is in our interest for as many people as possible to use email accounts. When all of these consequences are considered, it seems clear that spamming is wrong from a rule utilitarian perspective.

SOCIAL CONTRACT THEORY EVALUATION

Finally, let's look at spam from the point of view of social contract theory. We all have a right to free speech. You might think that implies you can send an email to anyone you want. On the other hand, your right to freedom of speech doesn't come with a requirement that anyone has to listen to what you're saying. We ought to think of email as a conversation. When you contact me, you tell me who you are and what you want. Then I can tell you if I am interested. You will be careful about whom you contact, because if you annoy 100 million people, you will not be able to handle the barrage of 100 million angry replies. However, this is not what spammers do. Spammers disguise their identities and often disguise their motives (the subject lines of their messages) in order to get people's attention. This is not above board. Spamming is wrong.

MAKING DIRECT EMAIL MORAL

We've considered the sending of spam from four different perspectives, and our conclusion has been the same each time: sending spam is wrong. You may think we have engaged in ethical overkill, but if we take the analysis one step further, we can figure out how direct email could be considered morally acceptable.

Certain characteristics of spam led us to the conclusion that sending spam is immoral. It is unsolicited, the email addresses of the marketers sending the messages are not correct, and the subject lines of the messages are misleading. Eliminating these characteristics would change our conclusion.

Suppose direct marketers who wished to advertise via email would not change their return addresses, would give their messages accurate subject lines, and would send email only to people who had requested it. From a Kantian perspective, sending email only to people who have requested to be put on a mailing list eliminates the problem of message recipients being treated merely as a means to an end. From a utilitarian perspective, these changes on the part of direct marketers would eliminate the harm done to the millions of people receiving unwanted email messages, while maintaining the benefit to the few people who are actually interested in reading the advertisements and possibly purchasing the products. From the point of view of social contract theory, the use of accurate subject lines and correct return addresses restores the honesty of the email "conversation" between the marketer and the recipient. We conclude that email marketers who use correct return addresses, give their messages accurate subject lines, and send email only to those people who requested to be put on the mailing list are acting morally.

3.3 Fighting Spam

3.3.1 Mail Abuse Prevention System

Mail Abuse Prevention System (MAPS) is a California-based not-for-profit organization dedicated to reducing the flow of spam through the Internet. MAPS maintains a Real-time Blackhole List (RBL), a list of networks that either forward spam or allow spam to be created, and it makes this list available to third parties. Network administrators can make use of the RBL to protect their mail relays from being used by spammers. When a mail relay receives an incoming mail message from an untrusted host, it can make a DNS query against the RBL. If the result of the query indicates the host is on the RBL, the mail relay can reject the email, bouncing it back to the sender.

How does MAPS decide if a network should be blacklisted? The organization has produced a set of guidelines it believes represent "best current practices" for sending bulk email. Here is a summary of the MAPS guidelines for organizations that manage mailing lists:

- They must be able to verify that the recipients of their messages have subscribed to their email lists. Potential subscribers verify their desire to receive marketing emails by replying to a verification message or visiting a Web site and filling in a form. Subscribers to one list should not be automatically added to another list.
- They must provide subscribers with simple ways to terminate their subscriptions, including at least one non-email communication mechanism (such as a telephone number).
- They must remove invalid email addresses from their mailing lists in a timely manner.
- They must take care not to overwhelm individual hosts or LANs.
- They must disclose to subscribers how their addresses will be used, including the frequency and subject matter of future mailings.

When MAPS becomes aware of an email marketer violating these standards, it contacts the marketer (or its ISP) with a warning that it may be placed on the Realtime

Blackhole List [16]. This is a significant threat, because many ISPs use the Realtime Blackhole List to weed out spam. MAPS has been sued at least five times by direct marketing organizations objecting to being put on the list. The American Civil Liberties Union and other groups have also protested the existence of the blacklist, saying that it restricts free speech.

They have a point. Innocent Internet users can be hurt when an ISP is put on the Realtime Blackhole List. If MAPS lists an ISP, then all who use that ISP for their email account—spammers and ordinary email users alike—are blocked from sending email to large swaths of the Internet. Ordinary email users may be denied the ability to send email to family and friends subscribing to ISPs that refuse all email from networks on the Realtime Blackhole List. On the other hand, some believe this is the most effective way to get offending ISPs to change their behavior. If the customers of an ISP complain about their email getting blocked, the ISP must choose between losing these customers and adhering to the policies set forth by MAPS.

3.3.2 Ethical Evaluations of Blacklisting by MAPS

Is MAPS doing anything wrong when it blacklists ISPs who do not conform to its standards for appropriate behavior with respect to forwarding email?

SOCIAL CONTRACT THEORY EVALUATION

The MAPS Web site presents a rights-based argument in defense of its actions. MAPS line of reasoning is predicated on the assumption that an email message ought to be of "direct and equal benefit to the sender and the recipient." This seems to be an extreme assumption. How likely is it that *any* email message has precisely the same benefit to both the sender and the receiver? Even two friends or family members exchanging messages may derive different amounts of benefit or pleasure from the emails, depending upon the amount of time they spend reading and writing them, the number of other people with whom they exchange messages, and so on.

Under MAPS's assumption that an email message should benefit both the sender and the recipient equally, no Internet user can claim a right to have an email message delivered, because that would allow email messages to be sent that had higher benefit to the sender than to the recipient. However, an email relay administrator *does* have the right to refuse to accept any piece of email, because handling a piece of email consumes network resources, and each network administrator should be able to decide how to use the resources of the network for which he or she is responsible.

MAPS does not force other ISPs to use its Realtime Blackhole List. It simply makes this information available to them. Other ISPs choose to use the Realtime Blackhole List in order to control how their network resources are being used.

UTILITARIAN EVALUATION

Let's consider the effects of creating the Realtime Blackhole List from a utilitarian perspective. If an ISP uses the RBL, the users of the ISP benefit in two ways. They do not receive as much spam, and their network performs better because it is not busy passing along spam. On the other hand, they are unable to receive useful email from innocent users of a blacklisted ISP, which reduces the utility of the Internet email system to them. As the number of blacklisted ISPs grows, the more the Internet is fragmented into a bunch of little fiefdoms that do not communicate with each other. The direct marketing companies and innocent users of an ISP on the RBL lose the ability to send email to many other domains on the Internet. They are harmed by the blacklisting. Whether the creation of the RBL is right or wrong depends largely on the net benefit to customers of the ISPs consulting the RBL and the number of innocent customers of blacklisted ISPs compared to the total number of Internet email users.

KANTIAN EVALUATION

From a Kantian viewpoint, the actions of MAPS are more difficult to justify. MAPS puts an offending ISP on its Realtime Blackhole List (and gives the list to all other ISPs that are interested) with the goal of changing the behavior of that ISP. MAPS is counting on the customers of the ISP to complain about being unable to send email to many other domains. Presumably the customers will threaten to change ISPs if the ISPs cannot guarantee their email will be delivered. Since the ISP cannot make its spammers happy (their emails are all bouncing), it must try to make its customers happy, or it will go out of business. This pressure may be sufficient for an ISP to drop its customers who send spam and focus on serving ordinary computer users and direct marketers who abide by the MAPS guidelines. What is happening here? MAPS is treating the innocent customers of an offending ISP as a means to an end (eliminating spam), rather than as ends in themselves. This violates the Categorical Imperative. Hence the creation of the Realtime Blackhole List is wrong from a Kantian perspective.

3.3.3 Proposed Solutions to the Spam Epidemic

What is the best way to halt the spam epidemic? Here are four proposed solutions.

1. *Require an explicit opt-in of subscribers to email lists.*

   A direct marketer cannot include subscribers' email addresses on a mailing list unless they have explicitly indicated that they desire to be included on the list. This proposal has two significant advantages. It would greatly increase the likelihood that people receiving commercial email will be interested in its content. It would reduce the volume of email being sent across the Internet. In July 2002 the European Union issued a directive that email lists contain only those subscribers who have officially opted in. Member countries had until October 2003 to pass legislation enforcing this directive [17].

2. *Require labeling of email advertising.*

   Some states passed laws requiring that all commercial emails contain the letters "ADV" in their subject line, making it easy for an individual to filter spam. These laws were preempted by the federal CAN SPAM Act, described in Section 3.3.4.

3. *Add a cost to every email that is sent.*

   The primary attraction of email as an advertising medium is its low cost—a fraction of a cent per email. Advocates of this approach suggest that a system of micropayments be established. If person A sends person B an email, person A must pay person B a few cents through an automatic system of micropayments. Marketers will no longer be able to afford to send out millions of emails, hoping that one in 100,000 persons responds.

   The problem with this "solution" is that spammers often hijack other people's computers and use them to send out their massive emailings. It's the innocent owners of these computers, not the spammers, who would be stuck with the charges.

4. *Ban unsolicited email.*

   In the late 1980s direct marketers began sending unsolicited advertisements to fax machines, tying up these machines and costing the owners time and money for supplies [18]. In 1991 the U.S. Congress passed the Telephone Consumer Protection Act, which contained a prohibition against unsolicited faxes, called **junk faxes.** In March 2003 the 8th Circuit U.S. Court of Appeals upheld the ban on junk faxes. The three-judge panel ruled that "there is substantial governmental interest in protecting the public from the cost shifting and interference caused by unwanted fax advertisements" [19].

   Some see a parallel between the junk fax epidemic of the late 1980s and today's spam epidemic. In fact, in February 2003 Mark Reinertson won a $539 judgment against Sears for sending him spam. The small claims court judge held that Sears had violated the federal junk fax law. After the trial, a legal expert contended that the ruling was unusual and would not be a strong precedent [20]. Even if the law banning junk faxes does not apply to spam, the U.S. Congress could have enacted new legislation banning unsolicited emails. Congress, however, decided to follow another path, as the next section reveals.

3.3.4 CAN SPAM Act of 2003

In December 2003 President Bush signed legislation regulating commercial email sent in the United States. The Controlling the Assault of Non-Solicited Pornography and Marketing (CAN SPAM) Act of 2003 went into effect January 1, 2004. According to a report by the law firm Gardner Carton & Douglas [21], the law divides email sent by businesses into three categories:

1. Transactional or relationship email messages related to a commercial transaction or ongoing business relationship that the sender and the recipient have already established;
2. Commercial email messages that the recipient has presumably consented to receive, either because the recipient explicitly requested to be put on the mailing list or because the recipient never requested to be removed from the mailing list; and
3. Unsolicited commercial email messages.

Emails falling into the transactional or relationship category must meet these requirements:

- The message header, sender, organization, and transmission information must be correct.
- The message must not disguise the identity of the computer from which the message was sent.

Emails in Category 2—commercial messages to recipients who have consented to receive them—must meet both of the above requirements, as well as the following:

- The message must inform the recipient that the recipient has the option of being removed from the mailing list.
- The message must provide an Internet-based mechanism allowing the recipient to opt out of the company's mailing list. The mechanism must work for at least 30 days from the time the message was sent.
- The message must contain the postal address of the sender.

Emails in Category 3—unsolicited commercial email messages—must meet all of the above requirements and one additional requirement:

- The message must provide "clear and conspicuous" notice that the message is an advertisement. If the message contains sexually explicit material, this notice must be in the subject line.

The CAN SPAM Act also includes a prohibition on the use of dictionary attacks to harvest email addresses. The criminal penalties for violating the CAN SPAM Act include fines of $250 per spam message, with a limit of $2 million that rises to $6 million for repeat offenders. The Act also specifies criminal penalties of up to five years in prison for those who hack into computers to turn them into spam launchers, those who use false information to create five or more email accounts from which they launch spam, and those who falsify header information in bulk email messages [22].

Oregon Senator Ron Wyden, one of the sponsors of the Senate anti-spam bill, said, "The CAN SPAM law will help the Internet remain open for business and keep Americans' in-boxes closed to inappropriate and unwanted spam email."

Some critics are calling the new law the "You CAN Spam Act" because it makes sending spam perfectly legal as long as the direct email advertisers follow the regulations set forth in the law. "This law does not prevent a single spam from being sent," said Scott Mueller, chairman of the Coalition Against Unsolicited Commercial EMail. "It gives a federal stamp of approval for every legitimate marketer in the U.S. to start using unsolicited email as a marketing tool" [23]. There are 24 million small businesses in the United States alone. If 1 percent of these businesses sent you one email a year, you would receive on average more than 650 emails a day [24].

While the CAN SPAM Act requires advertisers to give you the ability to "opt out" from future mailings, there are good reasons for you to be reluctant to attempt to opt out. If the mail was sent by a law-breaking spammer, clicking on the link simply verifies there is someone reading the email. It makes your email address more valuable and increases the probability you will receive more spam. If the email was sent by a law-abiding company, the firm must remove you from its list. But it could sell your email address to another company that wants to send you spam.

Critics also point out that the federal law is weaker than the state laws it preempted. For example, the CAN SPAM Act requires unsolicited commercial email messages to contain a notice that they contain advertising. This notice, however, does not need to appear in the message's subject line unless the message contains sexually explicit material. Many state laws had required a notification in the subject line—in some cases accompanied by the letters ADV—to make spam easier to filter. The Direct Marketing Association supported the preemption language in the CAN SPAM Act, saying that dozens of different state regulations made compliance too difficult for legitimate companies using email advertising [25].

Critics note the CAN SPAM Act became the law of the land on January 1, 2004, the same day that a tough new California law banning all unsolicited emails was going to take effect. "The federal proposals stick a fork in the eye of every Californian who's had their fill of spam," complained California State Senator Debra Bowen [23].

Some question whether any law, state or federal, can solve the spam problem. A report from the technology research firm Gartner, Inc., concluded that CAN SPAM would be ineffective. It noted that spammers can avoid prosecution by sending their email from Internet providers outside the United States, where U.S. laws cannot be enforced [25]. Three different studies performed in the first week of 2004 revealed that only between 1 and 10 percent of bulk email complied with all of the provisions of the CAN SPAM Act [26]. At the end of 2004 America Online reported that the number of spam emails it filtered daily had been cut in half since the passage of the CAN SPAM Act, from 2.4 billion to 1.2 billion. However, other ISPs did not report a decline in the volume of spam [12].

3.3.5 Emergence of "Spim"

Marketers are no longer relying strictly upon email to send unsolicited bulk messages. A February 2005 survey by the Pew Internet & American Life Project revealed that 30 percent of instant messaging users have gotten **spim**—unsolicted bulk instant messages [27]. The same month, an 18-year-old man was arrested for violating the CAN SPAM Act by allegedly sending more than 1.5 million instant messages to users of MySpace.com's instant messaging system [28].

3.4 The World Wide Web

3.4.1 Attributes of the Web

In the past decade the World Wide Web has become the world's most important information storage and retrieval technology. Its creator, Tim Berners-Lee, initially proposed the Web as a documentation system for CERN, the Swiss research center for particle physics, but the creation of easy-to-use Web browsers made the Web accessible to "ordinary" computer users as well [29]. The Web is a hypertext system: a flexible database of information that allows Web pages to be linked to each other in arbitrary fashion.

In Chapter 1 we examined the history of technological innovations that led to the creation of the Web. Here we focus on the attributes that have enabled the Web to become a global tool for information exchange.

1. *It is decentralized.*

   An individual or organization can add new information to the Web without asking for permission from a central authority.

2. *Every object on the Web has a unique address.*

   Any object can link to any other object by referencing its address. A Web object's address is called a **URL** (Uniform Resource Locator).

3. *It is based on the Internet.*

   Building on the Internet makes the Web accessible to people using a wide variety of different computers, such as Macintoshes, Windows systems, or Unix workstations.

The decentralized nature of the Web is one reason why it has grown so rapidly. It also makes the Web more difficult to control. Sometimes the lack of control is viewed as a good thing. For example, the existence of the Web makes it more difficult for an authoritarian government to control the flow of information. On the other hand, the lack of control is sometimes viewed as a weakness. An example of this is when parents attempt to shield their children from Web pages with violent or pornographic content.

3.4.2 How We Use the Web

Web browsers, with their point-and-click navigation and file transfer capabilities, have made the Internet accessible to people with little or no formal computer training. Today, millions of people use the Web for a wide variety of purposes. Here are just a few examples of how people are using the Web.

1. *We shop.*

   The Web enables us to view and order merchandise from the comfort of our homes. According to Accenture, a global consulting service, South Korea is the leader in home shopping, with an amazing 8.7 percent of retail sales occurring electronically [30].

2. *We promote our businesses.*

   Seattle baker Wendi Chocholak's specialty cake business was foundering until she started advertising on Google. Now everyone who searches for "Seattle wedding cake" sees a link to her Web site, and her customer inquiries have increased fourfold [31].

3. *We learn.*

   The University of Phoenix, a private corporation, has 140,000 students, making it the largest university in the United States in terms of total enrollment. About 60,000 students attend classes online, including 4,000 students from other countries [32].

   With more than one million entries, Wikipedia has become the world's largest encyclopedia, and its articles are available for free. However, critics wonder about the quality of a reference work that allows anyone with a Web browser to contribute [33].

4. *We explore our roots.*

   In the past, genealogists interested in accessing American immigration and census records had the choice between mailing in their requests and waiting for them to be processed or visiting the National Archives and examining the documents by hand. Now that the National Archives has put more than 50 million historical records online, the same searches can be performed remotely—and much more quickly—over the Internet [34].

5. *We play games, sometimes for cash prizes.*

   Adults can go online and win cash playing World War II first person shooter *Return to Castle Wolfenstein*. However, Arizona, Arkansas, Connecticut, Delaware, Florida, Iowa, Louisiana, Maryland, Nevada, Tennessee, and Vermont have passed laws against fee-based online gaming [35].

6. *We enter virtual worlds.*

   An **online game** is a game played on a computer network that supports the simultaneous participation of multiple players. A **persistent online game** is an online game in which each player assumes the role of a character in a virtual world and the attributes of the character and the world persist beyond a single gaming session.

   The hub of persistent online gaming is South Korea. The most popular game is Lineage, based on a popular Korean comic book. The creator of Lineage, NCsoft, has attracted millions of monthly subscribers, who take on the roles of knights, elves, and wizards fighting for castles in a medieval kingdom [36]. At times more than 180,000 people are playing the game simultaneously [37]. Cybercafés (called **PC bangs** in South Korea) have large-screen monitors enabling spectators to watch the gameplay, which is full of virtual violence and mayhem. Some children spend up to 10 hours a day playing games, hoping to turn professional. Kim Hyun Soo, chairman of the Net Addiction Treatment Center, complains that "young people are losing their ability to relate to each other, except through games" [37]. We'll discuss the topic of Internet addiction in Section 3.10.

7. *We pay our taxes.*

   Millions of Americans now file their federal income taxes through the Internal Revenue Service's Web site [38].

8. *We gamble.*

   Internet gambling is a $6 billion-a-year business, attracting more than 12 million gamblers from around the world. Running an Internet-based casino is illegal in most states. As a result, many American emigres are operating gambling Web sites from the Caribbean or Central America [39].

9. *We blog.*

   A **blog** (short for Web log) is a personal journal or diary kept on the Web. Used as a verb, the word blog means to maintain such a journal. Blogs may contain plain text, images, audio clips, or video clips [40].

3.4.3 Too Much Control or Too Little?

Internet access is severely limited in certain countries. The governments of Burma (Myanmar), Cuba, and North Korea make it difficult for ordinary citizens to use the Internet to communicate with the rest of the world [41, 42, 43].

In other countries Internet access is easier, but still carefully controlled. Saudi Arabians gained access to the Internet in 1999, after the government installed a centralized control center outside Riyadh. Virtually all Internet traffic flows through this control center, which blocks pornography sites, gambling sites, and many other pages deemed to be offensive to Islam or the government of Saudi Arabia [44]. Blocked sites and pages are from such diverse categories as non-Islamic religious organizations, women's health and sexuality issues, music and movies, gay rights, Middle Eastern politics, and information about ways to circumvent Web filtering.

In contrast to Saudi Arabia, the government of the People's Republic of China does not direct all Internet traffic through a single control center. Instead, it allows many Internet Service Providers to make their own connections outside China. However, all Internet Service Providers (including Yahoo!) have agreed to abide by a "self-discipline" agreement forbidding them from forwarding politically or morally objectionable Web pages [45]. Besides blocking many sites containing sexually explicit material, Chinese ISPs typically block sites concerned with Chinese dissidents, sites related to Taiwan and Tibet, and many news sites, such as BBC News and CNN [46]. The government monitors ISPs to ensure sites that should be blocked are, in fact, unavailable to Web surfers.

Filtering within China can affect its neighbors, too. Some ISPs send packets to China with destinations beyond China (North Korea, for example). However, since Chinese routers filter packets based on the IP addresses of the originating Web server, packets simply "passing through" China can be discarded before they reach their destinations.

Meanwhile, Western nations have different standards about what is acceptable and what is not. While Germany forbids access to any neo-Nazi Web site, Web surfers in the United States can access many such sites. The Aryan Nations West Coast Web site, for example, features an anti-Semitic quote from Adolf Hitler and a link to an online catalog selling a variety of "white power" items, including Nazi flags.

Political satire and pornography are easily available through American ISPs. Americans are used to political satire, but many citizens are concerned about the corrupting influence of pornography, particularly with respect to minors. Since 1996 the U.S. Congress has passed three laws aimed at restricting access of children to sexually explicit materials on the Web: the Communications Decency Act, the Child Online Protection Act, and the Children's Internet Protection Act. The first two laws were ruled unconstitutional by the U.S. Supreme Court; the third was upheld by the Supreme Court in June 2003.

3.5 Ethical Perspectives on Pornography

The government of Saudi Arabia attempts to block pages from any pornographic Web sites from reaching computers inside the kingdom. The U.S. government has insisted on the installation of Web filters in public libraries receiving federal funding for Internet access. Should a government restrict the availability of pornography to some or all of its citizens? Such action is legitimate if pornography is immoral *and* certain kinds of government censorship are moral. In this section we consider the first of these issues: the morality of producing and using pornography.

According to *The Encyclopedia of Ethics*, "pornography is the sexually explicit depiction of persons, in words or images, created with the primary, proximate aim, and reasonable hope, of eliciting significant sexual arousal on the part of the consumer of such materials" [47]. To ground our discussion, we'll consider pornography in the form of sexually explicit photographs of women.

3.5.1 Analyses Concluding Pornography Is Immoral

Some ethicists hold that any production and use of pornography is wrong. Kant wrote that "sexual love makes of the loved person an Object of appetite" [48]. In other words, sexual desire is focused on the body rather than the complete person. Under these circumstances it is easy to treat another person solely as a means to an end—the end being sexual gratification. According to Kant, the objectification of the sexual partner can only be avoided within marriage, where two people have already given their bodies to each other and become one. Any sexual gratification outside of marriage, including the viewing of pornography, is wrong. Furthermore, the model(s) depicted in the pornographic images are being exploited because they are being treated merely as sex objects rather than as complete persons.

Some utilitarians conclude that pornography is wrong based on five harmful consequences it has on society:

1. Pornography reduces the dignity of human life. Hence it harms everyone.
2. By portraying violence and sexual abuse, pornography helps create an environment in which true victims of these crimes are less likely to be dealt with sympathetically by the justice system.

3. Some people imitate crimes such as rape that they have seen portrayed in pornography.
4. Pornography offends most people. It is similar to pollution in that it "poisons the environment."
5. The existence of the pornography industry diverts resources from activities that would have more redeeming social value.

3.5.2 Analyses Concluding Adult Pornography Is Moral

While some ethicists conclude all forms of pornography are wrong, others divide pornography into different categories depending upon whether the persons being depicted and the persons being exposed to the material are adults or children.

Some utilitarians conclude it is moral for adults to view pornography depicting other adults. They cite three beneficial consequences of pornography:

1. Those involved in producing pornography (models, photographers, and proprietors) make money.
2. The consumer derives physical pleasure from viewing pornography.
3. Pornography provides the consumer with a harmless outlet for exploring sexual fantasies.

As soon as children are involved, the analysis becomes markedly different. There is near-universal agreement that any depiction of children in pornography is an exploitation of these children and immoral. There is also widespread agreement that parents have the right to protect their children from exposure to sexually explicit material.

3.5.3 Commentary

The debate over pornography provides a practical illustration of how difficult it is to perform the utilitarian calculus. First, some of the harms brought up by opponents of pornography contradict some of the benefits raised by its apologists. For example, does viewing pornography give men a harmless outlet for sexual fantasies, or does it make them more likely to commit the crime of rape? When so-called experts provide completely contradictory information, it can be challenging to sort the facts from the fairy tales. Second, even when the facts have been established as well as possible, many benefits or harms can be difficult to determine. For example, how do we measure the harm done to the people who find pornography offensive?

3.5.4 Summary

To summarize, some ethicists conclude that any production and use of pornography is wrong, while others hold that it is not wrong for adults to view pornography depicting adults. However, most agree that producing child pornography exploits the children who are involved and that parents have the right to protect their children from exposure to pornographic materials.

If parents ought to shield their children from pornography and other material that could harm them, what role should government play in assisting parents to fulfill their duties? We'll discuss the role of government in controlling the flow of information in the next section.

3.6 Censorship

Censorship is the attempt to suppress or regulate public access to material considered offensive or harmful. Historically, most censorship has been exercised by governments and religious institutions. For example, Roman censors banished the poets Ovid and Juvenal for their writings. During the Middle Ages the Inquisition suppressed the publication of many books, including the work of Galileo Galilei.

Censorship became a much more complicated issue with the invention of the printing press. The printing press broke the virtual monopoly held by governments and religious institutions on distributing material to a large audience, and the increase in printed material increased the number of literate people. For the first time, private individuals could broadcast their ideas to others on a wide scale.

In Western democracies, the gradual separation of church and state has left the government as the sole institution responsible for censorship. In other parts of the world, such as the Middle East, religious institutions continue to play a significant role in determining which material should be accessible to the public.

3.6.1 Direct Censorship

Direct censorship has three forms: government monopolization, prepublication review, and licensing and registration.

The first form of direct censorship is government monopolization. In the former Soviet Union, for example, the government owned all the television stations, radio stations, and newspapers. Private organizations could not even own a copier. Government monopolization is an effective way to suppress the flow of information. Modern computer and communication technology makes government monopolization much more difficult than it has been in the past.

Prepublication review is the second form of direct censorship. This form of censorship is essential for material the government wishes to keep secret, such as information about its nuclear weapons program. Most governments have laws restricting the publication of information that would harm the national security. In addition, autocratic governments typically block publication of material deemed injurious to the reputation of their rulers.

The third form of direct censorship is licensing and registration. This form of censorship is typically used to control media with limited bandwidth. For example, there are a limited number of radio and television stations that can be accommodated on the electromagnetic spectrum. Hence a radio or television station must obtain a license to broadcast at a particular frequency. Licensing invites censorship. For example, the U.S.

Federal Communications Commission has banned the use of certain four-letter words. This led to a challenge that went all the way to the U.S. Supreme Court, as we will see in Section 3.7.3.

3.6.2 Self-Censorship

Perhaps the most common form of censorship is self-censorship: a group deciding for itself not to publish material. In some countries a publisher may censor itself in order to avoid persecution. For example, after U.S.-led forces toppled the regime of Saddam Hussein in April 2003, CNN's chief news executive Eason Jordan admitted that CNN had suppressed negative information about the actions of the Iraqi government for more than a decade in order to keep CNN's Baghdad bureau open and protect Iraqi employees of CNN [49].

In other countries, publishers may want to maintain good relations with government officials. Publications compete with each other for access to information. Often this information is available only from government sources. Publishers know that if they offend the government, their reporters may not be given access to as much information as reporters for rival publications, putting them at a competitive disadvantage. This knowledge can lead a "free" press to censor itself.

Publishers have adopted ratings systems as a way of helping people decide if they (or their children) should access particular offerings. For example, television stations in the United States broadcast shows with "mature content" late in the evening. Voluntary rating systems help people decide if they (or their children) will see a movie, watch a television show, or listen to a record.

The Web does not have a universally accepted ratings system. Some Web sites practice a form of labeling. For example, the home page may warn the user that the site contains nudity and require the user to click on an "I agree" button to enter the site. However, other sites have no such warnings. People who stumble onto these sites are immediately confronted with images and text they may find offensive.

3.6.3 Challenges Posed by the Internet

Five characteristics of the Internet make censorship more difficult.

1. *Unlike traditional one-to-many broadcast media, the Internet supports many-to-many communications.*

   While it is relatively easy for a government to shut down a newspaper or a radio station, it is very difficult for a government to prevent an idea from being published on the Internet, where millions of people have the ability to post Web pages.

2. *The Internet is dynamic.*

   Millions of new computers are being connected to the Internet each year.

3. *The Internet is huge.*

   There is simply no way for a team of human censors to keep track of everything that is posted on the Web. While automated tools are available, they are fallible.

Hence any attempt to control access to material stored on the Internet cannot be 100 percent effective.

4. *The Internet is global.*

   National governments have limited authority to restrict activities happening outside their borders.

5. *It is hard to distinguish between children and adults on the Internet.*

   How can an "adult" Web site verify the age of someone attempting to enter the site?

3.6.4 Ethical Perspectives on Censorship

KANT'S VIEWS ON CENSORSHIP

As a thinker in the tradition of the Enlightenment, Kant's motto was "Have courage to use your own reason" [50]. Kant asks the rhetorical question "Why don't people think for themselves?" and answers it: "Laziness and cowardice are the reasons why so great a portion of mankind, after nature has long since discharged them from external direction, nevertheless remain under lifelong tutelage, and why it is so easy for others to set themselves up as their guardians. It is so easy not to be of age. If I have a book which understands for me, a pastor who has a conscience for me, a physician who decides my diet, and so forth, I need not trouble myself. I need not think, if I can only pay—others will readily undertake the irksome work for me" [50].

The Enlightenment was a reaction to the institutional control over thought held by the aristocracy and the Church. Kant believed he was living in a time in which the obstacles preventing people from exercising their own reason were being removed. He opposed censorship as a backward step.

MILL'S VIEWS ON CENSORSHIP

John Stuart Mill also championed freedom of expression. He gave four reasons why freedom of opinion, and freedom of expression of opinion, were necessary.

First, none of us is infallible. All of us are capable of error. If we prevent someone from voicing their opinion, we may actually be silencing the voice of truth.

Second, while the opinion expressed by someone may be erroneous, it may yet contain a kernel of truth. In general, the majority opinion is not the whole truth. We ought to let all opinions be voiced so that all parts of the truth are heard.

Third, even if the majority opinion should happen to be the whole truth, it is in the clash of ideas that this truth is rationally tested and validated. The whole truth left untested is simply a prejudice.

Fourth, an opinion that has been tested in the fire of a free and open discourse is more likely to have a "vital effect on the character and conduct" [51].

Hence Mill, like Kant, fundamentally supported the free exchange of ideas with the conviction that good ideas would prevail over bad ones. Applying their philosophy to the World Wide Web, it seems they would support the free exchange of opinions and oppose any kind of government censorship of opinions.

MILL'S PRINCIPLE OF HARM

However, a lack of government censorship can also lead to harm. Under what circumstances should the government intervene? Mill proposed the principle of harm as a way of deciding when an institution should intervene in the conduct of an individual.

Principle of Harm

"The only ground on which intervention is justified is to prevent harm to others; the individual's own good is not a sufficient condition" [51].

In other words, the government should not get involved in the private activities of individuals, even if the individuals are doing something to harm themselves. Only if individuals' activities are harming other people should the government step in.

The principle of harm can be used to explain the position of most Western democratic governments with respect to censoring pornographic material depicting adults. As our discussion in the previous section indicates, some ethicists conclude it is not wrong for adults to view pornography with adult models. Others hold that this activity is immoral. If the activity is immoral, it is more certain the harm is being done to the individual consumer; less certain is how much harm is being done to other people. Hence the principle of harm can be used as an argument why the government should not be trying to prevent adults from using pornography depicting adults.

3.7 Freedom of Expression

In the United States, freedom of expression is one of the most cherished—and most controversial—rights. In this section we explain the history behind the adoption of the First Amendment to the United States Constitution. We also explore why the freedom of expression has not been treated as an absolute right.

3.7.1 History

At the time of the American revolution, any criticism of government was seen as a threat to public order and could result in fines and/or imprisonment. Restrictions on freedom of speech in England date back to 1275 and a law called De Scandalis Magnatum. According to this law, a person could be imprisoned for spreading stories about the King that could have the effect of weakening the loyalty of his subjects. The scope of the law became much broader through numerous revisions over the next two centuries. Eventually it encompassed seditious words and words spoken against a wide variety of government officials, including justices [52].

The De Scandalis Magnatum was administered by the Court of Star Chamber, or "Star Chamber" for short. The Star Chamber reported directly to the King, and it did not have to obey traditional rules of evidence. Rulings of the Star Chamber demonstrated that a person could be convicted for making a verbal insult or for something written in a private letter. The Star Chamber was abolished in 1641, but the law continued to be enforced through Common Law Courts [52].

At the end of the eighteenth century, freedom of the press in England and its colonies meant freedom to print without a license. In other words, there were no **prior restraints** on publication. People could publish what they pleased. However, those who published material found to be seditious or libelous would face severe consequences [52].

The law against libel simply considered if the material printed was harmful; arguing that the information was true was not relevant to the proceedings and could not be used in a publisher's defense. Between 1760 and the end of the American Revolution, about 50 people were successfully prosecuted for libel. To prevent such prosecutions from continuing, most states adopted bills of rights after gaining independence from England [52].

In May 1787 delegates from the thirteen states gathered in Philadelphia to revise the Articles of Confederation. Soon they were drafting a completely new Constitution. Delegate George Mason, author of the Virginia Declaration of Rights, strongly opposed the proposed Constitution because it contained no declaration of the rights of the citizens. Patrick Henry and other political leaders shared Mason's objections [52].

While the proposed Constitution was ratified by all thirteen states, most state legislatures adopted the Constitution with the expectation that Congress would offer amendments addressing the human rights concerns brought up by the opponents of the Constitution. During the first Congress, James Madison proposed 12 such amendments. All 12 of these amendments were sent to the states for ratification. Of these 12 amendments, 10 were quickly ratified. Today, these 10 amendments are commonly known as the Bill of Rights. The first of these amendments, the one Madison considered most essential, was the one guaranteeing freedom of speech and freedom of the press [52].

FIRST AMENDMENT TO THE UNITED STATED CONSTITUTION

Congress shall make no law respecting an establishment of religion, or prohibiting the free exercise thereof; or abridging the freedom of speech, or of the press; or the right of the people peaceably to assemble, and to petition the government for a redress of grievances.

3.7.2 Freedom of Expression Not an Absolute Right

The primary purpose of the First Amendment's free speech guarantee is political. Free speech allows an open discussion of public issues. It helps make government responsive to the will of the people [53].

However, the First Amendment right to free expression is not limited to political speech. Nonpolitical speech is also covered. There are good reasons for protecting nonpolitical as well as political speech. First, it is sometimes hard to draw the line between the two. Asking a judge to make the distinction turns it into a political decision. Second, society can benefit from nonpolitical as well as political speech. Hence the free speech guarantee of the First Amendment also promotes scientific and artistic expression. For the same reason, the definition of "speech" encompasses more than words. Protected "speech" includes art and certain kinds of conduct, such as burning an American flag [54].

Decisions by the U.S. Supreme Court have made clear that freedom of expression is not an absolute right. Instead, the private right to freedom of expression must be balanced against the public good. Those who abuse this freedom and harm the public may be punished. For example, protection is not given to "libel, reckless or calculated lies, slander, misrepresentation, perjury, false advertising, obscenity and profanity, solicitation of crime, and personal abuse or 'fighting' words," because these actions do not serve the ends of the First Amendment [53].

Various restrictions on freedom of speech are justified because of the greater public good that results. For example, U.S. law prohibits cigarette advertising on television because cigarette smoking has detrimental effects on public health. Some cities use zoning laws to concentrate adult bookstores in a single part of town because the presence of adult bookstores lowers property values and increases crime.

3.7.3 *FCC v. Pacifica Foundation et al.*

To illustrate limits to First Amendment protections, we consider the decision of the U.S. Supreme Court in the case of *Federal Communications Commission v. Pacific Foundation et al.*

In 1973 George Carlin recorded a performance made in front of a live audience in California. One track on the resulting record is a 12-minute monologue called "Filthy Words." In the monologue Carlin lists seven words that "you couldn't say on the public, ah, airwaves, um, the ones you definitely wouldn't say, ever" [55]. The audience laughs as Carlin spends the rest of the monologue creating colloquialisms from the list of banned words.

On the afternoon of October 30, 1973, counterculture radio station WBAI in New York aired "Filthy Words" after warning listeners the monologue contained "sensitive language which might be regarded as offensive to some" [56]. A few weeks after the broadcast, the Federal Communications Commission (FCC) received a complaint from a man who had heard the broadcast on his car radio in the presence of his son. In response to this complaint, the FCC issued a declaratory order and informed Pacifica Foundation (the operator of WBAI) the order would be placed in the station's license file. The FCC warned Pacifica Foundation that further complaints could lead to sanctions.

Pacifica sued the FCC, and the resulting legal battle reached the U.S. Supreme Court. In 1978 the Supreme Court ruled, in a 5-4 decision, that the FCC did not violate the First Amendment [56]. The majority opinion states, "[O]f all forms of communication, it is broadcasting that has received the most limited First Amendment protection." There are two reasons why broadcasters have less protection than book sellers or theater owners:

1. *"Broadcast media have a uniquely pervasive presence in the lives of all Americans."*

   Offensive, indecent material is broadcast into the privacy of citizens' homes. Since people can change stations or turn their radios on or off at any time, prior warnings cannot completely protect people from being exposed to offensive material. While someone may turn off the radio after hearing something indecent, that does not undo a harm that has already occurred.

2. *"Broadcasting is uniquely accessible to children, even those too young to read."*

   In contrast, restricting children's access to offensive or indecent material is possible in bookstores and movie theaters.

The majority emphasized that its ruling was a narrow one and that the context of the broadcast was all-important. The time of day at which the broadcast occurred (2 p.m.) was an important consideration, because that affected the composition of the listening audience.

3.8 Children and the Web

Many parents believe they ought to protect their children from exposure to pornographic and violent materials on the Web. A large software industry has sprung up to meet these needs.

3.8.1 Web Filters

A **Web filter** is a piece of software that prevents certain Web pages from being displayed by your browser. While you are running your browser, the filter runs as a background process, checking every page your browser attempts to load. If the filter determines the page is objectionable, it prevents the browser from displaying it.

Filters can be installed on individual computers, or an ISP may provide filtering services for its customers. Programs designed to be installed on individual computers, such as Cyber Sentinel, eBlaster, and Spector PRO, can be set up to email parents as soon as they detect an inappropriate Web page [57]. America Online's filtering service is called AOL Guardian. It enables parents to set the level of filtering on their children's accounts. It also allows parents to look at logs showing the pages their children have visited.

Typical filters use two different methods to determine if a page should be blocked. The first method is to check the URL of the page against a "blacklist" of objectionable sites. If the Web page comes from a blacklisted site, it is not displayed. The second method is to look for combinations of letters or words that may indicate a site has objectionable content.

Neither of these methods is foolproof. The Web contains millions of pages containing pornography, and new sites continue to be created at a high rate; hence any blacklist of pornographic sites will be incomplete by definition. Some filters sponsored by conservative groups have blacklisted sites associated with liberal political causes, such as those sponsored by the National Organization of Women and gay and lesbian groups. The algorithms used to identify objectionable words and phrases can cause Web filters to block out legitimate Web pages.

3.8.2 Child Internet Protection Act

In March 2003 the Supreme Court weighed testimony in the case of *United States v. American Library Association*. The question: Can the government require libraries to install antipornography filters in return for receiving federal funds for Internet access?

More than 14 million people access the Internet through public library computers. About one-sixth of the libraries in the United States have already installed filtering software on at least some of their computers. The Child Internet Protection Act requires that libraries receiving federal funds to provide Internet access to its patrons must prevent children from getting access to visual depictions of obscenity and child pornography. The law allows adults who desire access to a blocked page to ask a librarian to remove the filter.

In his testimony before the Supreme Court, Solicitor General Theodore Olson argued that since libraries don't offer patrons X-rated magazines or movies, they should not be obliged to give them access to pornography over the Internet.

Paul Smith, representing the American Library Association and the American Civil Liberties Union, argued that in their attempt to screen out pornography, filters block tens of thousands of inoffensive pages. He added that requiring adults to leave the workstation, find a librarian, and ask for the filter to be turned off would be disruptive to their research and would stigmatize them.

In June 2003 the U.S. Supreme Court upheld CIPA, ruling 6-3 that antipornography filters do not violate First Amendment guarantees [58]. Chief Justice William Rehnquist wrote, "A public library does not acquire Internet terminals in order to create a public forum for Web publishers to express themselves, any more than it collects books in order to provide a public forum for the authors of books to speak . . . Most libraries already exclude pornography from their print collections because they deem it inappropriate for inclusion" [59].

3.8.3 Ethical Evaluations of CIPA

In this section we evaluate CIPA from the perspectives of Kantianism, act utilitarianism, and social contract theory.

KANTIAN EVALUATION

We have already covered Kant's philosophical position against censorship. He optimistically believed that allowing people to use their own reason would lead to society's gradual enlightenment. In this case, however, the focus is narrower. Rather than talking about censorship in general, let's look at CIPA in particular.

The goal of CIPA is to protect children from the harm caused by exposure to pornography. The way the goal is being implemented is through Web filters. Studies have demonstrated that Web filters do not block all pornographic material, but do block some nonpornographic Web pages. Some nonpornographic information posted on the Web will not be easily accessible at libraries implementing government-mandated Web filters. The people posting this information did not consent to their ideas being blocked. Hence the decision to require the use of Web filters treats the creators of non-offensive, but blocked Web pages solely as means to the end of restricting children's access to pornographic materials. This analysis leads us to conclude that CIPA is wrong.

ACT UTILITARIAN EVALUATION

Our second evaluation of CIPA is from an act utilitarian point of view. What are the consequences of passing CIPA?

1. While not all children access the Web in public libraries, and while Web filtering software is imperfect, it is probable that enacting CIPA will result in fewer children being exposed to pornography, which is good.
2. Because Web filters are imperfect, people will be unable to access some legitimate Web sites. As a result, Web browsers in libraries will be less useful as research tools, a harmful consequence.
3. Adult patrons who ask for filters to be removed may be stigmatized (rightfully or not) as people who want to view pornography, a harm to them.
4. Some blocked sites may be associated with minority political views, reducing freedom of thought and expression, which is harmful.

Whenever we perform the utilitarian calculus and find some benefits and some harms, we must decide how to weigh them. This is a good time to think about utilitarian philosopher Jeremy Bentham's seven attributes. In particular, how many people are in each affected group? What is the probability the good or bad event will actually happen? How soon is the event likely to occur? How intense will the experience be? To what extent is the pain not diluted by pleasure, or vice versa? How long will it last? How likely is the experience to lead to a similar experience? Actually performing the calculus for CIPA is up to each person's judgment. Different people could reach opposite conclusions about whether enacting CIPA is the right thing for the U.S. government to do.

SOCIAL CONTRACT THEORY EVALUATION

In social contract theory, morally binding rules are those rules mutually agreed to in order to allow social living [60]. Freedom of thought and expression is prized. According to John Rawls, "liberty of conscience is to be limited only when there is a reasonable expectation that not doing so will damage the public order which the government should maintain" [61].

It would be difficult to gain consensus around the idea that the private viewing of pornography makes social living no longer possible. For this reason, the private use of pornography is considered to be outside the social contract and nobody else's business. However, when we think about the availability of pornography in public libraries, the issue gets thornier.

Some argue that allowing people to view pornography in a public place demeans women, denying them dignity as equal persons [62]. On the other hand, we know that filtering software is imperfect. In the past it has been used to promote a conservative political agenda by blocking sites associated with other viewpoints [63, 64]. Hence it reduces the free exchange of ideas, limiting the freedoms of thought and expression. For some adults, public libraries represent their only opportunity to access the Web for no cost. In order to be treated as free and equal citizens, they should have the same Web access as people who have Internet access from their homes. If Web filters are in place, their access is not equal because they must ask for permission to have the filters disabled. Finally, while most people would agree that children should not be exposed to pornographic material, it would be harder to convince reasonable people that social living would no longer be possible if children happened to see pornography in a library.

Our analysis from the point of view of social contract theory has produced arguments both supporting and opposing the Children's Internet Protection Act. However, installing filters does not seem to be necessary to preserve the public order. For this reason, the issue is outside the social contract and freedom of conscience should be given precedence.

3.9 Breaking Trust on the Internet

3.9.1 Identity Theft

Just over nine million people were victims of identity theft in 2005, and about one-eighth of these cases were computer related [65]. How can this happen?

The email looks authentic (Figure 3.4). The message header is identical to the header on the PayPal Web site, and it includes the color logos of Visa, MasterCard, Discovery Card, and American Express. According to the message, my credit card is about to expire. It asks me to update my personal information to avoid an interruption in my PayPal service.

The message, however, is fraudulent. If I respond with the information requested, con artists will be able to assume my electronic identity and make purchases or get cash advances using my credit card.

The attempt to deceive Internet users into disclosing personal information through the use of official-looking emails or Web sites is called **phishing**, and it is a growing

PayPal®

Sign Up | Log In | Help

Welcome | Send Money | Request Money | Merchant Tools | Auction Tools

Welcome

► PayPal members

LOG IN

► New user?

SIGN UP

What's New

Sell with PayPal Seller Protection

Refer new merchants - Earn up to $100.00 USD

Community Spotlight

Dear Customer
Your credit card will expire soon To avoid any interruption to your service, please update your credit card expiration date by following the steps below. If you do not update your credit card expiration date you will no longer be able to fund payments with this card.

Your Address Information - You may only enter **English** characters during Sign Up. This does NOT include characters with accents. Please enter your name and address as they are listed for your credit card or bank account. Your primary currency is the currency in which you are expecting to send and receive the majority of your payments.

E-Mail #

Re-Enter E-Mail #

Password #

Re-Enter Password #

Full Name #

Full Address #

State #

Zip #

Credit Card #

Exp.Date(mm/yy) #: Month / Year

ATM PIN (4 digits) #:

Re-enter ATM PIN:

Security Measure: This must be completed to sign up for a PayPal account
Enter the characters as they are shown in the box below. This will further increase the security of your account and the PayPal network. This is not your password. Help

BS8 8MJV7

Log In

FIGURE 3.4 This official-looking email message is fraudulent. An authentic email from PayPal would address the customer by name and would not ask for personal information via return email.

problem. The messages appear to be from PayPal, Citibank, EarthLink, AOL, Best Buy, or other companies with which the recipient may have an account. Since many of the perpetrators of these scams are outside the United States, American law enforcement agencies have had a difficult time tracking them down and prosecuting them [66].

The stereotypical victim of identity theft is an elderly person who isn't computer savvy, but the facts speak otherwise. The average age of a victim of identity theft is 40. Many victims are experienced computer users who have become comfortable typing in their credit card information while online [66].

3.9.2 Chat-Room Predators

Instant messaging is a real-time communication between two or more people supported by computers and a telecommunications system. A **chat room** is similar to instant messaging, except that it supports discussions among many people. A large number of organizations sponsor chat rooms dedicated to a wide variety of topics. For example, in July 2005 America Online's "All Chats" page listed 67 chat rooms under the general headings of Hot Chats, Arts & Entertainment, Computers & Science, Family & Home, Health & Wellness, Hobbies & Interests, International, Lifestyles, Local, Love & Romance, News, Personal Finance, Sports & Recreation, Travel, and Workplace.

The popularity of instant messaging varies from country to country. According to Nielsen/NetRatings, the number of people who used instant messaging between January 2002 and March 2002 varied from 13 percent of all Internet users in Denmark to 43 percent in Spain [67]. Participation in chat rooms also varies from country to country. According to the same survey, the number of people with Internet accounts who participated in a chat room between January and March 2002 varied from 16 percent in the United Kingdom to 41 percent in Brazil. Conservatively estimating average use of instant messaging or chat rooms at 25 percent, the number of people worldwide who use this technology at least occasionally is about 150 million.

For many young people, instant messaging has replaced the telephone as their preferred technology for communicating with their peers. In 1995 Katie Tarbox, a 13-year-old swimmer from New Canaan, Connecticut, met a man in an AOL chat room [68]. He said his name was Mark and his age was 23. His grammar and vocabulary were good, and he made her feel special. Katie agreed to meet Mark at a hotel in Texas, where her swim team was competing. Soon after she entered his hotel room, he molested her. "Mark" turned out to be 41-year-old Francis Kufrovich from Calabasas, California, a man with a history of preying on children. In March 1998, Kufrovich was the first person in the United States to be sentenced for Internet pedophilia. After pleading guilty, he served 18 months in prison.

In 1999 the FBI investigated 1,500 crimes in which an alleged pedophile crossed a state line to meet and molest a child met through an Internet chat room [68]. Many say the problem is growing. Parry Aftab, executive director of Cyber Angels, says, "I know that I can go into a chat room as a 12-year old and not say anything, and be hit on and asked if I'm a virgin within two minutes" [68]. In New York a 42-year-old man was sentenced to 150 years in prison after being convicted of kidnapping a 15-year-old girl and raping her repeatedly over the course of a week. He met the girl in a chat room [69].

Police have begun entering chat rooms posing as young girls to lure pedophiles [70]. During a three-week-long sting operation in Spokane, Washington, a police detective posed as a 13-year-old girl in a chat room. In early March 2003 police arrested a 22-year-old man on charges of attempted second-degree rape of a child. Inside his car the officers found handcuffs, a large folding knife, and a condom. The suspect was still on parole for an earlier conviction for fourth-degree assault with sexual motivation. Police sergeant Joe Petersen asked, "What happens had it been a real girl?" [69]. Chat-room sting operations are leading to many arrests all over the United States [71, 72, 73, 74, 75, 76].

3.9.3 Ethical Evaluations of Police "Sting" Operations

Is it morally right for police detectives to entrap pedophiles by posing as children in chat rooms and agreeing to meet with them?

UTILITARIAN ANALYSIS

Let's consider the various consequences of such a sting operation. A person allegedly interested in having sex with an underage minor is arrested and charged with attempted child rape. Suppose the person is found guilty and must serve time in prison. The direct effects of the sting operation are the denial of one person's freedom (a harm) and an increase in public safety (a benefit). Since the entire public is safer and only a single person is harmed, this is a net good.

The sting operation also has indirect effects. Publicity about the sting operation may deter other chat-room pedophiles. This, too, is a beneficial result. It is harder to gauge how knowledge of sting operations influences innocent citizens. First, it may reduce citizens' trust in the police. Many people believe that if they are doing nothing wrong, they have nothing to fear. Others may become less inclined to provide information to the police when requested. Second, sting operations can affect everyone's chat-room experiences. They demonstrate that people are not always who they claim to be. This knowledge may make people less vulnerable to being taken advantage of, but it may also reduce the amount of trust people have in others. Sting operations prove that supposedly private chat-room conversations can actually be made public. If chat-room conversations lack honesty and privacy, people will be less willing to engage in serious conversations. As a result, chat rooms lose some of their utility as communication devices. How much weight you give to the various consequences of police sting operations in chat rooms determines whether the net consequences are positive or negative.

KANTIAN ANALYSIS

A Kantian focuses on the will leading to the action rather than the results of the action. The police are responsible for maintaining public safety. Pedophiles endanger innocent children. Therefore, it is the duty of police to try to prevent pedophiles from accomplishing what they intend to do. The will of the police detective is to put a pedophile in prison. This seems straightforward enough.

If we dig a level deeper, however, we run into trouble. In order to put a pedophile in prison, the police must identify this person. Since a pedophile is unlikely to confess on the spot if asked a question by a police officer, the police lay a trap. In other words, the will of the police detective is to deceive a pedophile in order to catch him. To a Kantian, lying is wrong, no matter how noble the objective. By collecting evidence of chat-room conversations, the police detective also violates the presumed privacy of chat rooms. These actions of the police detective affect not only the alleged pedophile, but also every innocent person in the chat room. In other words, detectives are using every chat-room occupant as a means to their end of identifying and arresting the pedophile. While police officers have a duty to protect the public safety, it is wrong for them to break other moral laws in order to accomplish this purpose. From a Kantian point of view the sting operation is morally wrong.

SOCIAL CONTRACT THEORY ANALYSIS

An adherent of social contract theory could argue that in order to benefit everyone, there are certain moral rules that people in chat rooms ought to follow. For example, people ought to be honest, and conversations ought to be kept confidential. By misrepresenting identity and/or intentions, the pedophile has broken a moral rule and ought to be punished. In conducting sting operations, however, police detectives also misrepresent their identities and record everything typed by suspected pedophiles. The upholders of the law have broken the rules, too. Furthermore, we have the presumption of innocence until proven guilty. What if the police detective, through miscommunication or bad judgment, actually entraps someone who is not a pedophile? In this case, the innocent chat-room users have not broken any rules. They were simply in the wrong place at the wrong time. Yet society, represented by the police detective, did not provide the benefits chat-room users expect to receive (honest communications and privacy). In short, there is a conflict between society's need to punish a wrongdoer and its expectation that everyone (including the agents of the government) will abide by its moral rules.

SUMMARY OF ETHICAL ANALYSES

To summarize our ethical evaluation of police sting operations, the actions of the police seem immoral from a Kantian point of view. Evaluations using the other ethical theories do not yield a clear-cut endorsement or condemnation of the stings. While the goals of the police are laudable, they accomplish their goals by deceiving other chat-room users and revealing details of conversations thought to be private. Sting operations are more likely to be viewed as morally acceptable by someone who is more focused on the results of an action than the methods used; in other words, a consequentialist.

3.9.4 False Information

The Web is a more open communication medium than newspapers, radio stations, or television stations. Individuals or groups whose points of view may never be published in a newspaper or broadcast on a television or radio show may create an attractive Web site. The ease with which people may get information out via the Web is one of the reasons the Web contains billions of pages. However, the fact that no one has to review a Web page before it is published means the quality of information available on the Web varies widely.

You can find many Web sites devoted to the American manned space program. You can also find many Web sites that provide evidence the moon landings were a hoax by NASA. Many Web sites describe the Holocaust committed by the Nazis before and during World War II. Other sites explain why the Holocaust could not have happened.

Disputes about commonly held assumptions did not begin with the Web. Some television networks and newspapers are well known for giving a forum to people who question information provided through government agencies. Twice in 2001 the Fox TV network aired a program called "Conspiracy Theory: Did We Land on the Moon?" The program concludes NASA faked the moon landing in the Nevada desert. Supermarket tabloids are notorious for their provocative, misleading headlines. Experienced consumers take into account the source of the information. Most people would agree *60 Minutes* on CBS is a more reliable source of information than *Conspiracy Theory* on Fox. Similarly, people expect information they find in *The New York Times* to be more reliable than the stories they read in a tabloid.

In traditional publishing, various mechanisms are put in place to improve the quality of the final product. For example, before Addison-Wesley published this book, an editor sent draft copies of the manuscript to a dozen reviewers who checked it for errors, omissions, or misleading statements. The author revised the manuscript to respond to the reviewers' suggestions. After the author submitted a revised manuscript, a copy editor made final changes to improve the readability of the text, and a proofreader corrected typographical errors.

Web pages, on the other hand, can be published without any review. As you're undoubtedly well aware, the quality of Web pages varies dramatically. Fortunately, search engines can help people identify those Web pages that are most relevant and of the highest quality. Let's take a look at how the Google search engine does this.

The Google search engine keeps a database of more than eight billion Web pages. Google uses a software algorithm to rank the quality of these Web pages. The algorithm invokes a kind of voting mechanism. If Web page A links to Web page B, then page B gets a vote. However, all votes do not have the same weight. If Web page A is itself getting a lot of votes, then page A's link to page B gives its vote more weight than a link to B from an unpopular page.

When a user makes a query to Google, the search engine first finds the pages that closely match the query. It then considers their quality (as measured by the voting algorithm) to determine how to rank the relevant pages.

3.10 Internet Addiction

3.10.1 Is Internet Addiction Real?

Using an Internet-enabled computer can be a lot of fun—the number of different things you can do online is staggering. Some psychologists warn about the dangers of Internet addiction. Are these fears justified?

In 1976, long before most computers were networked, Joseph Weizenbaum pointed out what attracts computer programmers to their machines: feelings of freedom and power. Because programmers deal with bits instead of physical objects, they are largely free to create whatever they can imagine, and computers execute their instructions without hesitation. The resulting thrill causes some programmers to have a "compulsion to program." In Weizenbaum's words:

> [B]right young men of disheveled appearance, often with sunken glowing eyes, can be seen sitting at computer consoles, their arms tensed and waiting to fire their fingers, already poised to strike, at the buttons and keys on which their attention seems to be as riveted as a gambler's on the rolling dice... They work until they nearly drop, twenty, thirty hours at a time. Their food, if they arrange it, is brought to them: coffee, Cokes, sandwiches. If possible, they sleep on cots near the computer. But only for a few hours—then back to the console or the printouts. Their rumpled clothes, their unwashed and unshaven faces, and their uncombed hair all testify that they are oblivious to their bodies and to the world in which they move. They exist, at least when so engaged, only through and for the computers [77].

Weizenbaum's observation is echoed by Maressa Orzack, who states "Computer addiction is real... As an impulse control disorder, computer addiction resembles pathological gambling" [78].

The traditional definition of addiction is the persistent, compulsive use of a chemical substance, or drug, despite knowledge of its harmful long-term consequences [79]. Today, however, Orzack and some other psychologists and psychiatrists have extended the definition of addiction to include any persistent, compulsive behavior that the addict recognizes to be harmful. According to their broader definition of addiction, people can be addicted to gambling, food, sex, long-distance running, and other activities, including computer-related activities [80].

Some people spend between 40 and 80 hours per week on the Internet, with individual sessions lasting up to 20 hours [81, 82]. Spending so much time online can have a wide variety of harmful consequences. Fatigue from sleep deprivation can lead to unsatisfactory performance at school or at work. Physical ailments include carpal tunnel syndrome, back strain, and eyestrain. Too many hours in front of a computer can weaken or destroy relationships with friends and family members [81].

Kimberly Young has created a test for Internet addiction. Using the diagnosis of pathological gambling in the *Diagnostic and Statistical Manual of Mental Disorders* as her starting point, Young has produced an eight-question screening instrument, which I reproduce verbatim [83]:

1. Do you feel preoccupied with the Internet (think about previous on-line activity or anticipate next on-line session)?
2. Do you feel the need to use the Internet with increasing amounts of time in order to achieve satisfaction?
3. Have you repeatedly made unsuccessful efforts to control, cut back, or stop Internet use?
4. Do you feel restless, moody, depressed, or irritable when attempting to cut down or stop Internet use?
5. Do you stay on-line longer than originally intended?
6. Have you jeopardized or risked the loss of significant relationship, job, educational or career opportunity because of the Internet?

7. Have you lied to family members, therapist, or others to conceal the extent of involvement with the Internet?
8. Do you use the Internet as a way of escaping from problems or of relieving a dysphoric mood (e.g., feelings of helplessness, guilt, anxiety, depression)?

Young considers patients who answer "yes" to five or more of these questions to be addicted to the Internet, unless "their behavior could not be better accounted for by a Manic Episode" [81].

Young's use of the phrase "Internet addiction" and her questionnaire are controversial. John Charlton points out that computer use, unlike drug use, is generally considered to be a positive activity. In addition, while drug addiction leads to an increase in criminal activity, the same level of societal harm is unlikely to occur even if the Internet is overused by some people. Charlton performed his own study of computer users and has concluded that Young's checklist approach is likely to overestimate the number of people addicted to the Internet. According to Charlton, some "people who are classified as computer-dependent or computer-addicted might often be more accurately said to be highly computer-engaged" [84].

Mark Griffiths holds a position similar to Charlton, stating that "to date there is very little empirical evidence that computing activities (i.e., Internet use, hacking, programming) are addictive" [82]. Richard Ries argues that it would be more accurate to call excessive use of the Internet a compulsion [85].

However, others share Young's perspective. Stanton Peele maintains that "people become addicted to experiences" [80]. In his broader view of addiction, nondrug experiences can be addictive. Peele has developed a model of addiction that extends "to all areas of repetitive, compulsive behavior" [80].

Our concern in this section is excessive Internet use that causes harm. The dispute over terminology is not important to our discussion. We will use the term "Internet addiction" rather than "Internet compulsion," since the former term appears to be more widely used by the press.

3.10.2 Contributing Factors

According to Peele, social, situational, and individual factors can increase a person's susceptibility to addiction. For example, peer groups play an important role in determining how individuals use alcohol and other drugs. People in stressful situations are more likely to become addicted, as are those who lack social support and intimacy, and those who have limited opportunities for "rewarding, productive activity" [80]. Individual factors that make a person more susceptible to addiction include a tendency to pursue an activity to excess, a lack of achievement, a fear of failure, and feelings of alienation.

Young's studies have led her to "believe that behaviors related to the Internet have the same ability to provide emotional relief, mental escape, and ways to avoid problems as do alcohol, drugs, food, or gambling" [81]. She notes that the typical Internet addict is addicted to a single application.

3.10.3 Ethical Evaluation of Internet Addiction

People who use the Internet excessively can harm themselves and others for whom they are responsible. For this reason, excessive Internet use is a moral issue.

Kantianism, utilitarianism, and social contract theory all share the Enlightenment view that individuals, as rational beings, have the capacity and the obligation to use their critical judgement to govern their lives [86]. Kant held that addiction is a vice, because it's wrong to allow your bodily desires to dominate your mind [48]. Mill maintained that some pleasures are more valuable than others and that people have the obligation to help each other "distinguish the better from the worse" [51].

Ultimately, people are responsible for the choices they make. Even if an addict is "hooked," the addict is responsible for choosing to engage in the activity the first time. This view assumes that people are capable of controlling their compulsions. According to Jeffrey Reiman, vices are "dispositions that undermine the sovereignty of practical reason. Dispositions, like habits, are hard but not impossible to overcome, and undermining something weakens it without necessarily destroying it entirely" [86].

Reiman's view is supported by Peele, who believes addicts can choose to recover from their addictions. "People recover to the extent that they (1) believe an addiction is hurting them and wish to overcome it, (2) feel enough efficacy to manage their withdrawal and life without the addiction, and (3) find sufficient alternative rewards to make life without the addiction worthwhile" [80].

While our analysis to this point has concluded that individual addicts are morally responsible for their addictions, it's also possible for a society to bear collective moral responsibility for the addictions of some of its members. We have already discussed how social conditions can increase a person's susceptibility to addiction, and Peele states an addict will not recover unless life without the addiction has sufficient rewards.

Addiction is wrong because it means voluntarily surrendering the sovereignty of your reason by engaging in a compulsion that has short-term benefits but harms the quality of your life in the long term. However, if somebody is living in a hopeless situation where any reasonable person would conclude there are no long-term prospects for a good life, then what is lost by giving in to the compulsion? Reiman believes that this is the case for many American inner-city drug addicts. "They face awful circumstances that are unjust, unnecessary, and remediable, and yet that the society refuses to remedy. Addiction is for such individuals a bad course of action made tolerable by comparison to the intolerable conditions they face. In that face, I think that moral responsibility for their strong addictions . . . passes to the larger society" [86].

Of course, the circumstances facing a typical suburban Internet addict are radically different from those facing a typical inner-city drug addict. For this reason, it is tempting to dismiss the notion that society could in any way be responsible for the Internet addiction of some of its members. However, some people use the Internet as a way to escape into their own world, because in the "real world" they suffer from social isolation [82]. Perhaps we should reflect on whether any of our actions or inactions make certain members of our community feel excluded.

Summary

The Internet has provided us with powerful new ways to interact. In this chapter we have explored three important Internet applications: email, chat rooms, and the Web. All of these technologies have had both positive and negative consequences.

Fifteen years ago, relatively few people had email accounts. Back then, email advertising was virtually unheard of. Email users did not have to delete large numbers of unwanted messages from their mailboxes. On the other hand, email was not too useful outside work, because most people didn't have it.

Today, over half a billion people have an email account. Most anyone you'd like to communicate with has an email address. However, the large number of email users has attracted the attention of direct marketing firms. In the past few years the volume of unsolicited bulk email (spam) has risen dramatically. Many believe the presence of spam has harmed the email system, and a variety of steps have been taken to reduce the amount of spam email users receive.

In the United States, the CAN SPAM Act took effect on January 1, 2004. Supporters of the law say it will reduce the problem of spam and keep the email system usable. Critics say the law will not reduce the problem of spam. In fact, it may encourage even more companies to generate spam.

The Web contains over eight billion pages. It contains images of sublime beauty and shocking cruelty, uplifting poetry and expletive-ridden hate speech, well-organized encyclopedias and figments of paranoid imaginations. In short, it is a reflection of the best and the worst of humanity.

Governments have responded to the Web in a variety of ways. The most repressive governments have simply made the Web inaccessible to their people. Other governments have instituted controls that prevent certain sites from being accessed. Still other governments allow their citizens nearly universal access to Web sites.

In the United States, there have been numerous efforts to make pornography inaccessible to children via the Web. The U.S. Congress has passed three laws attempting to make pornography less accessible to children via the Web. All of these laws have raised objections from civil libertarians, who call them an infringement on free speech rights. The U.S. Supreme Court has ruled the first two laws unconstitutional; it has upheld the third.

The Internet provides new ways for people to become misled. For example, chat rooms are a popular way for groups of like-minded people to come together to discuss a topic of mutual interest. Unfortunately, sexual predators have used chat rooms as a tool to contact children. In response, police have begun to set up "sting" operations to snare these predators. While sting operations may catch sexual predators, they also change the climate of chat rooms.

The Internet has facilitated e-commerce. Many people are comfortable purchasing items over the Internet. Submitting a credit card number and other identifying information is part of this process. In this environment the problem of identity theft is a growing concern. Hundreds of thousands of people have been conned into revealing their credit card numbers to scam artists who use this information to get cash advances or purchase goods using someone else's identity.

The Web provides a remarkably simple way for people to post and access information. People looking for answers can often get more information, and get it much more quickly, by retrieving what they want from the Web instead of searching printed encyclopedias, books, journals, and newspapers. Ordinary people can also use the Web to broadcast their ideas around the globe. There are many advantages to this information-rich environment. Unfortunately, because anybody can post information on the Web, incorrect information is mixed in with correct information. Web users cannot believe everything they read on the Web. Web search engines incorporate algorithms that attempt to steer people toward higher-quality sites.

A wide variety of enticing activities are available online, and some people exhibit a compulsion to spend extraordinarily long hours connected to the Internet. Numerous commentators have compared compulsive computer users to compulsive gamblers. Whether or not a compulsive online activity is a true addiction, excessive computer use can have harmful consequences. According to Kantianism, utilitarianism, and social contract theory, people must take responsibility for the voluntary choices they make, including the decision to go online. However, we should also remember that social and cultural factors can make people more susceptible to addictions.

Review Questions

1. What is the Internet?
2. Explain the meaning of the two parts of an email address.
3. Describe how email is transmitted from the sender to the recipient.
4. Describe similarities between email and ordinary physical mail delivered by the postal service.
5. What is spam?
6. What does a spam blocker do?
7. Explain how MAPS attempts to reduce the amount of spam people receive.
8. What is the difference between email and instant messaging?
9. What is a PC bang?
10. Describe five uses of the Web not covered in the text.
11. What is a URL?
12. Summarize the different forms of direct censorship.
13. According to the U.S. Supreme Court, why do broadcasters have the most limited First Amendment rights?
14. What characteristics of the Internet make censorship difficult?

15. How does the idea of "Internet addiction" stretch the traditional concept of addiction?
16. What is the Enlightenment view regarding responsibility for addiction?

Discussion Questions

17. Does spam include unsolicited bulk email from nonprofit organizations? Should non-profit organizations be regulated the same way as for-profit organizations with respect to their use of unsolicited bulk email?
18. Does the Mail Abuse Prevention System have too much authority? What are the risks inherent in widespread use of the Realtime Blackhole List?
19. Suppose a fee (an electronic version of a postage stamp) was required in order to send an email message. How would this change the behavior of email users? Suppose the fee was one cent. Do you think this would solve the problem of spam?
20. The original Senate version of the CAN SPAM Act required the FCC to create a national "do not email" list. People who did not wish to receive spam would have been able to put their email addresses on the list. It would have been illegal for direct email advertisers to send messages to the addresses on the list. Why do you think Congress took this provision out of the final bill?
21. Internet service providers monitor their chat rooms and expel users who violate their codes of conduct. For example, users can be kicked off for insulting a person or a group of people based on their race, religion, or sexual orientation. Is it wrong for an ISP to expel someone for hate speech?
22. Suppose you are the director of an ISP that serves the email needs of 10,000 customers. You receive dozens of complaints from them every week about the amount of spam they are receiving. Meanwhile, American spammers are hacking into computers in Jamborea (an East Asian country) and using them to mail spam back to the United States. You estimate that at least 99 percent of email originating from Jamborea is spam. A few of the messages, however, are probably legitimate emails. Should you do anything to restrict the flow of email messages from Jamborea to your customers?
23. Stockbrokers are now required to save all their instant messaging communications. Is having a record of everything you type good or bad? Do you think this requirement will change the behavior of brokers?
24. What are the benefits and harms of Internet censorship?
25. Discuss similarities and differences between the Web and each of these other ways that we communicate: the telephone system, physical mail, bookstores, movie theaters, newspapers, broadcast and cable TV. Should governments ignore the Web, or should they regulate it somehow? If governments should regulate the Web, should the regulations be similar to the regulations for one of the aforementioned communication systems, or should they be unique in significant ways?
26. Should children be prevented from accessing some Web sites? Who should be responsible for the actions of children surfing the Web?

27. You are in charge of the computers at a large, inner-city library. Most of the people who live in the neighborhood do not have a computer at home. They go to the library when they want to access the Internet. About two-thirds of the people surfing the Web on the library's computers are adults.

    You have been requested to install filtering software that would block Web sites containing various kinds of material deemed inappropriate for children. You have observed this software in action and know that it also blocks many sites that adults might legitimately want to visit. How should you respond to the request to install filtering software?

28. What is the longest amount of time you have ever spent in a single session in front of a computer? What were you doing?

29. The income of companies providing persistent, online games depends on the number of subscribers they attract. Since consumers have a choice of many products, each company is motivated to create the best possible experience for its customers. Role-playing adventures have no set length. When playing one of these games, it's easy to spend more time on the computer than originally planned. Some subscribers cause harm to themselves and others by spending too much time playing these games. Should the designers of persistent online games bear some moral responsibility for this problem?

In-Class Exercises

30. Divide the class into teams representing each of the following groups:
    - Small, struggling business
    - Large, established corporation
    - Internet service provider
    - Consumer

    Debate the value of direct email versus other forms of advertising, such as direct mail, television advertising, radio advertising, the Yellow Pages, and setting up a Web site.

31. A company uses pop-up advertising to market its software product, which blocks pop-ups from appearing when someone is surfing the Web. Debate the morality of the company's marketing strategy.

32. Ad-blocking software attachments to Web browsers enable a Web surfer to visit Web sites without having to view the pop-up advertisements associated with these Web pages. Debate this proposition: "People who use ad-blocking software are violating an implicit 'social contract' with companies that use advertising revenues as a means of providing free access to Web pages."

33. A female employee of a high-tech company receives on average 40 spam messages per day. About one-quarter of them are advertising pornographic Web sites and have explicit photographs of people engaged in a sexual activity. All of these emails pass through the company's email server. The woman sues the company for sexual harassment, saying that the company tolerates an atmosphere that is degrading to women. Is the company responsible for the pornographic spam reaching the computers of its employees?

34. In 2000 the Estonian parliament passed a law declaring Internet access to be a fundamental human right of its citizens. What do you think will be the positive and negative consequences of this law?

35. How do you determine the credibility of information you get from the Web? Does the source of the information make any difference to you? If so, how would you rank the reliability of each of the following sources of Web pages? Does the type of information you're seeking affect your ranking?
    - Establishment newspaper
    - Counterculture newspaper
    - Television network
    - Corporation
    - Nonprofit organization
    - Individual

36. Discuss the morality of Google's page-ranking algorithm. Does it systematically exclude Web pages containing opinions held only by a small segment of the population? Should every opinion on the Web be given equal consideration?

Further Reading

Francis Canavan. *Freedom of Expression: Purpose as Limit*. Carolina Academic Press and The Claremont Institute for the Study of Statesmanship and Political Philosophy, Durham, NC, 1984.

Edward A. Cavazos and Gavino Morin. *Cyberspace and the Law: Your Rights and Duties in the OnLine World*. The MIT Press, Cambridge, MA, 1994.

Edward G. Hudon. *Freedom of Speech and Press in America*. Public Affairs Press, Washington, DC, 1963.

Steven E. Miller. *Civilizing Cyberspace: Policy, Power, and the Information Superhighway*. ACM Press, New York, NY, 1996.

Stanton Peele. *The Meaning of Addiction: Compulsive Experience and Its Interpretation*. Lexington Books, Lexington, MA, 1985.

Frederick Schauer. *Free Speech: A Philosophical Enquiry*. Cambridge University Press, Cambridge, England, 1982.

Richard A. Spinello and Herman T. Tavani, editors. *Readings in CyberEthics*. Jones and Bartlett, Sudbury, MA, 2001.

Cass R. Sunstein. *Democracy and the Problem of Free Speech*. The Free Press, New York, NY, 1993.

References

[1] Steve Lohr. "Smallpox Researchers Seek Help from Millions of Computer Users." *The New York Times*, February 5, 2003.

[2] Associated Press. "Idle Computers Could Seek Smallpox Cure." *The New York Times*, February 5, 2003.

[3] Reuters. "Researchers Use Computing Grid in Smallpox Battle." *The New York Times*, February 5, 2003.

[4] "Largest Prime Number Ever Is Found." *NewScientist.com News Service*, December 2, 2003.

[5] Alok Jha. "Why Do People Keep on Looking for Ever Bigger Prime Numbers?" *The Guardian (Manchester, England)*, December 18, 2003.

[6] Frank Ahrens. "Spam Lowers Its Appeal." *washingtonpost.com*, December 30, 2004.

[7] "Global Internet Population Grows an Average of Four Percent Year-over-Year." *Nielsen/NetRatings*, February 20, 2003. www.nielsen-netratings.com.

[8] "How Many Online?" *Nua Internet Surveys*, 2003. www.nua.com/surveys.

[9] Brad Templeton. "Origin of the Term 'Spam' to Mean Net Abuse," July 8, 2005. www.templetons.com/brad/spamterm.html.

[10] Jonathan Krim. "Spam's Cost to Business Escalates." *The Washington Post*, March 13, 2003.

[11] Saul Hansell. "Internet Is Losing Ground in Battle against Spam." *The New York Times*, April 22, 2003.

[12] Mike Musgrove. "AOL Reports Decline in Spam in the Past Year." *washingtonpost.com*, December 28, 2004.

[13] Harry A. Valetk. "Spoofing: Spam Scammers Hit a New Low with Spoofed E-mail." *New York Law Journal*, September 16, 2002.

[14] Florence Olsen. "Fed Up with Spam: Irate Students and Professors Want Colleges to Crack Down, but Doing So Is Difficult." *The Chronicle of Higher Education*, September 27, 2002.

[15] "Spam-for-Hire Scheme Uncovered at Tufts." *IDG.net*, February 24, 2003.

[16] Carrie Kirby. "Spam Wars; Markets Battle with E-mail Activists over Your Inbox." *San Francisco Chronicle*, June 2, 2001.

[17] "Stopping Spam." *The Economist*, April 26, 2003.

[18] "As Spam Multiplies, So Do Its Costs to Consumers." *USA Today*, April 30, 2003.

[19] Jim Salter. "Court Rules against Unsolicited Fax Ads." *Associated Press*, March 21, 2003.

[20] "U.S. Junk Fax Law Stretched into $500 Spam Damages against Sears." *Washington Internet Daily*, February 28, 2003.

[21] Lisa M. Thomas and Ashok M. Pinto. "Bush Has Signed Federal Anti-spam Legislation to Take Effect January 1, 2004." Technical report, Chicago, IL, January 2004.

[22] Gregg Keizer. "U.S. Senate Approves CAN SPAM Act." *TechWeb News*, November 26, 2003.

[23] Doug Bedell. "Effectiveness of Suing Spammers Questioned." *The Dallas Morning News*, December 20, 2003.

[24] Alyson Ward. "Still Being Buried by Bulk E-mail—Despite New Federal Legislation?" *Forth Worth Star-Telegram*, January 13, 2004.

[25] Tim Lemke. "House Passes Spam Controls." *The Washington Times*, December 9, 2003.

[26] Grant Gross. "Is the CAN-SPAM Law Working?" *IDG News Service*, January 13, 2004. www.pcworld.com.

[27] Lee Rainie. "The Advent of Spim." Pew Internet & American Life Project, February 21, 2005.

[28] Paul Roberts. "New York Teen Charged with CAN-SPAM Violations, Attempted Extortion." *Network World*, February 28, 2005.

[29] Tim Berners-Lee. *Weaving the Web*. HarperCollins Publishers, New York, NY, 1999.

[30] Kel Belson with Matt Richtel. "America's Broadband Dream Is Alive in Korea." *The New York Times*, May 5, 2003.

[31] Kim Peterson. "How 'Search' Is Redefining the Web—and Our Lives." *The Seattle (WA) Times*, May 2, 2005.

[32] Otto Pohl. "Universities Exporting M.B.A. Programs via the Internet." *The New York Times*, March 26, 2003.

[33] Daniel H. Pink. "The Book Stops Here." *Wired*, page 125, March 2005.

[34] Associated Press. "50 Million Historical Documents Posted on Web." *CNN.com*, April 5, 2003.

[35] Associated Press. "Service to Offer Cash, Prizes for Online Gamers." *The Miami Herald*, March 18, 2003.

[36] Kim Tae-jong. "Lineage Craze Takes On New Dimensions." *The Korea Times*, April 15, 2003.

[37] Jimmy Yap. "Power Up!" *Internet Magazine*, February 2003.

[38] Associated Press. "IRS Online Filing Tops 2M Users." *The New York Times*, March 25, 2003.

[39] Bob Tedeschi. "Gambling Sites Adjust to Scrutiny." *The New York Times*, March 31, 2003.

[40] Paul Festa. "Dialing for Bloggers." *The New York Times*, February 25, 2003.

[41] Privacy International. "Silenced—Burma." September 21, 2003. www.privacy international.org.

[42] Stephen Gibbs. "Cuba Law Tightens Internet Access." *BBC News*, January 24, 2004.

[43] Rebecca MacKinnon. "Chinese Cell Phone Breaches North Korean Hermit Kingdom." *YaleGlobal Online*, January 17, 2005.

[44] Jonathan Zittrain and Benjamin Edelman. "Documentation of Internet Filtering in Saudi Arabia." Technical report, Harvard Law School, Cambridge, MA, September 12, 2002.

[45] Elizabeth Croad. "China Steps Up Web Censorship." *dot journalism*, April 28, 2003.

[46] Jonathan Zittrain and Benjamin Edelman. "Empirical Analysis of Internet Filtering in China." Technical report, Harvard Law School, Cambridge, MA, March 20, 2003.

[47] Donald Vandeveer. "Pornography." In *Encyclopedia of Ethics*. Routledge, New York, NY, 2001.

[48] Immanuel Kant. *Lectures on Ethics*. Cambridge University Press, 2001.

[49] Eason Jordan. "The News We Kept to Ourselves." *The New York Times*, April 11, 2003.

[50] Immanuel Kant. "What Is Enlightenment?" In *Foundations of the Metaphysics of Morals*, Upper Saddle River, NJ, 1997. Library of Liberal Arts.

[51] John Stuart Mill. "On Liberty." In *On Liberty and Utilitarianism*. Bantam Books, New York, NY, 1993.

[52] Edward G. Hudon. *Freedom of Speech and Press in America*. Public Affairs Press, Washington, DC, 1963.

[53] Francis Canavan. *Freedom of Expression: Purpose as Limit*. Carolina Academic Press, Durham, NC, 1984.

[54] Cass R. Sunstein. *Democracy and the Problem of Free Speech*. The Free Press, New York, NY, 1993.

[55] George Carlin. "Filthy Words." In *Occupation: Foole*. Atlantic Records, 1973.

[56] Supreme Court of the United States. *Federal Communications Commission v. Pacifica Foundation et al.*, 1978. 438 U.S. 726.

[57] "Spying on Kids' Internet Use." *CBS News*, February 2003.

[58] Associated Press. "Justices Uphold Use of Internet Filters in Public Libraries." *NY-Times.com*, June 23, 2003.

[59] Jeffrey Kosseff. "Libraries Should Bar Web Porn, Court Rules." *The Oregonian (Portland, Oregon)*, June 24, 2003.

[60] James Rachels. *The Elements of Moral Philosophy*. 4th ed. McGraw-Hill, Boston, MA, 2003.

[61] John Rawls. *A Theory of Justice, Revised Edition*. The Belknap Press of Harvard University Press, Cambridge, MA, 1999.

[62] Lorenne Clark. "Sexual Equality and the Problem of an Adequate Moral Theory: The Poverty of Liberalism." In *Contemporary Moral Issues*, Toronto, 1997. McGraw-Hill Ryerson.

[63] Langdon Winner. "Electronically Implanted 'Values'." *Technology Review*, 100(2), February/March 1997.

[64] Doug Johnson. "Internet Filters: Censorship by Any Other Name?" *Emergency Librarian*, 25(5), May/June 1998.

[65] Al Swanson. "Analysis: Most ID Theft Begins at Home." *United Press International*, January 28, 2005.

[66] Jason Gertzen. "Protect Your Finances from Online Fraudsters, Experts Warn." *The Milwaukee Journal Sentinel (Wisconsin)*, December 8, 2003.

[67] "Nielsen/NetRatings Finds E-mail Is the Dominant Online Activity Worldwide." *Nielsen/NetRatings*, May 9, 2002. www.nielsen-netratings.com.

[68] Lynn Burke. "Memoir of a Pedophile's Victim." *Wired News*, April 26, 2000. www.wired.com.

[69] Thomas Clouse. "Man Accused of Seeking Sex with 13-Year-Old Girl; Police Say Internet Sting Caught Suspect Who Had Handcuffs, Knife." *Spokane Spokesman-Review*, March 5, 2003.

[70] Shaila K. Dewan. "Who's 14, 'Kewl' and Flirty Online? A 39-Year-Old Detective, and He Knows His Bra Size." *The New York Times*, April 7, 2003.

[71] "Police Say Arkansas Man Made an Online Deal to Buy a Little Girl for Sex." *ZDNet UK*, September 3, 1999. news.zdnet.co.uk.

[72] "Chat Room Cops Nab Possible Predator." *Tech TV Inc.*, May 17, 2002. www.techtv.com.

[73] Paige Akin. "Man Arrested in Undercover Cyber Sex Sting." *Richmond Times Dispatch (Virginia)*, March 8, 2003.

[74] Suzannah Gonzales. "Sex Case May Lead to More Charges." *St. Petersburg Times (Florida)*, March 14, 2003.

[75] Jennifer Sinco Kelleher. "Arrests in Sex Chats with 'Girls.'" *Newsday*, March 15, 2003.

[76] Amy Klein. "Cops in the Chat Room; Detectives Play Teenagers to Bait Sexual Predators." *The Record (Bergen County, NJ)*, April 6, 2003.

[77] Joseph Weizenbaum. *Computer Power and Human Reason: From Judgment to Calculation*. W. H. Freeman and Company, San Francisco, CA, 1976.

[78] "Computer Addiction: Is It Real or Virtual?" *Harvard Mental Health Letter*, 15(7), January 1999.

[79] *Merriam-Webster's Collegiate Dictionary, Tenth Edition*. Merriam-Webster, Springfield, MA, 1994.

[80] Stanton Peele. *The Meaning of Addiction: Compulsive Experience and Its Interpretation*. Lexington Books, Lexington, MA, 1985.

[81] Kimberly S. Young. "Internet Addiction: Symptoms, Evaluation, and Treatment." *Innovations in Clinical Practice*, volume 17, edited by L. VandeCreek and T. L. Jackson. Professional Resource Press, Sarasota, FL, 1999.

[82] Mark Griffiths. "Does Internet and Computer 'Addiction' Exist? Some Case Study Evidence." *CyberPsychology and Behavior*, 3(2), 2000.

[83] Kimberly S. Young. "Internet Addiction: The Emergence of a New Clinical Disorder." *CyberPsychology and Behavior*, 1(3), 1998.

[84] John P. Charlton. "A Factor-Analysis Investigation of Computer 'Addiction' and Engagement." *British Journal of Psychology*, 99(3), August 2002.

[85] Aydrea Walden. "Center Helps Those Hooked on Internet." *The Seattle (WA) Times*, February 5, 2002.

[86] Jeffrey Reiman. *Critical Moral Liberalism: Theory and Practice*. Rowman & Littlefield Publishers, Lanham, MD, 1997.

AN INTERVIEW WITH

Jennifer Preece

Jennifer J. Preece, Ph.D., is a professor and Dean of the College of Information Studies at the University of Maryland, College Park. At the University of Maryland, her research focus is on how online communities are designed and managed. Dr. Preece earned an honors degree in Biology from the University of Ulster. She earned her Ph.D. degree from Open University, Institute of Educational Technology. After receiving her Ph.D., Dr. Preece taught at several universities, including the University of Maryland, Baltimore County, where she held the position of department chair. In January 2005 she moved to the University of Maryland at College Park. Dr. Preece has authored several books and many journal articles. Her two most recent books are: *Online Communities: Designing Usability, Supporting Sociability* (2000) and *Interaction Design: Beyond Human-Computer Interaction* (2002), which is collaboratively authored with Yvonne Rogers and Helen Sharp.

How do you define the term "online community"?

I define an online community as a group of people who come together online via Internet technology for a particular purpose and whose behavior is guided by policies that can take the form of rules or norms. There are thousands, probably millions, of online communities that bring together people with a common interest in a health problem, hobby, sport, political goal, or other topic. These kinds of communities are referred to as Communities of Interest (COIs) to distinguish them from Communities of Practice (COPs). Typically COIs develop in a 'grass roots' fashion using widely available software such as bulletin boards, wikis, chats, IM, and listservers. COPs, on the other hand, tend to be developed within companies or organizations. They often use a purpose-built software platform, have a clear mission statement, and follow the governance model of the parent organization, which is often hierarchical rather than involving shared governance as in many COIs.

How did you get interested in studying online communities?

I worked for fifteen years at the British Open University, a large, internationally renowned distance education university. During that time, I was involved in using and developing state-of-the-art computer-mediated communication learning environments that we would now call online communities. As I saw how these technologies were used, it became obvious to me that focusing on the technology alone was a losing strategy. For these online community spaces to be successful, developers needed to understand and nurture social interaction online.

Can you give a few examples of highly successful online communities?

Interestingly, some of the most successful online communities use widely available, fairly unsophisticated software. One community that I have studied for around seven years is Bob's ACL Knee Community, which is a community of people with knee injuries who communicate with each other primarily via a bulletin board with threaded messages.

Other highly successful communities include e-commerce sites that you may not even think of as online communities. eBay engenders such enthusiasm and loyalty among its sellers that thousands of subcommunities have formed throughout the US and the world. These communities meet face-to-face and online, and like all strong communities, their members proudly wear their tee-shirts and sport other forms of eBay branding. Amazon, too, is a successful online community. As well as buying books, an army of volunteers compete to be notable reviewers, generating tens and hundreds of reviews each. Other interesting examples include those supported by network software such as Linkedin, Ringo, and Orkut, which are designed to facilitate safe-meeting between friends and potential dating partners or business associates, depending on the community, through trusted friends and third-party participants. And, of course, there are thriving wiki communities and communities that exchange music. Podcasting is becoming another technology around which communities are developing. Then there are small communities of friends who coordinate their activities and keep track of each other's activities via cell-phone texting. There are many, many different kinds of online communities.

What makes an online community successful?

The answer to this question depends to some extent on the kind of community. Community developers and leaders have to nurture the community's growth. If nothing appears to be happening in a community, people will not come back. Having a clearly articulated purpose also encourages successful communities, because everyone who belongs knows the intention of the community. Another contributor to success is shared trust and a sense of responsibility, so that when antisocial events happen, the community takes action because its members feel responsible for each other's well-being.

How do successful online communities cope with antisocial members?

Some of the most successful communities that I have encountered have a shared governance model in which participants take care of each other and ensure that a few agreed upon codes of behavior are upheld. So, if one of the members is verbally attacked for holding a particular position or saying something a bit off-beat, other members jump in to their defense and reprimand the offending person for their unacceptable behavior. Of course, strong moderators may also take action, but it tends to be more effective coming from the community themselves.

How have online communities changed in the past decade?

The concept of an online community has developed strongly during the last decade. This is due to an increase in the number of people who belong to some kind of online community and to changes in technology. Different types of software tend to integrate more easily. For example, participants may shift from synchronous to asynchronous communication. Multimedia software enables participants to incorporate pictures, video, and sounds into their communication. In addition, a variety of new community software environments have been created to support different types of community activities, such as collaborative authoring in wikis, cross-cultural communication, and voice over IP.

How will emerging technologies (such as new input/output devices) affect online communities?

I expect that there will be even stronger integration of synchronous and asynchronous communication, better support for identity, and more ways of ensuring that individuals' privacy is protected, while at the same time ensuring that people are who they say they are and take responsibility for their actions. It will become even easier to integrate different media, such as pictures and video, with verbal and written comments. Improved translation and cross-cultural support will enable people who speak different languages and come from different cultures to communicate more effectively. Many of these developments will be driven by the e-commerce marketplace. However, just as globalization encourages collaboration and communication, it will also drive invention of technologies to prevent exploitation of children and protect individuals' rights.

Technology alone will not create breakthroughs. Breakthroughs come only when technology is designed for usability and sociability is sensitively supported and nurtured.

CHAPTER

4 Intellectual Property

Friends share all things.

–PYTHAGORAS

Today's pirates operate not on the high seas but on the Internet.

–RECORDING INDUSTRY ASSOCIATION OF AMERICA

4.1 Introduction

THE ABILITY TO STORE MUSIC IN DIGITAL FORM has combined with the high communication speed of the Internet to create a legal crisis. A poll by Ipsos-Reid says that more than 60 million Americans have used the Internet to download music [1]. The Recording Industry Association of America (RIAA) responds by seeking billions of dollars in damages from four college students who have let others download copyrighted music from their computers [2]. Meanwhile, the Electronic Frontier Foundation starts a "Let the Music Play" campaign (Figure 4.1) to convince Americans they should put pressure on Congress to change copyright laws [3].

At the beginning of a Britney Spears concert, thousands of teenagers hold up their cell phones so their friends can hear the concert, too [4]. Does an entertainer have the right to control who hears a live performance of her music?

About 95 percent of "Microsoft" products purchased in China are actually counterfeits made by Chinese entrepreneurs [5]. Microsoft does not receive revenues from

FIGURE 4.1 The Electronic Frontier Foundation is advocating a reform of the copyright laws in the United States. (Courtesy of Electronic Frontier Foundation)

these sales; it calls the entrepreneurs "software pirates." Should Microsoft have the right to charge licensing fees to those individuals and institutions using its software?

As a society we benefit from access to high-quality music, movies, computer programs, and other products of the human intellect. The value of these intellectual properties is much higher than the value of the media on which they are distributed. Creating the first copy is very expensive. Duplicates cost almost nothing. There is a strong incentive for people to make unauthorized copies. When this happens, producers of intellectual property do not receive all of the payments the law says they are entitled to. The legal system has responded by giving more rights to the creators of intellectual property. Are these changes in the best interests of our society, or are our politicians catering to special interest groups?

In this chapter we discuss how information technology is affecting our notions of intellectual property. We consider what makes intellectual property different from tangible property and how governments have created a variety of mechanisms to guarantee intellectual property rights. We examine what has been considered "fair use" of intellectual property created by others, and how new copy-protection technologies are eroding the notions of fair use. Meanwhile, peer-to-peer networks are making it easier than ever for consumers to get access to music and movies without purchasing them. We look at what the entertainment industry is doing to eliminate free access to copyrighted material. We also explore the evolution of intellectual property protection for computer software and the rise of the open-source movement, which advocates the distribution of source code to programs. Finally, we take a look at one organization's efforts to make it easier for artists, musicians, and writers to use the Internet as a vehicle for stimulating creativity and enhancing collaboration.

4.2 Intellectual Property Rights

4.2.1 What Is Intellectual Property?

Intellectual property is any unique product of the human intellect that has commercial value [6]. Examples of intellectual property are books, songs, movies, paintings, inventions, chemical formulas, and computer programs.

It is important to distinguish between intellectual property and its physical manifestation in some medium. If a poet composes a new poem, the poem itself is the intellectual property, not the piece of paper on which the poem is printed.

In modern Western democracies there is a widely accepted notion that people have the right to own property. Does this right extend to intellectual property as well? To answer this question, we need to examine the philosophical justification for a natural right to property.

4.2.2 Property Rights

The English philosopher John Locke (1632–1704) developed an influential theory of property rights. In *The Second Treatise of Government*, Locke makes the following case for a natural right to property. First, people have a right to property in their own person. Nobody has a right to the person of anybody else. Second, people have a right to their own labor. The work that people perform should be to their own benefit. Third,

FIGURE 4.2 According to John Locke, people have a natural right to the things they have removed from Nature through their own labor.

people have a right to those things that they have removed from Nature through their own labor [7].

For example, suppose you are living in a village in the middle of the woods, which are held in common. One day you walk into the woods, chop down a tree, saw it into logs, and split the logs into firewood (Figure 4.2). Before you cut down the tree, everyone had a common right to it. By the time you have finished splitting the logs, you have mixed your labor with the wood, and at that point, it has become your property. Whether you burn the wood in your stove, sell it to someone else, pile it up for the winter, or give it away, the choice of what to do with the wood is yours.

Locke uses the same reasoning to explain how a person can gain the right to a piece of land. Taking a parcel out of the state of Nature by clearing the trees, tilling the soil, and planting and harvesting crops gives people who performed these labors the right to call the land their property.

To Locke, this definition of property makes sense as long as two conditions hold. First, no person claims more property than he or she can use. In the case of harvesting a natural resource, it is wrong for someone to take so much that some of it is wasted. For example, people should not appropriate more land than they can tend. Second, when people remove something from the common state in order to make it their own property, there is still plenty left over for others to claim through their labor. If the woods are full of trees, I can chop a tree into firewood without denying you or anyone else the opportunity to do the same thing.

Locke's description of a natural right to property is most useful at explaining how virtually unlimited resources are initially appropriated. It is not as useful in situations where there are few or no resources left for appropriation.

4.2.3 Extending the Argument to Intellectual Property

Is there a natural right to intellectual property?

We can try to demonstrate such a right exists by extending Locke's theory of property rights to intellectual property. However, since Locke was talking about the ownership of physical objects and we are talking about the ownership of ideas, we must resort to an analogy. We'll compare creating a piece of intellectual property to making a belt buckle [8]. In order to make a belt buckle, a person must mine ore, smelt it down, and cast it. To write a play, a playwright "mines" words from the English language, "smelts" them into stirring prose, and "casts" them into a finished play.

Attempting to treat intellectual property the same as ordinary property leads to certain paradoxes, as Michael Scanlan has observed [8]. We will consider two of Scanlan's scenarios illustrating problems that arise when we extend Locke's natural rights argument to intellectual property.

Scenario A, Act 1

After a day of rehearsals at the Globe Theatre, William Shakespeare decides to have supper at a pub across the street. The pub is full of gossip about royal intrigue in Denmark. After his second pint of beer, Shakespeare is visited by the muse, and in an astonishing burst of energy, he writes *Hamlet* in one fell swoop.

If we apply Locke's theory of property to this situation, clearly Shakespeare has the right to own *Hamlet*. He mixed his labor with the raw resources of the English language and produced a play. Remember, we're not talking about the piece of paper upon which the words of the play are written. We're talking about the sequence of words comprising the play. The paper is simply a way of conveying them.

What should Shakespeare get from his ownership of *Hamlet?* Here are two ideas (you can probably think of more): He should have the right to decide who will perform the play. He should have the right to require others who are performing the play to pay him a fee.

So far, so good. But let's hear the end of the story.

Scenario A, Act 2

On the very same night, Ben Jonson, at a pub on the opposite side of London, hears the same gossip, is struck by the same muse, and writes *Hamlet*—exactly the same play!

Ben Jonson has mixed his intellectual labor with the English language to produce a play. According to Locke's theory of natural rights to property, he ought to own it. Is it possible for both Ben Jonson and William Shakespeare to own the same play (Figure 4.3)? No, not as we have defined ownership rights. It is impossible for both of them to have the exclusive right to decide who will perform the play. Both of them cannot have an exclusive claim to royalties collected when *Hamlet* is performed. We've uncovered a paradox: two people labored independently and produced only a single artifact.

Figure 4.3 Suppose both Ben Jonson and William Shakespeare simultaneously write down *Hamlet*. Who owns it?

We ended up with this paradox because our analogy is imperfect. If two people go to the same iron mine, dig ore, smelt it, and cast it into belt buckles, there are two belt buckles, one for each person. Even if the belt buckles look identical, they are distinct, and we can give each person ownership of one of them. This is not the case with *Hamlet*. Even though Jonson and Shakespeare worked independently, there is only one *Hamlet*. Whether we give one person complete ownership or divide the ownership among the two men, both cannot get full ownership of the play, which is what they ought to have if the analogy were perfect. Therefore, the uniqueness of intellectual properties is the first way in which they differ from physical objects.

A second paradox has to do with the copying of intellectual property. Consider a slightly different version of our story.

Scenario B

One evening William Shakespeare stays up all night in a pub writing *Hamlet* while Ben Jonson goes to a party. The next morning Shakespeare returns to the Globe Theatre, but he carelessly leaves a copy of *Hamlet* in the pub. Jonson stops by for a pint, sees the manuscript, transcribes it, and walks out the door with a copy of the play in his possession, leaving the original copy where it was.

Did Jonson steal *Hamlet*? Shakespeare still has his physical copy of the play, but he has lost exclusive control over who will read, perform, or hear the play. If you want to call this stealing, then stealing in the sense of intellectual property is quite different from stealing a physical object. When you steal someone's car, they can't drive it any more. When you steal someone's joke, both of you can tell it.

Certainly any creator of a piece of intellectual property has the right to keep his ideas a secret. After Shakespeare wrote *Hamlet*, he could have locked it in a trunk to prevent others from seeing it. Ben Jonson would not have the right to break into Shakespeare's trunk to get access to the play. Hence we can argue that there is a natural right to keep an idea confidential. Unfortunately, this is a weak right, because Shakespeare cannot perform the play while he is keeping it confidential. He must give up the confidentiality in order to put his creation to good use.

We began this section with the question, Is there a natural right to intellectual property? We have found no right other than the weak right to keep an idea confidential. In our quest for stronger rights, we have uncovered two important differences between tangible property and intellectual property. First, every intellectual property is one-of-a-kind. Second, copying a piece of intellectual property is different from stealing a physical object.

4.2.4 Benefits of Intellectual Property Protection

New ideas in the form of inventions and artistic works can improve the quality of life for the members of a society. Some people are altruistic and will gladly share their creative energies. For example, Benjamin Franklin (1706–1790) invented many useful items, including an improved wood stove, the lightning rod, the odometer, and bifocals. He did not patent any of them. Franklin said, "As we enjoy great advantages from the invention of others, we should be glad of an opportunity to serve others by any invention of ours; and this we should do freely and generously" [9]. However, most people find the allure of wealth to be a strong inducement for laboring long hours in the hope of creating something useful. Hence even if there are no natural rights to intellectual property, a society may choose to grant intellectual property rights to people because of the beneficial consequences.

The authors of the Constitution of the United States recognized the benefits society reaps by encouraging creativity. Article I, Section 8, of the U.S. Constitution gives Congress the power to "promote the Progress of Science and useful Arts by securing for limited Times to Authors and Inventors the exclusive Right to their respective Writings and Discoveries."

If a person has the right to control the distribution and use of a piece of intellectual property, there are many opportunities for that person to make money. For example, suppose you build a better mousetrap and the government gives you ownership of this design. You may choose to manufacture the mousetrap yourself. Anyone who wants the better mousetrap must buy it from you, because no other mousetrap manufacturer has the right to copy your design. Alternatively, you may choose to license your design to other manufacturers, who will pay you for the right to build mousetraps according to your design.

4.2.5 Limits to Intellectual Property Protection

Of course, society benefits the most when inventions are in the public domain and anyone can take advantage of them. Going back to the mousetrap example, we would like everyone in society who needs a mousetrap to get the best possible trap. If someone invents a superior mousetrap, the maximum benefit would result if all mousetrap manufacturers were able to use the better design. On the other hand, if the inventor of the superior mousetrap did not have any expectation of profiting from her new design, she

Artist	*Work*	*Previous Rental Fee*	*Year Became Public Domain*	*Purchase Price*
Ravel	Daphnis et Chloe Suite no. 1	$450.00	1987	$155.00
Ravel	Mother Goose Suite	540.00	1988	70.00
Ravel	Daphnis et Chloe Suite no. 2	540.00	1989	265.00
Griffes	The White Peacock	335.00	1993	42.00
Puccini	O Mio Babbino Caro	252.00	1994	26.00
Respighi	Fountains of Rome	441.00	1994	140.00
Ravel	Le Tombeau de Couperin	510.00	1995	86.00
Respighi	Ancient Aires and Dances Suite no. 1	441.00	1996	85.00
Elgar	Cello Concerto	550.00	1997	140.00
Holst	The Planets	815.00	1997	300.00
Ravel	Alborada Del Gracioso	360.00	1999	105.00

TABLE 4.1 Once a piece of classical music enters the public domain, it may be purchased for much less than it cost simply to rent the same piece of music for two performances when it was still under copyright protection. These prices assume the orchestra has an annual budget of $150,000 or less [10].

may not have bothered to invent it. Hence there is a tension between the need to reward the creators of intellectual property by giving them exclusive rights to their ideas and the need to disseminate these ideas as widely as possible.

The way the Congress has traditionally addressed this tension is through a compromise. It has granted authors and inventors exclusive rights to their writings and discoveries, but only for a finite period of time. At the end of that time period, the intellectual property enters the public domain. While creators have control over the distribution of their properties, use of the properties is more expensive, and the creators are rewarded. After properties enter the public domain, using them becomes less expensive, and everyone has the opportunity to produce derivative works from them.

Consider a community orchestra that wishes to perform a piece of classical music. It may purchase a piece of music from the public domain for far less money than it cost simply to rent the same piece of music while it was still protected by copyright (Table 4.1).

"Happy Birthday to You" is one of the most popular songs in the English language. Have you ever wondered why you almost never hear it sung on television? The reason is that the music publisher Clayton F. Summy Company copyrighted the song in 1935. Television networks must pay the Clayton F. Summy Company to air "Happy Birthday to You." The company collects about $2 million in royalties each year for public performances of "Happy Birthday to You" [11]. Under the Copyright Term Extension Act of 1998, the song will remain copyrighted until at least 2030!

4.3 Protecting Intellectual Property

While the U.S. Constitution gives Congress the right to grant authors and inventors exclusive rights to their creations, it does not elaborate on how these rights will be protected. Today, there are four different ways in which individuals and organizations protect their intellectual property: trade secrets, patents, copyrights, and trademarks/service marks.

4.3.1 Trade Secrets

A **trade secret** is a confidential piece of intellectual property that provides a company with a competitive advantage. Examples of trade secrets include formulas, processes, proprietary designs, strategic plans, customer lists, and other collections of information. The right of a company to protect its trade secrets is widely recognized by governments around the world. In order to maintain its rights to a trade secret, a company must take active measures to keep it from being discovered. For example, companies typically require employees with access to a trade secret to execute a confidentiality agreement.

A famous trade secret is the formula for Coca-Cola syrup. The formula, known inside the company as "Merchandise 7X," is locked in a bank vault in Atlanta, Georgia. Only a few people within the company know the entire formula, and they have signed nondisclosure agreements. The task of making the syrup is divided among different groups of employees. Each group makes only one part of the final mixture, so that nobody in these groups learns the complete recipe.

An advantage of trade secrets is that they do not expire. A company never has to disclose a trade secret. Coca-Cola has kept its formula secret for more than 100 years.

The value of trade secrets is in their confidentiality. Hence trade secrets are not an appropriate way to protect many forms of intellectual property. For example, it would make no sense for a company to make a movie a trade secret, because a company can only profit from a movie by allowing it to be viewed, which would make it no longer confidential. On the other hand, it would be appropriate for a company to make the *idea* for a movie a trade secret. Art Buchwald pitched Paramount Pictures a story called *King for a Day*, about an African prince who visits the United States. After the studio produced the movie *Coming to America*, starring Eddie Murphy, Buchwald successfully sued the studio for breach of contract, because he had made the studio sign a confidentiality agreement before he gave them the plot [12].

While it is illegal to steal a trade secret, there are other ways in which the confidentiality may be broken. "Reverse engineering" is one way in which a competing firm can legally gain access to information contained in a trade secret. If another company can purchase a can of Coca-Cola and figure out the formula, it is free to manufacture a soft drink that looks and tastes just like Coke.

Another way in which a competing firm can gain access to information contained in another company's trade secret is by hiring its employees. While a firm can require its employees to sign confidentiality agreements, it cannot erase the memories of an employee who starts working for a competing firm. Hence some "leakage" of confidential information may be inevitable when employees move from one company to another.

4.3.2 Trademarks and Service Marks

A **trademark** is a word, symbol, picture, sound, color, or smell used by a business to identify goods. A **service mark** is a mark identifying a service [13].

By granting a trademark or service mark, a government gives a company the right to use it and the right to prevent other companies from using it. Through the use of a trademark, a company can establish a "brand name." Society benefits from branding because branding allows consumers to have more confidence in the quality of the products they purchase.

When a company is the first to market a distinctive product, it runs the risk that its brand name will become a common noun used to describe any similar product. When this happens, the company may lose its right to exclusive use of the brand name. Some trademarks that have become generic are "yo yo," "aspirin," "escalator," "thermos," and "brassiere." Companies strive to ensure their marks are used as adjectives rather than nouns (Figure 4.4). Kimberly-Clark's advertisements refer to "Kleenex **brand** facial tissue." Remember Johnson & Johnson's jingle "I am stuck on Band-Aid **brand** 'cause Band-Aid's stuck on me"?

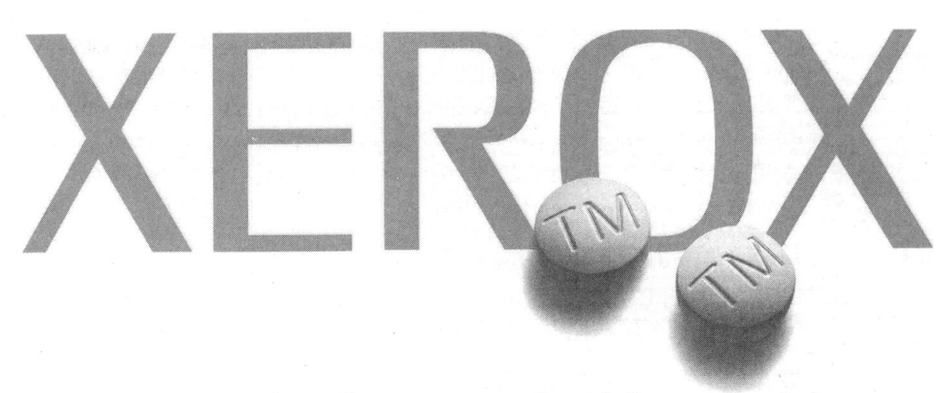

FIGURE 4.4 Xerox Corporation ran this advertisement in *The Chronicle of Higher Education* as part of a campaign to protect its trademark. (Courtesy of Xerox Corporation)

4.3.3 Patents

A **patent** is a way the U.S. government provides an inventor with an exclusive right to a piece of intellectual property. A patent is quite different from a trade secret because a patent is a public document that provides a detailed description of the invention. The owner of the patent can prevent others from making, using, or selling the invention for the lifetime of the patent, which is currently 20 years. After the patent expires, anyone has the right to make use of its ideas.

POLAROID V. KODAK

Dr. Edwin Land invented "instant" photography. The company he founded, Polaroid Corporation, had 10 patents protecting the invention of film that developed in 60 seconds. Polaroid did not license these patents to other firms, and for many years it was the only company to sell cameras and film allowing photographs to be developed in a minute.

When Kodak introduced its first instant camera in 1976, Polaroid sued Kodak [14]. In 1985 a court ruled that Kodak had infringed on seven of Polaroid's original ten patents; six years later Kodak paid Polaroid a $925 million settlement [15, 16].

SPARC INTERNATIONAL

Sometimes companies see an advantage in licensing their inventions. After Sun Microsystems invented the SPARC architecture, it wanted to maximize the number of SPARC-compliant computers being manufactured. For this reason, Sun transferred ownership of the SPARC specifications to an independent, nonprofit organization called SPARC International. SPARC International has licensed SPARC technology to a variety of other firms. In 2005 the list of companies manufacturing SPARC-based systems included Continuous Computing Corp., Force Computers, Fujitsu, Nature Worldwide Technology Corporation (NatureTech), Rave Computer Association, Inc., Sun Microsystems, Inc., Tadpole, Themis Computer, and Toshiba Corporation.

4.3.4 Copyrights

A **copyright** is how the U.S. government provides authors with certain rights to original works that they have written. The owner of a copyright has five principal rights:

1. The right to reproduce the copyrighted work
2. The right to distribute copies of the work to the public
3. The right to display copies of the work in public
4. The right to perform the work in public
5. The right to produce new works derived from the copyrighted work

Copyright owners have the right to authorize others to exercise these five rights with respect to their works. The owner of a copyright to a play may sell a license to a high school drama club that wishes to perform it. After a radio station broadcasts a song, it must pay the songwriter(s) and the composer(s) through a performance rights organization such as ASCAP, BMI, or SESAC.

Copyright owners have the right to prevent others from infringing on their rights to control the reproduction, distribution, display, performance, and production of works derived from their copyrighted work.

Several important industries in the United States, including the movie industry, music industry, software industry, and book publishing, rely upon copyright law for protection. "Copyright industries" account for over 5 percent of the United States gross domestic product, with over $500 billion in sales. About 5 million U.S. citizens work in these industries, which are growing at a much faster rate than the rest of the U.S. economy. With foreign sales and exports of $89 billion, copyright industries were the leading export sector in the United States in 2001 [17].

In this section we examine court cases and legislation that have helped define the limits of copyright in the United States.

By permission of John Deering and Creators Syndicate, Inc.

GERSHWIN PUBLISHING CORPORATION V. COLUMBIA ARTISTS MANAGEMENT, INC.

Columbia Artists Management, Inc. (CAMI) managed concert artists, and it sponsored hundreds of local, nonprofit community concert associations that arranged concert series featuring CAMI artists. CAMI helped the associations prepare budgets, select artists, and sell tickets. CAMI printed the programs and sold them to the community concert associations. In addition, all musicians performing at these concerts paid CAMI a portion of their fees.

On January 9, 1965, the CAMI-sponsored Port Washington (NY) Community Concert Association put on a concert that included Gershwin's "Bess, You Is My Woman Now" without obtaining copyright clearance from Gerwshwin Publishing Corporation. The American Society of Composers, Authors, and Publishers (ASCAP) sued CAMI to determine whether CAMI could be held liable for the copyright infringement.

CAMI argued that it was not responsible for the copyright infringement, since the concert was put on by the Port Washington Community Concert Association. However, the U.S. District Court for the Southern District of New York ruled that CAMI could be held liable for the copyright infringement. In 1971 the U.S. Court of Appeals for the Second Circuit upheld the ruling of the district court [18].

BASIC BOOKS V. KINKO'S GRAPHICS CORPORATION

In the 1980s Kinko's Graphics Corporation engaged in what it called the "Professor Publishing" business. It distributed brochures to university professors asking them to provide lists of readings they planned to use in their courses. Kinko's used these lists to produce packets of reading materials for students taking these classes. The packets typically contained chapters from books. In 1991 the U.S. District Court for the Southern District of New York ruled that when Kinko's produced these packets it infringed upon the copyrights held by the publishers. The judge ordered Kinko's to pay statutory damages of $510,000 to the plaintiffs, a group of eight book publishers [19]. Kinko's subsequently got out of the Professor Publishing business.

DAVEY JONES LOCKER

Richard Kenadek ran a computer bulletin board system (BBS) called Davey Jones Locker. Subscribers paid $99 a year for access to the BBS, which contained copies of more than 200 commercial programs. In 1994 Kenadek was indicted for infringing on the copyrights of the owners of the software. He pleaded guilty and was sentenced to six months' home confinement and two years' probation [20].

NO ELECTRONIC THEFT ACT

Another incident in 1994 led to further legislation protecting copyrights. David LaMacchia, an MIT student, posted copyrighted software on a public bulletin board he created on a university computer. According to prosecutors, bulletin board users downloaded more than a million dollars' worth of software in less than two months. However, the prosecutors were forced to drop charges against LaMaccia because he had made the programs available for free. Since he had not profited from his actions, he had not violated copyright law. To close this legal loophole, Congress passed the No Electronic Theft Act of 1997, which made it a criminal offense *simply to reproduce or distribute* more than $1,000 worth of copyrighted material in a six-month period.

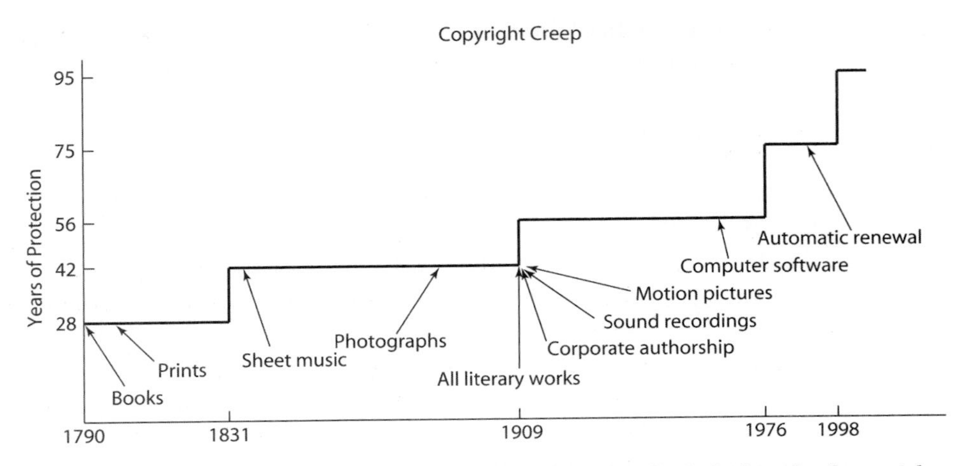

FIGURE 4.5 Since the first Copyright Act was passed in 1790, both the length of copyright protection and the kinds of intellectual property that can be copyrighted have grown dramatically.

COPYRIGHT CREEP

Currently, works created and published before January 1, 1978, are protected for 95 years. Works created on or after January 1, 1978, are protected for the author's lifetime plus 70 years after the author's death. If the work is a work made for hire, the length of protection is 95 years from the date of publication or 120 years from the date of creation, whichever is less.

According to Siva Vaidhyanathan, "in the early republic and the first century of American legal history, copyright was a Madisonian compromise, a necessary evil, a limited, artificial monopoly, not to be granted or expanded lightly" [21]. Over time, however, Congress has gradually increased both the term of copyright protection and the kind of intellectual properties that are protected by copyright (Figure 4.5). One reason has been the desire to have international copyright agreements. In order to complete these agreements, Congress has had to reconcile American copyright law with European law, which in general has had much stronger protections for the producers of intellectual property [21]. Another reason for "copyright creep" has been the introduction of new technologies such as photography, audio recording, and video recording.

For example, since 1831 music publishers have been able to copyright sheet music and collect royalties from musicians performing this music in public. In 1899 Melville Clark introduced the Apollo player piano, which played songs recorded on rolls of heavy paper. Apollo manufactured and sold piano rolls of copyrighted songs. White-Smith Music Company sued Apollo for infringing on its copyrights. In 1908 the Supreme Court ruled that Apollo had not infringed on White-Smith Music's copyrights. The court suggested Congress ought to change copyright law if it wanted owners of copyrights to have control over recordings such as piano rolls and phonograph records. Congress responded by revising the Copyright Act in 1909. The new copyright law recognized that player piano rolls and phonograph records could be copyrighted.

4.4 Fair Use

The right given to a copyright owner to reproduce a work is a limited right. Under some circumstances, called **fair use,** it is legal to reproduce a copyrighted work without the permission of the copyright holder. Examples of fair use include citing short excerpts from copyrighted works for the purpose of teaching, scholarship, research, criticism, commentary, and news reporting.

The United States Copyright Act does not precisely list the kinds of copying that are fair use. Instead, what is considered to be fair use has been determined by the judicial system. The courts have relied upon Section 107 of the Copyright Act, which lists four factors that need to be considered [22]:

1. *What is the purpose and character of the use?*

   An educational use is more likely to be permissible than a commercial use.

2. *What is the nature of the work being copied?*

   Use of nonfiction is more likely to be permissible than use of fiction. Published works are preferred over unpublished works.

3. *How much of the copyrighted work is being used?*

   Brief excerpts are more likely to be permissible than entire chapters.

4. *How will this use affect the market for the copyrighted work?*

   Use of out-of-print material is more likely to be permissible than use of a readily available work. A spontaneously chosen selection is better than an assigned reading in the course syllabus.

In the previous section on copyright we discussed the case against Kinko's. A number of factors led the judge to conclude that the reproductions made by Kinko's Professor Publishing business were not fair use. Kinko's is a commercial enterprise; it started the Professor Publishing business to make a profit. It copied significant portions of books to create the course reading packets. Some of the books were still in print; hence Kinko's negatively affected the market for the copyrighted work. Finally, the readings were not spontaneously chosen. Kinko's had time to contact publishers and gain permission to reproduce the materials, perhaps by paying a licensing fee.

Let's consider two scenarios in which copyrighted works are duplicated and determine if they made fair use of the material. These scenarios are closely modeled after situations presented on the Web site of CETUS, the Consortium for Educational Technology in University Systems (www.cetus.org).

Fair use Example #1

A professor puts a few journal articles on reserve in the library and makes them assigned reading for the class. Some students in the class complain that they cannot get access to the articles because other students always seem to have them checked out. The professor scans them and posts them on his Web site. The professor gives the students in the class the password they need to access the articles.

The first factor to consider is the purpose of the use. In this case the purpose is strictly educational. Hence this factor weighs in favor of fair use.

The second factor is the nature of the work being copied. The journal articles are nonfiction. Again this weighs in favor of fair use.

The third factor is the amount of material being copied. The fact that the professor is copying entire articles rather than brief excerpts weighs against a ruling of fair use.

The fourth factor is the effect the copying will have on the market for journal sales. If the journal issues containing these articles are no longer for sale, then the professor's actions cannot affect the market. The professor took care to prevent people outside the class from accessing the articles. Overall, this factor appears to weigh in favor of fair use.

Three of the four factors weigh in favor of fair use. The professor's actions probably constitute fair use of the copyrighted material.

FAIR USE EXAMPLE #2

An art professor takes slide photographs of a number of paintings reproduced in a book about Renaissance artists. She uses the slides in her class lectures.

The first factor to consider is the purpose of the copying. The professor's purpose is strictly educational. Hence the first factor weighs in favor of fair use.

The second factor is the type of material being copied. The material is art. Hence this factor weighs against a ruling of fair use.

The third factor is the amount of material copied. In this case the professor is displaying copies of the paintings in their entirety. Fair use almost never allows a work to be copied in its entirety. Note that even if the original painting is in the public domain, the photograph of the painting appearing in the art book is probably copyrighted.

The final factor is the effect the copying will have on the market. The determination of this factor would depend on how many images the professor took from any one book and whether the publisher is in the business of selling slides of individual images appearing in its book.

Overall, this professor's actions are less likely to be considered fair use than the actions of the professor in the first scenario.

4.4.1 *Sony v. Universal City Studios*

In 1975 Sony introduced its Betamax system, the first consumer VCR. People used these systems to record television shows for viewing later, a practice called **time shifting.** Some customers recorded entire movies onto videotape.

A year later, Universal City Studios and Walt Disney Productions sued Sony, saying it was responsible for copyright infringements performed by those who had purchased VCRs. The movie studios sought monetary damages from Sony and an injunction against the manufacturing and marketing of VCRs. The legal battle went all the way to the U.S. Supreme Court. The Supreme Court evaluated the case in light of the four fair use factors.

The first factor is the intended purpose of the copying. Since the purpose is private, not commercial, time shifting should be seen as fair use with respect to the first factor.

The second factor is the nature of the copied work. Consumers who are time shifting are copying creative work. This would tend to weigh against a ruling of fair use.

The third factor is the amount of material copied. Since a consumer copies the entire work, this weighs against a ruling of fair use.

The final factor is the affect time shifting will have on the market for the work. The Court determined that the studios were unable to demonstrate that time shifting had eroded the commercial value of their copyrights. The movie studios receive large fees from television stations in return for allowing their movies to be broadcast. Television stations can pay these large fees to the studios because they receive income from advertisers. Advertising rates depend upon the size of the audience; the larger the audience, the more a television station can charge an advertiser to broadcast a commercial. Time shifting allows people who would not ordinarily be able to watch a show to view it later. Hence it can be argued that VCRs actually increase the size of the audience, and since audience size determines the fees studios receive to have their movies broadcast on television, it is not at all clear whether the copying of these programs harms the studios.

The Supreme Court ruled, in a 5-4 decision, that time shifting television programs is a fair use of the copyrighted materials [23]. It said that the private, noncommercial use of copyrighted materials ought to be presumed fair use unless it could be shown that the copyright holder would be likely to suffer economic harm from the consumer's actions (Figure 4.6). Importantly, the Court also noted that the Sony Betamax VCR could be used to copy both copyrighted and noncopyrighted material, and that Sony should not be held accountable if some of the people who buy a VCR choose to use it to infringe on copyrights.

4.4.2 *RIAA v. Diamond Multimedia Systems Inc.*

Diamond Multimedia Systems introduced the Rio portable music player in 1998. About the size of an audio cassette, the Rio stored an hour of digitized music. The Recording Industry Association of America (RIAA) asked for an injunction preventing Diamond Multimedia Systems from manufacturing and distributing the Rio. The RIAA alleged that the Rio did not meet the requirements for the Audio Home Recording Act of 1992, because it did not employ a Serial Copyright Management System to prevent unauthorized copying of copyrighted material.

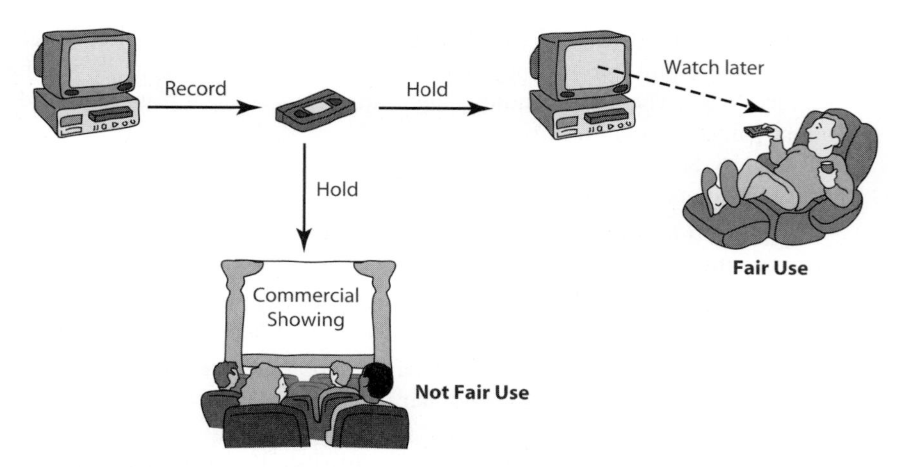

FIGURE 4.6 The Supreme Court ruled that videotaping television broadcasts for private viewing at a later time is fair use of the copyrighted material. This practice is called time shifting. Using videotaped material for a commercial purpose is not considered fair use.

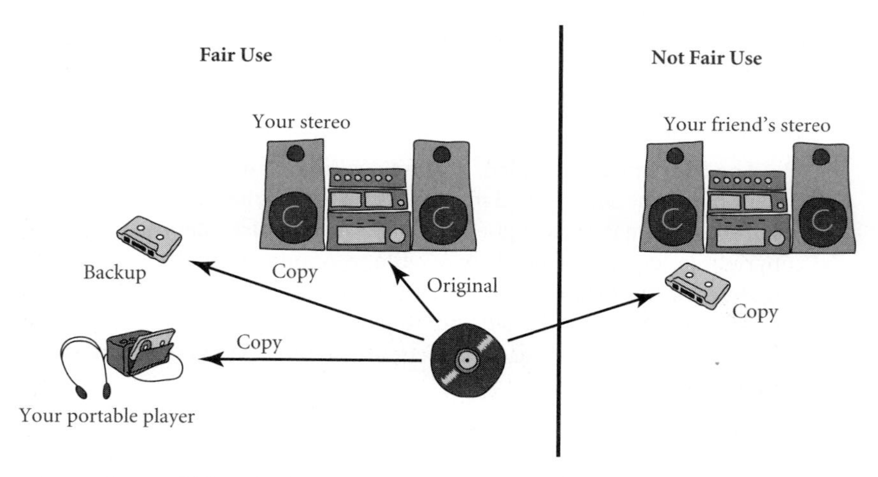

FIGURE 4.7 Space shifting is the creation of a copy for backup purposes or for use in a portable device, and it is considered fair use. Making a copy for a friend is not considered fair use.

The U.S. Court of Appeals, Ninth Circuit, upheld the ruling of a lower court that the Rio was not a digital audio recording device as defined by the Audio Home Recording Act. It denied the injunction on these technical grounds. In addition, the Court affirmed that **space shifting**, or copying a recording in order to make it portable, is fair use and entirely consistent with copyright law (Figure 4.7).

4.4.3 Digital Technology and Fair Use

In the not-so-distant past, music publishers distributed content on vinyl records, and some purchasers made copies onto cassette tapes. The copying process introduced hiss and distortions that significantly degraded the quality of the music. Trying to make a copy from a copy resulted in a nearly worthless tape. Music publishers focused on suing major violators of copyright law (those producing thousands of tapes) and ignored people who made a few copies of albums for their friends [24].

Three technological innovations disrupted the status quo. The first advance was the compact disc (CD). Initially, the introduction of CDs was a huge boon for the music publishing industry. The per-unit production costs of CDs was lower than vinyl albums or tapes, but their sound quality was higher, meaning companies could charge more for them. As a result, their profits swelled.

However, because a CD encodes music digitally—as a stream of ones and zeros—it can be copied perfectly. Now that CD burners are standard equipment on personal computers, millions of people have the ability to clone music CDs. Since each copy is perfect, a copy of a copy is as good as the original. Blank CDs cost only a few cents. CDs sold in stores often cost nearly 20 dollars. Respecting copyright law has become an expensive proposition.

Digital reproduction means people in any country can make perfect copies. Organized crime has entered the lucrative business of manufacturing and selling counterfeit software, CDs, and DVDs in a variety of countries, including Russia and Malaysia (Figure 4.8) [25].

FIGURE 4.8 Counterfeit CDs are destroyed in Thailand. (©Reuters/CORBIS)

The second advance was the creation of the MP3 standard for compressing audio signals. A **compression** algorithm reduces the number of bits needed to store a picture or sound. An MP3 music file is typically less than 10 percent the size of the original file, but it is difficult to hear the difference between the original and the compressed versions. The introduction of MP3 made it practical for people to exchange music over the Internet.

The third technological advance was the increase in the number of people with high-speed Internet connections. While a patient person with an ordinary dial-up connection to the Internet can download large files, connections that are dozens of times faster make file sharing much more practical. As more people got DSL or cable access to the Internet, the number of downloads soared [26].

While most of the focus in the media has been on the illegal sharing of music files, sharing video may soon become much more common. A **personal video recorder** (PVR) allows you to record TV onto your computer. Microsoft is introducing its Media Center PC. This system allows you to record TV shows, play them later without commercials, and burn them onto a DVD. Imagine putting every episode of Seinfield on a $100 hard disk [27]!

4.5 New Restrictions on Use

Legal and technological initiatives are restricting the ability of consumers to use CDs and DVDs, even for purposes that were previously considered fair use, such as making a backup copy. Larry Kenswil of Universal Music Group says, “What we really want to do is not to stop copying, simply to stop redistributing. But the technology available doesn’t distinguish between the two” [24].

4.5.1 Digital Millennium Copyright Act

The Digital Millennium Copyright Act (DMCA), passed by Congress in 1998, was the first major revision of United States copyright law since 1976. The primary purpose of the DMCA was to bring the United States into compliance with international copyright agreements it had signed [22]. Provisions in the DMCA significantly curtail fair use of copyrighted material. For example, the DMCA makes it illegal for consumers to make copies of any digitally recorded work for any purpose [28]. It is illegal to sell (or even discuss online) a software program designed to circumvent copy controls [29].

Online service providers that misuse copyrighted materials face severe penalties [29]. That means, for example, a university that knows students are exchanging MP3 files on the campus network and does nothing to stop them can be sued [30].

The DMCA extends the copyright protection to music broadcast over the Internet. It requires royalty payments to be made to copyright holders of music played over the Internet since October 1998. For example, a college Internet radio station would pay the larger of an annual fee of $500 or $0.0002 per listener per song for every song that it plays. Radio stations are having a hard time determining how much they owe, because most of them have not kept track of how many online listeners they have or the number of songs they have played [31].

4.5.2 Digital Rights Management

Digital rights management (DRM) can refer to any of a variety of actions owners of intellectual property may take to protect their rights. As Christopher May puts it, "all DRM technologies are aimed at tracking and controlling the use of content once it has entered the market" [32]. DRM technologies may be incorporated into a computer's operating system, a program, or a piece of hardware.

One approach to DRM is to encrypt the digital content so that only authorized users can access it. Another approach is to place a digital mark on the content so that a device accessing the content can identify the content as copy-protected.

4.5.3 Secure Digital Music Initiative

The Secure Digital Music Initiative (SDMI) was an effort to create copy-protected CDs and secure digital music downloads that would play only on SDMI-compliant devices. About 200 entertainment and technology companies joined the consortium, which worked for three years to develop "digital watermarks" that would make unauthorized copying of audio files impossible. The SDMI was unsuccessful for three reasons. First, before any copy-protection technologies could be put in place, the number of music files being copied on the Internet mushroomed. Second, some of the sponsors of the SDMI—consumer electronics companies—started making a lot of money selling devices that became more attractive to customers as access to free MP3 files got easier. Their sales could be hurt by restrictions on copying. Third, the digital watermarking scheme was cracked [33].

In September 2000 SDMI issued a "Hack SDMI" challenge. It released some digitally watermarked audio files and offered a $10,000 prize to the first person to crack them. Princeton computer science professor Edward Felten and eight colleagues picked up the gauntlet. Three weeks later the team had successfully read the audio files. The team declined to accept the cash prize. Instead, it wrote a paper describing how it broke the encryption scheme. It prepared to present a paper at the Fourth Annual Information Hiding Workshop at Carnegie-Mellon University in April 2001 [34]. At this point, the Recording Industry Association of America sent Dr. Felten a letter stating "Any disclosure of information gained from participating in the public challenge would be outside the scope of activities permitted by the agreement and could subject you and your research team to actions under the Digital Millennium Copyright Act" [35]. Fearing litigation, Dr. Felten agreed to withdraw the paper from the conference. However, that did not prevent the information from being leaked. Even before the conference, copies of the research paper and the letter from the RIAA were placed on a freedom-of-speech Web site [35]. Four months later Felten's group published the paper [36].

4.5.4 Encrypting DVDs

A DVD (Digital Versatile Disc) is capable of storing a full-length motion picture. DVDs are smaller than videotapes and have higher video and audio fidelity. People can view DVDs on DVD players attached to home entertainment systems; they can also watch DVDs on Windows and Macintosh computers equipped with DVD players.

To prevent unauthorized viewing of DVD movies, the contents of the discs are encrypted using a scheme called the Content Scramble System (CSS), developed by Matsushita and Toshiba. DVD players and DVD drives inside PCs and Macintoshes have a licensed copy of CSS including the decryption keys [37].

In 1999 16-year-old Norwegian Jon Johansen wrote a computer program called DeCSS that decoded the CSS encryption scheme. DeCSS enabled him to view DVD movies on a computer running the Linux operating system, which was not supported by CSS. Johansen distributed the program to others via the Internet.

2600 Magazine published the code and provided links to it. Eight major motion picture studios successfully sued the publisher of *2600 Magazine* for violating the Digital Millennium Copyright Act [38]. In November 2001 a federal appeals court upheld the ruling. The appeals court ruled that while a computer code is "speech," the code enjoys only limited First Amendment protection because its purpose is more "functional" than "expressive." The court held that the publisher's right to post the code on the Internet was outweighed by the potential harm the program could do in the form of increasing the illegal copying of digitally encoded motion pictures [39].

This decision is difficult to understand, since CSS encryption is designed to prevent the unauthorized *viewing* of DVDs, not the unauthorized *copying* of DVDs. Commercial DVDs do not have copy protection, but what is on the DVDs is encrypted. In order to view a DVD, you need to have access to a system that can decrypt its contents. Jon Johansen didn't make it possible to copy DVDs. What he made possible was viewing DVDs on Linux computers [40]. I found this great analogy on the OpenDVD.org Web site: "Look, it's like this—a DVD movie is basically just a message (the movie) written in secret code on a piece of paper. To read the message (watch the movie), you need a secret decoder ring. To be a pirate, you need a photocopier, but you don't need a decoder ring because you don't really care what the secret message is, as long as your photocopier makes nice, crisp copies that your client (who has a decoder ring) can read. All these guys did was make a decoder ring that works under Linux, because all the commercial decoder rings only run on Windows (or stand alone DVD players)."

Jon Johansen was also brought to trial in Norway for creating and distributing DeCSS, but in January 2003 an Oslo City Court acquitted Johansen. The court ruled he had the right to access information on a DVD that he had purchased. It noted the program Johansen developed to decrypt DVDs could be used for both legal and illegal purposes [38].

4.5.5 Making CDs Copyproof

The recording industry is working to develop copyproof CDs [4]. Several systems are being considered. The Israeli company Midbar Tech has developed the Cactus Data Shield. An audio file on a CD protected by the Cactus Data Shield cannot be converted (ripped) into an MP3 file. Sony is developing a CD protection system called key2audio. SunnComm of Phoenix has developed encryption software called MediaCloq [41].

These copyproofing schemes work by exploiting a crucial difference between CD-ROM drives and CD audio players. CD-ROM drives in PCs comply with one standard called Yellow Book. When a CD-ROM drive encounters a bad block of bits, it reads it again. It keeps rereading the block until the data appear to be correct. CD audio players follow another standard, called Red Book. When a CD player encounters a bad block of bits, it skips over it. By deliberately planting bad blocks of data onto a CD, you can make it unreadable by a CD-ROM drive but acceptable to a CD audio player [41].

Another strategy is being pursued by the developers of Macrovision's SafeAudio software. It doesn't prevent copying, but it encodes patterns into the audio data that, when decoded by a computer, translate into annoying sounds that ruin the quality of the song [41].

In their efforts to make it impossible to copy music CDs, entertainment companies face two significant obstacles. The first obstacle is the existence of a large community of programmers who are going to try to "crack" any encryption scheme that is invented. Programmers have already posted on the Internet instructions for breaking the SafeAudio and key2audio schemes.

The second obstacle is that people can record what they hear. As P. J. McNealy says, "Music is ultimately not secure because of the way it is delivered. No matter how secure the music is on a CD, it can always be hacked. All you have to do is put two microphones in front of your computer speakers" [4].

4.5.6 Criticisms of Digital Rights Management

The introduction of DRM technologies has been controversial. Here are four criticisms that have been raised against DRM.

Some experts suggest that any technological "fix" to the problem of copyright abuse is bound to fail. As we have seen in the previous examples, all prior attempts to create encryption or anti-copying schemes have been circumvented.

Others argue that DRM undermines the well-established principle of fair use. Under DRM, a consumer may not be able to make a private copy of a DRM-protected work without making an extra payment, even if he has the right to do so under traditional fair use standards. Selena Kim writes:

> In the analogue world, people go ahead and use the work if they believe themselves entitled to do so. It is only if users are sued for infringement that they invoke the relevant copyright exceptions as defence. In a digital world encapsulated by access control and embedded with copy control, a potential user of a work may have to ask for permission twice: once to access a work, and again to copy an excerpt. The

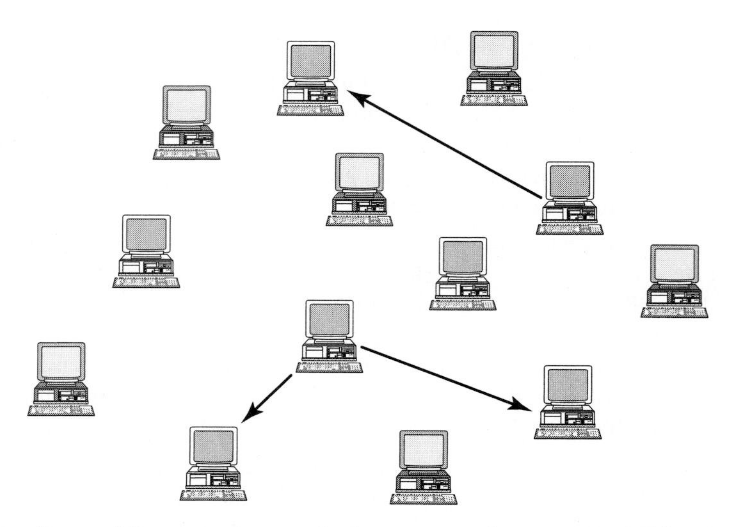

FIGURE 4.9 Some of the computers on the Internet run the same networking program to form a peer-to-peer network. The network supports multiple simultaneous file transfers. The files may contain digitized music, images, computer software, or other content.

> exception to copyright is not being put forward as a defence; it is put forward to show entitlement to use the work. [42]

A third criticism is that DRM technologies could reduce competition. A consortium of large electronic companies could develop a media player based on a new DRM technology and make it difficult for a start-up company to introduce competing product.

Finally, some DRM schemes prevent people from anonymously accessing content. Microsoft's Windows Media Player has an embedded globally unique identifier (GUID). The Media Player keeps track of all the content the user views. When the Media Player contacts Microsoft's central server to obtain titles, it can upload information about the user's viewing habits.

4.6 Peer-to-Peer Networks

On the Internet, the adjective **peer-to-peer** refers to a transient network allowing computers running the same networking program to connect with each other and access files stored on each other's hard drives (Figure 4.9). Peer-to-peer networks stimulate the exchange of data in three ways. First, they give each user access to data stored in many other computers. Second, they support simultaneous file transfers among arbitrary pairs of computers. Third, they allow users to identify those systems that will be able to deliver the desired data more rapidly, perhaps because they have a faster Internet connection or are fewer routing hops away.

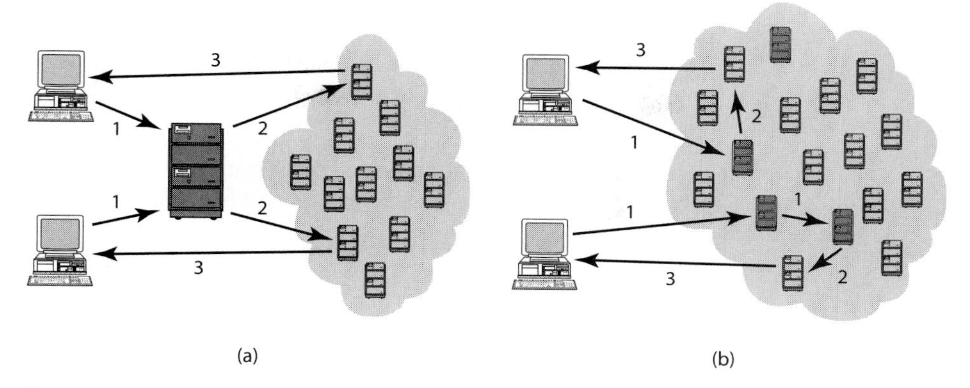

Figure 4.10 Comparison of the Napster and FastTrack implementations of peer-to-peer file sharing. (a) In Napster, a central server maintains the index of all files available for sharing. Retrieving a file is a three-step process: (1) making the request to the central server, (2) establishing a peer-to-peer connection between the sending and receiving computers, and (3) transferring the file. (b) In FastTrack, the index of available files is distributed among many "supernodes." Each supernode has information about files available for sharing on "nearby" computers. Different users connect with different supernodes.

4.6.1 Napster

Napster, which began operation in 1999, was a peer-to-peer network that facilitated the exchange of music files. In December 1999 the RIAA sued Napster for copyright infringement, asking for damages of $100,000 each time a Napster user copied a copyrighted song. In June 2000 the RIAA asked for a preliminary injunction to block Napster from trading any copyrighted content from major record labels. In February 2001 a federal appeals court ruled that Napster must stop its users from trading copyrighted material. Napster put in place file-filtering software that was 99 percent effective in blocking the transfer of copyrighted material. In June 2001 a district court judge ruled that unless Napster could block 100 percent of attempted transfers of copyrighted material, it must disable file transfers. This court order effectively killed Napster, which went off-line in July 2001 and officially shut down in September 2002 [43, 44, 45]. The following year, Napster reemerged as an online subscription music service and music store.

4.6.2 FastTrack

FastTrack is a second-generation peer-to-peer network technology developed by Scandinavians Niklas Zennström and Janus Friis. Because of its decentralized design, a FastTrack network may be more difficult to shut down than Napster [46, 47].

Figure 4.10 illustrates the differences between the Napster and FastTrack implementations of peer-to-peer file sharing. Napster relied upon a central computer to maintain a global index of all files available for sharing. The existence of this central index made it easy to eliminate the distribution of copyrighted files via Napster.

In contrast, FastTrack distributes the index of available files among a large number of "supernodes." Any computer with a high-speed Internet connection running FastTrack has the potential to become a supernode. The use of multiple supernodes makes searching for content slower, but it also makes it much more difficult for legal authorities to shut down the file-sharing network.

Popular peer-to-peer networks KaZaA and Grokster use the FastTrack technology. Morpheus, operated by StreamCast, is based on a different file-sharing technology called Neonet [48].

4.6.3 BitTorrent

For a computer with a broadband connection to the Internet, downloading a file from the network is about ten times faster than uploading a file to the network. A problem with FastTrack and other peer-to-peer networking protocols is that when one peer computer shares a file with another peer computer, the file is transferred at the slower, upload speed rather than the faster, download speed. To solve this problem, Bram Cohen developed BitTorrent [49].

BitTorrent divides a file into pieces about a quarter megabyte in length. Different pieces of a file can be downloaded simultaneously from different computers, avoiding the uploading bottleneck (Figure 4.11). As soon as a user has a piece of a file, the user can share this piece with other users. Since BitTorrent gives a priority for downloads to those users who allow uploading from their machines, users tend to be generous. As a result, downloading speeds increase as more peers get a copy of the file. Put another way, downloading speeds increase with the popularity of a title.

With its markedly higher downloading rates, BitTorrent has made practical the exchange of files hundreds of megabytes long. People are using BitTorrent to download copies of computer programs, television shows, and movies. Linspire, a Linux operating system developer, reduces demand on its servers (and saves money) by using BitTorrent to distribute its software [50]. BitTorrent was also the vehicle by which *Revenge of the Sith* became available on the Internet before it appeared in movie theaters [51].

4.6.4 RIAA Lawsuits

In April 2003 the RIAA warned Grokster and KaZaA users that they could face legal penalties for swapping files containing copyrighted music. The message read, in part:

> It appears that you are offering copyrighted music to others from your computer. . . . When you break the law, you risk legal penalties. There is a simple way to avoid that risk: DON'T STEAL MUSIC, either by offering it to others to copy or downloading it on a 'file-sharing' system like this. When you offer music on these systems, you are not anonymous and you can easily be identified [52].

The RIAA identified the IP addresses of the most active KaZaA supernodes, leading it to the ISPs of users who have stored large numbers of copyrighted files on their computers. Under the terms of the Digital Millennium Copyright Act, the RIAA subpoenaed

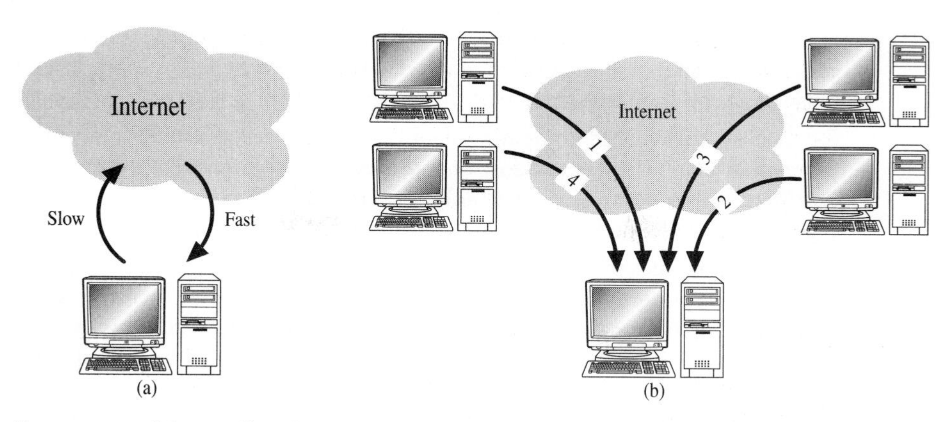

Figure 4.11 (a) Broadband Internet connections provide higher speeds for downloading than for uploading. (b) BitTorrent reduces downloading times by enabling a computer to download different pieces of a file simultaneously from many different peers.

Verizon, asking it to identify the names of customers suspected of running these KaZaA supernodes. Verizon resisted responding to the subpoenas, claiming that responding to the subpoenas would violate the privacy of its customers. In June 2003 a judge in Washington, D.C., ruled that Verizon had to release the names of these customers [53].

In September 2003 the RIAA sued 261 individuals for distributing copyrighted music over the Internet [54]. A month later the RIAA sent letters to 204 people who had downloaded at least 1,000 music files, giving them an opportunity to settle before being sued by the RIAA [55].

In December 2003 the RIAA suffered a setback when the U.S. Court of Appeals for the District of Columbia Circuit ruled that Verizon did not have to respond to the subpoenas of the RIAA and identify its customers [56]. Still, there is some evidence the RIAA lawsuits have reduced illegal file-swapping across the Internet. A survey from ComScore reported activity on KaZaA declined by 15 percent between November 2002 and November 2003 [57]. The Pew Internet & American Life Project reported that the percentage of Internet users who say they download music dropped from 32% in October 2002 to 22% in January 2005, and more than half of the January 2005 downloaders said that they purchased their music from an online service, such as iTunes. However, the report cautioned that because of the stigma associated with illegal downloading, fewer people may be willing to admit they do it. Interestingly, about half of music downloaders said they have gotten music from email, instant messages, or someone else's MP3 player or iPod [58].

4.6.5 MP3 Spoofing

In the meantime, the recording industry wants to make downloading MP3 files from peer-to-peer networks unreliable. Millions of times a day people download MP3 files onto their computers, only to discover that the files are spoofs; in other words, their contents do not match their titles. A spoofed track from a CD, for example, may contain noise or a repetition of a small part of the song.

A New York–based company called Overpeer is responsible for posting spoofs of more than 30,000 songs, videos, and games onto peer-to-peer networks. Dozens of Overpeer employees busily create and post altered MP3 files. The company believes its spoofs are downloaded more than 200 million times a month.

The companies producing peer-to-peer networking software are responding to the onslaught of spoofed files. BearShare and KaZaA software now incorporate user ratings that help people identify the spoofs. Still, the technology research firm Jupiter Research reports that MP3 spoofs do frustrate many users. The music industry hopes users would rather join an online music subscription service than go through the hassle of trying to find free copies of the songs they want [59].

4.6.6 Universities Caught in the Middle

Universities have found themselves at the center of the peer-to-peer file-sharing debate, because many institutions have high-speed Internet connections and plenty of storage space on their file servers, two essential ingredients for setting up a repository of music and video files [60].

In 2003 the RIAA sued students at Michigan Technical University, Princeton University, and Rensselaer Polytechnic Institute, charging them with using their universities' networks to distribute copyrighted music files. The RIAA said that the four students were offering from between 27,000 and a million songs to other students, and it asked for about $100 billion in damages [2]. Notably, the RIAA sued the students even though they were only sharing songs with other students on the same college network. They were not distributing them off campus. All four students settled the lawsuits out of court. None of the students admitted any wrongdoing, but all agreed they would stop distributing copyrighted music. One student agreed to pay the RIAA $12,000, two others agreed to pay $15,000, and the fourth student settled for $17,500 [61].

Universities have responded to the file-swapping problem in a variety of ways. In November 2002 the U.S. Naval Academy seized the personal computers of 92 students suspected of using the institution's Internet connection to download copyrighted materials. In April it punished 85 of these midshipmen. Navy commander Bill Spann said, "This boiled down to holding the next generation of the nation's combat leaders accountable for their actions" [62].

In April 2003 the New Jersey Institute of Technology banned students from using file-sharing programs on its networks after the student senate narrowly approved the policy. Dean of Students Jack Gentul wrote in a memo to students, faculty, and staff that file sharing "put not only the students, but also the university at risk for legal action" [63].

Pennsylvania State University decided to provide students a legal way to download copyrighted music. Using student fees, it signed a deal giving its students access to the

Napster 2.0 service. Beginning in 2003, Penn State students had access to more than a half million music tracks [64]. Since then, dozens of other universities have signed similar agreements with Napster [65].

4.6.7 *MGM v. Grokster*

A group of movie studios, recording companies, music publishers, and songwriters sued Grokster and StreamCast for the copyright infringements of their users. The plaintiffs (henceforth referred to as MGM) sought damages and an injunction against the defendants.

During the discovery phase of the litigation, the following facts were revealed:

- The defendants' networks were used to transfer billions of files every month.
- About 90 percent of the files available on Grokster's FastTrack network were copyrighted.
- Grokster and StreamCast promoted their networks to investors and potential customers as replacements for Napster.
- An internal StreamCast document revealed that StreamCast's executives wanted to have more copyrighted songs available on their network than on competing networks.
- Grokster sent its users a newsletter touting its ability to deliver popular copyrighted songs.
- Grokster and StreamCast provided technical support to users who were having difficulty locating or playing copyrighted content.

A U.S. District Court granted Grokster and StreamCast a summary judgment; that is, it made its decision without a trial based on the facts and evidence collected. According to the judge, "the defendants distribute and support software, the users of which can and do choose to employ it for both lawful and unlawful ends. Grokster and StreamCast are not significantly different from companies that sell home video recorders or copy machines, both of which can be and are used to infringe copyrights" [66]. The judge referred to *Sony v. Universal City Studios*, the Supreme Court's 1984 ruling on the legality of Sony's Betamax VCR. MGM appealed to the U.S. Court of Appeals for the Ninth Circuit, which upheld the ruling.

After another appeal, the U.S. Supreme Court *unanimously* reversed the decision of the lower courts in June 2005. Justice Souter wrote: "The question is under what circumstances the distributor of a product capable of both lawful and unlawful use is liable for acts of copyright infringement by third parties using the software. We hold that one who distributes a device with the object of promoting its use to infringe copyright, as shown by clear expression or other affirmative steps taken to foster infringement, is liable for the resulting acts of infringement by third parties" [67].

The Supreme Court made clear it was not reversing the Sony Betamax decision. Instead, it ruled that the "safe harbor" provided to Sony did not apply to Grokster and StreamCast. The Sony Betamax VCR was primarily used for time-shifting television shows, which the Court found to be a fair use. There was no evidence Sony had done anything to increase sales of its VCRs by promoting illegal uses. Therefore, Sony could not be found liable simply for selling VCRs.

The situation for Grokster and StreamCast was quite different. Both companies gave away their software, but made money by streaming advertisements to users. Advertising rates are higher when the number of users is greater. Hence both companies wanted to increase their user base. They realized the way to do this was to make sure their networks had the content people are interested in downloading. The opinion notes dryly, "Users seeking Top 40 songs, for example, or the latest release by Modest Mouse, are certain to be far more numerous than those seeking a free Decameron, and Grokster and StreamCast translated that demand into dollars... [T]he unlawful objective is unmistakable" [67].

According to the Supreme Court, the Ninth Circuit Court of Appeals erred when it cited *Sony v. Universal City Studios.* The more relevant precedent was *Gershwin Publishing Corporation v. Columbia Artists Management, Inc.* The Supreme Court remanded the case to the Court of Appeals, suggesting that a summary judgment in favor of MGM would be in order.

4.6.8 Legal Music Services on the Internet

Subscription music services, such as Napster, Rhapsody, pressplay, and MusicNet, are an alternative to illegal file-swapping [46]. In February 2003 America Online started an online music service for its 27 million customers. For a monthly fee of $8.95, customers can choose from an online collection of about 250,000 songs. They can download a song for 99 cents. The customer has the right to listen to the downloaded music on no more than two computers and does not have the right to copy the files to other devices or give them to other people [68].

The biggest player in the legal online music business, however, is Apple Computer. Apple announced its own music service, called iTunes Music Store, in April 2003. Rather than requiring consumers to pay a monthly subscription fee, it simply charges 99 cents to download a song. A customer has the right to copy each track to three computers, burn it onto CDs, and download it onto an iPod portable music player [46]. By September 2004 Apple Computer had sold 125 million songs, and its iTunes Music Store was handling about 70 percent of online music purchases [69].

4.7 Protections for Software

The two primary sources for the information in this section are the BitLaw Web site (www.bitlaw.com), created by Daniel A. Tysver of the law firm Beck & Tysver, and *Legal Protection of Digital Information* by Lee Hollaar [70].

In the early days of the computer industry, there was no strong demand for intellectual property protection for software. Most commercial software was produced by the same companies manufacturing computer hardware. They sold complete systems to customers, and the licensing agreements covered use of the software as well as the hardware. Interest in copyrighting software grew with the emergence of an independent software industry in the 1960s.

4.7.1 Software Copyrights

The first software copyrights were applied for in 1964. The Copyright Office allowed the submitted computer programs to be registered, reasoning that a computer program is like a "how-to" book. The Copyright Act of 1976 explicitly recognizes that software can be copyrighted.

When a piece of software gets copyright protection, what exactly is copyrighted? First, copyright protects the expression of an idea, not the idea itself. For example, suppose you develop a program for a relational database management system. You may be able to copyright your implementation of a relational database management system, but you cannot copyright the concept of using relational databases to store information.

Second, copyright usually protects the object (executable) program, not the source program. Typically, the source code to a program is confidential, in other words, a trade secret of the enterprise that developed it. The company only distributes the object program to its customers. The copyright also protects the screen displays produced by the program as it executes. This is particularly valuable for the developers of video games.

4.7.2 Violations of Software Copyrights

The holder of a copyright has a right to control the distribution of the copyrighted material. Obviously, this includes making copies of the program. The definition of what it means to make a copy of a program is broad. Suppose you purchase a program stored on a CD. If you transfer a copy of the program from the CD to a hard disk, you are making a copy of it. If you execute the program, it is copied from the hard disk of the computer into its random access memory (RAM). This, too, is considered making a copy of the program. The standard licensing agreement that comes with a piece of commercial software allows the purchaser of the product to do both of the above-mentioned copying operations.

However, doing any of the following actions without authorization of the copyright holder is a violation of copyright law:

1. Copying a program onto a CD to give or sell to someone else
2. Preloading a program onto the hard disk of a computer being sold
3. Distributing a program over the Internet

Another kind of copyright violation can occur when a company attempts to create software that competes with an existing product. Two court cases illustrate a copyright infringement and fair use of another company's product.

APPLE COMPUTER, INC. V. FRANKLIN COMPUTER CORP.

In the early 1980s Franklin Computer Corp. manufactured the Franklin ACE to compete with the Apple II. The Franklin ACE was Apple II compatible, meaning that programs sold for the Apple II would run on the Franklin ACE without modification. In order to ensure compatibility, the Franklin ACE contained operating systems functions directly copied from a ROM on the Apple II. Apple sued Franklin for infringing on its copyright. The U.S. Court of Appeals for the Third Circuit ruled in favor of Apple Computer, establishing that object programs are copyrightable.

SEGA V. ACCOLADE

Video game–maker Accolade wanted to port some of its games to the Sega Genesis console. Sega did not make available a technical specification for the Genesis console, so Accolade disassembled the object code of a Sega game in order to determine how to interface a video game with the game console. Sega sued Accolade for infringing on its copyright. In 1992 the U.S. Court of Appeals for the Ninth Circuit ruled in favor of Accolade, judging that Accolade's actions constituted fair use of the software. It noted that Accolade had no other way of discerning the hardware interface and that the public would benefit from additional video games being available on the Genesis console.

4.7.3 Software Patents

Until the early 1980s, the U.S. Patent and Trademark Office refused to grant patents for computer software. Its position was that a computer program is a mathematical algorithm, not a process or a machine.

However, a U.S. Supreme Court decision in 1981 forced the Patent and Trademark Office to begin considering software patents. In the case of *Diamond v. Diehr*, the Supreme Court ruled that an invention related to curing rubber could be patented. Even though the company's principal innovation was the use of a computer to control the heating of the rubber, the invention was a new process for rubber molding, and hence patentable.

After this decision, the Patent and Trademark Office began reviewing applications for software patents. In each case, it needed to sort out applications merely describing mathematical algorithms from those describing inventions.

One way the distinction can be made is by looking at the data manipulated by the program. If the software simply manipulates values, such as a program that sorts numbers, it is an expression of a mathematical algorithm and should not be patented. On the other hand, if the software manipulates data representing measurements made in the real world, it is more likely to be a patentable invention. An example of such a piece of software would be a program inside a pacemaker that interprets electrocardiograph signals and determines when to administer an electric shock to the heart.

One problem faced by patent examiners in the Patent and Trademark Office is knowing what the existing technical knowledge (prior art) in computer programming is. Patent examiners typically look at patents already issued to determine prior art. This works fine for other kinds of inventions, but it doesn't work well for software patents, because a significant amount of software was written before software patents were first granted. The consequence is that patent examiners issue many "bad patents"—patents that would not have been issued if the examiner knew about all of the prior art. While organizations such as the Software Patent Institute are trying to collect information about prior art for patent examiners to use, the Patent and Trademarks Office continues to issue bad patents.

Bad patents can lead to increased legal costs for software companies. A software developer accused of patent infringement by another firm holding a bad patent must do its own research into the prior art to demonstrate the patent is invalid. Also, the existence of bad patents makes software patents in general more suspect. A general skepticism about the validity of software patents increases the likelihood that the owner of a software patent will have to defend challenges to the patent mounted by another software developer.

4.7.4 Safe Software Development

An organization must be careful not to violate the copyrights held by its competitors. Even unconscious copying can have serious consequences. Years after hearing the song "He's So Fine," George Harrison wrote "My Sweet Lord." The owner of "He's So Fine" sued Harrison for copyright infringement and prevailed after a lengthy legal battle. Unconscious copying is a real concern in the software industry because programmers frequently move from one firm to another.

Suppose a company needs to develop a software product that duplicates the functionality of a competitor's product without violating the competitor's copyright. For example, in the 1980s companies developing IBM-compatible computers needed to develop their own implementations of the BIOS (Basic Input/Output System). A "clean room" software development strategy helps ensure a company's software program does not duplicate any code in another company's product.

In this strategy, two independent teams work on the project. The first team is responsible for determining how the competitor's program works. It may access the program's source code, if it is available. If it cannot get access to the source, it may disassemble the object code of the competitor's product. It also reads the product's user manuals and technical documentation. The first team produces a technical specification for the software product. The specification simply states how the product is supposed to function. It says nothing about how to implement the functionality.

The second team is isolated from the first team. Members of this team have never seen any code or documentation from the competitor's product. It relies solely on the technical specification to develop, code, and debug the software meeting the specification. By isolating the code developers from the competitor's product, the company developing the competing product can demonstrate that its employees have not copied code, even unconsciously.

4.8 Open-Source Software

In the early years of commercial computing, there was no independent software industry. Computer manufacturers such as IBM produced both the hardware and the software needed for the system to be usable. Well into the 1960s, software distributions included the source code. Customers who wanted to fix bugs in the programs or add new features could do so by modifying the source code and generating a new executable version of the program.

In the 1970s the number of computer applications expanded, and organizations recognized the increasing value of software. To protect their investments in software development, most companies decided to make their programs proprietary (owned).

Today, companies developing proprietary software tightly control the distribution of their intellectual property. Typically they do this by treating source code as a trade secret and distributing only the object code, which is not in human-readable form. In addition, they do not sell the object code. Instead, when people "purchase" the program, what they are actually buying is a license allowing them to run the program. Their rights to do other things with the code, such as make backup copies, are limited.

4.8.1 Consequences of Proprietary Software

Governments have given ownership rights to those who produce computer software because of the perceived beneficial consequences. A key benefit is the ability to profit from the licensing of the software. The assumption is that people will work harder and be more creative if they must compete with others to produce the best product. Those who produce the best products will have the opportunity to make money from them.

While most people point to the benefits of a system encouraging the development of proprietary software, some people have noted the harms caused by such a system. A well-known critic of proprietary software is Richard Stallman. According to Stallman, granting intellectual property rights to creators of computer software has numerous harmful consequences:

- The copyright system was designed for an era in which it was difficult to create copies. Digital technology has made copying trivial. In order to enforce copyrights in the digital age, increasingly harsh measures are being taken. These measures infringe on our liberties.
- The purpose of the copyright system is to promote progress, not to make authors wealthy. Copyrights are not promoting progress in the computer software field.
- It is wrong to allow someone to "own" a piece of intellectual property. Granting someone this ownership forces the users of a piece of intellectual property to choose between respecting ownership rights and helping their friends. When this happens, the correct action is clear. If a friend asks you for a copy of a proprietary program, you would be wrong to refuse your friend. "Cooperation is more important than copyright" [71].

The **open-source movement** is the philosophical position that source code to software ought to be freely distributed and that people should be encouraged to examine and improve each other's code. The open-source software movement promotes a cooperative model of software development.

4.8.2 Open-Source Definition

Open source is an alternative way of distributing software. Licenses for open-source programs have the following key characteristics (there are others) [72]:

1. There are no restrictions preventing others from selling or giving away the software.
2. The source code to the program must be included in the distribution or easily available by other means (such as downloadable from the Internet).
3. There are no restrictions preventing people from modifying the source code, and derived works can be distributed according to the same license terms as the original program.
4. There are no restrictions regarding how people can use the software.
5. These rights apply to everyone receiving redistributions of the software without the need for additional licensing agreements.
6. The license cannot put restrictions on other software that is part of the same distribution. For example, a program's open-source license cannot require all of the other programs on the CD to be open source.

Note there is nothing in these guidelines that says an open-source program must be given away for free. While people may freely exchange open-source programs, a company has the right to sell an open-source program. However, a company cannot stop others from selling it either. In order for a company to be successful selling open-source software that people can find for free on the Internet, it must add some additional value to the software. Perhaps it packages the software so that it is particularly easy to install. It may provide great manuals, or it may provide support after the sale.

The Open Source Initiative (www.opensource.org) is a nonprofit corporation that promotes a common definition of open source. In July 2005 its Web site listed the names of 58 software licenses that met its definition of open source.

4.8.3 Beneficial Consequences of Open-Source Software

Advocates of open-source software describe five beneficial consequences of open-source licensing.

The first benefit of open source is that it gives everyone using a program the opportunity to improve it. People can fix bugs, add enhancements, or adapt the program for entirely new uses. Software evolves more quickly when more people are working on it.

Rapid evolution of open-source software leads to the second benefit: new versions of open-source programs appear much more frequently than new versions of commercial programs. Users of open-source programs do not have to wait as long for bug fixes and patches [73].

A third benefit of open source is that it eliminates the tension between obeying copyright law and helping others. Suppose you legally purchased a traditional license to use a program, and your friend asks you for a copy. You must choose between helping your friend and conforming to the license agreement. If the program had an open-source license, you would be free to distribute copies of it to anyone who wanted it.

The fourth benefit is that open-source programs are the property of the entire user community, not just a single vendor. If a vendor selling a proprietary program decides not to invest in further improvements to it, the user community is stuck. In contrast, a user community with access to the source code to a program may continue its development indefinitely [73].

The fifth benefit of open source is that it shifts the focus from manufacturing to service, which can result in customers getting better support for their software [73]. As Eric Raymond puts it:

> Anybody who has studied software engineering knows that programmers do not actually spend most of their time originating software. They spend most of their time on service updates and maintenance. Nobody thinks about the implications of this: that the software industry is actually a service industry operating under the delusion that it is a manufacturing industry. Software producers are operating under a manufacturing model, under which the way you make money is building a product and getting it out the door. Because they have this model of themselves as a manufacturing industry, all the bright people go to production and the dumb people go to the support desk. That's why when you call a vendor support line you have to fight your way through three layers of idiots to get down to anyone who knows anything.
>
> As long as the software industry continues to misperceive itself as a manufacturing industry, instead of a service industry, reliability is going to be awful. But that shift is not going to happen until source is open. That's the difference between closed and open source.
>
> In the closed source world, your short-term profit incentive is to try and keep everything you do a trade secret and extract the absolute maximum rent from that trade secret in terms of initial cost of the software. And then your economic incentive is to put as little money as you can get away with into supporting the fiction that you support your software. OK? Now as a consumer do you want to live in that world, or do you want to live in a world where source is primarily open and the people competing for your dollars are service bureaus? [74]

4.8.4 Examples of Open-Source Software

Open-source software is a key part of the Internet's infrastructure. Here are a few examples of highly successful programs distributed under open-source licenses:

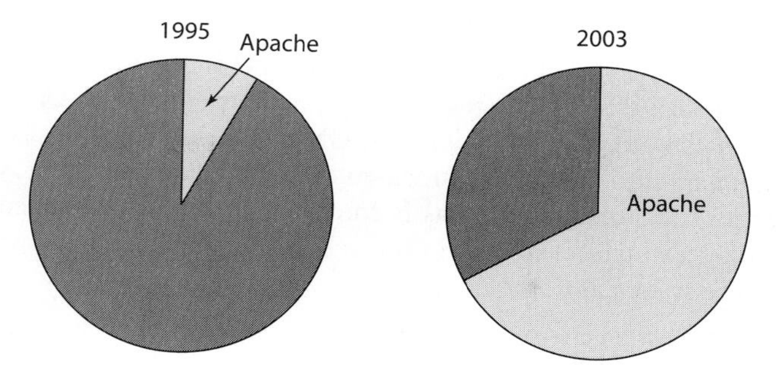

Figure 4.12 Two-thirds of Web servers are now based on Apache open-source software.

- BIND provides DNS (domain name service) for the entire Internet.
- Apache runs more than half of the world's Web servers (Figure 4.12).
- The most widely used program for moving email about the Internet is the open-source program sendmail.
- Perl is the most popular Web programming language.
- Other popular open-source programming languages and tools are Python, Ruby, TCL/TK, PHP, and Zope.
- Programmers have recognized the high quality of the GNU compilers for C, C++, Objective-C, Fortran, Java, and Ada.

Surveys indicate that the quality and dependability of open-source software is about the same as the quality of commercial software [75].

4.8.5 The GNU Project and Linux

The GNU Project and Linux are important success stories in the history of the open-source movement. Richard Stallman began the GNU Project in 1984. (GNU is pronounced "guh-new" with the accent on the second syllable. It's a tradition among hackers to invent recursive acronyms; GNU stands for "GNU's Not Unix.") The goal of the GNU Project was ambitious: to develop a complete Unix-like operating system consisting entirely of open-source software.

In order to be fully functional, a modern operating system must include text editors, command processors, assemblers, compilers, debuggers, device drivers, mail servers, and many other programs. During the late 1980s Stallman and others developed most of the necessary components. The GNU Project also benefited from open-source software previously developed by others, notably Donald Knuth's TEX typesetting system and MIT's X Window System. Most of the software developed as part of the GNU Project is distributed under the GNU Public License, an example of an open-source license. (For technical reasons some programs have been distributed as open-source software under other licenses.)

In 1991 Linus Torvalds began work on a Unix-like kernel he named Linux. (The kernel is the software at the very heart of an operating system.) He released version 1.0 of the kernel in 1994. Because the other major components of a Unix-like operating system had already been created through the GNU Project, Torvalds was able to combine all of the software into a complete, open-source, Unix-like operating system. To the obvious chagrin of Stallman, Linux has become the commonly accepted name for the open-source operating system based on the Linux kernel. (Stallman urges people to refer to the entire system as GNU/Linux [76].)

4.8.6 Impact of Open-Source Software

Andrew Leonard summarized the impact of Linux this way: "Linux is subversive. Who could have thought even five years ago that a world-class operating system could coalesce as if by magic out of part-time hacking by several thousand developers scattered all over the planet, connected only by the tenuous strands of the Internet?" [74].

As a reliable open-source alternative to Unix, Linux is putting price pressure on companies selling proprietary versions of Unix. Many corporations, including Morgan Stanley, Credit Suisse First Boston, Pixar, and the E*Trade Group, have replaced Sun file servers with less expensive "Lintel" boxes—servers running the Linux operating system on Intel-compatible CPUs [77]. A 2004 survey of 140 large North American firms by Forrester Research revealed that slightly more than half of them were using Linux for "mission-critical" applications, and slightly more than half of them were running new applications on Linux [78]. However, another Forrester Research survey resulted in the conclusion that despite the shift toward purchasing Linux servers, large companies would continue to maintain servers running proprietary operating systems [79].

Linux is also putting pressure on Microsoft and Apple, which sell proprietary operating systems for desktop computers. The cost of commodity, off-the-shelf hardware has gotten so low that the cost of a proprietary operating system is a significant portion of the selling price of low-end systems. Retailers such as Wal-Mart have begun offering Linux-equipped PCs for about $300.

While more than 90 percent of the personal computers on people's desktops run Microsoft operating systems, Microsoft is clearly worried about losing market share to Linux. In the summer of 2002, Microsoft sent an email message to senior managers urging them to hold on to government and large institutional accounts at all costs. If a negotiation to renew a licensing agreement looked hopeless, managers were authorized to draw from a special fund enabling them to offer the Microsoft software at large discounts, or even for free. "Under NO circumstances lose against Linux," the memo instructed [80].

4.8.7 Critique of the Open-Source Software Movement

The open-source movement has many detractors. They have raised the following criticisms of the open-source model of software development.

First, if a particular open-source project does not attract a critical mass of developers, the overall quality of the software can be poor [73].

Second, without an "owner," there is always the possibility that different groups of users will independently make enhancements to a software product that are incompatible with each other. The source code to a single program may fork into a multitude of irreconcilable versions. (In reality, this possibility hasn't materialized. Code forking would fragment the developer community, which is bad for everyone. Hence there are incentives to keep a single version of the source code. About 99 percent of Linux distributions have the same source code [73].)

Third, open-source software as a whole tends to have a relatively weak graphical user interface, making it harder to use than commercial software products. This is one explanation why to this point open-source systems have made greater inroads as servers than as desktop systems [73].

Fourth, open source is a poor mechanism for stimulating innovation. Currently, corporations invest billions of dollars developing new software products. By removing the financial reward for creating new software, companies will sharply curtail or even eliminate research and development. They will no longer be a fountain of new programs. The open-source movement has proven it is able to produce alternatives to proprietary programs (for example, StarOffice instead of Microsoft Office), but it has not demonstrated its ability to innovate completely new products.

4.9 Legitimacy of Intellectual Property Protection for Software

Licenses for proprietary software usually forbid you from making copies of the software to give or sell to someone else. These licenses are legal agreements. If you violate the license, you are breaking the law. In this section we are *not* discussing the morality of breaking the law. Rather, we are considering whether as a society we ought to give the producers of software the right to prevent others from copying the software they produce. In other words, should we give copyright and/or patent protection to software?

Rights-based and consequentialist arguments have been given for granting intellectual property protection to those who create software. Let's review and test the strength of these arguments. To simplify the discussion, we'll assume that a piece of software is written by a person. In reality, most software is created by teams, and the company employing the team owns the rights to the software the team produces. However, the logic is the same whether the software creator is an individual or a corporation.

4.9.1 Rights-Based Analysis

Not everyone can write good computer programs, and programming is hard work. Programmers who write useful programs that are widely used by others should be rewarded for their labor. That means they should own the programs they write. Ownership implies control. If somebody creates a piece of software, he or she has the right to decide who gets to use it. Software owners ought to be able to charge others for using their programs. Everybody ought to respect these intellectual property rights.

This line of reasoning is a variation of Locke's natural rights argument we discussed at the beginning of the chapter. It is based on the Lockean notion that mixing your labor with something gives you an ownership right in it.

Here are two criticisms of the "just deserts"[1] argument. First, why does mixing your labor with something mean that you own it? Doesn't it make just as much sense to believe that if you mix your labor with something you lose your labor? Robert Nozck gives this example: If you own a can of tomato juice and pour it in the ocean, mixing the tomato juice with the salt water, you do not own the ocean. Instead, you have lost your can of tomato juice. Certainly it would be unjust if someone else could claim ownership of something you labored to produce, but if there were no notion of property ownership, and everybody understood when they mixed their labor with something they lost their labor, it would be just.

Of course, we do live in a society that has the notion of ownership of tangible property. How can we justify giving a farmer the right to the crop he labors to produce while failing to give a programmer the right to the accounting program he produces for the benefit of the farmer?

Still, if we do want to give ownership rights to those who produce intellectual property, we run into the problems we discussed at the beginning of the chapter. The second criticism of the "just deserts" argument is that Locke's natural rights argument does not hold up well when extended to the realm of intellectual property. There are two crucial differences between intellectual property and tangible property. Each piece of intellectual property is unique, and copying intellectual property is different from stealing something physical.

4.9.2 Utilitarian Analysis

A second argument in favor of providing intellectual property protection for software producers is based on consequences. Failing to provide this protection would have net harmful consequences. The argument goes like this [81]: When software is copied, it reduces software purchases. If less software is purchased, less money will flow to the producers of software. As a result, less new software will be produced. As a whole, new software titles benefit society. When the number of new titles drops, society is harmed. Therefore, when software is copied, society is harmed. Copying software is wrong.

You can view this argument as a chain of consequences (Figure 4.13). Copying software causes software sales to drop, which causes the software industry to decline, which causes fewer products to be released, which causes society to be harmed. Logically, all of the links in the chain must be strong in order for the argument to be convincing. We will look at each of the links in turn, and we'll see that none of them are strong.

1. Pronounced with the accent on the second syllable. Think of the related word "deserve."

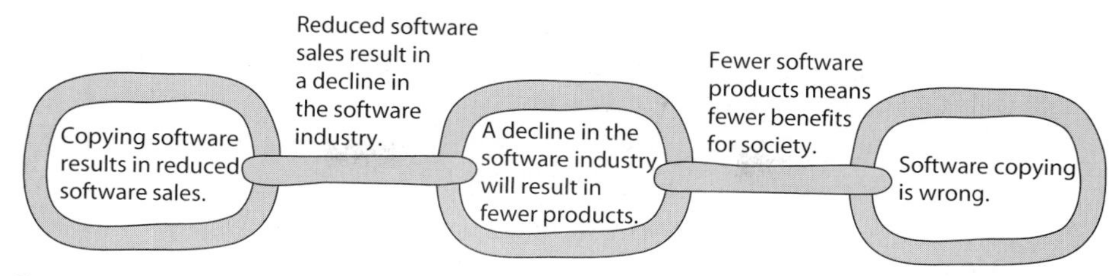

FIGURE 4.13 The chain of reasoning for a consequentialist argument why copying software is bad. (Beth Anderson)

The first claim is that copying software results in reduced sales of software. When talking about software piracy, the computer industry cites the dollar value of the copied software as if each instance of copying represents a lost sale. Obviously this is an exaggeration. Not everyone who gets a free copy of a computer game has the money or the desire to purchase the game for $50. In fact, sometimes software copying may lead to a sale. A person may not have been interested in buying a particular program. After trying it out for free, the person may decide it is so useful she is willing to buy a copy of the program in order to get access to all of the documentation, the technical support line, or another service provided to registered users of the program. It is fair to say that copying software sometimes results in reduced sales of software, but it is not always the case. Hence it is incorrect to make a universal statement.

The second claim is that reduced sales of software will result in a decline in the software industry. An argument against this claim is that Microsoft is flourishing, despite the fact that software counterfeiting is prevalent in some countries. A stronger argument against the claim is that it makes a strong cause-and-effect connection between the creation of software and financial renumeration. However, the open-source movement demonstrates many people are willing to create software without being rewarded financially. Some people write programs because they find it fun. Others are motivated by the desire to gain a good reputation by writing a program many people find useful. Advocates of open-source software, including Richard Stallman, suggest that the best way to stimulate innovation is to allow a free exchange of ideas and source code. From this point of view, allowing software producers to control the distribution of their code stifles, rather than promotes, innovation in the software industry.

Finally, the second claim assumes that software customers are solely responsible for the health of the software industry. In reality, other groups want to ensure that there are plenty of new software titles released. Intel, for example, makes its money from selling CPU chips. Every year the chips are faster. If a person owns a computer fast enough to run his current programs, he has little motivation to upgrade the hardware. However, if that same person purchases a new program that requires additional CPU cycles, he may be motivated to upgrade his computer. Hence it is in Intel's interest to encourage the development of ever more computationally intensive computer programs. Software customers are not solely responsible for promoting the growth of the software industry.

The third claim is that new software packages benefit society. This is a difficult claim to prove. Certainly some programs would benefit society more than others. Hence it's not the number of different programs that matters, it's what they can be used for. The utility of new software titles must be weighed against the utility of letting people give away copies of programs that would help their friends.

4.9.3 Conclusion

We have examined two arguments for why society ought to provide intellectual property protection to software creators. The first argument is based on the notion of just deserts. It is a variation of the natural rights argument we discussed at the beginning of the chapter. This argument is weak; it rests on the faulty assumption that a natural right to own property extends cleanly to intellectual property.

The second argument is based on consequences. It holds that denying intellectual property protection for software would have harmful consequences. It relies upon a chain of cause-and-effect relationships: copying leads to a loss of revenue, which leads to a decline in software production, which harms society. The strength of each of the links in the chain is debatable; taken as a whole, the argument is not strong.

Our conclusion is that the arguments for granting intellectual property protection for software are not strong. Nevertheless, our society *has* granted copyright protection to owners of computer programs. If you violate a licensing agreement by copying a CD containing a computer program and giving it to a friend, you are breaking the law. As we discovered in Chapter 2, from the viewpoint of Kantianism, rule utilitarianism, and social contract theory, breaking the law is wrong unless there is a strong overriding moral obligation.

4.10 Creative Commons

As we saw earlier in this chapter, some believe strong intellectual property protection stimulates creativity by dangling the prospect of financial reward in front of artists and inventors. Others believe that creativity is suppressed in such an environment. They argue that people are more creative when they are free to build on the work of others. Consider music, for example. It's not just rap musicians who sample the works of others to create new songs. Listen to the classical piece *Appalachian Spring* by Aaron Copland and you'll find that he used the Shaker hymn "Simple Gifts."

Information technology has created an environment in which an unprecedented amount of creativity could be unleashed. Never before has it been so inexpensive to record and mix music, combine photographs and computer-generated images, or tape and edit movies. Wouldn't it be great to take what others have done and add your own talents to produce even better works of art for everyone's enjoyment? Quoting the movie *Get Creative* on the Creative Commons Web site: "Collaboration across space and time. Creative co-authorship with people you've never met. Standing on the shoulders of your peers. It's what the Internet is all about" [82].

Strong intellectual property protection, however, stands in the way of this vision. Under current U.S. copyright law, works of intellectual property are copyrighted the moment they are made, even if the creator does not attach a copyright symbol © to the work. Since copyright is implicit, permission is required before use. The current system discourages people from building on the work of others.

Imagine the difficulty an art professor has trying to put together a Web site of images for an online course! She needs to request permission for every image she wishes to display on the Web site. Suppose there are three suitable images of Michelangelo's *Pieta*. It may be impossible for her to tell in advance which, if any, of the photographers would be willing to let her use the image. It would be better if there were an official way for a photographer to say, "It's fine if you use this photograph, as long as you give me credit for taking it."

Stanford law professor Lawrence Lessig realized there was a need for a system that would allow producers of intellectual property to indicate to the world the rights they wanted to keep. Lessig asks us to think about instances of the **commons,** a "resource to which anyone within the relevant community has a right without obtaining the permission of anyone else" [40]. Examples of the commons include public streets, parks, beaches, the theory of relativity, and the works of Shakespeare. Lessig says that "there is a benefit to resources held in common and the Internet is the best evidence of that benefit . . . [T]he Internet forms an *innovation commons*" [40]. The reason Lessig calls the Internet an innovation commons is because its control is decentralized: one person can introduce a new application or new content without getting anyone else's permission.

Lessig joined with Hal Abelson, James Boyle, Eric Eldred, and Eric Saltzman to found the nonprofit corporation Creative Commons in 2001. Creative Commons provides standard copyright licenses free of charge. Every license comes in three forms: human-readable, lawyer-readable, and computer-readable. With a Creative Commons license, you can retain the copyright while allowing some uses of your intellectual property under certain circumstances. Because you have published the circumstances under which your work may be used, others do not have to ask for permission before using your work [82].

How does the system work? Suppose you have taken a photograph and wish to post it on your Web site accompanied by a Creative Commons license. You visit the Creative Commons Web site, which gives you the following menu of choices (from the Creative Commons Web site www.creativecommons.org):

- Attribution. You let others copy, distribute, display, and perform your copyrighted work—and derivative works based upon it—but only if they give you credit.
- Noncommercial. You let others copy, distribute, display, and perform your work—and derivative works based upon it—but for noncommercial purposes only.
- No Derivative Works. You let others copy, distribute, display, and perform only verbatim copies of your work, not derivative works based upon it.

- Share Alike. You allow others to distribute derivative works only under a license identical to the license that governs your work. *Note:* A license cannot feature both the Share Alike and No Derivative Works options. The Share Alike requirement applies only to derivative works.

After you choose the combination of options that describes the rights you want to protect for your photograph, you get three electronic documents from Creative Commons: a plain-language summary of the options you have chosen (Figure 4.14), the official legal document matching your selection, and a computer-readable version of the license. You post your photograph on the Web along with these three documents.

Commercial artists may choose to use Creative Commons licenses to increase exposure to their work. For example, suppose you take a great photograph of the Golden Gate Bridge. You post it on your Web site with a Creative Commons license indicating the photograph may be used for noncommercial purposes as long as the user gives attribution to you. People from around the world think the image is stunning, and they copy it to their own personal Web sites, giving you credit for the photo. A travel agent in a foreign country sees the image and wants to put it on a travel poster. Since this is a commercial purpose, she must gain your permission before using the image. At that time you can negotiate a fair price for its use. Without the widespread distribution of the image through a Creative Commons license, the travel agent might never have seen it.

The computer-readable versions of the licenses are designed to make it easier for search engines to identify content based upon the particular criteria. For example, a history professor might use a search engine hoping to find an image of the Coliseum in Rome that he could include on his Web site. His purpose is noncommercial, and he is happy to credit the photographer, but he does not want to have to pay to display the image or write a letter asking for the photographer's permission. A search engine could return only those images that meet these criteria.

By March 2005 more than 10 million different pieces of intellectual property had been distributed using Creative Commons licenses. John Buckman has used Creative Commons licenses to create an online record label called Magnatune [83]. Magnatune puts complete MP3 albums online for potential customers to preview. Customers who wish to purchase an album choose how much they wish to pay for it (between $5 and $18), with half of the proceeds going to the artists. All of the music distributed under this label is covered by a Creative Commons license that allows listeners to download free music, share it with others, and create derivative works, such as samples and remixes, as long as these uses are noncommercial [84].

Summary

Intellectual property is any unique product of the human intellect that has commercial value. Because our society values property rights, simply calling products of the intellect "intellectual property" creates a bias toward ownership. Some believe the creators of intellectual property have a natural right to own what they create. However, paradoxes

Attribution-NonCommercial-ShareAlike 1.0

You are free:

- to copy, distribute, display, and perform the work
- to make derivative works

Under the following conditions:

Attribution. You must give the original author credit.

Noncommercial. You may not use this work for commercial purposes.

Share Alike. If you alter, transform, or build upon this work, you may distribute the resulting work only under a license identical to this one.

- For any reuse or distribution, you must make clear to others the license terms of this work.
- Any of these conditions can be waived if you get permission from the author.

Your fair use and other rights are in no way affected by the above.

This is a human-readable summary of the Legal Code (the full license).

Disclaimer

FIGURE 4.14 One of the electronic documents returned by the Creative Commons Web site is a plain-language summary of the rights you have chosen to reserve.

occur when we try to extend John Locke's theory of property rights to intellectual property. As we saw in our hypothetical scenarios involving William Shakespeare and Ben Jonson, intellectual property has two characteristics that make it significantly different from ordinary property. First, each creation is unique. That creates a problem when two people independently create the same work. Second, ideas are copied, not stolen. When

I take your idea, you still have it. These paradoxes illustrate that Locke's natural-rights argument for property does not extend to intellectual property. We conclude there are no strong arguments for a natural right to property.

Nevertheless, our society recognizes the value of intellectual property creation. In order to stimulate creativity in the arts and sciences, governments have decided to grant limited ownership rights in intellectual property to its creators. In the United States, there are four different ways in which individuals and organizations can protect their intellectual property: trade secrets, patents, copyrights, and trademarks/service marks.

A trade secret is a confidential piece of intellectual property that provides a company with a competitive advantage. The formula for Coca-Cola is a well-known trade secret. A company may keep a trade secret confidential indefinitely.

A trademark is a word, symbol, picture, sound, color, or smell used by a business to identify goods. A service mark is a mark identifying a service. Xerox is a well-known trademark identifying a brand of photocopy machine. Companies protect their marks to ensure they are used as adjectives rather than common nouns.

A patent gives an inventor the exclusive right to an invention for a period of 20 years. A patent is a public document, and after the patent expires, anyone has the right to make use of its ideas.

A copyright provides authors with certain rights to original works they have written: reproduction, distribution, public display, performance, and production of derivative works. Books, movies, sheet music, songs, and computer software are all protected by copyright. Industries producing products protected by copyright account for 5 percent of the U.S. economy, with over $500 billion in sales. Over time, both the length of copyright protection and the kinds of intellectual property that can be copyrighted have increased significantly. Works created today are protected for the author's lifetime plus 70 years.

The rights given copyright holders are limited. The fair use doctrine allows certain uses of copyrighted works without asking the copyright holder for permission. To determine whether a particular use is fair use, courts consider the purpose of the use (commercial versus noncommercial), the nature of the work being copied (fiction versus nonfiction), how much of the copyrighted work is being used, and how the use will affect the market for the copyrighted work. Two courts cases legitimized time shifting—recording a TV program for viewing later—and space shifting—copying a recording to make it portable.

The introduction of digital technology and the Internet have brought intellectual property issues to the forefront. For the first time, consumers can make perfect copies of audio recordings. In response, new laws and technologies have been put into place to restrict copying. The Digital Millennium Copyright Act has made illegal certain actions previously considered fair use, such as making backup copies of an original disc. Digital rights management technologies are being introduced in an attempt to prevent unauthorized copying or viewing of copyright-protected works. New copy-protection and encryption technologies are often subverted by programmers who break the codes and share their solutions with the world.

John Branch/San Antonio Express-News

Peer-to-peer networks enable people to swap files around the world. Many of these files contain copyrighted songs. Napster facilitated the exchange of music files until it was sued by the Recording Industry Association of America (RIAA). A judge shut down Napster after Napster indicated it could not block 100 percent of attempted transfers of copyrighted material. However, other free file-sharing services such as Grokster and StreamCast took Napster's place. The RIAA has sued individuals who allegedly have distributed large numbers of copyrighted songs via the Internet. These lawsuits have reduced the percentage of Internet users who illegally download music, or at least the percentage of Internet users who are willing to admit to doing it. A diverse group of movie studios, recording companies, music publishers, and songwriters sued Grokster and StreamCast. The U.S. Supreme Court ruled that Grokster and StreamCast could be held liable for the copyright infringements of their users since they had actively promoted these activities.

Until the mid-1960s, there was no intellectual property protection for computer software other than trade secrets. Now, software is protected by copyright law. The case of Apple Computer versus Franklin Computer demonstrates that object code as well as source code is protected by copyright. Some software-based inventions can be patented, but the large number of bad patents for software has limited the effectiveness of software patents.

The open-source movement is an alternative to the more conventional proprietary model of software development. A great deal of the software that keeps the Internet running is open-source software. Nevertheless, because this software is hidden from the ordinary user, most people have not paid much attention to the open-source movement. Even the emergence of the Linux operating system has been quiet, because a company can convert its servers to Linux without changing any computers sitting on employees' desks. The availability of Linux has put price pressure on companies selling proprietary versions of Unix. Linux is also putting pressure on Microsoft and Apple. The cost of off-the-shelf commodity hardware has gotten so low that the cost of a proprietary operating system is a significant portion of the selling price of low-end systems.

We examined the question, "Should we give intellectual property protection to software?" There are both rights-based and utilitarian arguments why we ought to give intellectual property protection to software. The first argument is based on the notion of just deserts. It relies upon a natural right to intellectual property, which as we have seen is a weak right at best. The second argument is based on a chain of consequences: copying leads to a loss of revenue, which leads to a decline in software production, which harms society. Taken as a whole, the second argument is not strong. In short, we concluded the arguments for providing intellectual property protection to software are weak.

The story of the GNU Project and Linux demonstrate how thousands of volunteers can work together to produce high-quality, industrial-strength software. Today, millions of people have access to personal computers, digital cameras, digital recording devices, and the Internet. Why can't the success of GNU/Linux be replicated? Imagine a culture that encouraged the production of new creative works from existing works, a culture in which songs would rapidly evolve, different versions of movies were exchanged and compared, and hypertext novels accumulated links to fan sites.

Existing intellectual property laws have made it difficult to reproduce the success of GNU/Linux in the entertainment arena. Little can be done with a copyrighted work without first asking for permission, a labor-intensive process that puts a drag on innovation. Creative Commons is an effort to streamline the process by allowing copyright holders to indicate up front the conditions under which they are willing to let other people use their work.

Review Questions

1. What is intellectual property? Give 10 examples of intellectual property.
2. Summarize John Locke's explanation why there is a natural right to property.
3. What paradoxes arise when we attempt to extend a natural right to property into the realm of intellectual property?
4. What are the ways in which an individual or firm may protect intellectual property in the United States?
5. What is the difference between a trademark and a trade secret?
6. What are the relative advantages and disadvantages of patents versus trade secrets?
7. When referring to copyrighted materials, what is meant by the term "fair use"?
8. Explain how advances in information technology have made it easier for consumers to violate copyright law.

9. How has the Digital Millennium Copyright Act curtailed fair use of copyrighted material by consumers?
10. What does the term "digital rights management" mean? Describe three different technologies that have been used or proposed for digital rights management.
11. How is it possible to create a CD that produces music in a CD player but is unreadable on a PC's CD-ROM drive?
12. What is a peer-to-peer network?
13. What property makes the peer-to-peer network FastTrack more difficult to shut down than Napster?
14. How does BitTorrent provide an order-of-magnitude increase in downloading speed, compared to KaZaA and Grokster?
15. Research your university's policy on bandwidth abuse and file sharing. Describe the kinds of activities that your university explicitly allows and forbids.
16. What is MP3 spoofing?
17. The U.S. Supreme Court ruled that Sony was not responsible for the copyright infringements of Betamax customers, but Grokster and StreamCast were responsible for the copyright infringements of those who used their peer-to-peer networks. Explain the differences in the two situations that led the Supreme Court to reach this conclusion.
18. Some companies are now providing services that allow consumers to download music legally from the Internet (for a fee). Research the performance of these companies. How many people are using these services? Are these companies making money?
19. Why are patents considered an unreliable way of protecting intellectual property rights in software?
20. Suppose company A wants to develop a program that duplicates the functionality of a program made by company B. Describe how company A may do this without violating the copyrights held by company B.
21. When describing a software license, what does the phrase "open source" mean?
22. How has Linux affected the market for proprietary software?
23. Suppose your band has recorded a song and posted it as an MP3 file on your Web site. How can you allow people to download your music for noncommercial purposes while retaining your copyright on the song?

Discussion Questions

24. Benjamin Franklin created many useful inventions without any desire to receive financial reward. Is intellectual property protection needed in order to promote innovation?
25. Any original piece of intellectual property you have created, such as a poem, term paper, or photograph, is automatically copyrighted, even if you did not label it with a copyright notice. Think about your most valuable piece of copyrighted material. Describe in detail the ownership rights you would like to claim on it.

26. How does the debate over digital music illuminate the differences among ethics, morality, and law?

27. What does the U.S. Supreme Court decision in *MGM v. Grokster* mean for the development of future peer-to-peer network technologies?

28. The current legal system allows both proprietary software and open-source software to be distributed. What are the pros and cons of maintaining the status quo?

29. Examine the analyses of Section 4.9 regarding the legitimacy of providing intellectual property protection for software. Do these arguments apply equally well to the question of providing intellectual property protection for music? Why or why not?

30. Should copyright laws protect musical compositions? Should copyright laws protect recordings of musical performances?

31. Which is more likely to be effective in protecting intellectual property in digital media such as CDs and DVDs: tougher copyright laws or new technologies incorporating more sophisticated anti-copying measures?

In-Class Exercises

32. A plane makes an emergency crash landing on a deserted tropical island. Two dozen survivors must fend for themselves until help arrives. All of them are from large cities, and none of them has camping experience. The survivors find it impossible to gather enough food, and everyone begins losing weight. One person spends a lot of time by himself and figures out how to catch fish. He brings fish back to camp. Others ask him to teach them how to catch fish. He refuses, but offers to share the fish he has caught with the other passengers as long as they take care of the other camp chores, such as hauling fresh water, gathering firewood, and cooking.

    Debate the morality of the bargain proposed by the fisherman. One group should explain why the fisherman's position is not morally justifiable. The other group should explain why the fisherman has done nothing wrong.

33. Survey 10 of your peers. How many own a computer? How many have MP3 files on their computer? How many of these MP3 files were purchased?

34. Is it right for Pennsylvania State University to use student fees to pay for the campuswide Napster license even though not all Penn State students are interested in using this service?

Further Reading

Justin Hughes. "The Philosophy of Intellectual Property." *The Georgetown Law Review*, 77, pp. 287–366, 1988.

Lawrence Lessig. *The Future of Ideas: The Fate of the Commons in a Connected World*. Random House, New York, NY, 2001.

Michael C. McFarland. "Intellectual Property, Information, and the Common Good." In *Readings in CyberEthics*, edited by Richard A. Spinello and Herman T. Tavani. Jones and Bartlett, Boston, MA, 2001, pp. 252–262.

Eric S. Raymond. *The Cathedral & the Bazaar: Musings on Linux and Open Source by an Accidental Revolutionary*. O'Reilly & Associates, Sebastopol, CA, 1999.

Paula Samuelson. "Good News and Bad News on the Intellectual Property Front." *Communications of the ACM*, 42(3), pp. 19–24, 1999.

Richard M. Stallman. *Free Software, Free Society: Selected Essays of Richard M. Stallman*. Edited by Joshua Gay. GNU Press, Boston, MA, 2002.

Siva Vaidhyanathan. *Copyrights and Copywrongs: The Rise of Intellectual Property and How It Threatens Creativity*. New York University Press, New York, NY, 2001.

Shelly Warwick. "Is Copyright Ethical? An Examination of the Theories, Laws, and Practices Regarding the Private Ownership of Intellectual Work in the United States." In *Readings in CyberEthics*, edited by Richard A. Spinello and Herman T. Tavani. Jones and Bartlett, Boston, MA, 2001, pp. 263–279.

References

[1] Phil Kloer. "Downloading Music Brings Issues of Law and Morality." *Cox News Service*, December 19, 2002.

[2] Scott Carlson. "Recording Industry Sues 4 Students for Allegedly Trading Songs within College Networks." *The Chronicle of Higher Education*, April 18, 2003.

[3] Adamson Rust. "'RIAA Hit by 'EFFing' Music Campaign." *The Inquirer IT*, June 30, 2003.

[4] David Kushner. "Music." *Technology Review*, November 2002.

[5] "Phonies Galore." *The Economist*, November 8, 2001.

[6] The University of Texas at Arlington, Office of Technology Transfer. "Intellectual Properties." www.uta.edu/tto/ip-defs.htm.

[7] John Locke. *Two Treatises of Government*. Cambridge University Press, Cambridge, England, 1988.

[8] Michael Scanlan. "Locke and Intellectual Property Rights." Technical report, Oregon State University, Philosophy Department, 2003.

[9] Edmund S. Morgan. *Benjamin Franklin*. Yale University Press, New Haven, CT, 2002.

[10] Randolph P. Luck. "Letter to The Honorable Senator Spencer Abraham." Luck's Music Library, June 28, 1996.

[11] snopes.com. "Happy Birthday, We'll Sue." *Urban Legends Reference Pages*, August 11, 2002. www.snopes.com.

[12] Neill A. Levy. "The Rights and Wrongs of Copyright." *CINAHLnews*, 15(1), Spring 1996.

[13] www.freeadvice.com. "What Is a Trademark?"

[14] John Case. "Snapshots in Legal Drama: Polaroid Inventor vs. Kodak." *The Christian Science Monitor*, October 21, 1981.

[15] David E. Sanger. "Kodak Infringed on Polaroid Patents." *The New York Times*, September 14, 1985.

[16] Lawrence Edelman. "Kodak Pays Polaroid $925m; Part of a Surprise Out-of-Court Settlement Ends 15-Year Legal Hassle." *The Boston Globe*, July 16, 1991.

[17] International Intellectual Property Alliance, Washington, DC. "IIPA Economic Study Reveals Copyright Industries Remain a Driving Force in the U.S. Economy." Press release, April 22, 2002.

[18] United States Court of Appeals for the Second Circuit. *Gershwin Publishing Corporation, Plaintiff-Appelle, v. Columbia Artists Management, Inc., Defendant-Appellant, and Community Concerts, Inc., Defendant*, May 24, 1971. 443 F.2d 1159.

[19] United States District Court for the Southern District of New York. *Basic Books, Inc. v. Kinko's Graphics Corporation*, March 28, 1991. 758 F. Supp. 1522.

[20] "Millbury Man Pleads Guilty in 'Davey Jones' Computer Case." *Worcester (MA) Telegram & Gazette*, December 16, 1994.

[21] Siva Vaidhyanathan. *Copyrights and Copywrongs: The Rise of Intellectual Property and How It Threatens Creativity*. New York University Press, New York, NY, 2001.

[22] Kathleen Amen, Tish Keogh, and Necia Wolff. "Digital Copyright." *Computers in Libraries*, May 2002.

[23] Supreme Court of the United States. *Sony Corporation of American et al. v. Universal City Studios, Inc., et al.*, January 17, 1984. 464 U.S. 417.

[24] "The Copyright Wars." *IEEE Spectrum*, pages 21–23, May 2003.

[25] "New Target for Copyright Enforcement: Organized Crime." *CNet News.com*, March 13, 2003.

[26] Steven Andersen. "How Piracy, Culture and High-Tech Hackers Brought the Recording Indsutry to Its Knees." *Corporate Legal Times*, November 2002.

[27] Farhad Manjoo. "Replay It Again, Sam." *Salon.com*, December 9, 2002.

[28] Roy Mark. "Boucher Introduced Fair Use Rights Bill." *internetnews.com*, January 8, 2003.

[29] Royal Van Horn. "The Digital Millenium Copyright Act and Other Egregious Laws." *Phi Delta Kappan*, November 2002.

[30] Kelly McCollum and Peter Schmidt. "How Forcefully Should Universities Enforce Copyright Law on Audio Files?" *The Chronicle of Higher Education*, November 11, 1999.

[31] Dan Carnevale. "Some College Radio Stations Struggle to Determine Webcasting Payments." *The Chronicle of Higher Education*, November 11, 2002.

[32] Christopher May. *First Monday*, 8(11), November 2003.

[33] Ron Harris. "Where's SDMI? Digital Music Protection Effort Flames Out." *Associated Press*, April 29, 2002.

[34] Kevin Coughlin. "Cyber Music Makers Seek to Gag Code-Breakers." *Newhouse News Service*, April 24, 2001.

[35] John Markoff. "Scientists Drop Plan to Present Music-Copying Study That Record Industry Opposed." *The New York Times*, April 27, 2001.

[36] "A Speed Bump vs. Music Copying; Master Cryptographer—and Code Cracker—Edward Felten Says Technology Isn't the Answer to Digital Copyright Violations." *Business Week Online*, January 9, 2002.

[37] Richard A. Spinello and Herman T. Tavani. "Notes on the DeCSS Trial." In *Readings in CyberEthics*, edited by Richard A. Spinello and Herman T. Tavani. Jones and Bartlett, Sudbury, MA, 2001.

[38] Gillian Law. "Defendant Acquitted in DVD Hacking Case." *IDG News Service*, January 7, 2003. www.pcworld.com.

[39] Donna Euben. "Talkin' 'bout a Revolution? Technology and the Law." *Academe*, May/June 2002.

[40] Lawrence Lessig. *The Future of Ideas: The Fate of the Commons in a Connected World*. Random House, New York, NY, 2001.

[41] Michael Jay Geier. "For Your Ears Only." *IEEE Spectrum*, pages 25–26, May 2003.

[42] S. Kim. "The Reinforcement of International Copyright for the Digital Age." *Intellectual Property Journal*, 16:93–122, 2003.

[43] Danielle Roy. "Napster Timeline." *IDG News Service, Boston Bureau*, April 2, 2001.

[44] Nathan Ruegger. "Napster Withers Away, but Peer-to-Peer Legacy Remains." *The Dartmouth*, October 3, 2002.

[45] Ron Harris. "Bankruptcy Judge Blocks Sale of Napster to Bertelsmann." *Associated Press*, September 4, 2002.

[46] "How to Pay the Piper." *The Economist*, May 1, 2003.

[47] "FastTrack." *Wikipedia*, July 1, 2005.

[48] John Borland. "Super-Powered Peer to Peer." *CNet News.com*, October 6, 2004.

[49] Clive Thompson. "The BitTorrent Effect." *Wired*, page 150, January 2005.

[50] Krysten Crawford. "BitTorrent as Friend, Not Foe." *CNN.com*, April 30, 2005.

[51] "Authorities Strike Back at 'Star Wars' Pirates." *Associated Press*, May 25, 2005.

[52] Reuters. "Music Industry Sends Warnings to Song Swappers." *NYTimes.com*, April 29, 2003.

[53] Phil Hardy. "Verizon Agrees to Give the RIAA the Names of Four Subscribers Alleged to be File-Sharing Copyrighted Works." *Music & Copyright*, June 11, 2003.

[54] Sara Calabro. "RIAA Lawsuits—Music Industry Mistakes Its Lawsuits for a PR Maneuver." *PR Week*, September 22, 2003.

[55] Phil Hardy. "Media Reaction to RIAA's New Round of Lawsuits Less Hostile but Grassroots Oppposition Grows." *Music & Copyright*, October 29, 2003.

[56] John Schwartz. "Record Industry May Not Subpoena Online Providers." *NYTimes.com*, December 19, 2003.

[57] Kevin Fitchard. "Verizon Gains Upper Hand in RIAA Subpoena Ruling." *Telephony*, January 12, 2004.

[58] Mary Madden and Lee Rainie. "Music and Video Downloading Moves beyond P2P." Pew Internet & American Life Project. Report, March 23, 2005. www.pewinternet.org/reports.

[59] David Kushner. "Digital Decoys." *IEEE Spectrum*, page 27, May 2003.

[60] Reuters. "Music Industry Drops Anti-piracy Pamphlets on Campus." *NYTimes.com*, March 27, 2003.

[61] Associated Press. "4 Students Settle File-Swapping Lawsuit." *NYTimes.com*, May 1, 2003.

[62] Vincent Kiernan. "Naval Academy Punishes 85 Students for Downloads, but Won't Describe Penalties." *The Chronicle of Higher Education (Daily News)*, April 16, 2003.

[63] Scott Carlson. "New Jersey Institute of Technology Prohibits File Sharing on Campus." *The Chronicle of Higher Education*, May 1, 2003.

[64] "Penn State Launches Napster Music Service." *Associated Press*, January 13, 2004.

[65] Jefferson Graham. "Students Score Music Perks as Colleges Fight Piracy." *USA Today*, August 24, 2004.

[66] John Borland. "Judge: File-Swapping Tools Are Legal." *CNet News.com*, April 25, 2003.

[67] Supreme Court of the United States. *Metro-Goldwyn-Mayer Studios Inc. et al. v. Grokster, Ltd., et al.*, June 27, 2005. 545 U.S.

[68] Sual Hansell. "E-music Settles on Prices. It's a Start." *NYTimes.com*, March 3, 2003.

[69] Frank Ahrens. "Microsoft Steps into the Ring." *washingtonpost.com*, September 4, 2004.

[70] Lee A. Hollaar. *Legal Protection of Digital Information*. The Bureau of National Affairs, Washington, DC, 2002.

[71] Richard P. Stallman. "Why Software Should Not Have Owners." *GNU Project Web Server*, June 17, 2003. www.gnu.org/philosophy/why-free.html.

[72] *The Open Source Definition*, 2003. www.opensource.org/docs/definition.php.

[73] Carolyn A. Kenwood. "A Business Case Study of Open Source Software." Technical report, The MITRE Corporation, Bedford, MA, July 2001.

[74] Andrew Leonard. "Let My Software Go!" *Salon.com*, April 14, 1998.

[75] Stephen Shankland. "Study Lauds Open-Source Code Quality." *Cnet News.com*, February 19, 2003.

[76] Richard P. Stallman. "Linux and the GNU Project." *GNU Project Web Server*, December 14, 2002. www.gnu.org/gnu/linux-and-gnu.html.

[77] Gary Rivlin. "McNealy's Last Stand." *Wired*, July 2003.

[78] Brad Day, Laura Koetzle, and Carey Schwaber. "Linux Crosses into Mission-Critical Apps." Technical report, Forrester Research, April 26, 2004. www.forrester.com.

[79] Brad Day, Frank E. Gillett, and Richard Fichera. "Firms Plan to Maintain Windows, Add Linux OS." Technical report, Forrester Research, June 18, 2004. www.forrester.com.

[80] Thomas Fuller. "How Microsoft Warded Off Rival." *International Herald Tribune*, May 15, 2003.

[81] Helen Nissenbaum. "Should I Copy My Neighbor's Software?" In *Computers, Ethics, & Social Values*, edited by Deborah G. Johnson and Helen Nissenbaum. Prentice Hall, Englewood Cliffs, NJ, 1995.

[82] Creative Commons, Stanford Law School. *Get Creative (movie)*. www.creativecommons.org.

[83] Ariana Eunjung Cha. "Creative Commons Is Rewriting Rules of Copyright." *Washington Post*, March 15, 2005.

[84] "What Is 'Open Music'?" *Magnatune (web site)*, July 7, 2005.

AN INTERVIEW WITH

Wendy Seltzer

Wendy Seltzer is an attorney and special projects coordinator with the Electronic Frontier Foundation, where she specializes in intellectual property and free speech issues. For the 2005–06 term, she is a visiting professor of law at the Brooklyn Law School. As a Fellow with Harvard's Berkman Center for Internet & Society, Wendy founded and leads the Chilling Effects Clearinghouse, helping Internet users to understand their rights in response to cease-and-desist threats. Prior to joining EFF, Wendy taught Internet Law as an Adjunct Professor at St. John's University School of Law, and practiced intellectual property and technology litigation with Kramer Levin in New York. Wendy speaks frequently on copyright, trademark, open source, and the public interest online. She has an A.B. from Harvard College and J.D. from Harvard Law School, and occasionally takes a break from legal code to program with Perl.

What motivated you to become involved with Internet law?

I had my first glimpse of the Internet in college—text-only, through a dial-up modem, but I was hooked. By the time I got to law school, the Internet had expanded to graphical Web browsers and streaming audio, and people were starting to ask whether off-line law could adapt. At Harvard's Berkman Center for Internet & Society, I got to examine first-hand questions of privacy in a networked environment or of copyright protection for material that could be copied at the click of a mouse. I wanted to help the courts understand the technologies they were increasingly asked to judge.

As a staff attorney for the Electronic Frontier Foundation, what are your principal activities?

EFF attorneys do a mix of litigation, policy work, and education, so I spend some of my time writing briefs and policy analyses, and some out at conferences explaining our positions. I also coordinate a few projects at the law-tech intersection, such as helping people to build high-definition digital television recorders so they can see what a broadcast flag rule would block (http://www.eff.org/broadcastflag/).

What is the most satisfying case you have worked on for the EFF?

Online Policy Group v. Diebold, the first case to win damages for misuse of the Digital Millennium Copyright Act's notice and takedown procedure (17 U.S.C. 512). When emails describing flaws in Diebold electronic voting machines leaked and voting activists began posting them on Websites, Diebold responded with copyright claims to the Internet service providers. Most providers, including colleges, took down the emails, but one ISP and a group of Swarthmore College students wanted to fight for their fair-use rights. In order to protect this democratic debate, we sued Diebold and won, setting a precedent that others can use to protect their own online speech: Companies can't use copyright to block public debate.

What is the purpose of the Chilling Effects Clearinghouse?

Chilling Effects, (http://www.chillingeffects.org/), aims to help people understand their online rights by collecting and annotating cease-and-desist letters sent regarding online activity, and by building a set of frequently asked questions and answers to explain the law in non-technical terms. We give people a place to turn when they get takedown demands, a resource to help them evaluate whether these are valid demands or just lawyers' threats attempting to chill lawful speech. We're also using the database of cease-and-desists to understand how law is used in the real world, outside the courtroom.

Given current U.S. laws as interpreted by the courts through May 2005, under what circumstances can I legally make a copy of a copyrighted music CD or movie DVD I have purchased?

Somewhat surprisingly, the answer differs between CDs and DVDs. For CDs, it is generally accepted that making a personal copy is fair use, so you can "rip" a CD to your hard drive or digital music player, or "burn" a copy to listen to in a car stereo. The Audio Home Recording Act even says that you can't be sued for non-commercial use of special audio recording media. But because DVDs are encrypted, they're covered by the Digital Millennium Copyright Act's "anticircumvention" prohibitions. Courts have held that using unlicensed software to "access" a DVD violates the DMCA—even if it would be to make fair use of the underlying content. The DMCA even makes it unlawful to play a DVD on a machine running GNU/Linux, because there's no licensed player for that operating system.

A few days after the release of The Revenge of the Sith in movie theaters (and its concurrent distribution on BitTorrent), the FBI and the U.S. Department of Homeland Security (DHS) shut down the Elite Torrents network. Please comment on this action.

We need to distinguish between technologies and their unlawful uses. BitTorrent as a program and peer-to-peer protocol has numerous lawful uses: I've downloaded many Linux distributions and independent videos via BitTorrent. But those using the software to distribute copyrighted movies they don't have the rights to are breaking the law and can expect to face suits (even if ultimately there are better ways than lawsuits of addressing new means of distribution). I can't say why this case was a DHS priority, but it does show that copyright enforcers don't need to shut down makers of general purpose software, as the RIAA and MPAA have been trying to do with their lawsuits against Grokster and Morpheus.

Isn't it important to have strong intellectual property protection in order to stimulate the creativity of artists and inventors?

It's important to have strong intellectual property protection balanced with strong protection for innovation and for public access to culture. When copyright law gets out-of-balance, it prevents technologists from developing video jukeboxes and stops home critics from adding movie clips to their reviews, rather than promoting "the progress of science and useful arts." A well-balanced copyright law recognizes both protections and their limits, like fair use.

Do you think U.S. copyright laws, or the interpretation of these laws by the courts, will change significantly in the next decade?

A decade may be too short a timeframe. I think the copyright lockdown will get worse before it gets better. A few developments give me long-term hope, though: Free Software licenses and Creative Commons help thousands of programmers and artists to say they encourage re-use of their works; cheaper computing power puts digital artistry into more hands; and the Internet's peer-to-peer distribution lets independent artists publish their own music and writing. Sure there will be lots of junk, but we'll also see some artistic gems that don't demand or depend on hundred-year exclusive rights.

CHAPTER

5 Privacy

Count not him among your friends who will retail your privacies to the world.

–Publilius Syrus (100 BCE)

5.1 Introduction

Computers and the Internet have accelerated the rate at which organizations can collect, exchange, combine, and distribute information about individuals. All these capabilities make it a challenge to preserve your privacy.

Are you familiar with Google's Phonebook service? If you visit the Google Web site and type my phone number in as a query, it returns a page giving my name and address. Click on a link, and the screen shows a map to my house. Phonebook can do this because it's easy for a computer to combine information from multiple sources, as long as they share a key. In this case the common key is my home address. Given my phone number, Phonebook accesses telephone directory records to learn my address. It can then consult a geographic information system to determine my home's location from its address [1].

Someone in campus security at Georgetown University accidentally sent out to the entire campus community an email crime report containing the names of three students. To protect these students, system administrators shut down the email system at Georgetown University for several hours and deleted the offending email from the mailboxes of the recipients [2]. This incident illustrates the power of modern communication networks to broadcast personal information at high speed. It's also a reminder that system administrators have the ability to read our email messages.

In 1993 Maryland created a database containing medical records of its residents. The purpose of the database was to help the state find ways to contain health care costs. A member of Maryland's public health commission, who happened to be a banker, had access to the database. He used this information to call in the loans of his customers who had cancer [3].

On the morning of July 18, 1989, actress Rebecca Schaeffer opened the door to her apartment and was shot to death by obsessed fan Robert Bardo. Bardo got Schaeffer's home address from a private investigator who purchased her driver's license information from the California Department of Motor Vehicles [4]. In response to this murder, the U.S. Congress passed the Driver's Privacy Protection Act in 1994. The law prohibits states from revealing certain personal information provided by drivers in order to obtain licenses. *It also requires states to provide this information to the federal government.*

In this chapter we focus on privacy issues related to the introduction of information technology. We begin by taking a philosophical look at privacy. What is privacy exactly? Do we have a natural right to privacy distinct from other rights, such as the right to property and the right to liberty? What about our need to know enough about others that we can trust them? How do we handle conflicts between the right to privacy and the right to free expression?

We then survey some of the ways that we leave an "electronic trail" of information behind us as we go about our daily lives. Both private organizations and governments construct databases documenting our activities. A variety of laws have been passed to regulate the collection and distribution of information gathered by private and public entities. We will study what these laws do—and don't do—to protect individual privacy.

With new technologies have come new ways for governments to intercept the communications of their citizens. We examine the history of covert electronic surveillance by the U.S. government, and how the Fourth Amendment to the Constitution has put boundaries around the surveillance activities of law enforcement organizations. Since 1968, Congress has passed a variety of laws allowing various forms of surveillance by law enforcement and intelligence agencies. Most notable is the USA PATRIOT Act, passed after the September 11, 2001, hijacking of four passenger airliners. Because the Patriot Act gives the government many new powers to collect information, it has generated controversy. We'll examine the major provisions of the Patriot Act and the concerns raised by its detractors.

Next, we take a look at data mining, an important tool for building profiles of individuals and communities. Companies use data mining to improve service and target product marketing to the right consumers. Governments use it to fight crime and enhance national security. The Total Information Awareness program is another response to the terrorist attacks of September 11. Like the Patriot Act, this program has raised the ire of privacy advocates.

Identity theft is an increasingly common crime. We describe a variety of ways in which thieves steal credit card numbers and other personal and financial information. The Social Security number has become a commonly used identifier; we study its weaknesses. Some have proposed the creation of a new national identification card for the Unites States. We consider the arguments in favor and against this idea, and we discuss the implications of the REAL ID Act of 2005.

How can people preserve their privacy in the Information Age? One powerful tool is encryption. New encryption technology allows individuals to send messages that are very difficult, if not impossible, for others to decipher. Another tool is digital cash, a technology enabling people to make anonymous transactions in the Information Age. We survey both of these technologies, as well as the attempts by the U.S. government to prevent strong encryption software from being exported to foreign countries.

5.2 Perspectives on Privacy

5.2.1 Defining Privacy

Philosophers struggle to define privacy. Discussions about privacy revolve around the notion of *access*, where access means either physical proximity to a person or knowledge about that person. There is a tug of war between the desires, rights, and responsibilities of a person who wants to restrict access to himself, and the desires, rights, and responsibilities of outsiders to gain access.

Edmund Byrne takes the point of view of the individual seeking to restrict access when he defines privacy as a "zone of inaccessibility" that surrounds a person [5]. You have privacy to the extent that you can control who has access into your zone of inaccessibility. For example, you exercise your privacy when you lock the door behind you when using the toilet. You also exercise your privacy when you choose not to tell the clerk at the video store your Social Security number. However, privacy is not the same thing as being alone. Two people can have a private relationship. It might be a physical relationship, in which each person lets the other person become physically close while excluding others. It might be an intellectual relationship, in which they exchange letters containing private thoughts.

When we look at privacy from the point of view of outsiders seeking access, the discussion revolves around where to draw the line between what is private and what is public (known to all). As Edward Bloustein has pointed out, stepping over this line and violating someone's privacy is an affront to that person's dignity [6]. You violate someone's privacy when you treat him or her as a means to an end. Put another way, some things ought not to be known. Suppose a friend invites you to see a cool movie trailer available on the Web. You follow him into the computer lab. He sits down at an available computer and begins to type in his login name and password. While it is his responsibility to keep his password confidential, it is also generally accepted that you ought to avert your eyes when someone is typing in their password. Another person's password is something that you should not know.

On the other hand, society can be harmed if individuals have too much privacy. Suppose a group of wealthy white, Anglo-Saxon, Protestant men forms a private club. The members of the club share information with each other that is not available to the general public. If the club facilitates business deals among its members, it may give them an unfair advantage over others in the community who are just as capable of fulfilling the contracts. In this way privacy can encourage social and economic inequities, and the public at large may benefit if the group had less privacy (or its membership were more diverse).

Here is another example of a public/private conflict, but this one focuses on the privacy of an individual. Most of us distinguish between a person's "private life" (what they do at home) and their "public life" (what they do at work). In general, we may agree that people have the right to keep outsiders from knowing what they do away from work. However, suppose a journalist learns that a wealthy candidate for high public office has lost millions of dollars gambling in Las Vegas. Does the public interest outweigh the politician's desire for privacy in this case?

In summary, privacy is a social arrangement that allows individuals to have some level of control over who is able to gain access to their physical selves and their personal information.

5.2.2 Harms and Benefits of Privacy

HARMS OF PRIVACY

Giving people privacy can result in harm to society. Some people take advantage of privacy to plan and carry out illegal or immoral activities. Ferdinand Schoeman observes that most wrongdoing takes place under the cover of privacy [7].

Edmund Leach suggests that increasing privacy has caused unhappiness by putting too great a burden on the nuclear family to care for all of its members. He notes that in the past, people received moral support not just from their immediate family, but also from other relatives and neighbors. Today, by contrast, families are expected to solve their own problems, which puts a great strain on some individuals [8].

On a related note, family violence leads to much pain and suffering in our society. Often, outsiders do not even acknowledge that a family is dysfunctional until one of its members is seriously injured. One reason dysfunctional families can maintain the pretense of normality as long as they do is because our culture respects the privacy of each family [9].

Humans are social beings. Most of us seek some engagement with others. The poor, the mentally ill, and others living on the fringes of society may have no problem maintaining a "zone of inaccessibility," because nobody is paying any attention to them. For outcasts, privacy may be a curse, not a blessing.

BENEFITS OF PRIVACY

According to Morton Levine, socialization and individuation are both necessary steps for a person to reach maturity. Privacy is necessary for a person to blossom as an individual [10].

Jeffrey Reiman has defined privacy as the way in which a social group recognizes and communicates to the individual that he is responsible for his development as a unique person, a separate moral agent [11]. Stanley Benn reinforces this point when he says that privacy is a recognition of each person's true freedom [12].

Charles Sykes argues that privacy is valuable because it lets us be ourselves, suggesting the following example [13]. Imagine you are in a park playing with your child. How would your behavior be different if you knew someone was carefully watching you, perhaps even videotaping you, so that he or she could tell others about your parenting skills? You might well become self-conscious about your behavior. Few people would be able to carry on without any change to their emotional state or physical actions.

In a related observation, Gini Graham Scott points out that privacy lets us remove our public persona [14]. Imagine a businessman who is having a hard time with one of his company's important clients. At work he must be polite to the client and scrupulously avoid saying anything negative about the client in front of any coworkers, lest he demoralize them, or even worse, lose his job. In the privacy of his home he can "blow off steam" by confiding in his wife, who lends him a sympathetic ear and helps motivate him to get through the tough time at work. If people did not have privacy, they would have to wear their public face at all times, which could be damaging to their psychological health.

Other philosophers have pointed out the ways in which privacy can foster intellectual activities. Constance Fischer has pointed out that privacy allows us to shut out the rest of the world so that we can focus our thoughts without interruption [15]. Robert Neville describes how privacy is needed to live a creative life [16]. Joseph Keegan argues that privacy is needed for spiritual growth, the opportunity to become intimate with the Absolute Being [17].

Charles Fried goes a step further, stating that privacy is the only way in which people can develop relationships involving respect, love, friendship, and trust. According to Fried, "privacy is not merely a good technique for furthering these fundamental relations; rather without privacy they are simply inconceivable" [18]. Fried refers to privacy as "moral capital." People use this capital to build intimate relationships. Taking away people's privacy means taking away their moral capital. Without moral capital, they have no means to develop close personal relationships.

James Rachels voices a similar sentiment, when he writes that "there is a close connection between our ability to control who has access to us and to information about us, and our ability to create and maintain different sorts of social relationships with different people" [19]. Charles Sykes echoes Rachels when he says that each person has a "ladder" of privacy [13]. At the top of the ladder is the person we share the most information with. For many people this person is their spouse. As we work our way down the ladder, we encounter people we would share progressively less information with. Here is an example of what someone's ladder of privacy might look like:

spouse
priest/minister/rabbi
brothers and sisters
parents
children
friends
in-laws
coworkers
neighbors
marketers
employers
government
news media
ex-spouses
potential rivals/enemies

On the other hand, Jeffrey Reiman is critical of suggestions that tie intimacy too closely to sharing information [11]. A woman might tell her psychoanalyst things she would not even reveal to her husband, but that does not imply she experiences deeper intimacy with her psychoanalyst than with her husband. Intimacy is not just about sharing information, it's also about caring. The mutual caring that characterizes a healthy marriage results in a greater level of intimacy than can be gained simply by sharing personal information.

SUMMARY

To summarize our discussion, allowing people to have some privacy has a variety of beneficial effects. Giving people privacy is one way that society recognizes them as adults and indicates they are responsible for their own moral behavior. Privacy allows people to develop as individuals and to truly be themselves. Privacy gives people the opportunity to shut out the world, be more creative, and develop spiritually. Privacy gives each of us the opportunity to create different kinds of relationships with different people. Privacy also has numerous harmful effects. Privacy provides people with a way of covering up actions that are immoral or illegal. If a society sends a message that certain kinds of information must be kept private, some people caught in abusive or dysfunctional relationships may feel trapped and unable to ask others for help. Weighing these benefits and harms, we conclude that allowing people at least some privacy is better than denying people any privacy at all. That leads us to our next question: Is privacy a natural right, like the right to life?

5.2.3 Is There a Natural Right to Privacy?

Most of us agree that every person has certain natural rights, such as the right to life, the right to liberty, and the right to own property. Many people also talk about our right to privacy. Is this a natural right as well?

LEVINE: PRIVACY RIGHTS EVOLVE FROM PROPERTY RIGHTS

Morton Levine has shown how our belief in a right to privacy grew out of our property rights [10]. Historically, Europeans have viewed the home as a sanctuary. The English common law tradition has been that "a man's home is his castle." No one—not even the King—can enter without probable cause of criminal activity.

In 1765 the British Parliament passed the Quartering Act, which required American colonies to provide British soldiers with accommodations in taverns, inns, and unoccupied buildings. After the Boston Tea Party of 1773, the British Parliament attempted to restore order in the colonies by passing the Coercive Acts. One of these acts amended the Quartering Act to allow the billeting of soldiers in private homes, breaking the centuries-old common law tradition and infuriating many colonists. It's not surprising, then, that Americans restored the principle of home as sanctuary in the Bill of Rights.

THIRD AMENDMENT TO THE UNITED STATES CONSTITUTION

No Soldier shall, in time of peace be quartered in any house, without the consent of the Owner, nor in time of war, but in a manner to be prescribed by law.

In certain villages in the Basque region of Spain, each house is named after the person who originally constructed it. Villagers refer to people by their house names, even if the family living in the house has no relation to the family originally dwelling there.

These examples show a strong link between a person and his property. From this viewpoint, privacy is seen in terms of control over personal territory, and privacy rights evolve out of property rights.

WARREN AND BRANDEIS: CLEARLY PEOPLE HAVE A RIGHT TO PRIVACY

We can see this evolution laid out in a highly influential paper, published in 1890, by Samuel Warren and Louis Brandeis. Samuel Warren was a Harvard-educated lawyer who became a businessman when he inherited a paper manufacturing business. His wife was the daughter of a U.S. Senator and a leading socialite in Boston. Her parties attracted the upper-crust of Boston society. They also attracted the attention of the *Saturday Evening Gazette*, a tabloid that delighted in shocking its readers with lurid details about the lives of the Boston Brahmins.[1] Fuming at the paper's coverage of his daughter's wedding, Warren enlisted the aid of Harvard classmate Louis Brandeis, a highly successful Boston attorney (and future U.S. Supreme Court justice). Together, Warren and Brandeis published an article in the *Harvard Law Review* called "The Right to Privacy" [20]. In their highly influential paper, Warren and Brandeis argue that political, social, and economic changes demand recognition for new kinds of legal rights. In particular, they write that it is clear that people in modern society have a right to privacy and that this right ought to be respected. To make their case, they focus on—you guessed it—abuses of newspapers.

According to Warren and Brandeis:

> The press is overstepping in every direction the obvious bounds of propriety and of decency. Gossip is no longer the resource of the idle and of the vicious, but has become a trade, which is pursued with industry as well as effrontery. To satisfy the prurient taste the details of sexual relations are spread broadcast in the columns of the daily papers . . . The intensity and complexity of life, attendant upon advancing civilization, have rendered necessary some retreat from the world, and man, under the refining influence of culture, has become more sensitive to publicity, so that solitude and privacy have become more essential to the individual; but modern enterprise and invention have, through invasions upon his privacy, subjected him to mental pain and distress, far greater than could be inflicted by mere bodily injury. [20]

Meanwhile, Warren and Brandeis argue, there are no adequate legal remedies available to the victims. Laws against libel and slander are not sufficient because they do not address the situation where malicious, but true, stories about someone are circulated. Laws addressing property rights also fall short because they assume people have control over the ways in which information about themselves is revealed. However, cameras and other devices are capable of capturing information about a person without that person's consent.

Warren and Brandeis pointed out that the right to privacy had already been recognized by French law. They urged the American legal system to recognize the right to privacy, which they called "the right to be let alone" [20]. Their reasoning was highly influential. Though it took decades, the right to privacy is now recognized in courts across America [21].

THOMSON: EVERY "PRIVACY RIGHT" VIOLATION IS A VIOLATION OF ANOTHER RIGHT

Judith Jarvis Thomson has a completely different view about a right to privacy. She writes: "Perhaps the most striking thing about the right to privacy is that nobody seems to have any very clear idea what it is" [22]. Thomson points out problems with defining privacy as "the right to be let alone," as Warren and Brandeis have done. In some respects, this definition of privacy is too narrow. Suppose the police use an X-ray device

1. To learn more about the Boston Brahmins. consult Wikipedia (www.wikipedia.org).

and supersensitive microphones to monitor the movements and conversations of Smith in his home. The police have not touched Smith or even come close to him. He has no knowledge they are monitoring him. The police have let Smith alone, yet people who believe in a right to privacy would surely argue that they have violated Smith's privacy. In other respects, the definition of privacy as "the right to be let alone" is too broad. If I hit Jones on the head with a brick, I have not let him alone, but it is not his right of privacy I have violated—it is his right to be secure in his own person.

Thomson argues that whenever the right to privacy is violated, another right is violated as well. For example, suppose a man owns a pornographic picture. He doesn't want anyone else to know he owns it, so he keeps it in a wall safe. He only takes it out of his safe when he has taken pains to prevent others from looking into his home. Suppose we use an X-ray machine to look into his home safe and view the picture. We have violated his privacy, but we have also violated one of his property rights—the right to decide who (if anybody) will see the picture.

Here is another example. Suppose a Saudi Arabian woman wishes to keep her face covered for religious reasons. When she goes out in public, she puts a veil over her face. If I should walk up and pull away her veil to see her face, I have violated her privacy. But I have also violated one of her rights over her person—to decide whether people should see her face.

According to Thomson, there are a cluster of rights associated with privacy, just as there are a cluster of rights associated with property and a cluster of rights associated with our physical self. However, every violation of a privacy right is also a violation of a right in some other cluster. Since this is the case, there is no need to define privacy precisely or to decide exactly where to draw the line between violations of privacy and acceptable conduct.

BENN AND REIMAN: AUTONOMOUS MORAL AGENTS NEED SOME PRIVACY

Instead of referring to privacy as a natural right, Stanley Benn proposes that privacy principles be based on the more fundamental principle that each person is worthy of respect [12]. We give each other privacy because we recognize privacy is needed if people are to be autonomous moral agents able to develop healthy personal relationships and act as free citizens in a democratic society.

Jeffrey Reiman expands on Benn's view. He writes:

> The right to privacy protects the individual's interest in becoming, being, and remaining a person. It is thus a right which *all* human individuals possess—even those in solitary confinement. It does not assert a right never to be seen even on a crowded street. It is sufficient that I can control whether and by whom my body is experienced in some significant places and that I have the real possibility of repairing to those places. It is a right which protects my capacity to enter into intimate relations, not because it protects my reserve of generally withheld information, but because it enables me to make the commitment that underlies caring as *my* commitment uniquely conveyed by *my* thoughts and witnessed by *my* actions. [11]

Note Reiman's fairly restricted view of privacy. He carefully points out areas where privacy is necessary. He does not argue that privacy is a natural right, nor does he suggest that a person has complete control over what is held private.

CONCLUSION: PRIVACY IS A PRUDENTIAL RIGHT

In conclusion, people disagree whether there is a natural right to privacy. Even if there is no natural right to privacy, most commentators cite the benefits of privacy as a reason why people ought to have some privacy rights. Alexander Rosenberg calls privacy a *prudential right*. That means rational agents would agree to recognize some privacy rights, because granting these rights is to the benefit of society [23].

APPLICATION: TELEMARKETING

Telemarketing provides a good example of how privacy is treated as a prudential right. After being sworn in as Chairman of the Federal Trade Commission (FTC) in 2001, Timothy Muris looked for an action that the FTC could take to protect the privacy of Americans. It did not take long for the FTC to focus on telemarketing. A large segment of the American population views dinner-time phone calls from telemarketers as an annoying invasion of privacy. In fact, Harris Interactive concludes that telemarketing is the reason why the number of Americans who feel it is "extremely important" to not be disturbed at home rose from 49 percent in 1994 to 62 percent in 2003 [24]. Responding to this desire for greater privacy, the FTC created the National Do Not Call Registry (www.donotcall.gov), a free service that allows people who do not wish to receive telemarketing calls to register their phone numbers. The public reacted enthusiastically to the availability of the Do Not Call Registry by registering more than 50 million phone numbers before it even took effect in October 2003 [25, 26].

The Do Not Call Registry will not eliminate 100 percent of unwanted solicitations. The regulations exempt political organizations, charities, and organizations conducting telephone surveys. Even if your phone number has been registered, you may still receive phone calls from companies with which you have done business in the past eighteen months. Still, the Registry is expected to keep most telemarketers from calling people who do not wish to be solicited. The creation of the Registry demonstrates that privacy is seen as a prudential right: the benefit of shielding people from telemarketers is judged to be greater than the harm caused by putting limits on telephone advertising.

Some telemarketing firms have challenged the establishment of the Do Not Call Registry in court. There is a chance that a court will rule the FTC regulations to be illegal or unconstitutional. Skeptics of the Do Not Call Registry say that people will continue to be bothered, just in different ways. They predict the decline in telemarketing will be offset by an increase in direct mail advertising (sometimes called "junk mail") [27].

5.2.4 Privacy and Trust

While many people complain about threats to privacy, it is clear upon reflection that we have more privacy than our ancestors did [28]. Only a couple of centuries ago our society was agrarian. People lived with their extended families in small homes. The nearest community center was the village, where everyone knew everyone else and people took a keen interest in each other's business. The Church played an important role in everyday life. In this kind of society there was a strong pressure to conform [14]. There was greater emphasis on the community and lesser emphasis on the individual.

Charles Sykes writes: "Over the past two centuries, the rise of the modern has been the rise of the individual" [13]. He points out that prosperity, the single-family home, the automobile, television, and computers have contributed to our privacy. The single-family home gives us physical separation from other people. The automobile allows us to travel alone instead of on a bus or train in the presence of others. The television brings entertainment to us inside the comfort of our homes, taking us out of the neighborhood movie theater. With a computer and an Internet connection, we can access information at home rather than visit the public library [13]. These are just a few examples of ways in which modern conveniences allow us to spend time by ourselves or in the company of a few family members or friends.

In the past, young people typically lived at home with their parents until they were married. Today, many young unmarried adults live autonomously. This lifestyle provides them with previously unthought-of freedom and privacy [28].

The consequence of all this privacy is that we live among strangers. Many people know little more about their neighbors than their names (if that). Yet when we live in a society with others, we must be able to trust them to some extent. How do we know that the taxi driver will get us where we want to go without hurting us or overcharging us? How do parents know that their children's teachers are not child molesters? How does the bank know that if it loans someone money, it will be repaid?

In order to trust others, we must rely on their reputations. This was easier in the past, when people didn't move around so much and everyone knew everyone else's history. Today, society must get information out of people to establish reputations. One way of getting information from a person is through an **ordeal,** such as a lie detector test or a drug test. The other way to learn more about individuals is to issue (and request) **credentials,** such as a driver's license, key, employee badge, credit card, or college degree. As Steven Nock puts it, "A society of strangers is one of immense personal privacy. Surveillance is the cost of that privacy" [28].

5.3 Disclosing Information

As we go about our lives, we leave behind an electronic trail of our activities, thanks to computerized databases. Databases record the purchases we make with credit cards, the groceries we buy at a discount with our loyalty cards, the videos we rent by showing our driver's licenses, the calls we make with our telephones, and much more. The companies collecting this information use it to bill us. They also can use this information to serve us better. For example, Amazon.com uses information about book purchases to build profiles of its customers. With a customer profile, Amazon.com can recommend other books the customer may be interested in buying.

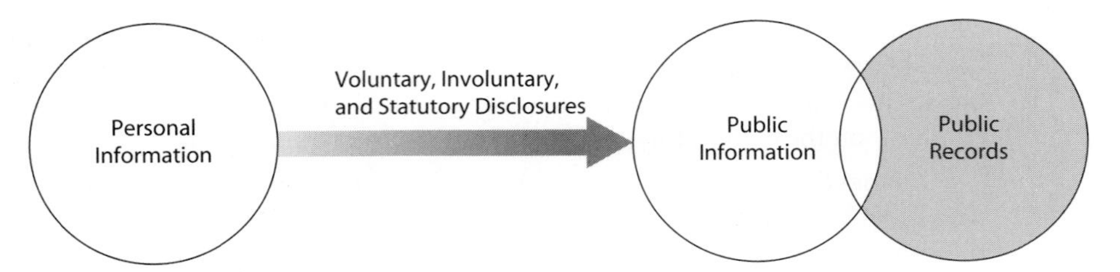

FIGURE 5.1 Personal information becomes public information or part of a public record as the result of a voluntary, involuntary disclosure, or statutory disclosure. The Privacy Act of 1974 puts some restrictions on access to information in public records. (We will discuss the Privacy Act in Section 5.6.4.)

It's important to distinguish between public information and public records [29]. A **public record** contains information about an incident or action reported to a government agency for the purpose of informing the public. Examples of public records are birth certificates, marriage licenses, motor vehicle records, criminal records, and deeds to property.

Public information is information you have provided to an organization that has the right to share it with other organizations. A good example of public information is a listing in a telephone directory. Most of us allow our name, address, and phone number to appear in telephone directories. By doing this, it is easier for our friends and acquaintances to call us or stop by our home. We judge this benefit to be worth the cost to us in the form of less privacy.

Personal information is information that is not public information or part of a public record. You may rightly consider your religion to be personal information. It remains personal information as long as you never disclose it to an organization that has the right to share it. However, if you do disclose your religious affiliation to such an organization, it becomes public information.

Personal information becomes public information or a public record through a voluntary, involuntary, or statutory disclosure (Figure 5.1).

Often people voluntarily make personal information public. Product registration forms and contest entries often ask consumers to reveal a great deal of personal information. I once received a product preference survey from Proctor & Gamble; it said, in part,

> Your opinions matter to us. That's why we've selected you to participate in one of the most important consumer research surveys we'll do this year. Whether or not you have completed one of our surveys in the past, you can help us continue to create the products that meet your needs. Simply answer the following questions, provide your name and address and mail it back to us. That way, we will be able to contact you if there are any special offers that might be of interest to you.

The questionnaire asked about my family's use of nasal inhalants, coffee, peanut butter, orange juice, laundry detergent, fabric softener, household cleaner, deodorant, toothpaste, detergents, skin care and hair care products, cosmetics, mouthwash, diapers, laxatives, and disposable briefs. It provided a list of 60 leisure activities, ranging from various sports to travel to gambling, and asked me to choose the three activities most important to my family. It also asked my date of birth, the sex and age of everyone living in my home, my occupation, the credit cards we used, and our annual family income. If I had returned the questionnaire (which I didn't), all of this information would have become public.

Sometimes you must disclose information in order to get something you want. If you want to fly on an airplane, you must allow others to search your luggage. You may even be subjected to a body search. You cannot refuse these searches if you want to travel by air. If you want to get a loan from a bank, you must provide the bank with your full name and Social Security number (so it can do a credit check), as well as detailed information about current income, your assets, and your liabilities. If you want to get married, you must fill out a marriage license and submit yourself to whatever tests are required by the local jurisdiction.

At other times, personal information becomes a public record without your consent. Police agencies and courts maintain records of arrests and convictions. Divorce records are public, and they can contain a significant amount of personal information.

Finally, information is sometimes gathered without our knowledge. There are more than a half million closed-circuit television cameras installed in public places in England. A resident of London may be captured on tape many times every day. A principal reason for installing these cameras is to reduce crime. However, detractors of this system point to abuses. Some allege that prosecutors have destroyed video footage that may have cleared a suspect. Others say that camera operators have acted like high-tech peeping Toms, using the cameras to watch people having sex [30].

5.4 Public Information

In this section we survey just a few of the many ways that personal information can become public information.

5.4.1 Rewards or Loyalty Programs

Rewards or loyalty programs for shoppers have been around for more than 100 years. Your grandparents may remember using S&H Green Stamps, the most popular rewards program in the United States from the 1950s through the 1970s. Shoppers would collect Green Stamps with purchases, paste them into booklets, and redeem the booklets by shopping in the Sperry and Hutchinson catalog for household items.

Today, many shoppers take advantage of rewards programs sponsored by grocery stores. Card-carrying members of the store's "club" save money on many of their purchases, either through coupons or instant discounts at the cash register. The most significant difference between the Green Stamps program and a contemporary shopper's club is that today's rewards programs are run by computers that record every purchase. Companies can use information about the buying habits of particular customers to provide them with individualized service.

For example, Safeway has unveiled computerized shopping carts at two of its stores in northern California. The shopping cart, called Magellan, has a small computer on the front handle and a card reader on the side. Customers identify themselves by swiping their Safeway Club card through the card reader. The computer taps into the database with the customer's buying history and uses this information to guide the customer to frequently purchased products. As the cart passes through the aisles, pop-up ads display items the computer predicts the customer may be interested in purchasing. It also lets customers purchase some products at sale prices unavailable to others [31].

Critics of grocery club cards say that the problem is not that card users pay less for their groceries, but that those who don't use cards pay more. They give examples of club-member prices being equivalent to the regular product price at stores without customer loyalty programs [32].

Some consumers respond to the potential loss of privacy by giving phony personal information when they apply for these cards. Others take it a step further by regularly exchanging their cards with those held by other people [33].

5.4.2 Body Scanners

Looking good is important to many, if not most, of us. Computer technology is making it possible for us to save time shopping and find clothes that fit us better (Figure 5.2).

In some stores in the United Kingdom, you can enter a booth, strip to your undergarments, and be scanned by a computer, which produces a three-dimensional model of your body. The computer uses this information to recommend which pairs of jeans ought to fit you the best. You can then sit in front of a computer screen and preview what various pairs of jeans will look like on you. When you have narrowed down your search to a few particular brands and sizes, you can actually try on the jeans.

Body scans are also being used to produce custom-made clothing. At Brooks Brothers stores in the United States, customers who have been scanned can purchase suits tailored to their particular physiques [34].

5.4.3 Digital Video Recorders

TiVo, Inc. manufactures a digital video recorder (DVR), which is similar to a VCR except that it records TV programs on a hard disk instead of videotape. TiVo also provides a service that allows its subscribers to more easily record programs they are interested in watching later. For example, with a single command a subscriber can instruct the TiVo to record every episode of a TV series. What many consumers may not know is that

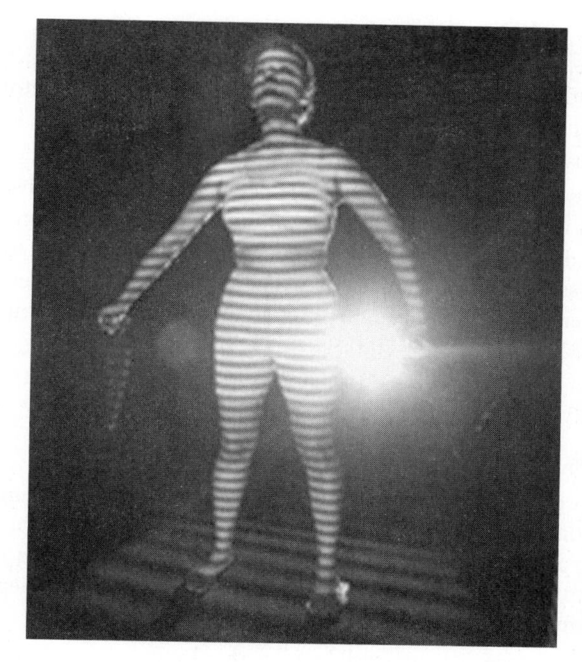

Figure 5.2 A computer takes a customer's measurements. (AP/Wideworld Photos)

TiVo sells detailed information about the viewing habits of its customers. Because the system monitors the activities of the user second by second, its data are more valuable than that provided by other services. For example, TiVo's records show that 54 percent of its customers skip commercials [35].

5.4.4 Automobile "Black Boxes"

You probably know about airplane flight data recorders, also called "black boxes," which provide information useful in postcrash investigations. Did you know that modern automobiles also come equipped with a "black box"? A microprocessor attached to the car's airbag records information about the speed of the car, the amount of pressure being put on the brake pedal, and whether the seat belts are connected. After a collision, investigators can retrieve the microprocessor from the automobile and view data collected in the five seconds before the accident [36].

5.4.5 Enhanced 911 Service

The U.S. Federal Communications Commission has passed an enhanced 911 mandate that requires cell phone providers to be able to track the locations of active cell phone users to within 100 meters. The safety benefit of enhanced 911 service is obvious. Emergency response teams can reach people in distress even if they are unable to speak or do not know exactly where they are.

The ability to identify the location of active cell phone users has other benefits. For example, it makes it easier for cell phone companies to identify where signal strength is weak and coverage needs to be improved. In the past, this information had to be gained by sending people into the field to check signal strength, à la the Verizon commercial ("Can you hear me now? Good!").

The downside to enhanced 911 service is a loss of privacy. Because it is possible to track the location of active cell phone users, what happens if information is sold or shared? Suppose you call your employer and tell him you are too sick to come into work. Your boss is suspicious, since this is the third Friday this winter you've called in sick. Your employer pays your cell phone provider and discovers that you made your call from a ski resort [37].

5.4.6 RFIDs

Imagine getting up in the morning, walking into the bathroom, and seeing a message on the medicine cabinet's computer screen warning you that your bottle of aspirin is close to its expiration date. Later that day, you are shopping for a new pair of pants. As you try them on, a screen in the dressing room displays other pieces of clothing that would complement your selection.

These scenarios are possible today thanks to a new technology called RFID, short for radio frequency identification. An RFID is a tiny wireless transmitter. Manufacturers are replacing bar codes with RFIDs, because they give more information about the product and are easier to scan. An RFID can contain specific information about the particular item to which it is attached (or embedded), and a scanner can read an RFID from six feet away. When barcodes are replaced by RFIDs, check-outs are quicker and companies track their inventory more accurately (Figure 5.3).

However, because RFIDs are not turned off when an item is purchased, the new technology has raised privacy concerns. Imagine a workplace full of RFID scanners. A scanner in your cubicle enables a monitoring system to associate you with the tags in your clothes. Another scanner picks up your presence at the water cooler. The next thing you know, your boss has called you in for a heart-to-heart talk about how many breaks you're taking. Some privacy advocates say consumers should have a way to remove or disable RFIDs in the products they purchase [38, 39].

The U.S. government plans to replace traditional passports with electronic passports equipped with RFID tags. The RFID tag would duplicate the passport's identifying information and include a digital photograph. By combining the RFID tag's information with new facial recognition technology, the government hopes to improve security at border crossings. Critics of this plan say that RFID tags can be read by anyone within 25 feet who has a powerful enough chip reader. They fear that these tags could make travelers more vulnerable to identity theft [40]. Others wonder if terrorists with powerful RFID tag readers might begin "scanning" foreign cafes, searching for locations with a high concentration of Americans [41]. Some experts, however, claim that these fears are exaggerated and that RFID tags are difficult to read at a distance [42].

Figure 5.3 Employees take inventory more quickly and make fewer errors when items are marked with RFID tags. (Courtesy Tibbett & Britten)

5.4.7 Implanted Chips

In Taiwan every domesticated dog must contain a microchip implant identifying its owner and residence [43]. The microchip, about the size of a grain of rice, is implanted into the dog's ear using a syringe.

Digital Angel Corporation has created a personal safety and location system that includes a clip-on monitor about the size of a pager, a temperature sensing watch, and a global positioning system. The system is designed to monitor an elderly person's location and environment, and to send an alert when necessary. For example, the system generates an alert if the subscriber falls down and doesn't get up. It also generates an alert if the subscriber wanders outside of a predetermined neighborhood.

Some people believe that parents should implant microchips in their children. They say that the life of a child is more important than any concerns about privacy [44].

5.4.8 Cookies

A **cookie** is a file placed on your computer's hard drive by a Web server. The file contains information about your visits to a Web site. Cookies can contain login names and passwords, product preferences, and the contents of virtual "shopping carts." Web sites use cookies to provide you with personalized services, such as custom Web pages. Instead of asking you to type in the same information multiple times, a Web site can retrieve that information from a cookie. Most Web sites do not ask for permission before creating a cookie on your hard drive. You can configure your Web browser to alert you when a cookie is being placed on your computer, or you can set your Web browser to refuse to accept any cookies. However, some Web sites cannot be accessed by browsers that block cookies.

5.4.9 Spyware

Spyware is a program that communicates over your Internet connection without your knowledge or consent. Spyware programs can monitor Web surfing, log keystrokes, take snapshots of your computer screen, summon pop-up advertisements, and send reports back to a host computer.

Free software downloaded from the Internet often contains spyware. A 2003 survey of 120 U.S. consumers with broadband Internet connections found that 91 percent of them had spyware on their computers.

Some ISPs are responding to the outbreak of spyware by releasing tools to help their customers protect their privacy. America Online plans to start including spyware-detecting programs with its software distribution [45].

5.5 U.S. Legislation

Reflecting public concerns about privacy, Congress has passed numerous laws regulating the collection and distribution of information gathered by private enterprises. We review six of these laws: the Fair Credit Reporting Act, the Family Education Rights and Privacy Act, the Video Privacy Protection Act, the Financial Services Modernization Act, the Children's Online Privacy Protection Act, and the Health Insurance Portability and Accountability Act.

5.5.1 Fair Credit Reporting Act

Credit bureaus and other consumer reporting agencies maintain information on your bill-paying record, whether you've been sued or arrested, and if you've filed for bankruptcy. They sell reports to other organizations that are trying to determine the creditworthiness of consumers who are applying for credit, applying for a job, or trying to rent an apartment. The Fair Credit Reporting Act, passed in 1970 and revised in 1996, was designed to promote the accuracy and privacy of information used by credit bureaus and other consumer reporting agencies to produce consumer reports. It also ensures that negative information does not haunt a consumer for a lifetime.

The three major credit bureaus are Equifax, Experian, and Trans Union. According to the Fair Credit Reporting Act, these credit bureaus may keep negative information about a consumer for a maximum of seven years. There are several exceptions to this rule. The two most important are that information about criminal convictions may be kept indefinitely, and bankruptcy information may be held for 10 years.

5.5.2 The Family Education Rights and Privacy Act

The Family Education Rights and Privacy Act (FERPA) provides students 18 years of age and older the right to review their educational records and to request changes to records that contain erroneous information. Students also have the right to prevent information in these records from being released without their permission, except under certain circumstances. For students under the age of 18, these rights are held by their parents or guardians. FERPA applies to all educational institutions that receive funds from the U.S. Department of Education; in other words, both public and private schools.

5.5.3 Video Privacy Protection Act

In 1988 President Ronald Reagan nominated Judge Robert Bork to the U.S. Supreme Court. Bork was a noted conservative, and his nomination was controversial. A Washington, D.C. video store provided a list of Bork's video rental records to a reporter for the *Washington City Paper*, which published the list. While the intention of the paper was most likely to embarrass Bork, it also had the effect of prompting Congress to pass the Video Privacy Protection Act of 1988. According to this law, videotape service providers cannot disclose rental records without the written consent of the customer. In addition, rental stores must destroy personally identifiable information about rentals within a year of the date when this information is no longer needed for the purpose for which it was collected.

5.5.4 Financial Services Modernization Act

The Financial Services Modernization Act (also called the Gramm-Leach-Bliley Act of 1999) contains dozens of provisions related to how financial institutions do business. One of the major provisions of the Act allows the creation of "financial supermarkets" offering banking, insurance, and brokerage services.

The law also contains some privacy-related provisions. It requires financial institutions to disclose their privacy policies to their customers. When a customer establishes an account, and at least once per year thereafter, the institution must let the customer know the kinds of information it collects and how it uses that information. These notices must contain an opt-out clause that explains to customers how they can request that their confidential information not be revealed to other companies. The law requires financial institutions to develop policies that will prevent unauthorized access of their customers' confidential information [46].

5.5.5 Children's Online Privacy Protection Act

The Children's Online Privacy Protection Act (COPPA), which went into effect in 2000, is designed to reduce the amount of public information gathered from children using the Internet. According to COPPA, online services must obtain parental consent before collecting any information from children 12 years old and younger.

5.5.6 Health Insurance Portability and Accountability Act

As part of the Health Insurance Portability and Accountability Act of 1996, Congress directed the Department of Health and Human Services (HHS) to come up with guidelines for protecting the privacy of patients. These guidelines went into effect in April 2003. They limit how doctors, hospitals, pharmacies, and insurance companies can use medical information collected from patients.

The regulations attempt to limit the exchange of information among health care providers to that information necessary to care for the patient. They forbid health care providers from releasing information to life insurance companies, banks, or other businesses without specific signed authorization from the person being treated. Health care providers must provide their patients with a notice describing how they use the information they gather. Patients have the right to see their medical records and to request corrections to errors they find in those records [47].

5.6 Public Records

The federal government maintains thousands of databases containing billions of records about the activities of U.S. citizens. In this section we consider the public-record-keeping activities of the Census Bureau, the Internal Revenue Service, and the FBI, paying particular attention to ways in which information collected for one purpose often has been used for another.

5.6.1 Census Records

In order to ensure each state has fair representation in the House of Representatives, the United States Constitution requires the government to perform a census every 10 years.

The first census of 1790 had six questions. It asked for the name of the head of the household and the number of persons in each of the following categories: free white males at least 16 years old; free white males under 16 years old; free white females; all other free persons (by sex and color); and slaves.

As time passed, the number of questions asked during the census increased. The 1820 census determined the number of people engaged in agriculture, commerce, and manufacturing. The 1840 census had questions regarding school attendance, illiteracy, and occupations. In 1850 census takers began asking questions about taxes, schools, crime, wages, and property values. The 1940 census is notable because for the first time statistical sampling was put to extensive use. A random sample of the population, about five percent of those surveyed, received a longer form with more questions. The use of sampling enabled the Census Bureau to produce detailed demographic profiles without substantially increasing the amount of data it needed to process.

According to federal law, the Census Bureau is supposed to keep confidential the information it collects. However, in times of national emergency the Census Bureau has revealed its information to other agencies. During World War I, the Census Bureau provided the names and addresses of young men to the military, which was searching for draft resistors. After the attack on Pearl Harbor, the Census Bureau provided the Justice Department with information from the 1940 census about the general location of Japanese-Americans. The Army used this information to round up Japanese-Americans and send them to internment camps.

5.6.2 Internal Revenue Service Records

The United States enacted a national income tax in 1862 to help pay for expenses related to the Civil War. In 1872 the income tax was repealed. Congress resurrected the national income tax in 1894, but a year later the Supreme Court ruled it unconstitutional. The 16th Amendment to the Constitution, ratified by the states in 1913, gives the United States government the power to collect an income tax. A national income tax has been in place ever since. The Internal Revenue Service (IRS) now collects more than $2 trillion a year in taxes.

Your income tax form may reveal a tremendous amount of personal information about your income, your assets, the organizations to which you give charitable contributions, your medical expenses, and much more. Every year the IRS investigates hundreds of employees for misusing their access to these records. In one notable case, a member of the Ku Klux Klan examined records of fellow Klan members, hoping to identify a suspected undercover agent in his group. The IRS has also misplaced hundreds of tapes and diskettes containing income tax data [48].

In 2003 five consumer protection groups complained to the U.S. Treasury Department that consumers using H&R Block's Web-based Free File tax filing service were being subjected to advertising for tax-related products and home mortgage loans [49]. For example, after a user entered mortgage interest in his tax form, a window popped up with this message:

> We noticed that you entered an itemized deduction for home mortgage interest. By refinancing your mortgage, you may be able to lower your monthly payments or pay off other debts. Now is a great time to take advantage of historically low interest rates. It's easy! Do you want to learn how refinancing your mortgage can help you?

The groups claimed that H&R Block was requiring everyone using its Free File service to consent to cross-marketing, even though that was against the law.

5.6.3 FBI National Crime Information Center 2000

The FBI National Crime Information Center 2000 (NCIC) is a collection of databases supporting the activities of federal, state, and local law-enforcement agencies in the United States, the United States Virgin Islands, Puerto Rico, and Canada [50]. Its predecessor, the National Crime Information Center, was established by the FBI in January 1967 under the direction of J. Edgar Hoover.

When it was first activated, the NCIC consisted of about 95,000 records in five databases: stolen automobiles, stolen license plates, stolen or missing guns, other stolen items, and missing persons. Today, NCIC databases contain more than 39 million records. The databases have been expanded to include such categories as wanted persons, criminal histories, people incarcerated in federal prisons, convicted sex offenders, unidentified persons, people believed to be a threat to the President, foreign fugitives, violent gang members, and suspected terrorists. More than 80,000 law enforcement agencies have access to these data files. The NCIC processes more than two million requests for information each day, with an average response time of less than one second.

The FBI points to the following successes of the NCIC:

- Investigating the assassination of Dr. Martin Luther King, Jr., the NCIC provided the FBI with the information it needed to link a fingerprint on the murder weapon to James Earl Ray.
- In 1992 the NCIC led to the apprehension of 81,750 "wanted" persons, 113,293 arrests, the location of 39,268 missing juveniles and 8,549 missing adults, and the retrieval of 110,681 stolen cars.
- About an hour after the April 19, 1995, bombing of the Alfred P. Murrah Federal Building in Oklahoma City, Oklahoma state trooper Charles Hanger pulled over a Mercury Marquis with no license plates. Seeing a gun in the back seat of the car, Hanger arrested the driver—Timothy McVeigh—on the charge of transporting a loaded firearm in a motor vehicle. He took McVeigh to the county jail, and the arrest was duly entered into the NCIC database. Two days later, when federal agents ran McVeigh's name through the NCIC, they saw Hanger's arrest record. FBI agents reached the jail just before McVeigh was released. McVeigh was subsequently convicted of the bombing.

Critics of the National Crime Information Center point out ways in which the existence of the NCIC has led to privacy violations of innocent people:

- Erroneous records can lead law enforcement agencies to arrest innocent persons.
- Innocent people have been arrested because their name is the same as someone listed in the arrest warrants database.
- The FBI has used the NCIC to keep records about people not suspected of any crime, such as opponents of the Vietnam War.
- Corrupt employees of law-enforcement organizations with access to the NCIC have sold information to private investigators and altered or deleted records.
- People with access to the NCIC have illegally used it to search for criminal records on acquaintances or to screen potential employees, such as baby-sitters.

5.6.4 Privacy Act of 1974

In the early 1970s, William Richardson, the Secretary of the U.S. Department of Health, Education, and Welfare, convened a group to recommend policies for the development of government databases that would protect the privacy of American citizens. The Secretary's Advisory Committee of Automated Personal Data Systems, Records, Computers, and the Rights of Citizens produced a report for Congress, which included the following "bill of rights" for the Information Age [51]:

CODE OF FAIR INFORMATION PRACTICES

1. There must be no personal data record-keeping systems whose very existence is secret.
2. There must be a way for a person to find out what information about the person is in a record and how it is used.
3. There must be a way for a person to prevent information about the person that was obtained for one purpose from being used or made available for other purposes without the person's consent.
4. There must be a way for a person to correct or amend a record of identifiable information about the person.
5. Any organization creating, maintaining, using, or disseminating records of identifiable personal data must assure the reliability of the data for their intended use and must take precautions to prevent misuses of the data.

Interestingly, the Richardson report had a greater impact in Europe than in the United States. Nearly every nation in Europe passed laws based on the Code of Fair Information Practices [52].

The Privacy Act of 1974 represents Congress's codification of these principles. While the Privacy Act does allow individuals in some cases to get access to federal files containing information about them, in other respects it has fallen short of the desires of privacy advocates. In particular, they say the Privacy Act has not been effective in reducing the flow of personal information into governmental databases, preventing agencies from sharing information with each other, or preventing unauthorized access to the data. They claim agencies have been unresponsive to outside attempts to bring them into alignment with the provisions of the Privacy Act. The Privacy Act has the following principal limitations [53]:

1. *The Privacy Act applies only to government databases.*

   Far more information is held in private databases, which are excluded. This is an enormous loophole, because government agencies can purchase information from private organizations that have the data they want.

2. *The Privacy Act only covers records indexed by a personal identifier.*

   Records about individuals that are not indexed by name or another identifying number are excluded. For example, a former IRS agent tried to gain access to a file containing derogatory information about himself, but the judge ruled he did not have a right to see the file, since it was indexed under the name of the IRS investigator, not the IRS agent.

3. *No one in the federal government is in charge of enforcing the provisions of the Privacy Act.*

Federal agencies have taken it upon themselves to determine which databases they can exempt. The IRS has exempted its database containing the names of taxpayers it is investigating. The Department of Justice has announced that the FBI does not have to ensure the reliability of the data in its NCIC databases.

4. *The Privacy Act allows one agency to share records with another agency as long as they are for a "routine use."*

   Each agency is able to decide for itself what "routine use" means. The Department of Justice has encouraged agencies to define "routine use" as broadly as possible.

5.7 Covert Government Surveillance

Section 5.4 gave a few examples of ways private organizations collect information about individuals. In most cases the goal of the organization's information gathering is to stimulate commerce, and for the most part the individual consumer voluntarily provides the information. Section 5.6 focused on the databases of public records created by the federal government. Some of these records are the result of transactions initiated by individuals, such as filing a tax return. In other cases individuals are required by law to provide the information, such as filling out a census form. Whether people are providing the information willingly or unwillingly, they are aware the government is collecting the data.

In this section we focus on ways in which the United States government has collected information in order to detect and apprehend suspected criminals or to improve national security. Because the individuals being observed are suspected of wrongdoing, they are not alerted or asked for permission before the surveillance begins.

Does covert surveillance violate any of the rights of a citizen? The most relevant statement in the U.S. Constitution is the Fourth Amendment:

FOURTH AMENDMENT TO THE UNITED STATES CONSTITUTION

The right of the people to be secure in their persons, houses, papers, and effects, against unreasonable searches and seizures, shall not be violated, and no Warrants shall issue, but upon probable cause, supported by Oath or affirmation, and particularly describing the place to be searched, and the persons or things to be seized.

Before the American Revolution, English agents in pursuit of smugglers made use of *writs of assistance*, which gave them authority to enter any house or building and seize any prohibited goods they could find. This activity drew the ire of the colonists. It is not surprising, then, that a prohibition against unreasonable searches and seizures appears in the Bill of Rights.

The position of the U.S. Supreme Court with respect to covert electronic surveillance has changed over time. Today, the Supreme Court's interpretation of the Constitution is that any kind of electronic surveillance without a warrant is unconstitutional. Let's see how the Supreme Court gradually came to this position.

5.7.1 Wiretaps and Bugs

Wiretapping refers to the interception of a telephone conversation. (The term is somewhat anachronistic, because many telephone conversations are no longer transmitted over wires.) Wiretapping has been taking place ever since the 1890s, when telephones became commonly used. In 1892 the State of New York made wiretapping a felony, but the police in New York City ignored the law and continued the practice of wiretapping. Until 1920 the New York City police listened to conversations between lawyers and clients, doctors and patients, and priests and penitents. On several occasions the police even tapped the trunk lines into hotels and listened to the telephone conversations of all the hotel guests [54].

OLMSTEAD V. UNITED STATES

Wiretapping was a popular tool for catching bootleggers during Prohibition (1919–1933). The most famous case involved Roy Olmstead, who ran a $2 million-a-year bootlegging business in Seattle, Washington. Without a warrant, federal agents tapped Olmstead's phone and collected enough evidence to convict him. Although wiretapping was illegal under Washington law, the state court allowed evidence obtained through the wiretapping to be admitted. Olmstead appealed all the way to the U.S. Supreme Court. His lawyer argued that the police had violated Olmstead's right to privacy by listening in on his telephone conversations. He also argued that the evidence should be thrown out because it was obtained without a search warrant [54, 55].

In a 5-4 decision, the Supreme Court ruled in *Olmstead v. United States* that the Fourth Amendment protected tangible assets alone. The federal agents did not "search" a physical place; they did not "seize" a physical item. Hence the Fourth Amendment's provision against warrantless search and seizure did not apply. Justice Louis Brandeis was one of the four judges siding with Olmstead. In his dissenting opinion, Brandeis argued that the protections afforded by the Bill of Rights ought to extend to electronic communications as well. He wrote:

> Whenever a telephone line is tapped, the privacy of the persons at both ends of the line is invaded, and all conversations between them upon any subject, and although proper, confidential, and privileged, may be overheard. Moreover, the tapping of one man's telephone line involves the tapping of the telephone of every other person whom he may call, or who may call him. As a means of espionage, writs of assistance and general warrants are but puny instruments of tyranny and oppression when compared with wiretapping. [56]

CONGRESS MAKES WIRETAPPING ILLEGAL

The public and the press were critical of the Supreme Court decision. Since the Court had ruled that wiretapping was constitutional, those interested in prohibiting wiretapping focused their efforts on the legislative branch. In 1934 the U.S. Congress passed the Federal Communications Act, which (among other things) made it illegal to intercept and reveal wire communications. Three years later, the Supreme Court used the Federal Communications Act to reverse its position on warrantless wiretaps. In *Nardone v. United States*, the Court ruled that evidence obtained by federal agents from warrantless wiretaps was inadmissible in court. In another decision, *Weiss v. United States*, it ruled that the prohibition on wiretapping applied to intrastate as well as interstate telephone calls. Subsequently, the Attorney General announced that the FBI would cease wiretapping [54, 55].

FBI CONTINUES SECRET WIRETAPPING

After World War II broke out in Europe, FBI Director J. Edgar Hoover pressed to have the ban on wiretapping withdrawn (Figure 5.4). The position of the Department of Justice was that the Federal Communications Act simply prohibited intercepting *and* revealing telephone conversations. In the Justice Department's view, it was permissible to intercept conversations as long as they were not revealed to an agency outside the federal government. President Roosevelt agreed to let the FBI resume wiretapping in cases involving national security, though he asked that the wiretaps be kept to a minimum and limited as much as possible to aliens [54].

FIGURE 5.4 Under the leadership of J. Edgar Hoover, the FBI engaged in illegal wiretapping. (©Bettmann/CORBIS)

Because it knew evidence obtained through wiretapping was inadmissible in court, the FBI began maintaining two sets of files: the official files that contained legally obtained evidence, and confidential files containing evidence obtained from wiretaps and other confidential sources. In case of a trial, only the official file would be released to the court [54].

The FBI was supposed to get permission from the Department of Justice before installing a wiretap, but in practice it did not always work that way. During his 48-year reign as Director of the FBI, J. Edgar Hoover routinely engaged in political surveillance, tapping the telephones of senators, congressmen, and Supreme Court justices. The information the FBI collected on these figures had great political value, even if the recordings revealed no criminal activity. There is evidence Hoover used information gathered during this surveillance to discredit Congressmen who were trying to limit the power of the FBI [54].

CHARLES KATZ V. UNITED STATES

A **bug** is a hidden microphone used for surveillance. In a series of decisions, the U.S. Supreme Court gradually came to an understanding that citizens should also be protected from all electronic surveillance conducted without warrants, including bugs. The key decision was rendered in 1967. Charles Katz used a public telephone to place bets. The FBI placed a bug on the outside of the telephone booth to record Katz's telephone conversations. With this evidence, Katz was convicted of illegal gambling. The Justice Department argued that since it placed the microphone on the outside of the telephone booth, it did not intrude into the space occupied by Katz [54]. In *Charles Katz v. United States*, the Supreme Court ruled in favor of Katz. Justice Stewart Potter wrote that "the Fourth Amendment protects people, not places" [57]. Katz entered the phone booth with the reasonable expectation that his conversation would not be heard, and what a person "seeks to preserve as private, even in an area accessible to the public, may be constitutionally protected" [57].

5.7.2 Operation Shamrock

During World War II the U.S. government censored all messages entering and leaving the country, meaning U.S. intelligence agencies had access to all telegram traffic. At the end of the war the censorship bureaucracy was shut down, and the Signal Security Agency (predecessor to the National Security Agency) needed to find a new way to get access to telegram traffic. It contacted Western Union Telegraph Company, ITT Communications, and RCA Communications, and asked them to allow it to make photographic copies of all foreign government telegram traffic that entered, left, or transited the United States. In other words, the Signal Security Agency asked these companies to break federal law in the interests of national security. All three companies agreed to the request. The Signal Security Agency gave this intelligence-gathering operation the name "Shamrock."

When the National Security Agency (NSA) was formed in 1952, it inherited Operation Shamrock. The sophistication of the surveillance operation took a giant leap forward in the 1960s, when the telegram companies converted to computers. Now the contents of telegrams could be transmitted electronically to the NSA, and the NSA could use computers to search for key words and phrases.

In 1961 Robert Kennedy became the new Attorney General of the United States, and he immediately focused his attention on organized crime. Discovering that information about mobsters was scattered piecemeal among the FBI, IRS, Securities and Exchange Commission (SEC), and other agencies, he convened a meeting in which investigators from all of these agencies could exchange information. The Justice Department gave the names of hundreds of alleged crime figures to the NSA, asking that these figures be put on its "watch list." Intelligence gathered by the NSA contributed to several prosecutions.

Also during the Kennedy administration, the FBI asked the NSA to put on its watch list the names of U.S. citizens and companies doing business with Cuba. The NSA sent information gathered from intercepted telegrams and international telephone calls back to the FBI.

During the Vietnam War the Johnson and Nixon administrations hypothesized that foreign governments were controlling or influencing the activities of American groups opposed to the war. They asked the NSA to put the names of war protesters on its watch list. Some of the people placed on the watch list included the Reverend Dr. Martin Luther King, Jr., the Reverend Ralph Abernathy, Black Panther leader Eldrige Cleaver, Dr. Benjamin Spock, Joan Baez, and Jane Fonda.

In 1969 President Nixon established the White House Task Force on Heroin Suppression. The NSA soon became an active participant in the war on drugs, monitoring the phone calls of people put on its drug watch list. Intelligence gathered by the NSA led to convictions for drug-related crimes.

Facing Congressional and press scrutiny, the NSA called an end to Operation Shamrock in May 1975 [58].

5.8 U.S. Legislation Authorizing Wiretapping

As we have seen, the Federal Communications Act of 1934 made wiretapping illegal, and rulings by the U.S. Supreme Court gradually closed the door to wiretapping and bugging performed without a warrant (court order). After the Katz decision, police were left without any electronic surveillance tools in their fight against crime.

Meanwhile, the United States was in the middle of the Vietnam War. In 1968 the country was rocked by violent antiwar demonstrations and the assassinations of Martin Luther King, Jr., and Robert F. Kennedy. Law-enforcement agencies pressured Congress to allow wiretapping under some circumstances.

5.8.1 Title III

Congress responded by passing Title III of the Omnibus Crime Control and Safe Streets Act of 1968. Title III allows a police agency that has obtained a court order to tap a phone for up to 30 days [54].

The government continued to argue that in cases of national security, agencies should be able to tap phones without a warrant. In 1972 the Supreme Court rejected this argument when it ruled that the Fourth Amendment forbids warrantless wiretapping, even in cases of national security [54].

5.8.2 Electronic Communications Privacy Act

Congress updated the wiretapping law in 1986 with the passage of the Electronic Communications Privacy Act (ECPA). The ECPA allows police to attach two kinds of surveillance devices to a suspect's phone line. If the suspect makes a phone call, a **pen register** displays the number being dialed. If the suspect gets a phone call, a **trap-and-trace device** displays the caller's phone number. While a court order is needed to approve the installation of pen registers and trap-and-trace devices, prosecutors do not need to demonstrate probable cause, and the approval is virtually automatic.

The ECPA also allows police to conduct **roving wiretaps**—wiretaps that move from phone to phone—if they can demonstrate the suspect is attempting to avoid surveillance by using many different phones [54].

5.8.3 Communications Assistance for Law Enforcement Act

The implementation of digital phone networks interfered with the wiretapping ability of the FBI and other organizations. In response to these technological changes, Congress passed the Communications Assistance for Law Enforcement Act of 1994 (CALEA), also known as the Digital Telephony Act. This law required networking equipment used by phone companies be designed or modified so that law-enforcement agencies can trace calls, listen in on telephone calls, and intercept email messages. CALEA thereby ensured that court-ordered wiretapping would still be possible even as new digital technologies were introduced.

CALEA left unanswered many important details about the kind of information the FBI would be able to extract from digital phone calls. The precise requirements were to be worked out between the FBI and industry representatives. The FBI asked for many capabilities, including the ability to intercept digits typed by the caller after the phone call was placed. This feature would let it catch credit card numbers and bank account numbers, for example. In 1999 the FCC finally issued the guidelines, which included this capability and five more requested by the FBI [59]. Privacy-rights organizations argued these capabilities went beyond the authorization of CALEA [60]. Telecommunications companies claimed that implementing these capabilities would cost them billions [61]. Nevertheless, in August 2005 the FCC gave voice over Internet Protocol (VoIP) and certain other broadband providers eighteen months to modify their systems as necessary so that law enforcement agencies could wiretap calls made using their services [62].

5.8.4 USA PATRIOT Act

BACKGROUND

On the morning of September 11, 2001, terrorists hijacked four passenger airliners in the United States and turned them into flying bombs. Two of the planes flew into New York's World Trade Center, a third hit the Pentagon, and the fourth crashed in a field in Pennsylvania. Soon after these attacks, which resulted in about 3,000 deaths and the destruction of the twin towers of the World Trade Center, the United States Congress passed the Uniting and Strengthening America by Providing Appropriate Tools Required to Intercept and Obstruct Terrorism (USA PATRIOT) Act of 2001, henceforth referred to as the Patriot Act [63].

PROVISIONS OF THE PATRIOT ACT

The Patriot Act amended many existing laws. Its provisions fall into four principal categories:

1. Providing federal law enforcement and intelligence officials with greater authority to monitor communications;
2. Giving the Secretary of the Treasury greater powers to regulate banks, preventing them from being used to launder foreign money;
3. Making it more difficult for terrorists to enter the United States; and
4. Defining new crimes and penalties for terrorist activity.

We focus on those provisions of the Patriot Act that most directly affect the privacy of persons living inside the United States.

The Patriot Act expands the kinds of information that law enforcement officials can gather with pen registers and trap-and-trace devices. It allows police to use pen registers on the Internet to track email addresses and URLs. The law does not require they demonstrate probable cause. To obtain a warrant, police simply certify that the information to be gained is relevant to an ongoing criminal investigation.

Law enforcement agencies seeking to install a wiretap or a pen register/trap-and-trace device have always been required to get a court order from a judge with jurisdiction over the location where the device was to be installed. The Patriot Act extends the jurisdiction of court-ordered wiretaps to the entire country. A judge in New York can authorize the installation of a device in California, for example. The act also allows the nationwide application of court-ordered search warrants for terrorist-related investigations.

The Patriot Act broadened the number of circumstances under which roving surveillance can take place. Previously, roving surveillance could only be done for the purpose of law enforcement, and the agency had to demonstrate to the court that the person under investigation actually used the device to be monitored. The Patriot Act allows roving surveillance to be performed for the purpose of intelligence, and the government does not have to prove that the person under investigation actually uses the device to be tapped. Additionally, it does not require that the law enforcement agency report back to the authorizing judge regarding the number of devices monitored and the results of the monitoring.

Under the Patriot Act, law enforcement officials wishing to intercept communications to and from a person who has illegally gained access to a computer system do not need a court order if they have the permission of the owner of the computer system.

The Patriot Act allows courts to authorize law enforcement officers to search a person's premises without first serving a search warrant when there is "reasonable cause to believe that providing immediate notification of the execution of the warrant may have an adverse affect." Officers may seize property that "constitutes evidence of a criminal offense in violation of the laws of the United States," even if that offense is unrelated to terrorism.

The Patriot Act makes it easier for the FBI to collect business, medical, educational, library, and church/mosque/synagogue records. To obtain a search warrant authorizing the collection of these records, the FBI merely needs to state that the records are related to an ongoing investigation. There is no need for the FBI to show probable cause. It is illegal for anyone supplying records to the FBI to reveal the existence of the warrant or tell anyone that they provided information to the government. The act does specifically prohibit the FBI from investigating citizens solely on the basis of activities protected by the First Amendment.

RESPONSES TO THE PATRIOT ACT

More than a hundred cities and several states have passed anti–Patriot Act resolutions [64]. Critics of the Patriot Act warn that its provisions give too many powers to the federal government. Despite language in the Patriot Act to the contrary, civil libertarians are concerned that law enforcement agencies may use their new powers to reduce the rights of law-abiding Americans, particularly those expressed in the First and Fourth Amendments to the United States Constitution.

We have seen that in the past, the FBI and the NSA used illegal wiretaps to investigate people who had expressed unpopular political views. Congressional investigations led to the termination of Operation Shamrock. Over time, however, Congress has gradually increased the surveillance options available to the police. By making court orders for various kinds of electronic surveillance easy to obtain, the Patriot Act is another step in this direction. If people know how easy it is for law enforcement personnel to monitor their activities, they may be less inclined to exercise their First Amendment rights. For example, knowledge that the Patriot Act allows the FBI to collect records from libraries and bookstores may make people less inclined to read certain books. In November 2003 the American Civil Liberties Union reported that public apprehension about the Patriot Act has led to a significant drop in attendance and donations at mosques [65].

Critics of the Patriot Act are also concerned that some of its provisions undermine rights guaranteed citizens under the Fourth Amendment to the Constitution:

- By revealing the URLs of Web sites visited by a suspect, a pen register is a much more powerful surveillance tool on the Internet than it is on a telephone network. The Patriot Act allows police to install Internet pen registers without demonstrating probable cause that the suspect is engaged in a criminal activity.
- Court orders authorizing roving surveillance do not "particularly describe the place to be searched."
- It allows law enforcement agencies, under certain circumstances, to search homes and seize evidence without first serving a search warrant.
- It allows the FBI to obtain—without showing probable cause—a warrant authorizing the seizure of business, medical, educational, and library records of suspects.

The Council of the American Library Association passed a resolution on the Patriot Act in January 2003. The resolution affirms every person's rights to inquiry and free expression. It "urges librarians everywhere to defend and support user privacy and free and open access to knowledge and information," and it "urges libraries to adopt and implement patron privacy and record retention policies" that minimize the collection of records about the activities of individual patrons [66].

FOLLOW-ON LEGISLATION

In February 2003 a draft copy of the Domestic Security Enhancement Act of 2003 was leaked to the press. The bill, dubbed "Patriot Act II," would have given the U.S. government sweeping new powers. Here are some of the provisions of the act:

- The government would have the ability to expatriate an American citizen "convicted of giving material support to a group that's designated a terrorist organization" [67].
- It would require the names of people being held on suspicion of terrorism to be kept secret.
- Law enforcement officials would be able to use administrative subpoenas to gain access to records held by ISPs, doctors, family members, or friends. An administrative subpoena does not require the approval of a judge unless the person being served the subpoena raises an objection.
- The act would make it simpler for police to gain access to credit reports.
- Police would have the right to collect DNA samples from suspected terrorists. The federal government would create a national DNA database. Federal, state, and local law enforcement agencies would be able to access the national database.
- Police would have the right to wiretap suspects and intercept their email for 15 days without obtaining a warrant.

Unlike the original Patriot Act, which passed Congress with little debate, many voices were raised against the follow-on bill. Congress adjourned at the end of 2003 without passing the Domestic Security Enhancement Act.

SUCCESSES AND FAILURES

According to Tom Ridge, Secretary of the Department of Homeland Security, the Patriot Act has helped the government in its fight against terrorism by allowing greater information-sharing among law enforcement and intelligence agencies, and by giving law enforcement agencies new investigative tools—"many of which have been used for years to catch mafia dons and drug kingpins" [68]. Terrorism investigations have led to charges being brought against 361 individuals in the United States. Of these, 191 have been convicted or pled guilty, including shoe-bomber Richard Reid, and John Walker Lindh, who fought with the Taliban in Afghanistan. More than 500 individuals linked to the September 11th attacks have been removed from the United States. Terrorist cells in Buffalo, Seattle, Tampa, and Portland (the "Portland Seven")[2] have been broken up [68].

Unfortunately, a few innocent bystanders have been affected by the war against terrorism. A notable example is Brandon Mayfield.

During the morning rush hour on March 11, 2004, ten bombs exploded on four commuter trains in Madrid, Spain, killing 191 people and wounding more than 2,000 others. The Spanish government retrieved a partial fingerprint from a bag of detonators, and the FBI linked the fingerprint to Brandon Mayfield, an attorney in Portland, Oregon [69].

Without revealing their search warrant, FBI agents secretly entered Mayfield's home multiple times, making copies of documents and computer hard drives, collecting ten DNA samples, removing six cigarette butts for DNA analysis, and taking 355 digital photographs. The FBI also put Mayfield under electronic surveillance [70]. On May 6, 2004, the FBI arrested Mayfield as a material witness and detained him for two weeks. After the Spanish government announced that it had matched the fingerprints to Ouhnane Daoud, an Algerian national living in Spain, a judge ordered that Mayfield be released. The FBI publicly apologized for the fingerprint misidentification [69].

Mayfield said his detention was "an abuse of the judicial process" that "shouldn't happen to anybody" [69]. The only evidence against Mayfield was a partial fingerprint match that even the Spanish police found dubious. Mayfield had not left the United States in more than a decade, and he had no connections with any terrorist organizations. Some civil rights groups suggest Mayfield was targeted by the FBI because of his religious beliefs. The affidavit that the FBI used to get an arrest warrant pointed out that Mayfield "had converted to Islam, is married to an Egyptian-born woman, and had once briefly represented a member of the Portland Seven in a child-custody case" [71].

PATRIOT ACT RENEWAL

Sixteen provisions of the Patriot Act authorizing certain forms of intelligence-gathering and surveillance activity were scheduled to expire at the end of 2005. The Bush Administration advocated making all of the provisions permanent. At the time this book

2. The "Portland Seven" included six American Muslim men accused of attempting to travel to Afghanistan to fight with the Taliban.

went to press, the House of Representatives had voted to make fourteen of these provisions permanent, although it put a 10-year limit on the two most controversial provisions. The first of these provisions allows roving wiretaps associated with a person rather than a particular phone number. The other allows the FBI to seize records from financial institutions, libraries, doctors, and businesses with approval from the secret Foreign Intelligence Surveillance Court. The Senate voted to renew these two provisions for only four years, and it added new restrictions to the FBI's ability to seize business records [72].

5.9 Data Mining

Data mining is the process of searching through one or more databases looking for patterns or relationships. Data mining is a way to generate new information by combining facts found in multiple transactions. It can also be a way to predict future events.

A record in a database records a single transaction, such as a particular purchase at a store. You can think of a database record as being a single snapshot of a person. It tells you something about the person, but in isolation its value is limited. By drawing upon large numbers of records, data mining allows an organization to build an accurate profile of an individual from a myriad of snapshots.

The first time you rent a movie from a video rental store, it takes a long time. You have to fill out an application asking for a lot of personal information, such as your name, address, and phone number. After the store has approved your application, renting movies is quick. You identify yourself to the clerk, and he accesses your file on the computer. The clerk scans your choices, takes your money, and you are on your way. The primary use of the information you provided was to allow you to rent movies from this particular video store.

Frequently, however, information is put to another purpose. This is called a **secondary use** of your data. Companies can look through a series of transactions in order to create more personal relationships with their customers [73]. For example, **collaborative filtering** algorithms draw upon information about the preferences of a large number of people to predict what an individual may enjoy. An organization performing collaborative filtering may determine people's preferences explicitly, through rankings, or implicitly, by tracking their purchases. The filtering algorithm looks for patterns in the data. Perhaps many people who purchase item X also purchase item Y. If a new customer selects X or Y, the collaborative filtering software will suggest the customer may also like the other item. Collaborative filtering software is used by on-line retailers and DVD-rental sites to make recommendations [74].

Information about customers has itself become a commodity. Organizations sell or exchange information with other organizations (Figure 5.5). This is another secondary use of data, and it is a common way for organizations to gather large databases of information they can mine.

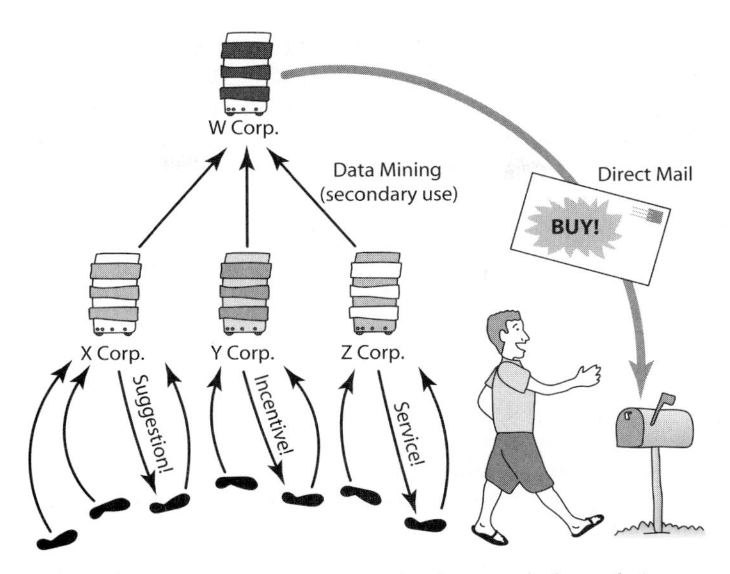

Figure 5.5 Companies use computers to record information about their customers and their buying habits. They analyze this information to suggest additional purchases, provide incentives, and deliver better service. They may also sell this information to other companies. By combining information from various sources, a company can build sophisticated profiles of individuals and target its direct mail advertising to those people most likely to be interested in its products.

For example, a company selling time-share condominiums purchases from a hotel chain the names and addresses of people who have vacationed in a resort area in the past two years. From another organization it purchases a database that gives the approximate annual household income of a family, based on that family's nine-digit ZIP code. Combining these lists allows the time-share agency to target people most likely to have both the interest and the financial resources to purchase a share of a vacation condominium. It uses direct mail to send brochures to these people.

Data mining can be surprisingly powerful. Suppose a government agency managing tollbooths sells information records of the form

⟨transponder number⟩ ⟨date⟩ ⟨time⟩ ⟨location⟩ ⟨charge⟩

The agency does not reveal the names of the owners of the cars, so it believes it is protecting their anonymity. However, many people have an account set up so that their tollbooth payments are automatically charged to their credit cards. If a credit card company buys these records from the tollbooth agency, it can match the date, time, and amount of the tollbooth payments with the date, time, and charge on its credit cards to determine the identity of the person driving a vehicle with a particular transponder number. Once this has been done, the credit card company can figure out which customers are driving the most miles and are likely to purchase new cars more frequently.

It can then sell this information to banks interested in soliciting automobile loan applications [37].

Advances in information technology have led to a drop in the cost of acquiring information. Meanwhile, the value of information continues to rise, as organizations refine their data mining techniques. The result of these trends is that organizations have increased incentives to acquire information, making it more difficult for individuals to protect their privacy [23]. Still, people can and do fight back when they feel an organization has gone too far. A case in point is Lotus Development's failure with its Marketplace: Households project.

5.9.1 Marketplace: Households

Lotus Development Corporation spent $8 million developing a CD with information on 120 million people, along with software that would help the purchaser produce mailing lists based on various criteria, such as household income. Lotus hoped to sell the CD, which it called "Marketplace: Households," to small businesses. When consumers found out about the CD, they complained loudly and vigorously, with more than 30,000 letters, phone calls, and emails. Lotus dropped plans to sell the CD [75].

5.9.2 IRS Audits

To identify taxpayers who have paid less than they owe, the IRS uses computer matching and data mining strategies. First, it matches information on the tax form with information provided by employers and financial institutions. This is a straightforward way to detect unreported income.

Second, the IRS audits a couple of million tax returns every year. Its goal is to select the most promising returns—those containing errors resulting in underpayment of taxes. The IRS uses a computerized system called the discriminant function (DIF) to score every tax return. The DIF score is an indicator of how many irregularities there are on a tax form, compared to carefully constructed profiles of correct tax returns. About 60 percent of tax returns audited by the IRS are selected due to their high DIF scores.

5.9.3 Syndromic Surveillance System

Another application of data mining by the government is protecting society from imminent dangers.

New York City has created the Syndromic Surveillance System, which analyzes more than 50,000 pieces of information every day, including 911 calls, visits to emergency rooms, and purchases of prescription drugs. The purpose of the system is to find patterns that might indicate the onset of an epidemic, an environmental problem leading to illnesses, or bioterrorism. In the fall of 2002, the system detected a surge in people seeking treatment for vomiting and diarrhea. These symptoms were the first signs of an outbreak of a Norwalk-type virus. The alert generated by the system allowed city officials to warn doctors about the outbreak and advise them to be particularly careful about handling the highly contagious body fluids of their affected patients [76].

5.9.4 Total Information Awareness

The Total Information Awareness (TIA) project, proposed by the Information Awareness Office of the U.S. Defense Advanced Research Projects Agency, is an example of data mining taken to a high level. The idea of the project is to detect potential terrorists by capturing everyone's "information signature" and using sophisticated computer algorithms to detect terrorist-like patterns of activity. TIA databases would contain financial, medical, communication, travel, and other records. The use of biometric technology to identify people from video images would allow TIA databases to incorporate person/location/time information as well [77].

The Total Information Awareness program received a skeptical reception from the U.S. Congress. In February 2003 Congress suspended funding for the domestic surveillance portion of the program [78]. Subsequently, DARPA changed the name of the program to Terrorist Information Awareness [79].

5.9.5 Criticisms of the TIA Program

In an open letter to John Warner and Carl Levin, ranking members of the Senate Committee on Armed Services, a group of computer scientists voiced objections to the Total Information Awareness program [80]. The letter, written on behalf of the U.S. Public Policy Committee of the Association for Computing Machinery (ACM), agrees with the notion that advances in information technology can contribute to public safety and national defense. However, the letter suggests that the Total Information Awareness program will have more harms than benefits.

The all-encompassing databases suggested by the TIA program would represent large security and privacy risks. They would contain a great deal of sensitive (and valuable) information about individuals. Hence the databases themselves would become targets for criminals and terrorists. In addition, tens of thousands of system administrators, law enforcement people, and intelligence officers would have access to the data, creating many opportunities for the security of the database to be compromised.

These databases would increase the risk of identity theft by putting in a single place a host of personal information. If terrorists could get access to the data, it would be easier for them to assume false identities.

Because the information in the TIA databases was secret, citizens would not have the ability to verify that the information stored about them is correct.

The TIA program may hurt the competitiveness of U.S. companies in the worldwide e-commerce. Non-Americans who do not wish their consumer profiles to become part of a TIA database may choose to purchase their goods from non-U.S. sources.

Identity theft is a growing problem. Transactions performed by people who have stolen the identities of others will introduce "noise" into the TIA database, giving a false impression of the activities of the actual person.

The goal of the TIA program is to identify patterns of behavior that indicate a person is a terrorist. It is inevitable that such a system will generate some "false positives"—labeling innocent persons as potential terrorists. Even a 99 percent accuracy rate could result in millions of Americans being incorrectly identified.

Finally, knowing about the existence of TIA will modify people's behavior. Terrorists will do everything possible to make sure their behavior looks "normal." Innocent people may avoid certain activities, even though they are legal, out of fear of being targeted by the system.

5.9.6 Who Should Own Information about a Transaction?

Does a person buying a product or service have the right to control information about the transaction? Does the seller have this right? Consider the following hypothetical example.

Dr. Knowitall, a computer science professor, takes his broken computer to the Computer Shop so that 18-year-old Andy can fix it for him. Dr. Knowitall is embarrassed that he can't fix the computer himself, and he doesn't want anybody to find out that he must pay someone to fix it. Dr. Knowitall certainly isn't going to tell anyone, but does he have the right to prevent Andy from telling anyone? Or maybe Andy wants to keep the transaction a secret, because he's embarrassed it took him so long to fix the computer and he doesn't want anyone to find out he was in over his head. Does Andy have the right to keep Dr. Knowitall from talking about it?

It seems that neither person can claim the right to control information about this transaction. Since information about the transaction becomes public information if either party discloses it, keeping the transaction private is more difficult (hence more valuable) than making it public.

If Dr. Knowitall wants to keep the transaction private, he should be willing to pay for it. He may tell Andy, "I'll give you an extra 20 bucks if you promise you'll not tell anybody that you fixed my computer." At this point Dr. Knowitall has purchased control over the information about this transaction. Andy is obliged to keep his mouth shut, not because of Dr. Knowitall's right to privacy, but because of his right to expect the agreement will be upheld.

5.9.7 Opt-in Versus Opt-out

What rules should govern the secondary use of information collected by organizations selling products or services? Two fundamentally different policies are called opt-in and opt-out.

The **opt-in** policy requires the consumer to explicitly give permission for the organization to share the information with another organization. Opt-in policies are preferred by privacy advocates.

The **opt-out** policy requires the consumer to explicitly forbid an organization from sharing information with other organizations. Direct marketing associations prefer the opt-out policy, because opt-in is a barrier for new businesses. New businesses do not have the resources to go out and collect all the information they need to target their mailings to the correct individuals. In an opt-out environment, most people will not go through the effort required to actually remove themselves from mailing lists. Hence it is easier for new businesses to get access to the mailing lists they need to succeed [81]. Another argument for opt-out is that companies have the right to control information about the transactions they have made. Information is a valuable commodity. An opt-in policy takes this commodity away from companies.

Some have suggested that the relationship between a consumer and a company is similar to the relationship between a patient and a doctor. A doctor is not supposed to reveal information about her patients. So-called **Hippocratic databases** contain rules about who should have access to the data and how long it should be stored. When users of a Hippocratic database enter information, they can see this information about the use and duration of the data and decide whether or not to actually submit the information [82].

5.9.8 Platform for Privacy Preferences (P3P)

The World Wide Web Consortium has developed an industry standard called Platform for Privacy Preferences (P3P) Project (www.w3.org/P3P). The goal of this project is to provide users with an automated way to control the use of personal information on the Web sites they visit. Participating sites disclose their privacy policies in a machine-readable format. The user's browser can compare these policies with the user's preferences. The user can then decide whether to visit the site.

Critics of P3P point out the P3P is a voluntary standard. Web sites are not required to disclose their privacy policies in a P3P-compatible format. Even worse, P3P cannot monitor whether a site is actually abiding by its stated policy.

5.10 Identity Theft

5.10.1 Background

Dorothy Denning defines **identity theft** as "the misuse of another person's identity, such as name, Social Security number, driver's license, credit card numbers, and bank account numbers. The objective is to take actions permitted to the owner of the identity, such as withdraw funds, transfer money, charge purchases, get access to information, or issue documents and letters under the victim's identity" [83].

The leading form of identity theft in United States is credit card fraud. Identity thieves either take out a new credit card in someone else's name or commandeer an existing account [84]. By changing the billing address of existing accounts, a thief can run up large debts before the victim becomes aware of the problem. These activities can blemish the target's credit history. As a result, victims of identity theft may have applications for credit cards, mortgage loans, and even employment denied. If the impostor shows false credentials to the police, the victim may even be saddled with a false criminal record or outstanding arrest warrants.

Financial institutions contribute to the problem of identity theft by making it easy for people to open up new accounts. Since information brokers on the Web are selling driver's license numbers, Social Security numbers, and credit card information, it's easy for an identity thief to gather a great deal of information about another person. Assuming another person's identity is made simpler by banks allowing people to open accounts online [85].

According to Privacy Rights Clearinghouse, more than 27 million Americans were victims of identity theft between 2000 and 2004. The frequency of identity theft is increasing; there were nearly 10 million identity theft victims in 2004. The average loss in 2004 was about $5,000 per victim [86].

Fortunately, United States law says that a consumer's liability for losses due to credit card fraud are limited to $50 if reported promptly. Many financial institutions do not even collect this amount [87]. However, victims of identity theft typically spend hundreds of hours cleaning up their financial records [86].

Most cases of identity theft are not the result of someone using computers to break into a database containing information about a target. Instead, identity thieves are much more likely to use low-tech methods to gain access to the personal information they need. Two popular sources of information are mailboxes and lost or stolen wallets. One in six cases of identity theft are traced to family members, friends, or coworkers [86].

Some identity thieves engage in **dumpster diving**—looking for personal information in garbage cans or recycling bins. Old bills, bank statements, and credit card statements contain a wealth of personal information, including names, addresses, and account numbers. Another simple way to get information is through **shoulder surfing**—looking over the shoulders of people filling out forms.

More recently, thieves have begun using **skimmers** (also called **wedges**) to steal credit card data. A skimmer is a small, battery-powered credit card reader. Identity theft rings use skimmers to collect hundreds of credit card numbers, then use these numbers to manufacture counterfeit credit cards. Credit card numbers are collected by waiters or store clerks, who match each legal swipe through a cash register with an illegal swipe through a skimmer. In one case, someone attached a skimmer to an ATM along with a sign requesting customers to use the "card cleaner" before putting their card in the ATM [88].

Some thieves send out spam messages designed to look like they originated from PayPal, eBay, or another well-known Internet-active business. Through these messages they hope to con unsuspecting recipients into revealing their credit card numbers or other personal information. Gathering financial information via spam is called **phishing** (pronounced "fishing").

Large institutions that store personal data are tempting targets for information thieves. Between 2003 and 2005 criminals used stolen passwords to access LexisNexis databases 59 times, retrieving the Social Security numbers and other financial information of more than 300,000 people [89]. ChoicePoint disclosed that it accidentally gave con artists access to about 150,000 personal financial dossiers, and Bank of America lost computer tapes containing information about more than a million federal employees

[90]. A hacker broke into a T-Mobile database and downloaded the photographs and personal information of at least 400 customers [90].

The Identity Theft and Assumption Act of 1998 makes identity theft a federal crime. In 2004 Congress passed the Identity Theft Penalty Enhancement Act, which lengthened prison sentences for identity thieves [91]. A variety of law enforcement agencies investigate alleged violations of this law: the U.S. Secret Service, the FBI, the U.S. Postal Inspection Service, and the Office of the Inspector General of the Social Security Administration [92]. Unfortunately, the probability that a particular case of identity theft will result in an arrest is about 1 in 700 [93].

The rapid increase in the number of identity theft victims is prompting new questions about how people identify themselves. In the United States, the Social Security number is a common form of identification, even though that was not how it was originally conceived.

5.10.2 History and Role of the Social Security Number

The Social Security Act of 1935 established two social insurance programs in the United States: a federal system of old-age benefits to retired persons, and a federal-state system of unemployment insurance. Before the system could be implemented, employers and workers needed to become registered. The Social Security Board contracted with the U.S. Postal Service to distribute applications for Social Security cards. The post office collected the forms, typed the Social Security cards, and returned them to the applicants. In this way, over 35 million Social Security cards were issued in 1936–1937 [94].

The U.S. government initially stated that Social Security numbers (SSNs) would be used solely by the Social Security Administration, and not as a national identification card. In fact, from 1946 to 1972, the Social Security Administration put the following legend on the bottom of the cards it issued: "FOR SOCIAL SECURITY PURPOSES—NOT FOR IDENTIFICATION." However, use of the SSN has gradually increased. President Roosevelt ordered in 1943 that federal agencies use SSNs as identifiers in new federal databases. In 1961 the Internal Revenue Service began using the SSN as the taxpayer identification number. Because banks report interest to the IRS, people must provide their SSN when they open a bank account. The SSN is typically requested on applications for credit cards. Motor vehicle departments and some other state agencies received permission to use SSNs as identification numbers in 1976. Many universities use the SSN as an identification number for faculty and students. The IRS now requires parents to provide the SSNs of their children over one year old on income tax forms in order to claim them as dependents. For this reason, children now get a SSN soon after they are born. Many private organizations ask people to provide SSNs for identification. The SSN has become a de facto national identification number in the United States.

Unfortunately, the SSN has serious defects that make it a poor identification number. The first problem with SSNs is that they are not unique. When Social Security cards were first issued by post offices, different post offices accidentally assigned the same SSN to different people. In 1938 wallet manufacturer E. H. Ferree included sample Social Security cards in one of its products. More than 40,000 people purchasing the wallets from

Woolworth stores thought the cards were real and used the sample card's number as their SSN [95].

A second defect of SSNs is that they are rarely checked. Millions of Social Security cards have been issued to applicants without verifying that the information provided by the applicants is correct. Many, if not most, organizations asking for a SSN do not actually require the applicant to show a card, making it easy for criminals to supply fake SSNs.

A third defect of SSNs is that they have no error-detecting capability, such as a check digit at the end of the number. A check digit enables computer systems to detect common data entry errors, such as getting one digit wrong or transposing two adjacent digits. If someone makes one of these mistakes, the data-entry program can detect the error and ask the person to retype the number. In the case of SSNs, if a person accidentally types in the wrong number, there is a high likelihood that it is a valid SSN (albeit one assigned to a different person). Hence it is easy to contaminate databases with records containing incorrect SSNs [96]. Similarly, without check digits or another error-detection mechanism, there is no simple way for a system to catch people who are simply making up a phony SSN.

5.10.3 Debate over a National ID Card

The events of September 11, 2001, resurrected the debate over the introduction of a national identification card for Americans.

Proponents of a national identification card point out numerous benefits to its adoption:

1. *Currently we are relying upon second-rate identification methods, such as SSNs and driver's licenses.*

   It would be better to rely on an identification card that was harder to forge. A modern card could incorporate a photograph as well as a thumbprint or other biometric data.

2. *A national ID card would make it much more difficult for people to enter the United States illegally.*

   This would help prevent terrorist attacks.

3. *Requiring employers to check the ID card would prevent illegal aliens from working in the United States.*
4. *Giving police the ability to positively identify people would reduce crime.*
5. *Many democratic countries already use national ID cards, including Belgium, France, Germany, Greece, Luxembourg, Portugal, and Spain.*

Opponents of a national identification card suggest these harms may result from its adoption:

1. *A national identification card does not guarantee that the apparent identity of an individual is that person's actual identity.*

Driver's licenses and passports are supposed to be unique identifiers, but there are many criminals who produce fake driver's licenses and passports. Even a hard-to-forge identification card system may be compromised by insiders. For example, a ring of motor vehicle department employees in Virginia was caught selling fake driver's licenses [97].

2. *It is impossible to create a biometric-based national identification card that is 100 percent accurate.*

   All known systems suffer from false positives (erroneously reporting that the person does not match the ID) and false negatives (failing to report that the person and ID do not match). Biometric-based systems may still be beaten by determined, technology-savvy criminals [97].

3. *There is no evidence that institution of a national ID card actually leads to a reduction in crime.*

   In fact, the principal problem faced by police is not the inability to make positive identifications of suspects, but the inability to obtain evidence needed for a successful prosecution.

4. *A national identification card makes it simpler for government agencies to perform data mining on the activities of its citizens.*

   According to Peter Neumann and Lauren Weinstein, "The opportunities for over-zealous surveillance and serious privacy abuses are almost limitless, as are opportunities for masquerading, identity theft, and draconian social engineering on a grand scale . . . The road to an Orwellian police state of universal tracking, but actually *reduced* security, could well be paved with hundreds of millions of such [national identification] cards" [97].

5. *While most people may feel they have nothing to fear from a national identification card system, since they are law-abiding citizens, even law-abiding people are subject to fraud and the indiscretions and errors of others.*

   Suppose a teacher, a doctor, or someone else in a position of authority creates a file containing misleading or erroneous information. Files created by people in positions of authority can be difficult to remove [98].

   In a society with decentralized record-keeping, old school or medical records are less likely to be accessed. The harm caused by inaccurate records is reduced. If all records are centralized around national identification numbers, files containing inaccurate or misleading information could haunt individuals for the rest of their lives.

5.10.4 The REAL ID Act

In May 2005 President George W. Bush signed the REAL ID Act, which significantly changes driver's licenses in the United States. The motivation for passing the REAL ID Act was to make driver's licenses a more reliable form of identification. Critics, however, say the act is creating a de facto national ID card in the United States.

The REAL ID Act requires that every state issue new driver's licenses by the end of 2008. These licenses will be needed in order to open a bank account, fly on a commercial airplane, or receive a government service, such as a Social Security check. The new law will make it more difficult for impostors to get driver's licenses by requiring applicants to supply four different kinds of documentation and requiring state employees to verify these documents using federal databases. If, as expected, the driver's license contains a biometric identifier, such as a fingerprint, it will be a stronger credential than current licenses [99].

While each state will be responsible for issuing new driver's licenses to its own citizens, these licenses must meet federal standards. One of these standards is that the licenses include data in machine-readable form. The Department of Homeland Security will determine the data to be included on a driver's license and the data-storage technology to be used, most likely an RFID chip [99].

Supporters of the measure say making the driver's license a more reliable identifier will have numerous benefits. Law enforcement is easier when police can be more certain that a driver's license correctly identifies the individual carrying it. Society is better off when parents ducking child support and criminals on the run cannot change their identities by crossing a state border and getting a new driver's license under a different name [100].

Some critics fear having machine-readable information on driver's licenses will aggravate problems with identity theft. American Civil Liberties Union lawyer Timothy Sparapani said, "We will have all this information in one electronic format, in one linked file, and we're giving access to tens of thousands of state DMV employees and federal agents" [101].

Proponents of the bill say such fears are unjustified. They suggest that the personal information actually available on the new driver's license is relatively insignificant compared with all the other personal information circulating around cyberspace [100].

5.11 Encryption

Encryption is the process of transforming a message in order to conceal its meaning. In an age in which information is easily captured and rebroadcast, encryption is a valuable tool for maintaining privacy. Even if someone should get a copy of an encrypted message, it is worthless unless the person can decode it.

5.11.1 Symmetric Encryption

In a traditional **symmetric encryption scheme,** a single key is needed to encrypt and decrypt a message. Suppose Smith wants to send a message to Jones. Smith and Jones are the only two people to know the key. Smith uses the key to encrypt the message into cipher. Jones uses the key to decrypt the cipher back into the message. Since no one else has the key, even if the cipher should fall into the wrong hands, it cannot be read. The weakness of symmetric encryption schemes is that it does not solve the problem of how

Smith gets the key to Jones. If a hostile outsider should get a copy of the key as it is transmitted, the security of the system is broken.

5.11.2 Public-Key Cryptography

The key transmission problem was solved by Whitfield Diffie and Martin Hellman, who published an alternative scheme, called **public-key cryptography,** in 1976. Public-key encryption is an example of **asymmetric encryption** because it uses two keys instead of one. Each person has a public key and a private key. A message encrypted with the public key can only be decrypted with the private key. That means everybody who wants to *receive* encrypted messages announces their public keys. If Smith wants to send an encrypted message to Jones, he uses *Jones's* public key. After Jones receives the cipher, he uses his private key to decrypt it. Public-key cryptography eliminates the Achilles heel of symmetric encryption schemes, because no longer is there a need for people to exchange keys. Figure 5.6 illustrates how the RSA public-key encryption algorithm works.

There is a mathematical relationship among the public and private keys, and it is theoretically possible to determine the private key from the public key. The time needed to determine the private key increases with the length of the key. We say encryption is weak if a computer can guess the private key from the public key in a reasonably small amount of time. In contrast, encryption is strong if the amount of time needed by a computer to guess the private key is so long that decryption is essentially impossible. For example, it may be possible for a computer to decipher a strongly encrypted message in 2,000 years, but by that time it probably will make no difference. Strong encryption is possible by choosing a long enough public key.

As we have seen, various federal agencies perform surveillance operations to fight crime and maintain the national security. The work of these agencies is simplified if they have the ability to read domestic and foreign messages. If messages are weakly encrypted, their work is still possible because they have high-speed computers that can decipher the messages. However, if messages are strongly encrypted, their work is made much more difficult. The traditional policy of the United States has been to regulate the use of cryptography within the United States and to forbid the exportation of strong encryption technology.

5.11.3 Pretty Good Privacy

In 1991 the U.S. Senate debated Senate Bill 266, a crime-fighting measure. Buried in the bill was a statement that all manufacturers of communications devices using cryptography would have to provide a "back door" enabling government agencies to read the ciphers. Phil Zimmerman was alarmed by this statement. He wrote:

> If privacy is outlawed, only outlaws will have privacy. Intelligence agencies have access to good cryptographic technology. So do the big arms and drug traffickers. So do defense contractors, oil companies, and other corporate giants. But ordinary people and grassroots political organizations mostly have not had access to affordable military grade public-key cryptographic technology. Until now. [102]

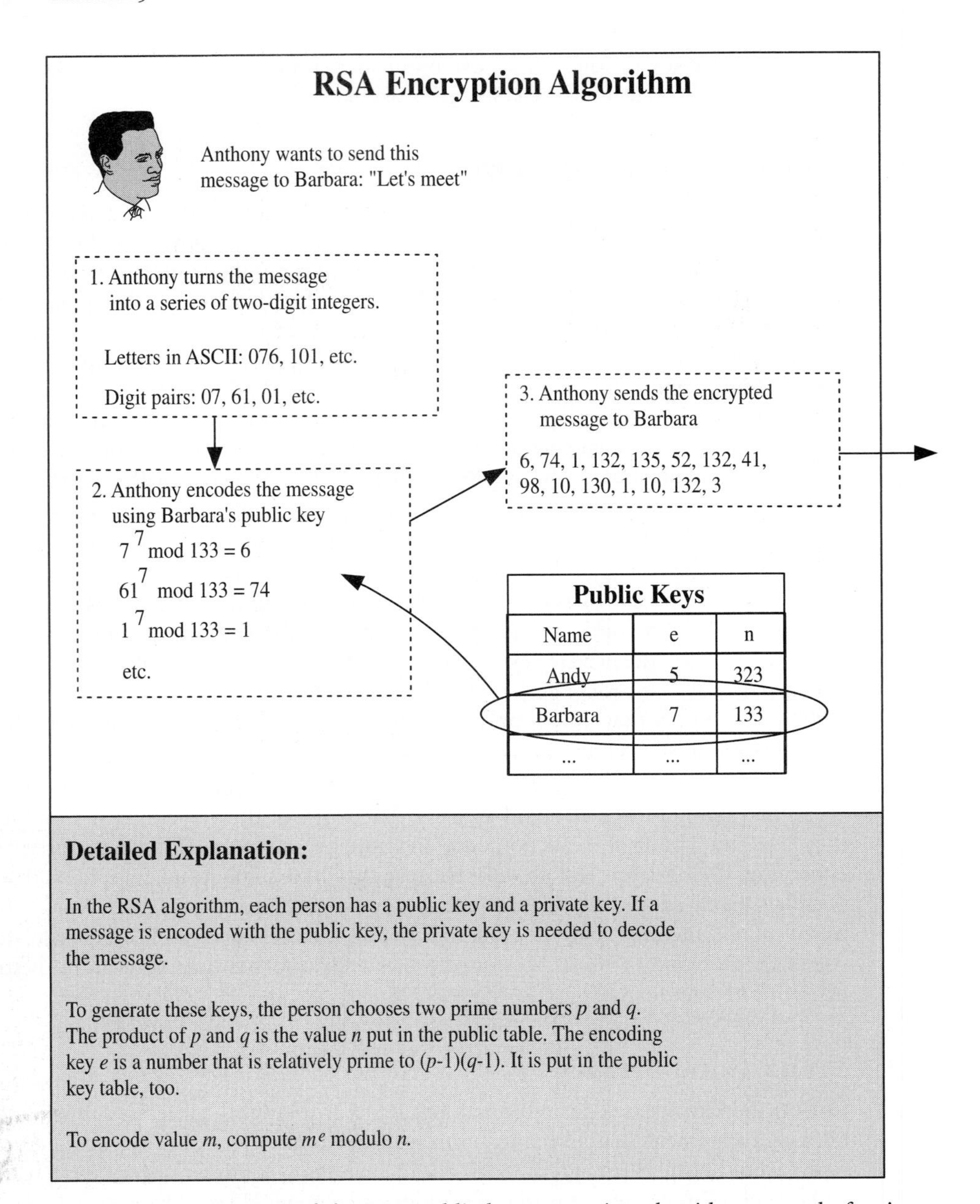

FIGURE 5.6 Illustration of the RSA public-key encryption algorithm, named after its inventors, Ron Rivest, Adi Shamir, and Leonard Adleman.

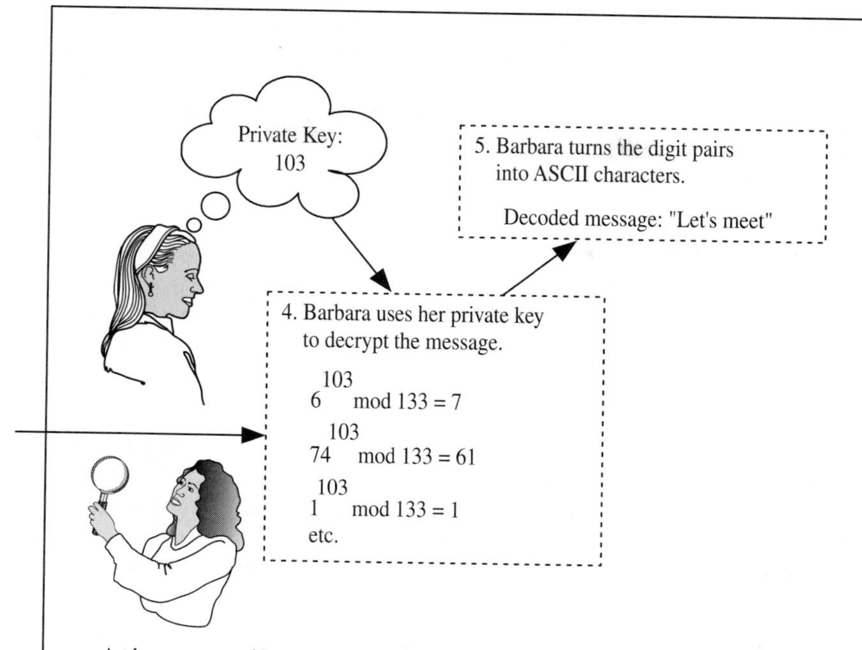

Anthony can *send* Barbara a message without knowing her private key, but an eavesdropper can't *read* the message without knowing the private key. Since Barbara doesn't have to reveal her private key to anyone, it can't be intercepted by an eavesdropper.

In this simple example Barbara has chosen $p = 11$ and $q = 13$. Since $n = pq$, n has the value 133. The encrypting key $e = 7$, which is relatively prime to $(p-1)(q-1)$, or 120.

The private, decoding key d is the unique number such that $de = 1$ modulo $(p-1)(q-1)$. Here, the decrypting key $d = 103$, since 7 times 103 is 721, which has a remainder of 1 when divided by $(p-1)(q-1)$, or 120.

To decode value m, compute m^d modulo n.

In the real world, primes p and q should be hundreds of digits long, making it impractical to discover the private key, even with the aid of a supercomputer. (Note: This explanation ignores other important, but more technical, considerations.)

FIGURE 5.6 *(continued)*

Zimmerman created a public key cryptography program called PGP, which stands for Pretty Good Privacy. He made the program freely available on several Internet sites in the United States. Many people, both inside and outside the United States, downloaded the program. For the next several years the U.S. government threatened legal action against Zimmerman for violating laws against exporting encryption technology. The government's view was that posting the code on the Internet was equivalent to exporting it.

5.11.4 Clipper Chip

In September 1992 AT&T announced plans to sell for $1,100 a telephone encryption device called the Surity 3600. The FBI and NSA were concerned that this device would compromise their ability to listen in on telephone calls. The Justice Department approached AT&T and suggested that it replace its own encryption scheme with NSA's encryption technology. The code name for this technology was called "Clipper."

The reason the U.S. government wanted AT&T to use the Clipper chip was that the government had the Clipper's mathematical decryption key. With the key in hand, the government would be confident it could listen in on conversations encrypted with the Clipper. AT&T agreed to the Justice Department's proposal, hoping that the U.S. government would make Clipper a national standard for telephone encryption and giving AT&T an edge over its competitors.

In March 1993 President Clinton made a decision to go ahead with the Clipper plan. Two weeks later he publicly announced his support for making the Clipper a national standard. The Justice Department ordered 9,000 Clipper-equipped phones from AT&T. It issued guidelines for control of the keys: two government agencies connected with law enforcement and the intelligence community would have copies of the keys. While the agencies were not supposed to release the keys without proper authorization, there were no penalties for improper release of the keys. The Justice Department did not provide citizens a way to object to the release of the keys or to suppress information gathered, even if the law enforcement or intelligence agency had violated the guidelines.

When the American public became aware of what the government was proposing, the reaction was overwhelmingly negative. A Time/CNN poll showed that 80 percent of the public was opposed to the Clipper proposal [103]. In February 1994 the Clinton administration retreated, proposing that Clipper encryption be named a voluntary standard. The Department of Justice and the NSA continued to urge manufacturers to use Clipper technology rather than other encryption schemes. Incredibly, the NSA also attempted to persuade other countries to make Clipper their standard encryption technology! Needless to say, foreign governments were more than a little reluctant to use an encryption scheme for which the NSA held the decryption key. In the end, the U.S. government's effort to standardize around Clipper failed.

5.11.5 Effects of U.S. Export Restrictions

While the U.S. government prevented American companies from exporting products using strong encryption, this ban did not prevent companies in other countries from developing products that did use strong encryption. Soon after its release on the Internet, the source code to PGP was downloaded to computers outside the United States.

It didn't take long for thousands of international software packages to incorporate PGP encryption. American software companies were allowed to manufacture software products using strong encryption for sale inside the United States. Eventually pirated versions of these programs became available outside the United States.

The U.S. State Department ban on selling software with strong encryption outside the United States resulted in an additional burden on the software industry. Each software maker was faced with a difficult choice. The more expensive alternative was to create and maintain two versions of each software product: the strong-encryption version for sale in the United States and the weak-encryption version for export. The less expensive alternative was to develop only a single version of the software product, the one based on weak encryption, and sell this inferior version both in the United States and internationally.

The ban reduced the international competitiveness of United States companies. U.S. companies forbidden from selling products with strong encryption lost sales to foreign companies that were able to use this technology.

In 1999 and 2000 two different federal appeals courts ruled that the export restrictions violated freedom of speech. In one of the rulings, the court lauded encryption as a way of protecting privacy. Since these rulings, the U.S. State Department has dropped export restrictions on encryption technology.

For decades the National Security Agency has been associated with high-speed computing to fulfill its mission of creating unbreakable codes and breaking the codes of other organizations. According to James Bamford, the NSA now has computers capable of performing a quintillion (1,000,000,000,000,000,000) operations a second [104]. Sara Baase raises an interesting conjecture: Perhaps the U.S. State Department dropped its export restrictions because by 1999 the NSA finally had computers fast enough to decrypt ciphers created using strong encryption [105].

5.11.6 Digital Cash

Glyn Davies defines **money** as "anything that is widely used for making payments and accounting for debts and credits" [106]. Various societies have used cattle, beads, precious metals, coins, paper currency, and many other objects as money. Today, money is often represented electronically. For example, banks routinely use electronic funds transfer to settle debts with each other. Electronic money is an alternative to physical coins and currency. Various systems have been devised that allow people to store electronic money in smart cards or on computer disks.

Electronic money relies upon public-key encryption. Earlier we stated that in a public-key encryption system, a message encrypted with a public key can be decrypted with the associated private key. It's also true that a message encrypted with a private key can be decrypted with the associated public key. When issuing electronic money, a bank signs it with its private key. Customers and merchants can use the bank's public key to verify that the money is authentic. In a similar way, bank customers can use their private key to withdraw funds. The bank uses the customer's public key to verify the identity of the customer.

Implementations of electronic money fall into two categories. In an **identified electronic money** system, the bank can trace the use of the money back to the person who withdrew it from the bank. In an **anonymous electronic money** system, there is no way for a bank to determine what the person who withdrew the money used it for. Anonymous electronic money is also called **digital cash.** Like the physical coins and currency we are familiar with, digital cash allows people to preserve their privacy by conducting transactions without leaving an electronic trail behind.

A digital cash system relies upon a **blind signature** protocol that prevents the bank from putting its mark on the electronic cash it issues a customer. Figure 5.7 illustrates a physical analogy created by Bruce Schneier that demonstrates how a bank can confidently sign a purchase order that it has never seen [107].

Digital cash systems can be divided into online and off-line systems. In an **online** system, the merchant communicates with the bank at the time of the sale, as most modern credit card transactions are handled. In an **off-line** system, there is no requirement that the merchant communicate with the bank at the time of the sale. Instead, the merchant collects information from many transactions and contacts the bank periodically to have the funds transferred into its account.

As we have seen in earlier chapters, digital information is simple to duplicate. Suppose Ann has $500 of digital cash. What keeps her from making 1,000 perfect copies of it and going on a wild spending spree?

An online system would catch Ann as soon as she attempted to spend the money the second time. The bank keeps a database of digital cash serial numbers. It knows which digital cash is still out in circulation and which has been spent. When the merchant seeks authorization from the bank, the bank will refuse payment, and the merchant can identify Ann as the person attempting to spend duplicated digital cash.

Even off-line cash systems can prevent people from spending the same cash twice. An observer chip can be placed in the smart card containing the digital cash. The observer chip maintains its own database of digital cash serial numbers, and it can prevent a person from spending the same money twice. Tampering with the observer chip results in the loss of all the information on the chip (i.e., the destruction of the digital cash).

Another off-line digital cash protocol does not prevent a person from spending the same money twice, but it does allow banks to catch those responsible for duplicating digital cash. This protocol has the amazing characteristic that it allows a bank to positively identify someone who has spent two copies of the same digital cash, while it maintains the anonymity of honest people who have spent their digital cash only once. This protocol is more complicated than the one illustrated in Figure 5.7. Bruce Schneier describes it in his book [107].

Finally, digital cash systems can be divided into electronic coins and electronic checks. **Electronic coins** have a fixed value. Combinations of electronic coins allow people to purchase items of arbitrary value. **Electronic checks** are good for purchases up to the value of the check. The unspent portion must be refunded to the purchaser.

Proponents of digital cash point to its privacy benefits. People can use their smart cards or digital computers to spend digital cash with the same anonymity associated

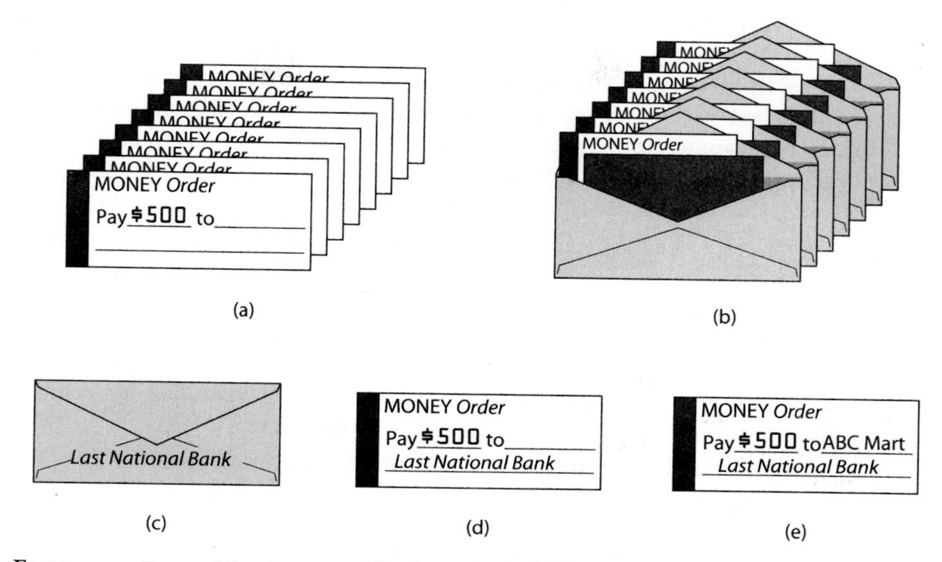

FIGURE 5.7 Bruce Schneier uses this physical analogy to explain how a blind signature protocol works: (a) Ann prepares 100 postal orders for $500, leaving the payee field blank. (b) Ann assembles 100 sealed envelopes. Each contains a postal order and a piece of carbon paper. She takes the envelopes to her bank. (c) The bank opens 99 of the sealed envelopes and sees that each contains a blank money order for $500. The bank is 99 percent sure that the last envelope also contains a money order for $500. It deducts $500 from Ann's account and signs the outside of the last envelope. The carbon paper transfers the signature to the money order. (d) Ann removes the signed money order from the last envelope. (e) Ann uses the money order to purchase $500 worth of goods from ABC Mart. ABC Mart takes the money order to the bank, which credits $500 to its account. The bank has never seen the money order before, so it cannot trace it back to Ann. However, the bank does know the money order is valid, because it recognizes its unique signature. What keeps Ann from trying to swindle the bank by putting $500,000 on one of the purchase orders? She knows there is only a 1 percent chance that the $500,000 purchase order will not be opened by the bank. She also knows that if any of the envelopes opened by the bank contain a different amount, she will be prosecuted. The high chance of punishment, combined with the slim chance of success, deters Ann from cheating the bank [107].

with physical coins and currency. Digital cash enables people to conduct electronic transactions without leaving an electronic trail.

Opponents of digital cash note that it will facilitate illicit transactions. Digital cash will make it easier for criminals to launder money. Law enforcement agencies and the IRS will find it more difficult to detect and prosecute criminals.

Consumers and merchants failed to embrace digital cash in pilot projects run by Citibank and Chase Manhattan Bank in New York in the late 1990s. For the time being, customers seem comfortable with their current options: cash, credit cards, and debit cards.

Summary

This chapter has focused on privacy issues brought to the forefront by the introduction of information technology. The issues of privacy and intellectual property are similar in the sense that both issues relate to how information ought to be controlled. Modern information technology makes it much easier to collect and transmit information, whether it be a song or a Social Security number. Information has become a valuable commodity.

Privacy can be seen as a balancing act between the desires of the individual and the needs of society. The individual seeks to restrict access. Society must decide where to draw the line between what ought to be private and what should be public. While privacy has both costs and benefits, the benefits of providing people at least some privacy exceed the costs. People do not have a natural right to privacy. Instead, it is a prudential right. We choose to give each other some privacy for our mutual good.

There is a tension between privacy and trust. We desire privacy, but we also want to be able to trust those we interact with each day. We trust those with good reputations. Good reputations are established through ordeals and credentials, which require that people reveal information about themselves.

Information can be put into three categories: personal information, public information, and public records. A public record is a piece of information collected by a government agency. Public information is information possessed by an organization that has the right to share it with other organizations. Personal information is information about an individual that is not yet public information or part of a public record. Sometimes public information is gained through voluntary disclosures. At other times people make information public as part of a commercial transaction. Certain activities, such as getting arrested or buying a house, result in the creation of a public record. Finally, organizations can collect information covertly.

People make information about themselves public in a variety of ways. The U.S. government has responded to the general desire for privacy by passing a variety of laws that regulate the collection and distribution of information gathered by private enterprises. The most important of these laws are the Fair Credit Reporting Act, the Family Education Rights and Privacy Act, the Video Privacy Protection Act, the Financial Services Modernization Act, the Children's Online Privacy Protection Act, and the Health Insurance Portability and Accountability Act.

Three federal agencies have collected a great deal of information about American citizens: the Census Bureau, the Internal Revenue Service, and the FBI. Concern about abuses of public record-keeping systems led to a report containing the Code of Fair Information Practices. Many European countries enacted laws incorporating the principles of this code. Congress's response to the code was the Privacy Act of 1974. The Privacy Act is supposed to protect the privacy of U.S. citizens by giving individual citizens access to public records about themselves, limiting the use of public records, and requiring that the records be accurate. However, the law has so many loopholes that many privacy advocates feel it is a weak piece of legislation.

Overt data collection by the U.S. government has been complemented by covert surveillance by various law enforcement organizations and the National Security Agency. In the name of national security or crime prevention, these agencies have routinely violated federal law in order to eavesdrop on telegraph and telephone conversations. The U.S. Supreme Court has ruled that electronic eavesdropping without a court order violates the Fourth Amendment to the Constitution.

The USA PATRIOT Act, passed in response to the terrorist attacks of September 11, 2001, amended many existing laws. It allows the use of pen registers and trap-and-trace devices to monitor Internet activity, allows a single judge to authorize wiretaps across the country, and broadens the number of circumstances under which roving surveillance can take place. The Patriot Act also makes it easier for the FBI to collect records from businesses, doctors, schools, libraries, and religious institutions. Citing concerns about possible violations of the Bill of Rights, more than a hundred cities and several states have passed anti–Patriot Act resolutions. A follow-on bill granting law enforcement agencies even greater powers was circulated in 2003, but it was not passed by Congress. In 2005 sixteen provisions of the Patriot Act were due to expire, so Congress began debating whether to make them permanent, as the Bush administration advocated. At the time this book went to press, it appeared that Congress would vote to make permanent all of these provisions except the two most controversial: one allowing roving wiretaps, and the other allowing the FBI to seize records from financial institutions, libraries, doctors, and businesses.

Data mining allows an organization to create a complex profile of a person from a collection of individual facts. It is a powerful tool for commerce. Companies use data mining to direct advertising to the most promising customers. Governments also use data mining. The IRS has used it for years to identify suspicious tax returns. The Department of Defense wants to create a sophisticated data-mining system to detect patterns of transactions indicating probable terrorist activity.

Data mining is possible because organizations handling transactions have the right to sell information about these transactions to other organizations. Some people believe that such transactions should be private by default. In other words, if Smith buys something from Acme Corp., Acme cannot reveal anything about the transaction to another company without Smith's permission. This is called an opt-in policy. Others believe it ought to be the other way around. Acme should have the right to share or sell information about the transaction unless Smith explicitly forbids Acme from doing so. This is called an opt-out policy.

Identity theft is a rapidly growing crime in the United States. The leading form of identity theft is credit card fraud. Typically, thieves use fairly simple methods to get access to credit card numbers. They look through mailboxes, garbage cans, or lost or stolen wallets. Some thieves get credit card information through fraudulent telephone calls or email messages. Others have targeted the computerized databases of large institutions, gaining access to hundreds of thousands of personal financial records.

The rapid rise in identity theft has highlighted the inadequacies of the Social Security number (SSN) as an identifier. Originally, use of the SSN was restricted to the Social

Security Administration. Over time, however, it has become an important identification number in the United States. Unfortunately, it has many flaws that make it a poor choice for an identification number. Some have proposed the option of a new, national identification card for the United States. They suggest it will reduce illegal immigration and criminal activity. Others say any identification card can be forged, and the creation of a national identification card will make data mining simpler.

The United States has created a new standard for driver's licenses. Even though they will be issued by the states, every state's license must meet the national standard, and the driver's license databases of the individual states will be connected, deterring fraud. The new driver's licenses, to be issued by 2008, will be required for opening bank accounts, traveling on commercial airplanes, and receiving federal services. They are likely to become the most trusted form of identification in the United States, a de facto national identification card.

Encryption is one way people can prevent eavesdroppers from reading the messages they send. In the past, only governments and large businesses had access to powerful cryptography systems. Public-key cryptography systems, such as PGP, have given small businesses and individuals this power. For many years, the U.S. State Department enforced a ban on selling software with strong encryption outside the United States, even though PGP had already been downloaded to computers around the world. In 1999 and 2000 two different federal appeals courts ruled that the export restrictions were unconstitutional, and the State Department dropped its ban.

Digital cash is an anonymous electronic money system that combines the strengths of credit cards and physical currency. Digital cash would be stored on a small, credit card–shaped smart card. A consumer would present this card to pay for goods or services. However, the transaction would be completely anonymous, like a cash transaction. To date, digital cash systems have not been well received. Consumers are comfortable using cash, credit cards, and debit cards.

Review Questions

1. How is Google's Phonebook service able to produce a map to a person's home, given only that person's phone number?
2. Give an example of a piece of information that a person should not have to reveal to anyone else. Give an example of a piece of information that society should be able to demand that a person reveal.
3. Is privacy a negative right or a positive right?
4. What right is guaranteed by the Third Amendment to the U.S. Constitution?
5. What does it mean when the author says privacy is a prudential right?
6. Give three examples of ways in which an inhabitant of New York City in 2003 has more privacy than an inhabitant of New York City in 1903.
7. What is the difference between a public record and public information?

8. Which is easier for a person to control: public records or public information?
9. List five pieces of information about a person that are public records.
10. Provide an example (not already given in the book) of a situation where people must disclose personal information in order to get something they want.
11. Provide an example (not already given in the book) of a situation where people must reveal personal information, whether or not they consent.
12. Provide an example (not already given in the book) of a situation where information about people is gathered without their knowledge.
13. What is the difference between a digital video recorder and a traditional VCR?
14. Why does enhanced 911 service raise new concerns about privacy?
15. What is spyware? Why does it raise privacy concerns?
16. The Fair Credit Report Act says that information which may negatively affect an individual's credit rating must be removed after seven years. What are two exceptions to this guideline?
17. What are the rights provided by the Family Education Rights and Privacy Act?
18. How does the Video Privacy Protection Act enhance privacy?
19. Summarize the major provisions of the Financial Services Modernization Act.
20. What is the purpose of the Children's Online Privacy Protection Act?
21. Describe the privacy protections resulting from the Health Insurance Portability and Accountability Act.
22. Give two examples of the Census Bureau illegally revealing census data to other federal agencies.
23. Why did consumer groups complain about H&R Block's Web-based Free File tax filing service?
24. Name two notable successes claimed by the National Crime Information Center.
25. Robert Bellair has said, "The Privacy Act, it turns out, is no protection at all. You can drive a truck through the Privacy Act" [53]. Explain why Bellair and other privacy advocates feel the Privacy Act of 1974 is a weak piece of legislation.
26. What right is guaranteed by the Fourth Amendment to the U.S. Constitution?
27. Explain the significance of the U.S. Supreme Court decision in *Katz v. United States*.
28. What are the key provisions of the Patriot Act?
29. What are we referring to when we talk about a secondary use of data?
30. What is collaborative filtering? Who uses it?
31. Give two examples of government data mining projects currently in operation.
32. What is the Platform for Privacy Preferences? What are the strengths and the weaknesses of this system?
33. What is the most common kind of identity theft?
34. Can a private company legally ask you for your Social Security number?

35. What are the problems with using the Social Security number as an identification number?
36. Name two benefits and two harms that may result from the passage of the Real ID Act.
37. What is the most significant difference between a public-key encryption scheme and a traditional symmetric encryption scheme?
38. Why did the United States government attempt to stop the distribution of the Pretty Good Privacy (PGP) program?
39. Why did the Justice Department advocate adoption of the Clipper chip?
40. Name three ways that criminals can steal personal information at ATMs.
41. What is digital cash? How does it differ from a credit card? How does it differ from ordinary cash?
42. What is a blind signature protocol? Why are blind signatures needed in a digital cash system?
43. Which pieces of legislation discussed in this chapter increased personal privacy rights? Which laws gave the government greater surveillance powers?

Discussion Questions

44. Section 5.1 describes the case of the Maryland banker who used information from a state medical records database to identify customers who had cancer. He then called in the loans of these customers. The banker broke no laws and received no legal penalties for his action. Did the banker do anything wrong? Why or why not?
45. Warren and Brandeis argued that it is a violation of a person's privacy to take their photograph without their consent. Do you agree with their position? Why or why not?
46. Critics of grocery club cards give examples of card-member prices being equal to the regular product price at stores without customer loyalty programs. In other words, customers who want to get food at the regular price must use the card. Customers pay extra if they don't want to use the card. Is it fair for a store to charge us more if we don't want to use its loyalty card? Explain your reasoning.
47. Some consumers give phony personal information when they apply for rewards or loyalty cards at stores. Others take it a step further by regularly exchanging their cards with those held by other people. Are these people doing anything wrong? Why or why not?
48. If you voluntarily have your body scanned at a department store, who should own that information, you or the store? Should the store have the right to sell your body measurements to other businesses? Explain your reasoning.
49. TiVo keeps detailed information about the television viewing habits of customers who subscribe to its service.
    a. Should your television viewing habits be private information?
    b. Do you care if anyone else knows what television shows or pay-per-view movies you have watched in the past year?

c. Do voters have the right to know the viewing habits of people running for elected office?

50. You are sitting on a jury. A driver of a car has been charged with manslaughter for killing a pedestrian. The prosecution presents evidence collected from the automobile's "black box" that indicates the car was traveling at 45 miles per hour before the accident. The defense presents four eyewitnesses to the accident, all of whom testify that the car could not have been going faster than 30 miles per hour. Are you more inclined to believe the eyewitnesses or the data collected from the "black box"? Explain your reasoning.

51. Enhanced 911 service allows cell phone companies to track the locations of active cell phone users within 100 meters.
    a. Who should have access to location information collected by cell phone companies?
    b. How long should this information be kept?
    c. If this information could be used to help you establish an alibi, would you want the cell phone company to be able to release it to the police?
    d. How would you feel about the cell phone company releasing compromising information about your whereabouts to the police?
    e. Should the police be able to get from the cell phone company the names of all subscribers using their phones close to a crime scene around the time of the crime?

52. Should parents implant microchips in their children to make them easier to identify in case they are lost or kidnapped? Why or why not?

53. Florida, Missouri, Ohio, and Oklahoma have passed laws that require lifetime monitoring of some convicted sex offenders after they have been released from prison. The offenders must wear electronic ankle bracelets and stay close to small GPS transmitters, which can be carried on a belt or in a purse. Computers monitor the GPS signals and alert law enforcement officials if the offenders venture too close to a school or other off-limits area. Police interested in the whereabouts of a monitored person can see his location, traveling direction, and speed plotted on a map [108].

    Do these laws represent an unacceptable weakening of personal privacy, or are they sensible public safety measures? Should they be repealed? Should people convicted of other crimes also be monitored for life? Would there be less crime if everyone in society were monitored?

54. Why do you think AOL would want to provide its customers with software tools enabling them to detect and eliminate spyware?

55. Before offering a job candidate a position, some potential employers do a criminal background check of the candidate. What are the pros and cons of this policy?

56. You are applying for an account at a video rental store. The clerk asks you to fill out the application form completely. One of the fields asks for your Social Security number. You leave that field blank. The clerk refuses to accept your application without the field filled in. You ask to speak to the manager, and the clerk says the manager is not available. Would it be wrong in this situation to fill in a fake Social Security number? Explain your reasoning.

57. A company discovers that some of its proprietary information has been revealed in Internet chat rooms. The disclosure of this information results in a substantial drop in

the price of the company's shares. The company provides Internet service providers with the screen names of the people who posted the confidential information. It asks the ISPs to disclose the actual identities of these people. Should the ISPs comply with this request? Explain your reasoning. (This scenario is adapted from an actual event [109].)

58. Think about what you do when you get up in the morning. How would you act differently if you knew you were being watched? Would you feel uncomfortable? Do you think you would get used to being watched?

59. In a recent study, people in subway stations were offered a cheap pen in return for disclosing their passwords. About 90 percent offered their passwords in return for the pen [110]. Do people really value privacy?

In-class Exercises

60. What does your "ladder of privacy" look like? How does it compare to those of your classmates?

61. Canadian science fiction author Robert Sawyer argues that we need privacy because we have "silly laws" which attempt to make people feel ashamed for indulging in certain harmless activities. He suggests that if there were no privacy, people would insist these laws be overturned [44]. Do you agree with Sawyer's position? Why or why not?

62. Do you agree with the author that it is more difficult to know whom to trust in modern society than it was in a small village of a few centuries ago? Why or why not?

63. Divide the class into two groups. The first group should come up with reasons supporting the proposition, "We live in a global village." The second group should come up with reasons supporting the proposition, "We live in a world of strangers."

64. When you purchase a product or service using a credit card, the merchant has information linking you to the transaction. Divide the class into two groups (pro and con) to debate the proposition that merchants should be required to follow an opt-in policy. Such a policy would require the consumer to explicitly give permission before a merchant could share information about that consumer with another organization.

65. The Code of Fair Information Practices applies only to government databases. Divide the class into two groups to debate the advantages and disadvantages of extending the Code of Fair Information Practices to private databases managed by corporations.

66. While the cost of automobile insurance varies from person to person, based on the driving record of each individual, health insurance premiums are typically uniform across groups of people, such as all of the employees of a company. However, a majority of health care costs are incurred by a minority of the population.

    Today it is possible to take a blood sample from a person and to extract a genetic profile that will reveal that person's disposition to certain diseases. Debate the proposition that health insurance rates should be tailored to reflect each individual's propensity to illness.

67. Divide the class into two groups (pro and con) to debate the proposition that every citizen of the United States ought to carry a national identification card.

68. The Department of Homeland Security is interested in using computers to identify suspected terrorists operating within the United States. It would like to mine databases containing information about purchases and travel to detect patterns that may identify individuals who are engaged in, or at least planning, terrorist activity. It asks a panel of computer scientists to determine the feasibility of this project. A panel member says the most difficult problem will be determining what patterns of transactions to look for. He suggests it might be possible to construct a computer program that uses artificial intelligence to mimic a terrorist organization. The program would determine the actions needed to execute a terrorist act. Once these actions were determined, it would be possible to search database records to find evidence of these actions.

    Debate the morality of developing a computer program capable of planning the steps needed to execute an act of terror.

Further Reading

James Bamford. *The Puzzle Palace: A Report on America's Most Secret Agency*. Penguin Books, New York, NY, 1983.

Whitfield Diffie and Susan Landau. *Privacy on the Line: The Politics of Wiretapping and Encryption*. The MIT Press, Cambridge, MA, 1998.

Steven L. Nock. *The Costs of Privacy: Surveillance and Reputation in America*. Aldine de Gruyter, New York, NY, 1993.

George Orwell. *1984*. Knopf, New York, NY, 1992.

Ellen Frankel Paul, Fred D. Miller, Jr., and Jeffrey Paul, editors. *The Right to Privacy*. Cambridge University Press, Cambridge, England, 2000.

Priscilla M. Regan. *Legislating Privacy*. The University of North Carolina Press, Chapel Hill, NC, 1995.

Bruce Schneier. *Applied Cryptography: Protocols, Algorithms and Source Code in C*. 2nd ed. John Wiley & Sons, New York, NY, 1996.

Ferdinand David Schoeman, editor. *Philosophical Dimensions of Privacy: An Anthology*. Cambridge University Press, Cambridge, England, 1984.

Charles J. Sykes. *The End of Privacy*. St. Martin's Press, New York, NY, 1999.

References

[1] Amy Harmon. "Some Search Results Hit Too Close to Home." *The New York Times*, April 13, 2003.

[2] Scott Carlson. "To Guard 3 Students' Privacy, Georgetown U. Expunges Thousands of E-mail Messages." *The Chronicle of Higher Education*, February 7, 2003.

[3] Noah Robischom. "Rx for Medical Privacy." *Netly News*, September 3, 1997.

[4] Jamie Prime. "Privacy vs. Openness." *Quill*, 82(8), October 1994.

[5] Edmund F. Byrne. "Privacy." In *Encyclopedia of Applied Ethics*, volume 3, pages 649–659. Academic Press, 1998.

[6] Edward J. Bloustein. "Privacy as an Aspect of Human Dignity: An Answer to Dean Prosser." In *Philosophical Dimensions of Privacy: An Anthology*, edited by Ferdinand David Schoeman, pages 156–202. Cambridge University Press, Cambridge, England, 1984.

[7] Ferdinand Schoeman. "Privacy: Philosophical Dimensions of the Literature." In *Philosophical Dimensions of Privacy: An Anthology*, edited by Ferdinand David Schoeman, pages 1–33. Cambridge University Press, Cambridge, England, 1984.

[8] Edmund Ronald Leach. *A Runaway World?* British Broadcasting Corporation, London, England, 1967.

[9] Marie Hartwell-Walker. "Why Dysfunctional Families Stay That Way." *Amherst Bulletin*, January 28, 1994.

[10] Morton H. Levine. "Privacy in the Tradition of the Western World." In *Privacy: A Vanishing Value?* edited by William C. Bier, S.J., pages 3–21. Fordham University Press, New York, NY, 1980.

[11] Jeffrey H. Reiman. "Privacy, Intimacy, and Personhood." *Philosophy & Public Affairs*, 6(1):26–44, 1976. Reprinted in *Philosophical Dimensions of Privacy: An Anthology*, ed. F. D. Schoenan, Cambridge University Press, 1984.

[12] Stanley I. Benn. "Privacy, Freedom, and Respect for Persons." In *Philosophical Dimensions of Privacy: An Anthology*, edited by Ferdinand David Schoeman, pages 223–244. Cambridge University Press, Cambridge, England, 1984.

[13] Charles J. Sykes. *The End of Privacy*. St. Martin's Press, New York, NY, 1999.

[14] Gini Graham Scott. *Mind Your Own Business: The Battle for Personal Privacy*. Insight Books / Plenum Press, New York, NY, 1995.

[15] Constance T. Fischer. "Privacy and Human Development." In *Privacy: A Vanishing Value?* edited by William C. Bier, S.J., pages 37–45. Fordham University Press, New York, NY, 1980.

[16] Robert C. Neville. "Various Meanings of Privacy: A Philosophical Analysis." In *Privacy: A Vanishing Value?* edited by William C. Bier, S.J., pages 22–33. Fordham University Press, New York, NY, 1980.

[17] Joseph G. Keegan, S.J. "Privacy and Spiritual Growth." In *Privacy: A Vanishing Value?* edited by William C. Bier, S.J., pages 67–87. Fordham University Press, New York, NY, 1980.

[18] Charles Fried. "Privacy: A Moral Analysis." *Yale Law Review*, 77:475–493, 1968. Reprinted in *Philosophical Dimensions of Privacy: An Anthology*, ed. F. D. Schoeman, 1984.

[19] James Rachels. "Why Privacy Is Important." *Philosophy & Public Affairs*, 4(4):323–333, 1975. Reprinted in *Philosophical Dimensions of Privacy: An Anthology*, ed. F. D. Shoeman, Cambridge University Press, 1984.

[20] Samuel D. Warren and Louis D. Brandeis. "The Right to Privacy." *Harvard Law Review*, 4(5), December 15, 1890.

[21] William L. Prosser. "Privacy: A Legal Analysis." *California Law Review*, 48:338–423, 1960. Reprinted in *Philosophical Dimensions of Privacy: An Anthology*, F. D. Shoeman, ed., Cambridge University Press, 1984.

[22] Judith Jarvis Thomson. "The Right to Privacy." *Philosophy & Public Affairs*, 4(4):295–314, 1975. Reprinted in *Philosophical Dimensions of Privacy: An Anthology*, ed. F. D. Schoeman, Cambridge University Press, 1984.

[23] Alexander Rosenberg. "Privacy as a Matter of Taste and Right." In *The Right to Privacy*, edited by Ellen Frankel, Jr., Fred D. Miller, and Jeffrey Paul, pages 68–90. Cambridge University Press, Cambridge, England, 2000.

[24] Humphrey Taylor. "Most People Are 'Privacy Pragmatists' Who, While Concerned About Privacy, Will Sometimes Trade It Off for Other Benefits." HarrisInteractive, March 19, 2003. The Harris Poll #17.

[25] James Toedtman. "Court Unblocks Do Not Call Registry in Latest Ruling." *Sun-Sentinel (Fort Lauderdale, FL)*, October 8, 2003.

[26] Heather Fleming Phillips. "Consumers Can Thank Do-Not-Underestimate FTC Chairman for Do-Not-Call Peace." *San Jose (CA) Mercury News*, January 2, 2004.

[27] Dave DeWitte. "Bulk Mailers Await Boost from 'Do Not Call' Registry." *The Gazette (Cedar Rapids, IA)*, November 10, 2003.

[28] Steven L. Nock. *The Costs of Privacy: Surveillance and Reputation in America*. Aldine de Gruyter, New York, NY, 1993.

[29] Michael L. Sankey and Peter J. Weber, editors. *Public Records Online: The National Guide to Private & Government Online Sources of Public Records*. 4th ed. Facts on Demand Press, Tempe, AZ, 2003.

[30] Jeffrey Rosen. "Being Watched: A Cautionary Tale for a New Age of Surveillance." *The New York Times on the Web*, October 7, 2001.

[31] Michael Liedtke. "New Shopping Technology Could Breed Supermarket Bias." *Kansas City Star*, December 1, 2002.

[32] John Vanderlippe. "Supermarket Cards: An Overview of the Pricing Issues." *Consumers Against Supermarket Privacy Invasion and Numbering*, 2003. www.nocards.org/overview.

[33] Elizabeth Weise. "Identity Swapping Makes Privacy Relative." *USA Today*, June 6, 2000.

[34] Amy Tsao. "So, We'll Take It In . . . " *Retail Traffic*, May 1, 2003.

[35] Amy Harmon. "TiVo Plans to Sell Information on Customers' Viewing Habits." *NYTimes.com*, June 2, 2003.

[36] Ian Austen. "Your Brake Pads May Have Something to Say (by E-mail)." *NYtimes.com*, March 27, 2003.

[37] Jay Warrior, Eric McHenry, and Kenneth McGee. "They Know Where You Are." *IEEE Spectrum*, pages 20–25, July 2003.

[38] Charles J. Murray. "Privacy Concerns Mount over Retail Use of RFID Technology." *Electronic Engineering Times*, (1298), December 1, 2003.

[39] Meg McGinty. "RFID: Is This Game of Tag Fair Play?" *Communications of the ACM*, 47(1):15–18, January 2004.

[40] Thomas Wailgum. "Is Big Brother Coming to Your Wallet?" *CIO*, July 1, 2005.

[41] Kristi Heim. "New Computerized Passport Raises Safety Concerns." *The Seattle (WA) Times*, January 3, 2005.

[42] "American Passports to Get Chipped." *Wired News*, October 21, 2004.

[43] "Owners of Dogs Lacking Implants Face Fines." *The China Post*, September 1, 2000.

[44] Robert J. Sawyer. "Privacy: Who Needs It?: We're Better Off without It, Argues Canada's Leading Sci-Fi Writer." *Maclean's (Toronto Edition)*, page 44, October 7, 2002.

[45] Edward C. Baig. "Keep Spies from Skulking into Your PC." *USA Today*, January 22, 2004.

[46] Privacy Rights Clearinghouse. "Fact Sheet 24: Protecting Financial Privacy," July 14, 2005. www.privacyrights.org.

[47] Department of Health of Human Services, USA. "Protecting the Privacy of Patients' Health Information," April 14, 2003. www.hhs.gov/news.

[48] United States General Accounting Office. "IRS Systems Security: Although Significant Improvements Made, Tax Processing Operations and Data Still at Risk," December 1998. GAO/AIMD-99-38.

[49] Jean Ann Fox, Chi Chi Wu, Edmund Mierzwinski, Chris Hoofnagle, and Shelley Curran. *Letter to Ms. Pamela F. Olson, Assistant Secretary, U.S. Treasury Department*. Consumer Federation of America, Consumers Union, Electronic Privacy Information Center, National Consumer Law Center, and U.S. Public Interest Research Group, March 24, 2003.

[50] Stephanie L. Hitt. "NCIC 2000." *FBI Law Enforcement Bulletin*, 69(7), July 2000.

[51] U.S. Department of Health, Education and Welfare. *Secretary's Advisory Committee of Automated Personal Data Systems, Records, Computers, and the Rights of Citizens*, 1973.

[52] Simson Garfinkel. "Privacy and the New Technology." *Nation*, 270(8), February 28, 2000.

[53] William Petrocelli. *Low Profile: How to Avoid the Privacy Invaders*. McGraw-Hill, New York, NY, 1981.

[54] Whitfield Diffie and Susan Landau. *Privacy on the Line: The Politics of Wiretapping and Encryption*. The MIT Press, Cambridge, MA, 1998.

[55] Priscilla M. Regan. *Legislating Privacy*. The University of North Carolina Press, Chapel Hill, NC, 1995.

[56] Supreme Court of the United States. *Dissenting Opinion in* Olmstead v. United States, 1928. 277 U.S. 438.

[57] Supreme Court of the United States. *Katz v. United States*, 1967. 389 U.S. 347.

[58] James Bamford. *The Puzzle Palace: A Report on America's Most Secret Agency*. Penguin Books, New York, NY, 1983.

[59] Nancy Gohring. "FCC Inflates CALEA." *Telephony*, 237(10), September 6, 1999.

[60] Charlotte Twight. "Conning Congress." *Independent Review*, 6(2), Fall 2001.

[61] Kirk Laughlin. "A Wounded CALEA Is Shuttled Back to the FCC." *America's Network*, 104(15), October 1, 2000.

[62] Federal Communications Commission. "FCC Requires Certain Broadband and VoIP Providers to Accommodate Wiretaps," August 5, 2005. www.fcc.gov.

[63] "USA Patriot Act: Major Provisions of the 2001 Antiterrorism Law." *Congressional Digest*, 82(4), April 2003.

[64] David Sarasohn. "Patriots vs. the Patriot Act." *Nation*, 277(8):23, September 22, 2003.

[65] American Civil Liberties Union, New York, NY. "PATRIOT Act Fears Are Stifling Free Speech, ACLU Says in Challenge to Law," November 11, 2003. www.aclu.org.

[66] American Library Association. "Resolution on the USA PATRIOT Act and Related Measures That Infringe on the Rights of Library Users," January 29, 2003. 2002–2003 CD #20.1, 2003 ALA Midwinter Meeting, www.ala.org.

[67] Lawrence Morahan. "'Patriot 2' Raises Concerns for Civil Liberties Groups." *CNSNews.com*, February 13, 2003.

[68] Tom Ridge. "Using the PATRIOT Act to Fight Terrorism." *Congressional Digest*, pages 266–268, November 2004.

[69] Ben Jacklet and Todd Murphy. "Now Free, Attorney Brandon Mayfield Turns Furious." *Washington Report on Middle East Affairs*, 23(6), July/August 2004.

[70] Dan Eggen. "Flawed FBI Probe of Bombing Used a Search Warrant." *The Washington Post*, April 7, 2005.

[71] Andrew Murr, Michael Isikoff, Eric Pape, and Mike Elkin. "The Wrong Man." *Newsweek*, 143(23), June 7, 2004.

[72] "Senate Approves Partial Renewal of Patriot Act." *The Washington Post*, July 30, 2005.

[73] L. A. Lorek. "Data Mining Extracts Online Gold; Stores Collect Information about Web Customers to Target Future Sales Pitches." *San Antonio Express-News*, December 15, 2002.

[74] "United We Find." *The Economist*, March 10, 2005.

[75] Ann Cavoukian and Don Tapscott. *Who Knows: Safeguarding Your Privacy in a Networked World*. McGraw-Hill, New York, NY, 1996.

[76] Richard Peŕez-Peña. "An Early Warning System for Diseases in New York." *NYTimes.com*, April 4, 2003.

[77] "Total Information Awareness Program." *Congressional Digest*, 82(4), April 2003.

[78] Jonathan Riehl. "Lawmakers Likely to Limit New High-Tech Eavesdropping." *CQ Weekly*, 61(7):406–407, February 15, 2003.

[79] Nikki Swartz. "Controversial Surveillance System Renamed." *Information Management Journal*, 37(4):6, July 2003.

[80] Barbara Simons and Eugene H. Spafford. *Letter to The Honorable John Warner and the Honorable Carl Levin*. U.S. ACM Public Policy Committee (USACM), Association for Computing Machinery, January 23, 2003.

[81] Carolyn Hirschman. "Congress Sticks Its Nose into Online Privacy." *Telephony*, 241(7), August 13, 2001.

[82] Martyn Williams. "IBM Researcher Eyes Databases with a Conscience." *InfoWorld Daily News*, August 27, 2002.

[83] Dorothy E. Denning. *Information Warfare and Security*. Addison-Wesley, Boston, MA, 1999.

[84] States News Service. "FTC Testifies on Identify Theft, Impact on Seniors." July 18, 2002.

[85] Matt Richtel. "Financial Institutions May Facilitate Identity Theft." *NYTimes.com*, August 12, 2002.

[86] Privacy Rights Clearinghouse. *How Many Identity Theft Victims Are There? What IS the Impact on Victims?* May 15, 2005. www.privacyrights.org.

[87] Tom Shean. "Damage Done by Identity-Theft Ring Demonstrates Need for Consumer Care." *Virginian-Pilot (Chesapeake, Virginia)*, November 27, 2002.

[88] Roy Furchgott. “In a Single Swipe, a Wealth of Data (Beware of Thieves).” *NY-Times.com*, March 13, 2003.

[89] “LexisNexis Uncovers More Consumer Data Breaches.” *Reuters*, April 12, 2005.

[90] Bruce Schneier. “Risks of Third-Party Data.” *Communications of the ACM*, 48(5):136, May 2005.

[91] David McGuire. “Bush Signs Identity Theft Bill.” *washingtonpost.com*, July 15, 2004.

[92] Federal Trade Commission. “Take Charge: Fighting Back Against Identity Theft,” February 2005. www.ftc.gov/bcp/conline.pubs.

[93] Gartner, Inc. “Gartner Says Identity Theft Is Up Nearly 80 Percent,” July 21, 2003.

[94] Social Security Administration, USA. “A Brief History of Social Security,” August 2000.

[95] Social Security Administration, USA. “Social Security Cards Issued by Woolworth.” www.ssa.gov/history/ssn/misused.html.

[96] Office of Inspector General, Department of Health and Human Services, USA. “Extent of Social Security Number Discrepancies,” January 1990. OAI-06-89-01120.

[97] Peter G. Neumann and Lauren Weinstein. “Risks of National Identity Cards.” *Communications of the ACM*, page 176, December 2001.

[98] Richard Turner. Letter to the editor. *The Times (London)*, September 7, 2001.

[99] Declan McCullagh. FAQ: How Real ID will affect you. *The New York Times*, May 6, 2005.

[100] Dennis Bailey. “Debating Barry Steinhardt’s UNREAL ID,” August 7, 2005. www.opensocietyparadox.com.

[101] Joseph Menn. “Federal ID Act May Be Flawed.” *The Los Angeles Times*, May 31, 2005.

[102] Phil Zimmerman. “Why Do You Need PGP?” www.pgpi.org/doc/whypgp/en/.

[103] Philip Elmer Dewitt. “Who Should Keep the Keys?” *Time*, March 14, 1993.

[104] James Bamford. *Body of Secrets: Anatomy of the Ultra-Secret National Security Agency*. Anchor Books, New York, NY, 2002.

[105] Sara Baase. *A Gift of Fire, Second Edition*. Prentice Hall, Upper Saddle River, NJ, 2003.

[106] Glyn Davies. *A History of Money: From Ancient Times to the Present Day*. University of Wales Press, Cardiff, Wales, 1994.

[107] Bruce Schneier. *Applied Cryptography: Protocols, Algorithms, and Source Code in C*. 2nd ed. John Wiley & Sons, New York, NY, 1996.

[108] David A. Lieb. “States Move on Sex Offender GPS Tracking.” *Associated Press*, July 30, 2005.

[109] Stewart Deck. “Legal Thumbs-Up for Raytheon Employee Suit; Privacy Groups Chilled by ISP Subpoenas.” *Computerworld*, April 12, 1999.

[110] John Leyden. “Office Workers Give Away Passwords for a Cheap Pen.” *The Register*, April 17, 2003. www.theregister.co.uk.

AN INTERVIEW WITH

Ann Cavoukian

Dr. Ann Cavoukian is recognized as one of the world's leading privacy experts. She oversees the operations of the freedom of information and privacy laws in Ontario, Canada, in her role as Information and Privacy Commissioner (IPC). Dr. Cavoukian joined the Office of the IPC as its first Director of Compliance in 1987. Prior to joining the IPC, she headed the Research Services Branch for the provincial Attorney General. She received her M.A. and Ph.D. in Psychology from the University of Toronto, where she specialized in criminology and law, and lectured on psychology and the criminal justice system.

Her published works include *Who Knows: Safeguarding Your Privacy in a Networked World*, with Don Tapscott (McGraw-Hill, 1997), and *The Privacy Payoff: How Successful Businesses Build Consumer Trust*, with Tyler Hamilton (McGraw-Hill Ryerson, 2002).

What is information privacy?

Information privacy essentially revolves around personal control—an individual's right to control the collection, use, and disclosure of his or her personal information. Freedom of choice is vital. Personal information is information that relates to an "identifiable individual." Organizations that collect, use, and disclose personal information can protect an individual's right to privacy by implementing what are commonly referred to as "fair information practices." Fair information practices are a set of common standards that balance an individual's right to privacy with the organization's legitimate need to collect, use, and disclose personal information. In Canada, fair information practices are set out in the *Canadian Standards Association Model Code for the Protection of Personal Information* (the *CSA Code*).

The *CSA Code* consists of ten principles. First, it requires the designation of at least one individual who is accountable for the organization's compliance with the other nine principles (**Accountability**). The organization must specify the purposes for which it collects personal information, at or before the time when the information is collected (**Identifying Purposes**). The consent of the individual must be obtained for the collection, use, or disclosure of personal information, except where it is not appropriate to obtain consent (**Consent**). The collection of personal information must be limited to that which is necessary to fulfill the specified purposes (**Limiting Collection**). Personal information must not be used or disclosed for purposes other than those for which it was collected, unless the individual consents or as required by law (**Limiting Use, Disclosure, and Retention**). Personal information must be as accurate, complete, and up-to-date as necessary for the purposes for which it is to be used (**Accuracy**). The organization must implement security safeguards that are appropriate for the level of sensitivity of the personal information (**Safeguards**). The organization must make readily available specific information about its policies and practices relating to the management of personal information (**Openness**). Individuals have a right to access and request correction of their own personal information (**Individual Access**). Finally, individuals must be able to challenge an organization's compliance with the privacy principles (**Challenging Compliance**).

People often sacrifice information privacy for the sake of convenience. Is information privacy really important?

Consumers are sometimes willing to provide personal information in exchange for some benefit or service. For example, consumers are sometimes willing to register personal information on a Web site in exchange for useful information or a discount on merchandise. It is up to each individual consumer to weigh the cost and benefits of providing personal information in any given situation—it's his or her choice. As long as this collection of personal information takes place with the knowledge and consent of the individual, it is not an invasion of privacy—it is a matter of personal choice and control which is central to the concept of privacy. When information is collected, used, or disclosed without the knowledge or consent of the individual, then privacy becomes an issue.

How has 9/11 affected people's attitudes toward information privacy?

Immediately after 9/11, people seemed willing to sacrifice civil liberties, and privacy if necessary, in order to feel secure. There was a surge of support, particularly in the United States, for increasingly invasive security measures and expanding public surveillance. However, as time passed, heads cooled and the public began to think more rationally about these issues and whether or not the invasive security measures that were being implemented and contemplated would actually having the desired impact on national security. The public began to question whether the privacy sacrifice that we were all being asked to make was actually worth it. This may need to be revisited after the events of July 7th, 2005, in London.

The public's interest in protecting consumer privacy, however, did not diminish in the post-9/11 period. If anything, the value of trusted business relationships has increased.

Information about customers is a valuable commodity. Why should a business be concerned about protecting the privacy of its customers?

In Canada, it happens to be the law for private-sector organizations, but a simple answer to the above question is that "privacy is good for business!" This assertion is supported by a Harris/Westin poll where in November 2001 and February 2002 it was found that if consumers had confidence in a company's privacy practices, they were much more likely to increase volume of business and frequency of business with that company. Conversely, they were likely to stop doing business with a company if it misused personal information. Further, The Information Security Forum reported in 2004 that a company's privacy breaches can cause major damage to brand and reputation. Robust privacy policies and staff training were viewed as keys to avoiding privacy problems.

Do you favor opt-in policies over opt-out policies?

As a general rule, opt-in policies are viewed as being more privacy-protective than opt-out policies. However, the type of consent that an organization should obtain (i.e., opt-in versus opt-out) depends on the circumstances in which personal information is being collected, used, and disclosed. When it is reasonable in the circumstances to infer implied consent, an opt-out type of consent may be appropriate, particularly where the information is not considered to be sensitive. For example, the individual's name and address may not be considered to be sensitive, and the collection of this information for specified purposes may take place with an opt-out type of consent. On the other hand, opt-in consent should generally be obtained whenever sensitive personal information, such as medical information or financial information, is being collected, used, or disclosed.

Is Canada ahead of the United States with respect to ensuring fair information practices?

Canada has more comprehensive privacy and data-protection laws and statutory oversight/enforcement agencies. By contrast, the U.S. has a multitude of specialized, sectoral laws, regulations, and self-regulation and more scope for private rights of action and financial penalties. There is strong evidence that Canadian organizations are more aware of privacy and much more likely to apply privacy principles throughout their operations than U.S. firms.

For example, in a benchmark study conducted by my office and the Ponemon Institute—a Tucson-based think tank dedicated to the advancement of responsible information management practices within business and government—we compared the corporate privacy practices of Canadian and U.S businesses. Some of the key findings of the study were that in comparison to U.S. companies, Canadian companies:

- are more likely to have a dedicated privacy officer and a privacy program with a clearly articulated mission,
- are more likely to have a formal redress process for customers to respond to queries and concerns about privacy,
- are more open to providing customers with the right to access and correct personal information,
- offer more choice to customers and consumers in terms of opting out (or opting in) to secondary uses and disclosures of personal information,
- are less likely to sell customer data,
- are more likely to offer privacy training or awareness programs for employees and contractors who handle sensitive personal information,
- hold their vendors and other third parties to higher standards or due diligence requirements,
- have a more aggressive data-control orientation when collecting and retaining sensitive personal information,
- are more concerned about insider misuse than external penetration,
- require more rigorous data-quality controls and monitoring requirements for transacting and moving of personal information about employees and customers, especially when the application involves transborder movement, and
- are more likely to have strict policies that protect the privacy of employees.

As you ponder new threats to information privacy, are there any emerging technologies you find particularly troubling?

A qualified answer would include Radio Frequency Identification (RFID) tags; biometric and other forms of authentication and identification; data mining; and video surveillance. If these emerging technologies are designed and implemented with fair information practices in mind, then they can enhance and enrich our lives immeasurably. But if used surreptitiously and without regard for privacy, they only hold the promise of a dreadful scenario of ever-present surveillance and discrimination, the proverbial "Orwellian nightmare." It all depends on the design and configuration of a particular technology.

CHAPTER

6 Computer and Network Security

A ship in harbor is safe, but that's not what ships are for.

–JOHN SHEDD

6.1 Introduction

AS COMPUTERS HAVE INCREASED IN POWER AND DECREASED IN PRICE, we have found more ways to use them, both at work and at home. Today, we rely upon computers for such activities as sending and receiving email, surfing the Web, shopping, managing our calendars, and keeping track of personal information. The utility of our computers and the information they hold makes computer security an important issue.

This chapter focuses on threats to computer security. We begin by surveying computer viruses, worms, and Trojan Horses. Through these mechanisms unauthorized programs can enter our computers. When executed, they can steal personal information, destroy data, and even launch attacks on other computers. System administrators play a key role in defending computers from outside threats. We review some of the ways administrators increase the security of the systems for which they are responsible.

Our focus then shifts from unauthorized programs to unauthorized people. People who use computers without authorization are called hackers. We look at the hacker culture of MIT in the 1950s and 1960s, as well as the origins of phone phreaking. We see how the electronic (and sometimes physical) trespassing of hackers and phreaks into telecommunications systems, combined with the public distribution of information related to break-ins, led to what Bruce Sterling calls the "hacker crackdown."

In the past few years, denial-of-service attacks have temporarily disabled Internet-based servers managed by many organizations. We examine some popular denial-of-service attack strategies and ways of combating them.

The controversy surrounding the 2000 Presidential election in Florida has raised the issue of online voting. Would voting over the Internet be superior to our present methods? We consider the benefits and risks associated with online voting.

6.2 Viruses, Worms, and Trojan Horses

There are a variety of ways in which undesired programs can become active on your computer. If you are lucky, these programs will do nothing other than consume a little CPU time and some disk space. If you are not so lucky, they may destroy valuable data stored in your computer's file system. An invading program may allow outsiders to seize control of your computer. Once this happens, they may use your computer as a depository for stolen credit card information, a Web server dishing out pornographic images, or a launch pad for spam or denial-of-service attacks on a corporate server.

"Computer pathologists" classify destructive programs as viruses, worms, or Trojan horses. In this section we describe these invasive programs and summarize technical means of defending against them.

6.2.1 Viruses

HOW VIRUSES WORK

A **virus** is a piece of self-replicating code embedded within another program called the **host** [1]. Figure 6.1 illustrates how a virus replicates within a computer. When a user executes a host program infected with a virus, the virus code executes first. The virus finds another executable program stored in the computer's file system and replaces the program with a virus-infected program. After doing this, the virus allows the host program to execute, which is what the user expected to happen. If the virus does its work quickly enough, the user may be unaware of the presence of the virus.

Because a virus is attached to a host program, you may find viruses anywhere you can find program files: hard disks, floppy disks, CD-ROMs, email attachments, and so on. Viruses can be spread from machine to machine via diskettes or CDs. They may also be passed when a person downloads a file from the Internet. Sometimes viruses are attached to free computer games that people download and install on their computers. A 2003 study revealed that 45 percent of the executable files people downloaded from KaZaA contained viruses or Trojan horses (which we will cover a little later) [2].

Today, many viruses are spread via email (Figure 6.2). An **attachment** is a file accompanying an email message. Attachments may be executable programs, or they may be word processing documents or spreadsheets containing macros, which are small pieces of executable code. If the user opens an attachment containing a virus, the virus takes

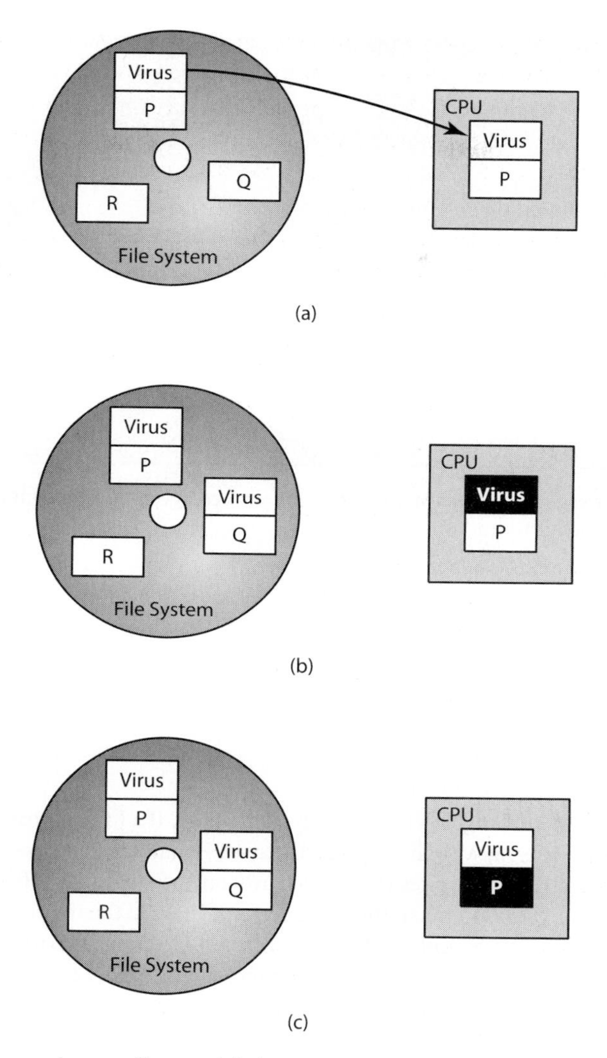

Figure 6.1 How a virus replicates. (a) A computer user executes program P, which is infected with a virus. (b) The virus code begins to execute. It finds another executable program Q and creates a new version of Q infected with the virus. (c) The virus passes control to program P. The user, who expected program P to execute, suspects nothing.

control of the computer, reads the user's email address book, and uses these addresses to send virus-contaminated emails to others, as illustrated in Figure 6.3.

Some viruses are fairly innocent; they simply replicate. These viruses occupy disk space and consume CPU time, but the harm they do is relatively minor. Other viruses are malicious and can cause significant damage to a person's file system.

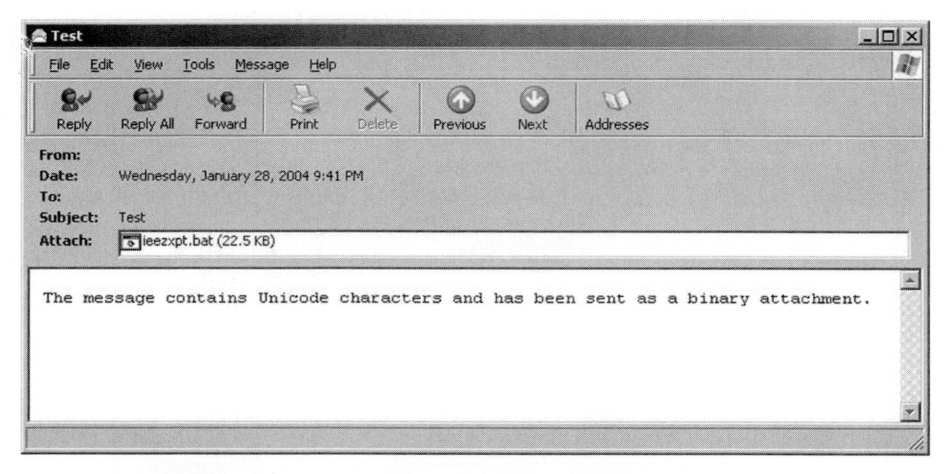

FIGURE 6.2 The attachment to this email message probably contains a virus. (The author didn't open it to find out.)

WELL-KNOWN COMPUTER VIRUSES

The Brain virus (c. 1986) was the first virus to move from one IBM PC to another. The virus was written by the owners of a Pakistani computer store called Brain Computer Services. They said their purpose was to determine the level of software piracy in Pakistan. The virus spread internationally, but it was not malicious and caused no significant harm to the PCs it infected [3].

The Michelangelo virus dates back to 1991. If a PC user executes a program infected with the virus on March 6, the birthday of Renaissance painter and sculptor Michelangelo, the virus overwrites critical records on the boot disk. If the boot disk is the user's hard drive, the contents of the drive are lost. In 1992 the media widely reported estimates that as many as five million PCs would be affected by the virus. As it turns out, only a few thousand computers were infected. Some say the whole episode was a classic example of media hype [4]. Others say the extensive media publicity encouraged institutions to perform checks that would not have been done otherwise. According to them, the outbreak on March 6 was not significant because institutions had already removed the virus [5].

The Melissa virus (c. 1999) lurks inside a macro in a Word document attached to an email message. When a user activates the virus by opening the infected attachment, Melissa sends an email message with the attachment to the first 50 people in the user's address book. When Melissa first appeared, email containing the virus flooded the Internet, crashing many email servers worldwide. It infected about 100,000 computers in its first weekend. David L. Smith of New Jersey pled guilty to posting the virus at an alt.sex.usenet group using a stolen AOL account [3]. In May 2001 Smith was sentenced to 20 months in federal prison plus 100 hours of community service. He was also fined $5,000 [6].

The Love Bug (c. 2000) is another virus lurking inside an email message. Unlike Melissa, which limits itself to the first 50 people in a victim's address book, the Love Bug

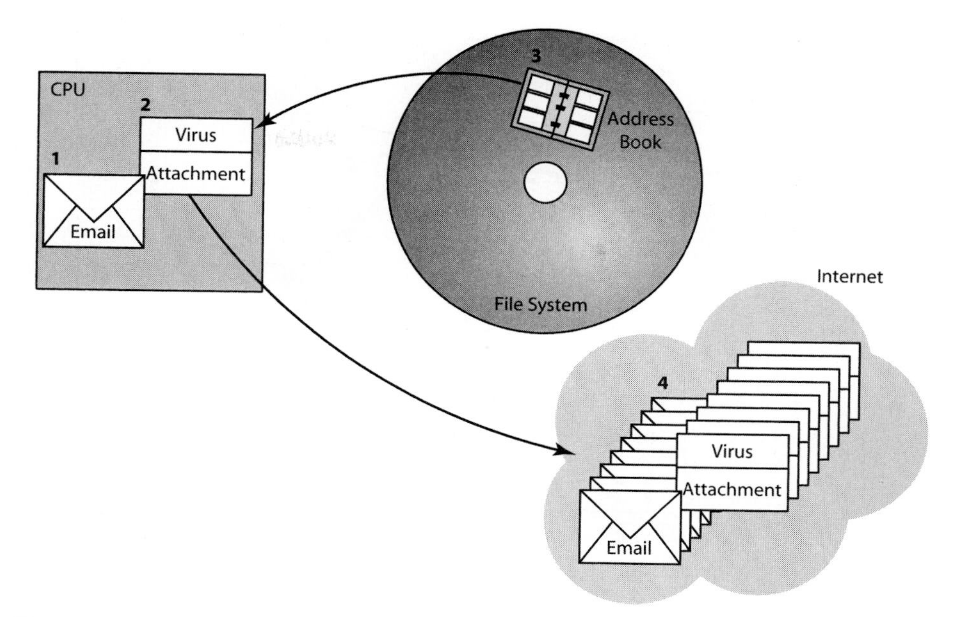

Figure 6.3 How an email virus spreads. A computer user reads an email with an attachment (1). The user opens the attachment, which contains a virus (2). The virus reads the user's email address book (3). The virus sends emails with virus-containing attachments (4).

creates email messages for everyone in the address book. It deletes some kinds of media files stored on the user's hard disk, and it also collects passwords and emails them to several different accounts in the Philippines. The creator of The Love Bug was a 23-year-old Filipino computer science student. When he created the virus, the Philippines had no laws against computer hacking, and he was not prosecuted [3].

VIRUSES TODAY

Commercial antivirus software packages allow computer users to detect and destroy viruses lurking on their computers. To be most effective, users must keep them up-to-date by downloading patterns corresponding to the latest viruses from the vendor's Web site.

There is evidence few people are diligent about keeping their computers virus-free. When students returned to Oberlin College in August of 2003, they were required to have their computers checked for viruses. System administrators found viruses in 90 percent of the computers running the Windows operating system [7].

6.2.2 Worms

A **worm** is a self-contained program that spreads through a computer network by exploiting security holes in the computers connected to the network (Figure 6.4). The technical term "worm" comes from *The Shockwave Rider*, a 1975 science fiction novel written by John Brunner [8].

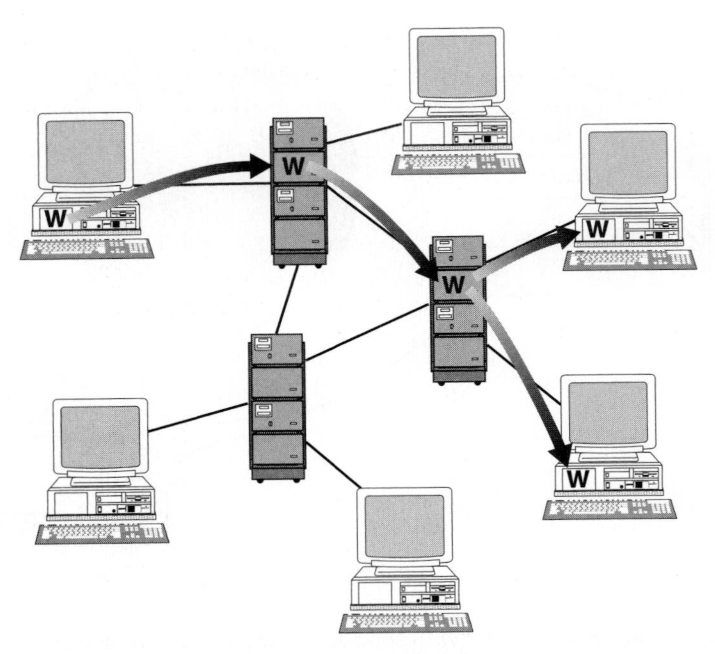

FIGURE 6.4 A worm spreads to other computers by exploiting security holes in computer networks.

WANK WORM

In October 1989, NASA scientists prepared for a Space Shuttle mission that would launch a probe to Jupiter. The robot probe, named Galileo, was fueled with radioactive plutonium. Antinuclear protestors created a worm that infiltrated a NASA network. Those who logged onto an infected computer were greeted with this banner:

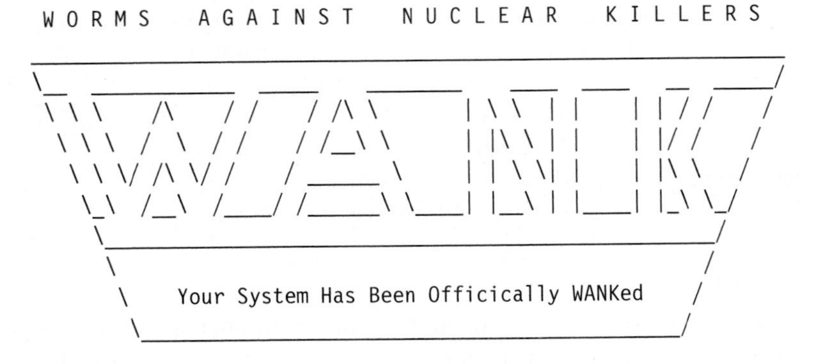

You talk of times of peace for all, and then prepare for war.

The WANK worm took a lot of system-administrator time to eradicate, but it did not delay the launch of the Space Shuttle. It is an example of **cyberterrorism**: a politically motivated attack against the information technology resources of a government or its people in order to inflict damage, disrupt services, or generate fear.

CODE RED

The Code Red worm, launched on July 19, 2001, exploited a bug in Microsoft's Internet Information Services (IIS) software to spread among Windows Web servers. If U.S. English was the default language on the server, the worm would change the server's local home page to the following message:

```
HELLO! Welcome to http://www.worm.com!

Hacked by Chinese!
```

Based on the day of the month, the Code Red worm either (1) attempted to propagate to other computers, (2) launched a denial-of-service attack against www.whitehouse.gov, or (3) slept. (We cover denial-of-service attacks in Section 6.4.) The Code Red worm spread to more than 359,000 hosts in less than 14 hours [9].

SAPPHIRE (SLAMMER)

The Sapphire worm, also known as Slammer, was released on January 25, 2003. Sapphire is notable for being the fastest-spreading computer worm in history. The number of hosts it infected doubled every 8.5 seconds, and within 10 minutes 90 percent of the vulnerable hosts were infected. The worm ended up affecting at least 78,000 computers worldwide [10].

Sapphire exploited a bug found in both Microsoft's SQL Server and SQL Server Desktop Engine. While it carried no malicious payload, the Sapphire worm overloaded networks and made database servers inaccessible. It resulted in cancelled airline flights, unavailable ATMs, and failures of emergency 911 service [11].

BLASTER

The Blaster worm appeared on August 11, 2003. It exploited a bug on Windows 2000 and Windows XP computers. Blaster infected hundreds of thousands of PCs worldwide. Besides spreading to as many computers as possible, the purpose of the Blaster worm seemed to be to launch a denial-of-service attack against windowsupdate.com, the Microsoft Windows Update Web server. The apparent goal of the worm was to prevent Microsoft customers from accessing the server to download the patch needed to fix the bug. It turns out windowsupdate.com was a shortcut to the actual Web site. Microsoft thwarted the attack by deleting the shortcut [12].

However, the Blaster worm did have the effect of slowing down some computer systems. It disrupted the signaling of CSX freight trains and Amtrak passenger trains in the Northeast, leading to service delays [13].

SASSER

The Sasser worm, launched in April 2004, exploited a previously identified security weakness with Windows computers. Computers with up-to-date software were safe from the worm, but it infected about 18 million computers worldwide nonetheless. The effects of the worm were relatively benign; infected computers simply shut themselves down shortly after booting. Still, the worm made millions of computers unusable and disrupted operations at Delta Airlines, the European Commission, Australian railroads, and the British coast guard [14].

After Microsoft offered a €250,000 award, a fellow student pointed the finger at German teenager Sven Jaschan, who confessed to the crime, and then began working for German computer security firm Securepoint. Because he was 17 when he released the worm, Jaschan was tried in a juvenile court, which sentenced him to one-and-a-half years' probation and 30 hours of community service [14, 15, 16].

INSTANT MESSAGING WORMS

Two early worms to strike instant messaging systems were Choke and Hello, which appeared in 2001. Worms were less devastating back then, because only about 141 million people used instant messaging. Today, more than 800 million people rely on instant messaging, so the impact of worms can be much greater. In April 2005 the appearance of the Kelvir worm forced the Reuters news agency to remove 60,000 subscribers from its Microsoft-based instant messaging service for 20 hours [17].

6.2.3 The Internet Worm

The Internet Worm was the first worm to affect thousands of computers. The primary source for this narrative is the excellent biography of Robert Morris in *Cyberpunk: Outlaws and Hackers on the Computer Frontier*, written by Katie Hafner and John Markoff [18].

BACKGROUND OF ROBERT TAPPAN MORRIS, JR

Robert Tappan Morris, Jr., began learning about the Unix operating system when he was still in junior high school. His father was a computer security researcher at Bell Labs, and young Morris was given an account on a Bell Labs computer that he could access from a teletype at home. It didn't take him long to discover security holes in Unix. In a 1982 interview with Gina Kolata, a writer for *Smithsonian* magazine, Morris admitted he had broken into networked computers and read other people's email. "I never told myself that there was nothing wrong with what I was doing," he said, but he acknowledged that he found breaking into systems challenging and exciting, and he admitted that he continued to do it.

As an undergraduate at Harvard, Morris majored in computer science. He quickly gained a reputation for being the computer lab's Unix expert. After his freshman year, Morris worked at Bell Labs. The result of his work was a technical paper describing a security hole in Berkeley Unix.

While at Harvard, Morris was responsible for several computer pranks. In one of them, he installed a program that required people logging in to answer a question posed by "the Oracle" and then ask the Oracle another question. (The Oracle program worked by passing questions and answers among people trying to log in.)

DESIGNING THE WORM

Morris entered the graduate program in computer science at Cornell University in the fall of 1988. He became intrigued with the idea of creating a computer worm that would exploit bugs he had found in three Unix applications: `ftp`, `sendmail`, and `fingerd`. Morris's worm used a **buffer overflow attack** to take control of a target computer (Figure 6.5). His "wish list" for the worm had about two dozen goals, including:

- Infect three machines per local area network.
- Only consume CPU cycles if the machines are idle.
- Avoid slow machines.
- Break passwords in order to spread to other computers.

The goal of the worm was to infect as many computers as possible. It would not destroy or corrupt data files on the machines it infected.

LAUNCHING THE WORM

On November 2, 1988, Morris learned that a fix for the `ftp` bug had been posted to the Internet, meaning his worm program could no longer take advantage of that security hole. However, nobody had posted fixes to the other two bugs Morris knew about. After making some last-minute changes to the worm program, he logged in to a computer at the MIT Artificial Intelligence Lab and launched the worm at about 7:30 p.m.

The worm quickly spread to thousands of computers at military installations, medical research facilities, and universities. Unfortunately, due to several bugs in the worm's programming, computers became infected with hundreds of copies of the worm, causing them to crash every few minutes or become practically unresponsive to the programs of legitimate users.

Morris contacted friends at Harvard to discuss what ought to be done next. They agreed that Andy Sudduth would anonymously post a message to the Internet. Sudduth's message is below. Harvard's computers were not affected (the security holes had already been patched), and you can tell from the last sentence that Sudduth was having a hard time believing Morris's story:

```
A Possible virus report:

There may be a virus loose on the internet.
Here is the gist of a message I got:

I'm sorry.
```

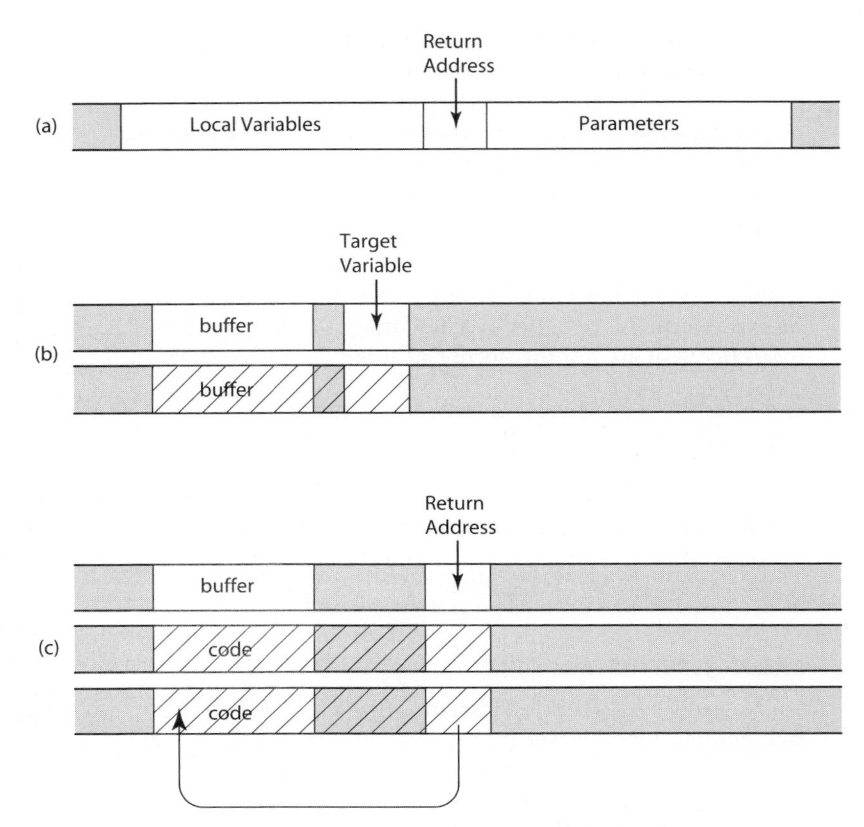

FIGURE 6.5 A **buffer overflow attack** is a common way to take control of a computer. (a) When a computer program makes a function call, the function's parameters and local variables are stored on the run-time stack, along with the return address—the address to which control should return when the function completes. The local variables occupy lower memory addresses than the return address. (b) In a **variable attack,** the goal of the intruder is to change the value of a key variable. The program expects the user to input some characters. It has allocated a buffer to store the characters. However, the string of characters provided by the intruder is too long, and it overflows the buffer, changing the value of the target variable. (c) In a **stack attack,** the goal of the intruder is to change the value of the return address. Again, a longer-than-expected string is input, overflowing the buffer and changing the value of the return address to point to the start of the buffer. When the function returns, code inserted by the intruder begins to execute, taking control of the computer. One way to prevent these attacks is to add checks to ensure array bounds are not exceeded. Another prevention measure is to modify the operating system so that it will not execute instructions stored on the run-time stack.

```
Here are some steps to prevent further
transmission:

1) don't run finger, or fix it to not
overrun its stack when reading arguments.
2) recompile sendmail w/o DEBUG defined
3) don't run rexed

Hope this helps, but more, I hope it is a hoax.
```

Sudduth's email was supposed to get routed through a computer at Brown University. However, computers at Brown were already infected with the virus and did not have spare cycles to route the message. Also, the email did not have a subject line, which made it less likely to be read during a crisis. The result is that the message was read too late to be of any help to those fighting the worm.

System administrators at various universities worked frantically to stop the spread of the worm. Within a day they had examined the worm's code, discovered the bugs in `sendmail` and `fingerd`, and published fixes to the Internet community. In all, about 6,000 Unix computers had been infected with the worm.

After some sleuthing by reporter John Markoff, *The New York Times* named Robert Tappan Morris, Jr. as the author of the worm. Morris was suspended from Cornell University. A year later, he was the first person to receive a felony conviction under the U.S. Computer Fraud and Abuse Act. He was sentenced to 3 years' probation, 400 hours of community service, and fined $10,000. His legal fees and fines exceeded $150,000.

ETHICAL EVALUATION

Was Robert Morris, Jr., wrong to unleash the Internet worm?

A Kantian evaluation must focus on Morris's will. Did Morris have good will? His stated goal was to see how many Internet computers he could infect with the worm. While Morris did not want to crash these computers or destroy any data stored on them, his motivation was fundamentally selfish: he wanted the thrill of seeing his creation running on thousands of computers. He used others because he gained access to their machines without their permission. There is also evidence Morris knew he was using others: he took measures designed to prevent people from discovering that he was the author of the worm. From a Kantian point of view, Morris's action was wrong.

From a social contract point of view, Morris's action was also wrong. He violated the property rights of the individuals and organizations whose computers were infected by the worm. They had the right to determine who would use their computers, and they attempted to enforce this right by requiring people to identify themselves by user name and password. Morris took advantage of security holes in these computers to gain unauthorized access to them. When his worm caused these computers to become unresponsive or crash, he denied access to the legitimate users of these computers.

A utilitarian evaluation of the case focuses on the benefits and harms resulting from the spread of the worm. The principal benefit of the Internet worm was that organizations managing these Unix computers discovered there were two significant security holes in their systems. They received the instructions they needed to patch these holes before a truly malicious intruder took advantage of them to enter their systems and do a lot of damage to their data. Of course, Morris could have produced the same beneficial result simply by contacting the system administrators at UC Berkeley and informing them of the security holes he had found.

The Internet worm had numerous harmful consequences. A large amount of time was spent by system administrators as they defended their machines from further attacks, tracked down the problem, installed patches, and brought machines back on line. There was a disruption in email and file exchange traffic caused by computers being taken off the network. About 6,000 computers were unavailable for a day or two. During this time, many thousands of people were less productive than they could have been had the systems been up and running. Morris himself was harmed by his actions. He was suspended from Cornell and sentenced to three years of probation and 400 hours of community service. His fines and legal fees exceeded $150,000. From a utilitarian viewpoint, Morris was wrong to have released the Internet worm.

In conclusion, Morris may not have been acting maliciously, but he was acting selfishly. If he had wanted to experiment with worms, he probably could have gotten permission to try out his creations on a local area network detached from the Internet, so that even if his worm had multiplied out of control, there would have been no fallout to the rest of the computer community. Instead, he chose to use the entire Internet as his experimental laboratory, inconveniencing thousands of people.

6.2.4 Trojan Horses

A **Trojan horse** is a program with a benign capability that conceals another, sinister purpose. When the user executes a Trojan horse, the program performs the expected beneficial task. However, the program is also performing actions unknown to, and not in the best interests of, the user.

Here are a few examples of the kinds of malicious tasks performed by Trojan horse programs:

- opening an Internet connection that allows an outsider to gain access to files on the user's computer;
- logging the keystrokes of the user and storing them in a file that the attacker can peruse to learn confidential information, such as passwords;
- looking for passwords stored on the computer and emailing them to the attacker's address;
- destroying files on the user's computer;
- launching a denial-of-service attack on a Web site;
- turning the user's computer into a proxy server that can be used to launch spam or stash information gained from illegal activities (such as stolen credit card numbers).

A **remote access Trojan (RAT)** is a Trojan horse program that gives the attacker access to the victim's computer. Two well-known RATs are Back Orifice and SubSeven. SubSeven is notable because of its easy-to-use point-and-click user interface. SubSeven consists of a client program running on the attacker's computer, and a server program running on the victim's computer. The attacker is able to capture images from the victim's monitor, record keystrokes, read and write files, watch traffic on the victim's local area network, and even control the mouse.

In order to gain access to another person's computer, the attacker must trick that person into downloading the RAT server. The most popular way to do this is to hide it inside a file posted to a Usenet newsgroup specializing in erotica. The attacker advertises the file as containing sexually explicit videos or photos. Those who download the file bring the RAT into their computer.

6.2.5 Defensive Measures

The ability of a computer network to withstand the attacks of viruses, worms, and Trojan horses depends to a great extent on the skill and dedication of its system administrators, as well as the cooperation of the network's users.

System administrators should set up reasonable authorization and authentication mechanisms. **Authorization** is the process of determining that a user has permission to perform a particular action. For example, a system administrator has authorization to reboot a computer, but a typical user does not. An ordinary user should not be able to examine the email messages of another user. Most operating systems create unique *user identifiers*, or uids, for its users. With each uid is information about the user's privileges. The system administrator should set user privileges appropriately to prevent one user from violating the privacy of another.

Computer security also depends upon **authentication**: determining that a person is who he claims to be. There are a variety of authentication mechanisms. The most common type is knowledge-based authentication, such as a password. Another authentication mechanism is the use of tokens, such as an identification card or smart card. A third authentication mechanism uses biometric data, such as a fingerprint or retinal scan. It is common for highly secure computer systems to use two different authentication schemes.

The most common knowledge-based authentication scheme is the password. System administrators should install automatic password checking software that prevents users from selecting passwords that are easily guessed, such as the login name, the reverse of the login name, or a circular shift of the login name. To foil a **dictionary attack**—an automated intruder attempting to guess a password by trying every word in the dictionary—a user should always have at least one nonalphabetic character in the password.

A sure-fire way to prevent a network from being attacked by an external virus or worm is to detach it from the Internet. If it is important that the computers on the network be able to communicate with the Internet, installing a firewall is the next best thing. A **firewall** is a computer, positioned between a local network and the Internet, that monitors the packets flowing in and out. One type of firewall is a packet filter, which accepts packets only from certain trusted computers on the Internet. Another use of a firewall is to limit the number of services external computers may access. For example, many attacks have taken advantage of the *finger* program. A way to prevent such attacks is simply to not provide *finger* service to outside computers.

An important responsibility of the system administrator is to keep the operating system up-to-date with the latest patches. When the provider of an operating system announces a security patch, the announcement also informs malicious persons that a vulnerability exists. Sometimes a new worm is launched well after the patch has been made available. Up-to-date systems are not vulnerable to attacks by these worms.

A system administrator can install filters on mail servers that screen out much unwanted mail, including spam and virus-laden email. Still, some contaminated email messages are likely to get through to individual users. Virus filters associated with email readers can check incoming messages for viruses. When such a message is found, it is deleted or put in a quarantine area.

6.3 Phreaks and Hackers

Telephone and computer systems are powerful technologies, prompting some curious people to invest a lot of time and energy into learning more about how they work. A few of these experts use the knowledge they have gained to enter systems without authorization. Once inside these systems, their actions vary widely, from simply "nosing around" to copying sensitive information to rerouting phone calls. In this section we examine two subcultures of techno-explorers: phreaks and hackers. This section relies upon three principal sources: *Cyberpunk: Outlaws and Hackers on the Computer Frontier* by Katie Hafner and John Markoff [18], *Hackers: Heroes of the Computer Revolution* by Steven Levy [19], and *The Hacker Crackdown* by Bruce Sterling [20].

6.3.1 Hackers

ORIGINAL DEFINITION OF "HACKER"

In its original meaning, a **hacker** is an explorer, a risk-taker, someone who is trying to make a system do something it has never done before. Hackers in this sense of the word abounded at MIT's Tech Model Railroad Club in the 1950s and 1960s. The Club constructed and continuously improved an enormous HO-scale model train layout. Members of the Signals and Power Subcommittee built an elaborate electronic switching system to control the movement of the trains. Wearing chino pants, short-sleeved shirts, and pocket protectors, the most dedicated members would drink vast quantities of Coca-Cola and stay up all night to improve the system. To them, a "hack" was a newly constructed piece of equipment that not only served a useful purpose, but also demonstrated its creator's technical virtuosity. Calling someone a hacker was a sign of respect; hackers wore the label with pride. In 1959, after taking a newly created course in computer programming, some of the hackers shifted their attention from model trains to electronic computers [19].

After extensive interviews with MIT hackers, Steven Levy has summarized the “hacker ethic” with these precepts, which I quote verbatim [19]:

- Access to computers—and anything which might teach you something about the way the world works—should be unlimited and total. Always yield to the Hands-On Imperative!
- All information should be free.
- Mistrust Authority—Promote Decentralization.
- Hackers should be judged by their hacking, not bogus criteria such as degrees, age, race, or position.
- You can create art and beauty on a computer.
- Computers can change your life for the better.

Computer security expert Dorothy Denning has observed that the will of the hacker is to make an improvement—a hacker is not malicious. A hacker is not out to destroy data or equipment. A hacker does not commit fraud for personal profit [21].

HACKING ON THE PDP-1

The story of MIT’s PDP-1 minicomputer illustrates some of the many ways that the hacker ethic translated into particular deeds.

Digital Equipment Corporation (DEC) donated the second PDP-1 it made to MIT in the summer of 1961. The PDP-1 was DEC’s first product, and it came with very little software. To help remedy this deficiency, six hackers put in about 250 man-hours in a single weekend to convert an assembler for MIT’s TX-0 computer to PDP-1 machine language. In one weekend they produced a program that would have taken a commercial enterprise months to complete.

Steve Russell came up with the idea of writing a shoot-em-up game for the PDP-1 that would utilize its programmable graphics display. He worked on it for over half a year, with help from other MIT hackers. In February 1962 he unveiled Spacewar, the first video game. Two players maneuvered space ships that shot torpedoes (dots) at each other. The game was an instant hit, but rather than commercialize it, the MIT group freely distributed copies of the program to other PDP-1 users.

Hackers also programmed the PDP-1 to produce the sounds needed to activate telephone switching equipment. With this capability, they were able to navigate the international telephone system. However, their excursions were simply for the sake of exploration, not for the purpose of defrauding AT&T. In fact, they reported problems they uncovered to the proper telephone service groups.

Stewart Nelson thought adding a new hardware instruction to the PDP-1 would make it better. Students had been expressly forbidden from working on the computer hardware itself, but they also knew that waiting for permission to modify the hardware would take months. Nelson decided not to ask for permission. One night, he and a few cohorts opened up the cabinet of the PDP-1 and did some rewiring. They tested the computer, and they thought they had increased the capability of the PDP-1 without affecting its other functionality. However, their testing was incomplete. The next morning, a legitimate user of the PDP-1 discovered that her program, an important weather simulation code, no longer worked. Adding a new instruction had caused another instruction to malfunction.

On another occasion, Nelson was making an unauthorized, middle-of-the night adjustment to the power supply on an MIT computer. Needing a large screwdriver, he took one from the locked cabinet of the machine shop craftsman. In the process of making the adjustment, Nelson accidentally shorted out a circuit, melting the screwdriver's handle. When the craftsman came to work the next morning, he opened the cabinet and saw the ruined screwdriver with this sign attached: USED UP.

ETHICAL EVALUATION

Was Stewart Nelson wrong to modify the PDP-1 hardware without permission? Let's evaluate his action.

A Kantian evaluation focuses on the will behind the action, rather than its results. We might be tempted to state that Stewart Nelson's will was to improve the PDP-1, but Kant writes that we should avoid a characterization that allows an expected result to provide the motivation for an action [22]. If we ignore the expected result, what do we have left? He appears to have been acting under the maxim, "Take advantage of every opportunity to demonstrate your technical skills." In his desire to demonstrate his technical prowess, Nelson made modifications to the PDP-1 without authorization. He disregarded the instructions issued by the person with legitimate authority to control access to the machine. He also disregarded the needs of the PDP-1's legitimate users, whose work depended upon the reliability of the computer. Hence Nelson treated other human beings as means to an end, and his action was wrong.

From the point of view of social contract theory, this moral problem is similar to the case of Robert Tappan Morris, Jr. By modifying a system he did not own, Nelson violated the rights of the legitimate owners and users of the computer. Hence his action was wrong.

A rule utilitarian analysis considers what would happen if everyone engaged in such behavior. Suppose everyone who had an idea about improving a system went ahead and made the change without asking permission. Perhaps most changes would make systems run better, but inevitably some people would accidentally make changes that made the system perform worse. A few supposed improvements would result in systems being broken, perhaps for long periods of time. You can also imagine situations where two different changes are being made to the same system. Either one of the changes, when made in isolation, would improve the system, but when both changes are made, they interfere with each other and make the system unusable. If changes are not systematically recorded, the missing documentation could make systems much harder to maintain. People who simply want to use the systems would not be able to predict when they would be available and when they would not, so another long-term consequence of such actions would likely be a lowering of productivity. In the long term, allowing people to make unauthorized changes would result in less reliable systems that no one understands. We conclude that Nelson's action was wrong from a rule utilitarian point of view.

Finally, let's evaluate Nelson's action from an act utilitarian point of view. The affected persons were Nelson, the PDP-1 administrator, and the computer's users. By modifying the PDP-1, Nelson learned more about computer engineering, a benefit. We know at least one computer user was harmed as a result of Nelson's failed modification: She spent a lot of time tracking down the problem, and she could not continue with her work until the computer was fixed. In order to repair the computer, it would have to be made unavailable to its programmers, another harm. Fixing the computer had an associated cost, measured in terms of labor and/or equipment. This cost is another negative effect. Nelson's deed most likely cost the PDP-1 administrator time and stress as he interacted with unhappy programmers and oversaw the repair job. While we have not assigned particular values to the benefit and the harms, it is likely a complete analysis would indicate Nelson's action was wrong.

It is worth considering how our analysis would change if Nelson's midnight modification of the PDP-1 had been successful and the system had worked even better after he operated on it. The Kantian, social contract, and rule utilitarian analysis did not take into account the actual result of Nelson's action, so even if his hacking had been successful, they would still have concluded that he did the wrong thing.

However, the act utilitarian analysis would be completely different. Nelson would have benefitted from learning more about computer engineering. The programmers of the computer would have benefitted from a more powerful instruction set. With no interruptions in the daily use of the computer, no one would have been harmed. If Nelson's hack had worked, you could conclude he did a good thing, from an act utilitarian point of view. At this point it's fair to ask: What good is an ethical theory if it can only tell you *afterward* whether your action was right or wrong? Does act utilitarianism encourage people to take morally dubious actions and then hope for the best outcome? Would you like to live in a world where everyone lived by the maxim, "Better to ask forgiveness than permission"?

DUMPSTER DIVING AND SOCIAL ENGINEERING

In the 1983 movie *WarGames*, a teenage hacker breaks into a military computer and nearly causes a nuclear Armageddon. After seeing the movie, a lot of teenagers were excited at the thought that they could prowl cyberspace with a home computer and a modem. A few of them became highly proficient at breaking into government and corporate computer networks.

Typically, you need a login name and password to access a computer system. Sometimes a hacker can guess a valid login name/password combination, particularly when system administrators allow users to choose short passwords or passwords that appear in a dictionary. Two other effective techniques for obtaining login names and passwords are dumpster diving and social engineering.

Dumpster diving means looking through garbage for interesting bits of information. Companies typically do not put a fence around their dumpsters. In midnight rummaging sessions hackers have found user manuals, phone numbers, login names, and passwords.

Social engineering, a term coined by hacker Kevin Mitnick, refers to the manipulation of a person inside the organization to gain access to confidential information. Social engineering is easier in large organizations where people do not know each other very well. For example, a hacker may identify a system administrator and call that person, pretending to be the supervisor of his supervisor and demanding to know why he can't access a particular machine. In this situation, a cowed system administrator, eager to please his boss's boss, may be talked into revealing or resetting a password [23].

MALICIOUS HACKERS

In the modern use of the word, "hacking" has come to include computer break-ins accompanied by malicious behavior, such as destroying databases or stealing confidential personal information. An example of this use of the word is a story in *Computerworld* describing how people hacked into *USA Today's* Web site on July 11, 2002, and inserted fabricated news stories [24].

6.3.2 Phone Phreaking

A **phone phreak** is someone who manipulates the telephone system in order to communicate with others without paying for the call. The prototypical phone phreaking activity is an hours-long, coast-to-coast conference call charged to the account of a large corporation.

Historically, phone phreaks used a variety of methods to access long-distance service:

1. *Stealing long-distance telephone access codes.*

   The easiest way to do this is by "shoulder surfing" at an airport, train station, or other public place. A phreak simply looks over people's shoulders as they key in their long distance access codes.

2. *Guessing long-distance telephone access codes.*

   Phreaks learned how to program a computer to try different codes. Running a computer all night typically resulted in about a dozen hits.

3. *Using a "blue box" to get free access to long-distance lines.*

   A "blue box" mimicked the telephone system's own access signal, a high-pitched tone of 2600 hertz.

In the 1980s phreaks used certain computer bulletin board systems (BBSs) called "pirate boards" to share stolen long-distance access codes and credit card numbers with each other.

In response to these activities, telecommunications firms installed software to detect overuse of particular long distance telephone codes. They also installed equipment to detect and trace attempts to guess access codes. The introduction of digital networks has made 2600-hertz blue boxes obsolete.

6.3.3 The Cuckoo's Egg

Clifford Stoll was a physics Ph.D. who took a job as a system administrator at Lawrence Berkeley Laboratory so he could stay in California. When Stoll was still new in the position, he was asked to reconcile a 75-cent discrepancy between two accounting systems that charged users for computer time. He carefully searched for the missing 75 cents and discovered, to his chagrin, that an unauthorized user was logging onto Lawrence Berkeley Lab's computer. Even worse, the hacker was using LBL computers as a staging point from which to jump to computers at military installations.

Stoll observed the intruder searching these systems for files with information about such topics as the Strategic Defense Initiative and stealth technology. Eventually investigators from the FBI, the CIA, the National Security Agency, the Air Force Office of Special Investigations, and the Defense Intelligence Agency joined Stoll in the search for the hacker. The trail led to a group of West German hackers who had sold various programs, but apparently no classified information, to the KGB, the intelligence service of the Soviet Union [25].

6.3.4 Legion of Doom

Plovernet was a popular phreak/hacker BBS operated in New York and Florida; more than 500 people subscribed to it. In 1984 "Lex Luthor" created an invitation-only BBS called Legion of Doom and recruited the sharpest phreaks from Plovernet. He also created a phreak/hacker group of the same name. According to Luthor, very few users of the Legion of Doom BBS were Legion of Doom members [26]. He took the name Legion of Doom straight out of the comic books, but the authorities did not think the group's activities were the least bit humorous.

One of the ways the Legion of Doom made a name for itself was by publishing *The Legion of Doom Technical Journal*, an obvious poke at AT&T's *Bell Labs Technical Journal*. This electronic publication contained articles of interest to phreaks and hackers. All of the articles were published under pseudonyms, of course.

The introduction to a Lex Luthor article appearing in the first issue, "Identifying, Attacking, Defeating, and Bypassing Physical Security and Intrusion Detection Systems," reveals something about the interests of Legion of Doom members as well as their attitude toward the establishment:

> The reasons for writing this article are twofold:
>
> 1. To prevent the detection and/or capture of various phreaks, hackers and others, who attempt to gain access to: phone company central offices, phone closets, corporate offices, trash dumpsters, and the like.

2. To create an awareness and prove to various security managers, guards, and consultants how easy it is to defeat their security systems due to their lack of planning, ignorance, and just plain stupidity.

In September 1988 Legion of Doom member Robert Riggs (a.k.a. "The Prophet") broke into a BellSouth computer known as an Advanced Information Management System. The computer contained employee email, documents, and databases. Because the system had no dial-up lines, BellSouth thought the system was hidden from the public and provided minimal security for it. It did not even ask users for passwords. Rummaging around the system, Riggs found a document called "Bell South Standard Practice 660-225-104SV Control Office Administration of Enhanced 911 Services for Special Services and Major Account Centers dated March 1988" (the E911 Document). He copied the E911 Document to his personal computer.

Five months later, Riggs sent a copy of the E911 Document to Craig Neidorf (a.k.a. "Knight Lightning"), a pre-law student at the University of Missouri. Neidorf was the publisher of *Phrack*, an electronic magazine widely distributed over BBSs. Both Riggs and Neidorf had something to gain from the publication of the E911 Document. Riggs would be able to brag about the trophy he had bagged from a BellSouth computer. Neidorf would be able to demonstrate the power of the hacker underground and thumb his nose at the telecommunications companies. Still, neither wanted to get caught. They edited the E911 Document heavily, deleting the document's NOT FOR USE OR DISCLOSURE warning, phone numbers of BellSouth employees, and other identifying and sensitive information. By the time they were done, they had removed nearly half the material from the report. On February 25, 1989, *Phrack* published the document under the pseudonym "The Eavesdropper."

6.3.5 Fry Guy

On June 13, 1989, all calls to the Palm Beach County Probation Department in Delray Beach, Florida were picked up by a phone-sex hotline in New York State. Phone phreaks thought it was a hilarious practical joke, but BellSouth was not amused. It immediately began a high-intensity, around-the-clock search for evidence of tampering with its computerized phone switching equipment. Investigators discovered that intruders had created new telephone numbers for themselves, manipulated proprietary databases, and reprogrammed diagnostic functions so that they could eavesdrop on conversations. If intruders could do these things, BellSouth reasoned, they could also reprogram 911 service. What if everyone dialing 911 were connected to a phone-sex hotline?

Within a matter of weeks, police investigating the phone-sex switcheroo got a lucky break. Someone called Indiana Bell to brag about the terrible things his friends in the Legion of Doom were about to do to the telephone system, including bring the entire network down the next Fourth of July. Indiana Bell traced the call back to its source, and the Secret Service installed pen registers at his home. The pen registers revealed long-distance telephone access code fraud. The Secret Service obtained a warrant, and on July 22 it seized all the equipment and notes of an Indiana 16-year-old with the nickname "Fry Guy."

Fry Guy had earned his nickname by using a password stolen from a local McDonald's manager to log into a McDonald's mainframe and give raises to some of his friends. He had moved on to stealing long-distance access codes and credit card numbers. He had used these stolen credit card numbers to purchase goods and get cash advances from Western Union.

The U.S. Attorney charged Fry Guy with 11 counts of computer fraud, unauthorized computer access, and wire fraud. In September 1990 he was sentenced to 44 months' probation and 400 hours of community service.

By Secret Service standards, a 16-year-old hacker was small fry. They were after his heroes, the members of the Legion of Doom, who were instigating all sorts of illegal activity through their publication of *The Legion of Doom Technical Journal*.

On January 15, 1990—Martin Luther King, Jr., Day—AT&T's long distance service failed. Sixty thousand people lost all their telephone service, and about 70 million telephone calls could not be completed. As we will see in Chapter 7, the crash was the result of a software bug in the switching equipment used to route long-distance calls. It took AT&T engineers about nine hours to understand the general cause of the crash. A few weeks later, they found the bug.

Despite this information from AT&T, law enforcement officials had their own theories about what had caused the crash. After all, they had interviewed numerous hackers who had claimed that the Legion of Doom could bring down the nationwide telephone switching system. It seemed too great a coincidence that the system should collapse on a national holiday, just as Fry Guy had predicted. The U.S. Attorney's Office in Chicago and the Secret Service decided it was time to take serious action against hackers and phreaks.

6.3.6 *U.S. v. Riggs*

Three days after the collapse of AT&T's long distance system, two U.S. Secret Service agents visited Craig Neidorf and accused him of causing the failure. They also confronted him with the stolen E911 Document. Neidorf cooperated with the Secret Service agents. He admitted that he had received the document from Riggs, and he also admitted that he knew the document had been taken from a BellSouth computer. The next day, Secret Service appeared at Neidorf's fraternity house with a warrant, searched his room, and seized his computer.

The U.S. Attorney in Chicago charged Riggs and Neidorf with wire fraud, interstate transportation of stolen property valued at $79,449, and computer fraud. Robert Riggs pleaded guilty to wire fraud for his unauthorized access of the BellSouth computer; he ended up serving time in a federal prison. Neidorf pleaded innocent to all charges, and the case went to trial in Chicago in July 1990.

The trial was short, lasting only four days. The defense quickly established that the information in the E911 Document was in the public domain. BellSouth was actually selling to the public two documents containing more detailed information about enhanced 911 service. These documents, which could be ordered by calling a toll-free number, sold for $13 and $21, respectively, belying BellSouth's contention that the E911 Document was worth $79,449. In light of this new information, the prosecution moved to dismiss the indictments against Neidorf. The judge agreed to the motion, dismissed the jury, and declared a mistrial.

The trial against Craig Neidorf is notable for a couple of reasons. First, it demonstrates how the long history of break-ins at telecommunications companies, the posting of information on BBSs about the inner workings of phone switches, and the collapse of AT&T's long distance service all combined to created an atmosphere in which the justice system was eager to "do something" about phone phreaking and computer hacking. In its zeal to prosecute, the government uncritically accepted AT&T's inflated valuation of the E911 Document. When the true value of the document was revealed, the government's case against Neidorf collapsed.

Second, the prosecution was careful to depict Neidorf as a thief, rather than a publisher. They could do this because Neidorf's "newsletter" was completely electronic. Viewing him as a publisher would have brought up a variety of First Amendment issues they were eager to avoid. In the early 1970s *The New York Times* and the *Washington Post* had published the Pentagon Papers, documents Daniel Ellsberg had stolen from the Pentagon describing government policies regarding the Vietnam war. The government never prosecuted these newspapers for publishing the documents. Should *Phrack* have been entitled to the same protection as *The New York Times*?

6.3.7 Steve Jackson Games

Another victim of the "hacker crackdown" was Steve Jackson Games (SJG) of Austin, Texas. SJG produces and sells role-playing games. In the late 1980s SJG operated a small BBS called Illuminati that provided various kinds of support to its customers, including email. Loyd Blankenship, a.k.a. "The Mentor" and an outspoken member of the Legion of Doom, happened to be a professional game designer and managing editor at SJG. Blankenship had already published the stolen E911 Document on his own BBS, called Phoenix Project.

On March 1, 1990, the Secret Service entered Blankenship's home and SJG. It seized four computers, including the one running the Illuminati BBS. According to the search warrant, which was only unsealed months later, the authorities had expected to find a copy of the stolen E911 Document on the Illuminati BBS. There was no copy of the document on any of the seized computers, and no charges were ever filed against SJG. Four months after the raid, the government returned most (but not all) of the hardware it had seized. The disruption in business caused by the Secret Service raid forced SJG to lay off half of its employees in order to survive.

The Secret Service raid of SJG is one of the key events that led to the creation of the Electronic Frontier Foundation, a nonprofit organization that speaks out for the Constitutional rights of Americans in cyberspace. With the financial backing of the Electronic Frontier Foundation, SJG and four Illuminati BBS users sued the Secret Service. The case went to trial in 1993. The court ruled that the Secret Service had violated the Electronic Communications Privacy Act when it seized, read, and (in some cases) deleted email on the Illuminati BBS without a court order. The judge noted that investigators simply could have logged on to the Illuminati BBS to determine if the E911 Document had been posted there. He awarded SJG $50,000 in damages plus over $250,000 in attorney's fees.

6.3.8 Retrospective

In *The Hacker Crackdown* Bruce Sterling writes:

> Hackers perceive hacking as a "game." This is not an entirely unreasonable or sociopathic perception. You can win or lose at hacking, succeed or fail, but it never feels "real." It's not simply that imaginative youngsters sometimes have a hard time telling "make-believe" from "real life." Cyberspace is *not real!* "Real" things are physical objects, such as trees and shoes and cars. Hacking takes place on a screen. Words aren't physical, numbers (even telephone numbers and credit card numbers) aren't physical. Sticks and stones may break my bones, but data will never hurt me. Computers *simulate* reality, such as computer games that simulate tank battles or dogfights or spaceships. Simulations are just make-believe, and the stuff in computers is *not real*.
>
> Consider this: If "hacking" is supposed to be so serious and real-life and dangerous, then how come *nine-year-old kids* have computers and modems? You wouldn't give a nine-year-old his own car, or his own rifle, or his own chainsaw—those things are "real."
>
> People underground are perfectly aware that the "game" is frowned upon by the powers that be. Word gets around about busts in the underground. Publicizing busts is one of the primary functions of pirate boards, but they also promulgate an attitude about them, and their own idiosyncratic ideas of justice. The users of underground boards won't complain if some guy is busted for crashing systems, spreading viruses, or stealing money by wire fraud. They may shake their heads with a sneaky grin, but they won't openly defend these practices. But when a kid is charged with some theoretical amount of theft: $264,846.14, for instance, because he sneaked into a computer and copied something, and kept it in his house on a floppy disk—this is regarded as a sign of near insanity on the part of prosecutors, a sign that they've drastically mistaken the immaterial game of computing for their real and boring everybody world of fatcat corporate money.[1] [20]

I quote Sterling at length because there are parallels between this viewpoint and the mentality of the millions of people who download MP3 files containing copyrighted music. The first parallel is the attitude that intellectual property is overvalued by the establishment. How can an AT&T technical document be worth $79,000? How can distributing songs over the Internet be a $100 billion offense? The second parallel is the

1. From *The Hacker Crackdown* by Bruce Sterling, copyright © 1992 by Bruce Sterling. Used by permission of Bantam Books, a division of Random House, Inc.

use of technology as a joyride: "Hey, I can make a long-distance phone call without getting a bill!" "Hey, I can make a music CD that costs me 17 cents instead of 17 bucks!" The knowledge that actions are wrong actually makes them more fun [27]. The third parallel is the idea that breaking certain laws is not that big a deal. There is the assumption that the chance of actually getting caught is small.

There are also parallels between the response of the Secret Service to the BBSs that posted information about hacking and phreaking, and the response of the Recording Industry Association of America (RIAA) to those who made available large number of MP3 files.

On May 9, 1990, in Operation Sundevil, the Secret Service shut down 25 BBSs for posting stolen long-distance telephone access codes and facilitating the exchange of stolen credit card numbers. A press release stated:

> Today, the Secret Service is sending a clear message to those computer hackers who have decided to violate the laws of this nation in the mistaken belief that they can successfully avoid detection by hiding behind the relative anonymity of their computer terminals . . .
>
> Underground groups have been formed for the purpose of exchanging information relevant to their criminal activities. These groups often communicate with each other through message systems between computers called "bulletin boards."
>
> Our experience shows that many computer hacker suspects are no longer misguided teenagers, mischievously playing games with their computers in their bedrooms. Some are now high tech computer operators using computers to engage in unlawful conduct. [20]

On September 8, 2003, the RIAA announced that its member companies had filed 261 federal lawsuits against what it called "major offenders," each of whom on average had been distributing more than 1,000 copyrighted music files through peer-to-peer networks. RIAA President Cary Sherman said:

> Nobody likes playing the heavy. There comes a time when you have to stand up and take appropriate action . . . We've been telling people for a long time that file sharing copyrighted music is illegal, that you are not anonymous when you do it, and that engaging in it can have real consequences . . . We hope that today's actions will convince doubters that we are serious about protecting our rights. [28]

The message from the Secret Service and the RIAA is consistent: cyberspace *is* real, those who break the law can be tracked down, and illegal actions in cyberspace can have severe consequences.

6.3.9 Penalties for Hacking

Under U.S. law, the maximum penalties for hacking are severe. The Computer Fraud and Abuse Act criminalizes a wide variety of hacker-related activities, including:

- transmitting code (such as a virus or worm) that causes damage to a computer system;

- accessing without authorization any computer connected to the Internet, *even if no files are examined, changed, or copied*;
- transmitting classified government information;
- trafficking in computer passwords;
- computer fraud;
- computer extortion.

The maximum penalty imposed for violating the Computer Fraud and Abuse Act is 20 years in prison and a $250,000 fine.

Another federal statute related to computer hacking is the Electronic Communications Privacy Act. This law makes it illegal to intercept telephone conversations, email, or any other data transmissions. It also makes it a crime to access stored email messages without authorization.

The use of the Internet to commit fraud or transmit funds can be prosecuted under the Wire Fraud Act and/or the National Stolen Property Act. Adopting the identity of another person to carry out an illegal activity is a violation of the Identity Theft and Assumption Deterrence Act.

6.3.10 Recent Incidents

Despite potentially severe penalties for convicted hackers, computer systems continue to be compromised by outsiders. Many break-ins are orchestrated by individuals or groups with a high degree of expertise, but others are committed by ordinary computer users who simply take advantage of a security weakness.

In 2003 a hacker broke into computers at the University of Kansas and copied the personal files of 1,450 foreign students. The files contained names, Social Security numbers, passport numbers, countries of origin, and birthdates. The University of Kansas had collected the information in one place in order to comply with a Patriot Act requirement that it report the information to the Immigration and Naturalization Service [29]. In a similar incident two years later, an intruder broke into a University of Nevada, Las Vegas computer containing personal information on 5,000 foreign students [30].

Another recent case demonstrates the time and effort sometimes required to identify those responsible for computer break-ins. In April 2004 several American supercomputer installations reported that hackers had broken into computers connected to a high-speed network called TeraGrid. Before the culprits could be apprehended, they had broken into thousands of computers at American research laboratories and military installations. The hackers also accessed computers at Cisco Systems and stole some of that company's software. Security experts, FBI agents, and Swedish police worked for more than a year to identify the European culprits and bring the break-ins to an end [31].

In March 2005 someone discovered a security flaw in the online-admissions software produced by ApplyYourself and used by six business schools. The discoverer posted instructions on a *Business Week* online forum explaining how business school applicants could circumvent the software security system and take a look at the status of their applications. It took ApplyYourself only nine hours to fix the flaw, but in the interim period hundreds of eager applicants had exploited the bug and peeked at their files. A week later, Carnegie Mellon University, Harvard University, and the Massachusetts Institute of Technology announced that they would not admit any of the applicants who had accessed their computer systems without authorization [32].

6.4 Denial-of-Service Attacks

A **denial-of-service (DoS) attack** is an intentional action designed to prevent legitimate users from making use of a computer service [33]. A DoS attack may involve unauthorized access to one or more computer systems, but the goal of a DoS attack is not to steal information. Instead, the aim of a DoS attack is to disrupt a computer server's ability to respond to its clients. Interfering with the normal use of computer services can result in significant harm. A company selling products and services over the Internet may lose business. A military organization may find its communications disrupted. A nonprofit organization may be unable to get its message out to the public.

A DoS attack is an example of an "asymmetric" attack, in which a single person can harm a huge organization, such as a multinational corporation or even a government. Since terrorist organizations specialize in asymmetric attacks, some fear that DoS attacks will become an important part of the terrorist arsenal [34, 35].

During the week of February 7–11, 2000, a 15-year-old initiated DoS attacks that disabled many Web sites, including Amazon.com, eBay, Yahoo, CNN.com, and Dell. The teenager, who went by the nickname "Mafiaboy," was sentenced to eight months in a juvenile detention center and a year of probation [36].

In October 2002 a DoS attack was launched against the Internet's 13 root servers, which act as the Internet's ultimate authority with respect to matching domain names to IP addresses [37].

Recently, many DoS attacks have focused on blacklist services, used by ISPs to shield their customers from spam. "We're usually under attack from 5,000 to 10,000 servers at once," says Steve Linford, CEO of Spamhaus [38].

The Cooperative Association for Internet Data Analysis at the University of California estimates that 4,000 Web sites suffer DoS attacks each week [39].

In this section we describe a variety of kinds of DoS attacks and some of the defensive measures that organizations can take to guard themselves against such attacks. Attackers do not want to give themselves away by initiating attacks from their own systems. Instead, they identify other computers they can use to launch their attacks. For this reason, all system administrators, not just those at targeted organizations, play a role in preventing DoS attacks.

6.4.1 Attacks that Consume Scarce Resources

The most common DoS attack is against a target system's network connection. A low-tech but effective way to do this is to cut the physical connection between the target

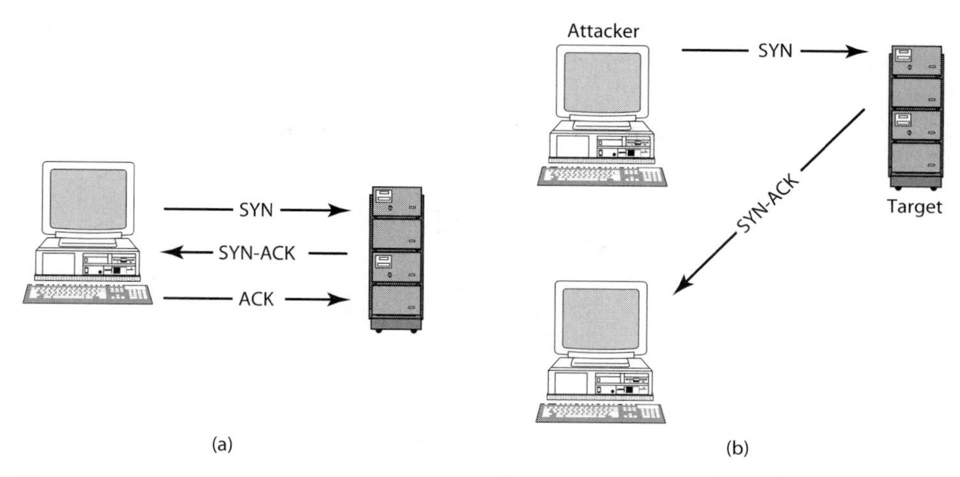

FIGURE 6.6 How a SYN flood attack works. (a) In a normal client-server connection, the client sends the server a SYN message, the server responds with a SYN-ACK message, and the client follows up with an ACK message. At this point the connection is established, and the client and server can interact. (b) In a SYN flood attack, the client sends the server a SYN message with a spoofed IP address. The server replies to a client that is unable to respond to the SYN-ACK message. Eventually, the server will stop waiting for the ACK message, but in the meantime the connection remains half-open, depriving legitimate clients of that connection.

computer and its network. Hence it is important that organizations provide their servers with adequate physical security.

The rest of the DoS attacks we are going to describe are electronic attacks on the server or its network.

Two Internet processes establish a TCP communication link by following a precise series of steps called a "three-way handshake" (Figure 6.6a). The three-way handshake assures each process that the other process is ready to communicate. Suppose process X wishes to communicate with process Y. Process X initiates the handshake by sending Y a SYN message. If Y agrees to communicate with X, it replies with a SYN-ACK message, acknowledging receipt of X's SYN message. At this point the communication channel is half open. In the third step of the handshake, X sends an ACK message to Y, acknowledging receipt of Y's SYN-ACK message. At this point the connection between X and Y is open.

In a **SYN flood attack,** the attacker's computer uses IP spoofing to send the target computer a SYN message from a phony client (Figure 6.6b). When the target computer receives this message, it sets up its side of the connection and replies with a SYN-ACK message. This message travels to the phony client, which cannot respond to the SYN-ACK message. While the target computer waits for the ACK message, the connection remains half-open. The attacker sends the target many such spoofed SYN messages.

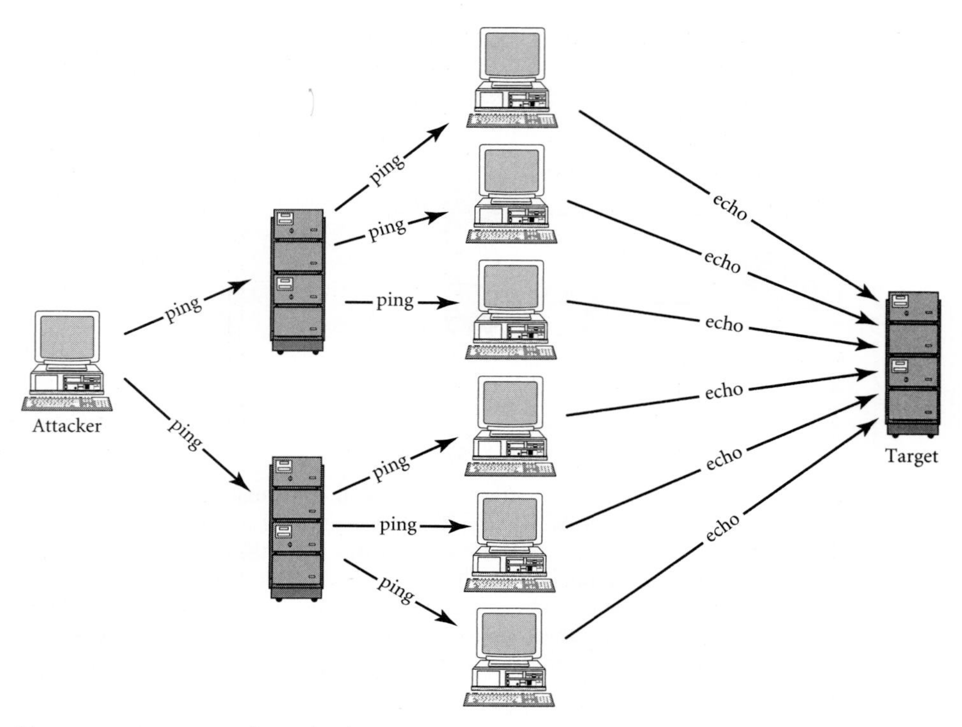

FIGURE 6.7 In a smurf attack, the attacker's computer "pings" many amplifier networks, which broadcast incoming messages. The attacker has spoofed the packet's IP address to appear to be the address of the target computer. The echoes of the "pinged" computers are routed to the target computer, consuming the target's network bandwidth.

Since a server can handle only so many clients at a time, it may turn away legitimate users while it waits futilely for connections to complete [40].

Another form of network attack consumes all the bandwidth on the target's network by generating a large number of messages directed to that network. The **smurf attack** is an example of this form of DoS attack (Figure 6.7). The attacker first identifies routers that support broadcasting of messages to all of the computers on their local area networks. The attacker sends "ping" messages to these routers, which multiply them. A computer receiving a "ping" message is supposed to echo it. In this case, the attacker has spoofed the IP address, making it look as if the ping came from the target computer. All of the computers receiving the ping message send an echo to the target computer. In a successful attack, the flood of incoming messages saturates the target server's network.

In a third kind of DoS attack, the attacker attempts to fill all of the available space on the target computer's disk. Here are three ways to fill a target computer's disk:

1. In **email bombing,** the attacker sends the target a flood of email messages. The target computer stores these email messages on its disk. By sending very long messages, the

attacker can quickly fill the target's disk drive. Email bombing is usually combined with email spoofing (changing the email address of the sender) to disguise the identify of the attacker from the target.

2. The attacker creates a worm that intentionally generates a very long stream of errors. Since the target computer logs errors in a data file, eventually the disk fills up.
3. The attacker breaks in to the target computer and copies over files from another site.

Most computers have a limit on the number of processes that may be active at one time. An attacker can disable the target's computer by penetrating it with a worm program that quickly replicates. (This is how Morris's Internet worm crashed many of the computers it infected.) Even if the target computer does not crash, the presence of many active processes can significantly degrade the performance of the computer's CPU.

Another form of DoS attack crashes the target computer by sending it unexpected data, such as an oversized IP packet.

6.4.2 Defensive Measures

System administrators can take a variety of defensive measures to reduce the threat of DoS attacks throughout the Internet.

Ensuring the physical security of a server is an important defensive measure. Beyond the server itself, physical security encompasses the network access point, the wiring closet, and the air conditioning and power systems.

System administrators should benchmark the performance of their computer systems in order to establish baselines. Once the baselines are known, it is easier to detect aberrations that may indicate a breach of security.

Disk quota systems are another good security measure. If single users have limits on the amount of disk space they may use, then it is tougher for an intruder to create files that eat up all the disk space.

Disabling unused network services is another prudent policy. Reducing available services reduces the options given potential attackers.

Another security measure is turning off the amplifier network capability of routers, taking a weapon out of the hands of those who wish to launch a smurf attack.

Companies have begun to create pattern-recognition software to detect DoS attacks. The software is used to discard requests for service that are coming from "clients" that have proven to be unreliable.

6.4.3 Distributed Denial-of-Service Attacks

In a **distributed denial-of-service (DDoS) attack**, the attacker gains access to thousands of computers. He installs software enabling them to launch a simultaneous attack on target servers. At the time of the attack, the attacker sends a command to the hijacked computers, and they launch their attack. Typically a DDoS attack is a smurf attack, except that now the initial “pings” are being sent from thousands of computers, so there are thousands of times more responses being echoed to the target system.

To defend against DDoS attacks, system administrators must be able to secure their computers to keep them from being hijacked. They can also install filters that check outgoing messages for forged IP addresses. An outgoing message packet should have a “from” address matching one of the local machines. If it does not, then the packet has been forged and should not be forwarded. Filtering outgoing messages means that even if someone has gotten into a machine, he can’t use it for an attack that depends on spoofing the addresses of IP packets.

6.4.4 SATAN

In 1995 computer-security expert Dan Farmer released a program called Security Administrator Tool for Analyzing Networks (SATAN). System administrators could use SATAN to probe their computers for security weaknesses. Farmer said, “SATAN was written because we realize that computer systems are becoming more and more dependent on the network, and more vulnerable to attack” [41]. In the first few days after its release, tens of thousands of copies were downloaded.

Critics fretted that SATAN, with its easy-to-use interface, would turn relatively unskilled teenagers into computer hackers. A security official noted it would be easy to create a script that would enable a hacker to probe hundreds of sites and report on their security holes [42]. Farmer admitted that SATAN was “a two-edged sword that can be used for good and evil.”

As it turns out, a flood of SATAN-enabled computer break-ins never materialized. Apparently, it served its purpose: helping system administrators, particularly novices, identify and fix security problems with their networks.

Still, nearly two years after the release of SATAN, Dan Farmer used it to survey the security of more than 2,200 Web sites. Farmer reported that more than 60 percent of the sites were vulnerable to break-ins. About half of these sites had major security problems, even though all of the security holes probed by SATAN had been publicized by the Computer Emergency Response Team (CERT) [43].

6.5 Online Voting

6.5.1 Motivation for Online Voting

The 2000 Presidential election was one of the closest contests in U.S. history. Florida was the pivotal state; without Florida’s electoral votes, neither Democrat Al Gore nor Republican George W. Bush had a majority of votes in the Electoral College. After a manual recount of the votes in four heavily Democratic counties, the Florida Secretary of State declared that Bush had received 2,912,790 votes to Gore’s total of 2,912,253. Bush’s margin of victory was incredibly small: less than 2 votes out of every 10,000 votes cast.

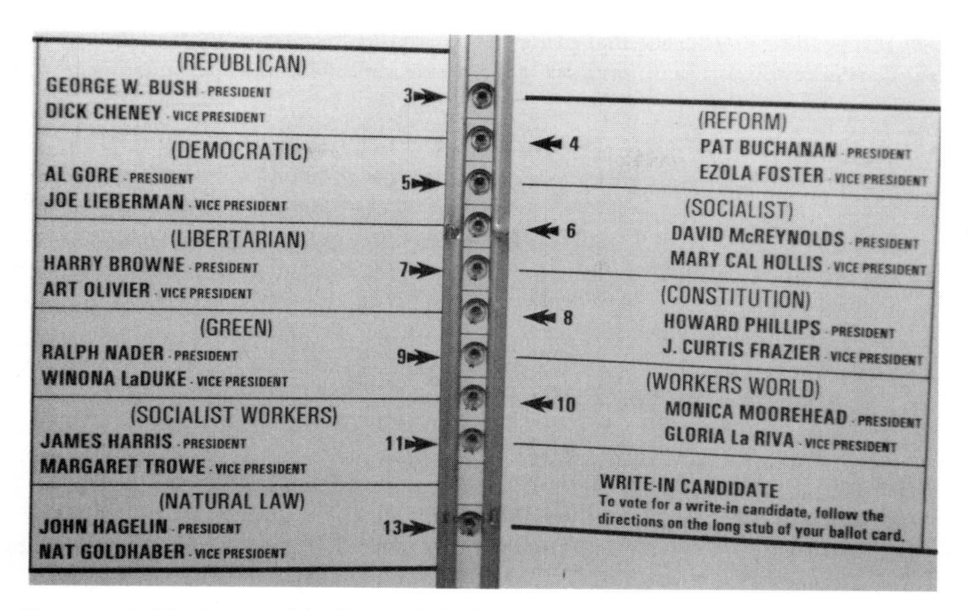

Figure 6.8 The layout of the "butterfly ballot" apparently led thousands of Palm Beach County, Florida voters supporting candidate Al Gore to punch the hole associated with Pat Buchanan by mistake. (AP/Wideworld Photos)

Most of these counties used a keypunch voting machine in which voters select a candidate by using a stylus to poke out a hole in a card next to the candidate's name. Two voting irregularities were traced to the use of these machines. The first irregularity was that sometimes the stylus doesn't punch the hole cleanly, leaving a tiny, rectangular piece of card hanging by one or more corners. Votes with "hanging chad" are typically not counted by automatic vote tabulators. The manual recount focused on identifying ballots with hanging chad that ought to have been counted. The second irregularity was that some voters in Palm Beach County were confused by its "butterfly ballot" and mistakenly punched the hole corresponding to Reform Party candidate Pat Buchanan rather than the hole for Democratic candidate Al Gore (Figure 6.8). This confusion may have cost Al Gore the votes he needed to win Florida [44].

6.5.2 Proposals

The problems with the election in Florida have led to a variety of actions to improve the reliability of voting systems in the United States. Many people have suggested that voting via the Internet be used, at least as a way of casting absentee ballots. In fact, online voting is already a reality. It was used in the 2000 Alaska Republican Presidential preference poll and the 2000 Arizona Democratic Presidential primary [45]. Local elections in the United Kingdom used online voting in 2001. One hundred thousand Americans in the military and living overseas were going to have the opportunity to vote over the Internet in the 2004 Presidential primaries as part of the Secure Electronic Registration and Voting Experiment, until the government cancelled the experiment at the last minute [46].

6.5.3 Ethical Evaluation

In this section we make a utilitarian evaluation of the morality of online voting by weighing its benefits and risks. The discussion assumes that online voting would be implemented via a Web browser, though similar arguments could be made if another technology were employed.

BENEFITS OF ONLINE VOTING

Advocates of online voting say it would have numerous advantages: [47]:

Online voting would give people who ordinarily could not get to the polls the opportunity to cast a ballot from their homes.

Votes cast via the Internet could be counted much more quickly than votes cast on paper.

Electronic votes will not have any of the ambiguity associated with physical votes, such as hanging chad, erasures, etc.

Elections conducted online will cost less money than traditional elections.

Online voting will eliminate the risk of somebody tampering with a ballot box containing physical votes.

While in most elections people vote for a single candidate, other elections allow a person to vote for multiple candidates. For example, a school board may have three vacancies, and voters may be asked to vote for three candidates. It would be easy to program the voting form to prevent people from accidentally overvoting—choosing too many candidates.

Sometimes a long, complicated ballot results in undervoting—where a voter accidentally forgets to mark a candidate for a particular office. A Web form could be designed in multiple pages so that each page had the candidates for a single office. Hence online voting could reduce undervoting.

RISKS OF ONLINE VOTING

Critics of online voting have pointed to numerous risks associated with casting ballots over the Web [47]:

Online voting is unfair because it gives an unfair advantage to those who are financially better off. It will be easier for people with computers and Internet connections at home to vote.

The same system that authenticates the voter also records the ballot. This makes it more difficult to preserve the privacy of the voter.

Online voting increases the opportunities for vote solicitation and vote selling. Suppose person X agrees to vote for candidate Y in return for getting a payment from Z. If person X votes from his personal computer, he could allow person Z to watch as he cast his vote for Y, proving that he fulfilled his end of the bargain. This is much less likely to occur at an official polling place monitored by election officials.

A Web site hosting an election is an obvious target for a DDoS attack. Unlike corporate Web sites, which have attracted the attention of teenage hackers, a national election Web site could attract the attention of foreign governments or terrorists trying to disrupt the electoral process. What happens if the Web site is unavailable and people are not able to access it before the election deadline?

If voting is done from home computers, the security of the election depends on the security of these home computers. The next few paragraphs describe ways in which the security of home computers could be compromised.

A virus could change a person's vote without that person even suspecting what had happened. Many people have physical access to other people's computers, giving them the opportunity to install voter-deceiving applications in the weeks leading up to the election. Alternatively, a rogue programmer or group of programmers within Microsoft, AOL, or another consumer software company could sneak in a vote-tampering virus.

A remote access Trojan such as SubSeven lurking in a voter's computer could allow a person's vote to be observed by an outsider. A RAT could even allow an outsider to cast a ballot in lieu of the rightful voter.

An attacker could fool a user into thinking he was connected to the vote server when in actuality he was connected to a phony vote server controlled by the attacker. For example, the attacker could send an email telling voters to click on a link to reach the polling site. When voters did so, they would be connected to the phony voting site. The attacker could ask for the voter's credentials, then use this information to connect to the real voter site and cast a vote for the candidate(s) desired by the attacker.

UTILITARIAN ANALYSIS

A utilitarian analysis must add up the positive and negative outcomes to determine whether allowing online voting is a good action to take. Recall from Section 2.6.2 that not all outcomes have equal weight. We must consider the probability of the outcome, the value of the outcome on each affected person, and the number of people affected.

Sometimes this calculation is relatively straightforward. For example, one of the benefits of online voting is that people who voted online would not have to travel to a polling place and wait in line. Suppose online voting replaced polling places in the United States. This change would affect about 50 percent of adult Americans (the ones who actually vote) [48]. We can estimate that the average voter spends about an hour traveling to a polling place, waiting in line, and traveling back. The average annual salary in the United States is about $37,000, or about $18.00 per hour [49]. We could compute, then, that the time savings associated with replacing polling places with online voting would be worth about $18.00 times one-half the adult population, or $9.00 for every adult.

It is more difficult to come up with reasonable weights for other outcomes. For example, a risk of online voting is that a DDoS attack may prevent legitimate voters from casting their votes before the deadline. While an election result that does not reflect the will of the voters is a great harm, the weight of this harm is reduced by three probabilities: the probability that someone would attempt a DDoS attack, the probability that a DDoS attack would be successful, and the probability that a successful DDoS attack would change the outcome of the election. Experts could have vastly different estimates of these probabilities, allowing the scales of the utilitarian evaluation to tip one way or the other.

KANTIAN ANALYSIS

A Kantian analysis of any voting system would focus on the principle that the will of each voter should be reflected in that voter's ballot. The integrity of each ballot is paramount. For this reason, every vote should leave a paper record, so that in the event of controversy a recount can be held to ensure the correctness of the election result. Eliminating paper records in order to achieve the ends of saving time and money or boosting voter turnout is wrong from a Kantian perspective.

CONCLUSIONS

We have surveyed the potential benefits and risks of holding elections online, and we have examined the morality of online voting from a utilitarian and a Kantian point of view.

Are we holding computers up to too high a standard? After all, existing voting systems are imperfect. There are two key differences, however, between existing mechanical or electromechanical systems and the proposed online system.

Existing systems are highly localized. A single person may be able to corrupt the election process at a few voting places, but it is impossible to taint the election results across an entire state. A Web-based election system would make it much easier for a single malicious person to taint the process on a wide scale.

The second difference is that most current systems produce a paper record of the vote. Where paper records do not exist, there is a push to make them mandatory [50]. When all else fails, the hard copy can be consulted to try to discern the intent of the voters. A Web-based voting system would not have paper records verified by citizens as true representations of their votes.

There is already evidence of tampering in online elections. In April 2002 Vivendi Universal, a Paris media conglomerate, held an online vote of its shareholders. Hackers caused ballots of some large shareholders to be counted as abstentions [47]. If a private election can draw the attention of a hacker, imagine how much more attractive a target a California election Web site will be!

Bruce Schneier has written, "A secure Internet voting system is theoretically possible, but it would be the first secure networked application ever created in computing history" [51].

Any election system that relies upon the security of personal computers managed by ordinary citizens will be vulnerable to electoral fraud. For this reason alone, there is a strong case to be made that a government should not allow online voting to be conducted in this way.

Summary

As computers become more fully integrated into our lives, the issue of computer security becomes more important. This chapter has described ways in which programs or people can gain unauthorized access into computer systems.

Unauthorized programs are categorized as viruses, worms, or Trojan horses. A virus is a piece of self-replicating code embedded within another program. Viruses can be found anywhere programs can be found. People can spread viruses by exchanging floppy disks or CDs or sharing files on peer-to-peer networks. A worm is a self-contained program that takes advantage of security holes to spread throughout a network. A worm is more autonomous than a virus. Once launched, a worm can spread without any human assistance. A Trojan horse is an apparently benign program that conceals a malicious purpose. Remote access Trojan horses (RATs) are often concealed inside files containing sexually explicit videos or photos. Once downloaded, a RAT enables the attacker to access the victim's computer. System administrators play an important role in securing systems against these external threats.

A person who accesses a computer without authorization is called a hacker. A phreak is someone who manipulates the phone system in order to make free calls. As telecommunications companies began computerizing their equipment in the 1980s, the line between hackers and phreaks got blurry. A well-known group of hackers in the 1980s was the Legion of Doom. Its members wrote "how-to" articles for hackers and phreaks. These stories were widely published on BBSs. In 1990 the U.S. Justice Department and the Secret Service made a number of widely publicized raids to curtail the activities of hackers and phreaks. Many hackers and phreaks served prison sentences for their activities. However, the Secret Service violated the Electronic Communications Privacy Act when it shut down the BBS of Steve Jackson Games.

Denial-of-service (DoS) attacks prevent legitimate users from making use of a computer service. There are different kinds of DoS attacks, including physical attacks on a server, attacks that tie up a server's memory or disk space, attacks that consume all the network bandwidth to the server, and attacks that attempt to "crash" the server. In the past few years, distributed denial-of-service (DDoS) attacks have become a significant new threat to prominent Web sites. Again, system administrators can take a variety of actions to ensure the computers they are responsible for do not contribute to DoS attacks.

Online voting has been suggested as one way of eliminating problems associated with traditional voting systems, and experiments in online voting have already begun. While online elections would result in some benefits, the risks are extensive. In particular, a networked application is only as strong as its weakest link. If people are allowed to vote from their home computer, that is likely to be a weak link that could be exploited by those determined to affect the outcome of an election.

Review Questions

1. What is a computer virus?
2. What is a computer worm?
3. What is the difference between a virus and a worm?
4. Soon after the Internet worm was released, Andy Sudduth sent out an email explaining how to stop the worm. Why was this email of no help to the system administrators fighting the spread of the worm?
5. What are the two reasons why a fast-moving worm is usually more dangerous than a worm that replicates more slowly?
6. In what way could a slow-moving worm be more dangerous than a fast-moving worm?
7. Name one virus launched by a computer science student. Name one worm launched by a computer science student.
8. What is a Trojan horse?
9. Why is it dangerous for an email program to open attachments automatically, without waiting for the user to select them?
10. Explain the origins of the terms "hacker" and "phreak."
11. What was the first major network to get hacked?
12. What parallels does the author draw between hackers/phreaks and those who download MP3 files of copyrighted music?
13. What is a denial-of-service attack?
14. In what way is email bombing like spamming? In what ways are they different?
15. Explain how computer worms are used in certain DoS attacks.
16. Why can't the administrator of a Web server stop a DoS attack by configuring the server so that it refuses to accept any packets from the attacker's computer?
17. Why is the filtering of outgoing Internet traffic an important tool in the fight against DoS and DDoS attacks?
18. Explain two different ways a vote thief could cast multiple votes in an online election.
19. Why does the author conclude it is a bad idea for a government to allow online voting from home computers?
20. What is the quickest and safest way to make a computer secure?

Discussion Questions

21. Email viruses are typically launched by people who modify header information to hide their identity. Brightmail's Enrique Salem says that in the future, your email reader will authenticate the sender before putting the message in your inbox. That way, you will know the source of all the emails you read. Alan Nugent of Novell says, "I'm kind of a fan of eliminating anonymity if that is the price for security" [52]. Will eliminating anonymity make computers more secure?
22. Are there conditions under which the release of a worm, virus, or Trojan horse would be morally justifiable?
23. When his worm program did not perform as expected, Robert Morris, Jr., contacted two old friends from Harvard to decide what to do next. One of them, Andy Sudduth, agreed to email an anonymous message apologizing for the worm and describing how to protect computers from it, without disclosing Morris as the creator of the worm [18]. Was this the right thing for Sudduth to do?
24. Oberlin College in Ohio requires that every computer brought to campus by a student be inspected for viruses. System administrators remove all of the viruses from the students' computers. Students whose computers subsequently pick up and spread a virus may be fined $25, whether they knew about the virus or not [7]. Is this a morally justifiable policy?
25. One person is curious about how the phone system works and finds a way to make free calls to random telephone operators around the globe. He uses this information to create a complex diagram showing the layout of the international telephone switching network. Another person wants to keep in touch with friends and relatives living in another country. He finds a way to make free international telephone calls to them. Is there a moral difference between the actions of these two people?
26. Adam and Charlene are good friends. Both attend East Dakota State University. One day, when Adam is off campus interviewing for a part-time job, someone asks him how many credit hours of computer science courses he has completed. Adam calls Charlene and asks her to access his student records by logging into the campus mainframe as if she were Adam. He provides Charlene with his student identification number and password so that she can do this. Is it wrong for Adam to share this information with Charlene? Is it wrong for Charlene to retrieve this information for Adam?
27. On a PC, a **port** is a software connection to the Internet. Different ports are used for different kinds of communications. For example, one port is used for email messages. Another port is used for Web traffic. A third port is used for transferring files. Since worms work their way into computers through the ports, security experts agree that unneeded ports should be kept closed.

    Apple's Mac OS X ships with no ports open to the Internet. Microsoft ships its Windows XP operating system with five ports open. Microsoft's decision means it is easier for users to set up home networks, but it also makes Windows XP computers more vulnerable to attacks from Internet worms [53]. Which policy is better? Why?
28. Should a person's use of a pseudonym be used as evidence that the person is aware he or she is doing something illegal?

29. Carnegie Mellon University, Harvard University, and the Massachusetts Institute of Technology denied admission to more than 100 business school applicants because they took an online peek at the status of their applications. These students learned how to circumvent the program's security, and they used this knowledge to view their files and see if they had been accepted. Students could see information about their own application, but could not view the status of other students' applications. In many cases the students learned that no admission decision had yet been made. Do you feel the response of these universities was appropriate?

30. Millions of American homes are equipped with wireless networks. If the network is not made secure, any nearby computer with a wireless card can use the network. The range of home wireless networks often extends into neighboring homes, particularly in apartment complexes. If your neighbor's wireless network extends into your home, is it wrong to use the network to get free Internet access?

In-Class Exercises

31. Debate this proposition: "Those who create nondestructive viruses and worms are doing the computer industry a favor because the patches created to block them make computers more secure. To use an analogy, each virus has the effect of strengthening the immune systems of the computers it targets."

32. The University of Calgary offered a senior-level computer science course called "Computer Viruses and Malware." The course taught students how to write viruses, worms, and Trojan horses. It also discussed the history of computer viruses and taught students how to block attacks. All course assignments were done on a closed computer network isolated from the Internet. Some computer security experts criticized the University for offering the course. One researcher said, "No one argues criminology students should commit a murder to understand how a murderer thinks" [54]. Debate whether the University of Calgary was wrong to offer the course.

33. Divide the class into two groups to debate the proposition: "It is wrong for a company to hire a former malicious hacker as a security consultant."

34. When we performed an act utilitarian evaluation of Stewart Nelson's unsuccessful attempt to add an instruction to the PDP-1 computer, we concluded that his action was wrong. However, we decided that his action would have been good from an act utilitarian point of view if his hack had actually worked.

    How much value does an ethical theory have if it can only tell you *after the fact* whether your decision was right or wrong? Does it lead people to undertake risky activities in the hope that they will turn out all right? Put another way, does act utilitarianism encourage people to follow the adage, "It is better to ask for forgiveness than permission"?

35. A distributed denial-of-service attack makes the Web site for a top electronic retailer inaccessible for an entire day. As a result of the attack, nearly a million customers were inconvenienced, and the retailer lost millions of dollars of sales to its competitors. Law enforcement agencies apprehend the person who launched the attack. Should the punishment be determined strictly by considering the crime that was committed, or

should the identity of the culprit be taken into account? If the identity of the perpetrator should be taken into account, what punishment do you think would be appropriate if he were

- a teenager who launched the attack out of curiosity;
- an adult dedicated to fighting the country's overly materialistic culture;
- a member of a terrorist organization attempting to harm the national economy.

36. East Dakota has decided to allow its citizens to vote over the Web in the Presidential election, if they so desire. Thirty percent of the eligible voters choose to cast their ballots over the Web. The national election is so closely contested that whoever wins the electoral votes of East Dakota will be the next President. After the election, state elections officials report the vote tally and declare Candidate X to be the winner.

    Two weeks after the inauguration of President X, state officials uncover evidence of massive electoral fraud. Some voters were tricked into connecting to a phony voting site. The organization running the phony site used the credentials provided by the duped voters to connect to the actual voting site and cast a vote for Candidate X.

    State officials conclude the electoral fraud may have changed the outcome of the election, but they cannot say for sure. They have no evidence that Candidate X knew anything about this scheme to increase his vote tally.

    Divide the class into groups representing President X, the other Presidential candidates, citizens of East Dakota, and citizens of other states to discuss the proper response to this revelation. For guidance, consult Article II, Section 1, and Amendment XII to the United States Constitution.

Further Reading

John Perry Barlow. "Crime and Puzzlement." *Whole Earth Review*, Fall 1990 (also on the Web at www.eff.org/Publications).

John Brunner. *The Shockwave Rider*. Harper & Row, New York, NY, 1975.

Dorothy E. Denning. *Information Warfare and Security*. Addison-Wesley, Boston, MA, 1999.

David Ferbrache. *A Pathology of Computer Viruses*. Springer-Verlag, London, England, 1992.

Katie Hafner and John Markoff. *Cyberpunk: Outlaws and Hackers on the Computer Frontier*. Simon & Schuster, New York, NY, 1991.

Steven Levy. *Hackers: Heroes of the Computer Revolution*. Penguin Books, New York, NY, 1994.

Bruce Sterling. *The Hacker Crackdown: Law and Disorder on the Electronic Frontier*. Bantam Books, New York, NY, 1992.

Clifford Stoll. *The Cuckoo's Egg: Tracking a Spy Through the Maze of Computer Espionage*. Doubleday, New York, NY, 1989.

Brett C. Tjaden. *Fundamentals of Secure Computer Systems*. Franklin, Beedle, & Associates, Wilsonville, OR, 2004.

References

[1] David Ferbrache. *A Pathology of Computer Viruses*. Springer-Verlag, London, England, 1992.

[2] Kim Zetter. "Kazaa Delivers More Than Tunes." *Wired News*, January 9, 2004. www.wired.com.

[3] Martin Sargent. "The Five Most Famous Computer Viruses." *TechTV*, June 6, 2001. www.techtv.com.

[4] S. Webb and B. Fox. Michelangelo Disappoints the Virus Hunters. *New Scientist*, March 14, 1992.

[5] Robert Slade. *Robert Slade's Guide to Computer Viruses: How to Avoid Them, How to Get Rid of Them, and How to Get Help*. 2nd ed. Springer-Verlag, New York, NY, 1996.

[6] U.S. Department of Justice Press Release. *Creator of Melissa Computer Virus Sentenced to 20 Months in Federal Prison*, May 1, 2001. www.usdoj.gov/criminal/cybercrime.

[7] "Colleges Crack Down on Viruses." *Associated Press*, September 4, 2003.

[8] John Brunner. *The Shockwave Rider*. Harper & Row, New York, NY, 1975.

[9] David Moore, Colleen Shannon, and Jeffery Brown. "Code-Red: A Case Study on the Spread and Victims of an Internet Worm." In *Proceedings of the Second ACM Internet Measurement Workshop*, 2002.

[10] Duncan Graham-Rowe. "Computer Antivirus Strategies in Crisis." *New Scientist*, September 3, 2003.

[11] Charles Choi. "The Fastest Worm Ever." *Science Now*, February 5, 2003.

[12] Byron Acohido. "Microsoft Thwarts Blaster Worm Attack." *USA Today*, August 18, 2003.

[13] Gregory Richards. "Train Signals Affected by Computer Virus at CSX Corp." *The Florida Times-Union*, August 21, 2003.

[14] "Worm Turns for Teenager Who Befuddled Microsoft." *The Times (London)*, July 6, 2005.

[15] "Hacker Behind Sasser, Netsky Worms Gets Job with German Security Company." *San Jose (CA) Mercury News*, September 28, 2004.

[16] John Leyden. "Sasser Suspect Walks Free." *The Register*, July 8, 2005. www.theregister.do.uk.

[17] Celeste Biever. "Instant Messaging Falls Prey to Worms." *New Scientist*, May 14, 2005.

[18] Katie Hafner and John Markoff. *Cyberpunk: Outlaws and Hackers on the Computer Frontier*. Simon & Schuster, New York, NY, 1991.

[19] Steven Levy. *Hackers: Heroes of the Computer Revolution*. Anchor Press/Doubleday, Garden City, NY, 1984.

[20] Bruce Sterling. *The Hacker Crackdown: Law and Disorder on the Electronic Frontier*. Bantam Books, New York, NY, 1992.

[21] Dorothy E. Denning. "Concerning Hackers Who Break into Computer Systems." In *Proceedings of the 13th National Computer Security Conference*, 1990.

[22] Immanuel Kant. *Foundations of the Metaphysics of Morals*. The Liberal Arts Press, 1959. Translated by Lewis White Beck.

[23] Marcia Savage. "Mitnick Turns Gamekeeper." *TechWeb.com*, October 30, 2000.

[24] Dan Verton. "Corporate America Is Lazy, Say Hackers; Vandalism of USA Today Site a Warning." *Computerworld*, July 22, 2002.

[25] Clifford Stoll. *The Cuckoo's Egg: Tracking a Spy Through the Maze of Computer Espionage*. Doubleday, New York, NY, 1989.

[26] Lex Luthor. "The History of LOD/H, Revision #3." *The LOD/H Technical Journal*, (4), May 20, 1990.

[27] Saint Augustine. *Confessions*. Oxford University Press, Oxford, England, 1991. Translated by Henry Chadwick.

[28] RIAA. "Recording Industry Begins Suing P2P File Sharers Who Illegally Offer Copyrighted Music Online." www.riaa.com/news/newsletter.

[29] Michael Arnone. "Hacker Steals Personal Data on Foreign Students at U. of Kansas." *The Chronicle of Higher Education*, January 24, 2003.

[30] Sara Lipka. "Hacker Breaks into Database for Tracking International Students at UNLV." *The Chronicle of Higher Education*, March 21, 2005.

[31] John Markoff and Lowell Bergman. "Internet Attack Called Broad and Long Lasting by Investigators." *The New York Times*, May 10, 2005.

[32] Dan Carnevale. "Harvard and MIT Join Carnegie Mellon in Rejecting Applicants Who Broke into Business-School Networks." *The Chronicle of Higher Education*, March 9, 2005.

[33] CERT Coordination Center. "Denial of Service Attacks," June 4, 2001. www.cert.org/tech_tips/ denial_of_service.html.

[34] Mike Toner. "Cyberterrorism Danger Lurking." *The Atlanta Journal and Constitution*, November 2, 2001.

[35] Toni O'Loughlin. "Cyber Terrorism Reaches New Heights." *Australian Financial Review*, April 4, 2003.

[36] Linda Rosencrance. "Hacker 'Mafiaboy' Sentenced." *Computerworld*, September 14, 2001.

[37] Steven M. Cherry. "Took a Licking, Kept on Ticking." *IEEE Spectrum*, page 49, December 2002.

[38] "Saboteurs Hit E-mail Spam's Blockers." *The Boston Globe*, August 28, 2003.

[39] Jonathan Casey. "Protect and Survive: Network Monitoring Tools, Rather Than Traditional Security Measures of Firewalls and IDSs (Intrusion Detection Systems), Provide the Strongest Protection Against 'Denial of Service' Attacks." *Business and Management Practices*, 36(8), August 2002.

[40] Brett Tjaden. *Fundamentals of Secure Computer Systems*. Franklin, Beedle & Associates, Wilsonville, OR, 2004.

[41] Vic Sussman. "The Devil of the Internet." *U.S. News & World Report*, 118(15), April 17, 1995.

[42] Sebastian Rupley. "Satan: Good or Evil?" *PC Magazine*, 14(11), June 13, 1995.

[43] Mark Ward. "Web Sites Are a Hacker's Heaven." *New Scientist*, January 18, 1997.

[44] A. Agresti and B. Presnell. "Misvotes, Undervotes, and Overvotes: the 2000 Presidential Election in Florida." *Statistical Science*, 17(4):436–440, 2002.

[45] Aviel D. Rubin. "Security Considerations for Remote Electronic Voting." *Communications of the ACM*, 45(12):39–44, December 2002.

[46] Sam Hananel. "Thousands to Cast Ballots by Web in 2004." *Associated Press*, July 12, 2003.

[47] Rebecca Mercuri. "A Better Ballot Box?" *IEEE Spectrum*, pages 46–50, October 2002.

[48] Thomas E. Patterson. *The Vanishing Voter: Public Involvement in an Age of Uncertainty*. Alfred A. Knopf/Random House, New York, NY, 2002.

[49] U.S. Department of Labor. "Quarterly Census of Employment and Wages," August 2, 2005. www.bls.gov.

[50] Todd R. Weiss. "N.J. to Get E-voting Paper Trail, but Not Until 2008; a Legal Battle Continues to Try to Put the Law into Effect Sooner." *Computerworld*, July 15, 2005.

[51] Bruce Schneier. "Technology Was Only Part of the Florida Problem." *Computerworld*, December 18, 2000.

[52] "Fighting the Worms of Mass Destruction." *The Economist*, pages 65–67, November 29th, 2003.

[53] Rob Pegoraro. "Microsoft Windows: Insecure by Design." *The Washington Post*, August 24, 2003.

[54] Brock Read. "How to Write a Computer Virus, for College Credit." *The Chronicle of Higher Education*, January 16, 2004.

AN INTERVIEW WITH

Matt Bishop

Matt Bishop received his Ph.D. in computer science from Purdue University, where he specialized in computer security, in 1984. He was a research scientist at the Research Institute for Advanced Computer Science and was on the faculty at Dartmouth College before joining the Department of Computer Science at the University of California at Davis. He teaches courses in computer security, operating systems, and programming.

His main research area is the analysis of vulnerabilities in computer systems, including modeling them, building tools to detect vulnerabilities, and ameliorating or eliminating them. This includes detecting and handling all types of malicious logic. He is active in the areas of network security, the study of denial of service attacks and defenses, policy modeling, software assurance testing, and formal modeling of access control. He also studies the issue of trust as an underpinning for security policies, procedures, and mechanisms.

He is active in information assurance education, is a charter member of the Colloquium on Information Systems Security Education, and led a project to gather and make available many unpublished seminal works in computer security. His textbook, *Computer Security: Art and Science*, was published in December 2002 by Addison-Wesley Professional.

What led you to focus your research on system vulnerabilities?

I became interested in this area because of the ubiquity of the problem. We have been designing and building computer systems since the 1950s, and we still don't know how to secure systems in practice. Why not? How can we find the existing vulnerabilities and improve the security of those existing systems?

Also, there are parallels with nontechnical fields. I find those parallels fascinating, and I enjoy learning and studying other fields to see if any of the methods and ideas from those fields can be applied to analyzing systems and improving their security. Some fields, like military science, political science, and psychology, have obvious connections. Others, such as art and literature, have less obvious connections. But all emphasize the importance of people to computer and software security.

Do you have an example of what can happen when security is treated as an add-on, rather than designed into a system from the beginning?

Yes. Consider the Internet. When it was first implemented (as the old ARPANET), the protocols were not developed to supply the security services that are now considered important. (The security services that were considered important were various forms of robustness, so that the network would provide connectivity even in the face of multiple failures of systems in the network and even of portions of the network itself. It supplied those services very well.) As a result, security services such as authentication, confidentiality of messages, and integrity of messages, are being treated as add-ons rather than the protocols being redesigned to provide those services inherently. So today we have security problems in the descendent of the ARPANET, the Internet.

How can the choice of programming language affect the security of the resulting program?

In two ways. The more obvious one is that some programming languages enforce constraints that limit unsafe practices. For example, in Java, the language prevents indexing beyond the end of an array. In C, the language does not. So you can get buffer overflows in C, but it's much harder to get buffer overflows in Java. The less obvious one is that the language controls how most programmers think about their algorithms. For example, a language that is functional matches some algorithms better than one that is imperative. This means the programmer will make fewer mistakes, and the mistakes he or she makes will tend to be at the implementation level rather than the conceptual or design level—and mistakes at the implementation level will be *much* easier to fix.

What can be done about the problem of viruses, worms, and Trojan horses?

These programs run with the authority of the user who triggers them; worms also spread autonomously through the network and most often take advantage of vulnerabilities to enter a system and spread from it. So, several things can ameliorate the situation:

1. Minimize the number of network services you run. In particular, if you don't need the service, disable it. This will stop the spread of many worms.
2. Don't run any attachments you receive in the mail unless you trust the person who sent them to you. Most viruses and many worms spread this way. In particular, some mailers (such as Outlook) can be set up to execute and/or unpack attachments automatically. This feature should be disabled.
3. The user should not be able to alter certain files, such as system programs and system configuration files. If the user must be able to alter them, confirmation should be required. This will limit the effect of most viruses to affecting the user rather than the system as a whole or other users on the system.

Many personal computer users do not update their systems with the latest operating system patches. Should computer manufacturers be given the ability (and the obligation) to keep up-to-date all of their customers' Internet-connected computers?

I question the wisdom of allowing vendors to update computers remotely. The problem is that vendors do not know the particular environment in which the computers function. The environment determines what "security" means. So, a patch that improves security in one realm may weaken it in another.

As an example, suppose a company disallows any connections from the network except through a virtual private network (VPN). Its systems were designed to start all servers in a particular directory that contains all network servers. So to enforce this restriction, all network servers *except* the VPN are removed from the systems. This prevents the other servers from being started.

The system vendor discovers a security vulnerability in the email server and the login procedure. It fixes both, and sends out a patch that includes a new login program and a new email server. The patch installs both, and reboots the system so the new login program and email server will be used immediately.

The problem here is that by installing the new email server (which improves security in most systems), the company's systems now are nonsecure, as they can be connected to via a port other than those used for the VPN (for example, the email port, port 25). The vendor's patch may therefore damage security.

We saw this with Windows XP SP2. It patched many holes, but also broke various third party applications, some of them very important to their users.

So, I believe vendors should be obligated to work with their customers to provide security patches and enhancements, but should not be given the ability to keep the systems up-to-date unless the customer asks for it. Vendors should also provide better configuration interfaces, and default configurations, that are easy to set up and change, as well as (free) support to help customers use them.

Do you expect personal computers a decade from now to be more secure than they are today?

In some ways yes, and in other ways no. I expect that they will provide more security services that can be configured to make the systems more secure in various environments—not all environments, though! I also expect that the main problem for securing systems will be configuration, operation, and maintenance, though, and those problems will not be overcome in a decade, because they are primarily people problems and not technical problems.

What advice can you offer students who are seriously interested in creating secure software systems?

Focus on all aspects of the software system. Identify the specific requirements that the software system is to solve, develop a security policy that the software system is to meet (and that will meet the requirements), design and implement the software correctly, and consider the environment in which it will be used when you do all this. Also, make the software system as easy to install and configure as possible, and plan that the users will make errors. People aren't perfect, and any security which depends upon them doing everything correctly will ultimately fail.

CHAPTER

7 Computer Reliability

The major difference between a thing that might go wrong and a thing that cannot possibly go wrong is that when a thing that cannot possibly go wrong goes wrong it usually turns out to be impossible to get at or repair.

–DOUGLAS ADAMS, *Mostly Harmless*

7.1 Introduction

COMPUTER DATABASES TRACK MANY OF OUR ACTIVITIES. What happens when a computer is fed bad information, or when someone misinterprets the information they retrieve from a computer? We are surrounded by devices containing embedded computers. What happens when a computer program contains an error that causes the computer to malfunction?

Sometimes the effects of a computer error are trivial. You are playing a game on your PC, do something unusual, and the program crashes, forcing you to start over. At other times computer malfunctions result in a real inconvenience. You get an incorrect bill in the mail, and you end up spending hours on the phone with the company's customer service agents to get the mistake fixed. Some software bugs have resulted in businesses making poor decisions that have cost them millions of dollars. On a few occasions, failures in a computerized system have even resulted in fatalities.

In this chapter we examine various ways in which computerized systems have proven to be unreliable. Systems typically have many components, of which the computer is just one. A well-engineered system can tolerate the malfunction of any single component without failing. Unfortunately, there are many examples of systems in which the computer was the weakest link and a computer error led to the failure of the entire system. The failure may have been due to a data-entry or data-retrieval error, poor design, or inadequate testing. Through a variety of examples, you will gain a greater appreciation for the complexity of building a reliable computerized system.

We also take a look at computer simulations, which are playing an increasingly important role in modern science and engineering. We survey some of the uses to which these simulations are put and describe how those who develop simulations can validate the underlying models.

Software engineering arose out of the difficulties organizations encountered when they began constructing large software systems. Software engineering refers to the use of processes and tools that allow programs to be created in a more structured manner. We describe the software development process and provide evidence that more software projects are being completed on time and on budget.

At the end of the chapter we take a look at software warranties. Software manufacturers typically disclaim any liability for lost profits or other consequential damages resulting from the use of their products. We discuss how much responsibility software manufacturers ought to take for the quality of their products. Some say software should be held to the same standards as other products, while others say we ought to have a different set of expectations when it comes to the reliability of the software we purchase. We explore the controversy surrounding the proposed Uniform Computer Information Transaction Act (UCITA), which would allow software manufacturers to disclaim all liability for defects in their products.

7.2 Data-Entry or Data-Retrieval Errors

Sometimes computerized systems fail because the wrong data have been entered into them or because people incorrectly interpret the data they retrieve. In this section we give several examples of wrong actions being taken due to errors in data entry or data retrieval.

7.2.1 Disfranchised Voters

In the November 2000 general election, Florida disqualified thousands of voters because pre-election screening identified them as felons. The records in the computer database, however, were incorrect; the voters had been charged with misdemeanors. Nevertheless, they were forbidden from voting. This error may have affected the outcome of the Presidential election [1].

7.2.2 False Arrests

As we saw in Chapter 5, the databases of the National Crime Information Center (NCIC) contain a total of about 40 million records related to stolen automobiles, missing persons, wanted persons, suspected terrorists, and much more. There have been numerous stories of police making false arrests based on information they retrieved from the NCIC. Here are three.

Sheila Jackson Stossier, an airline flight attendant, was arrested at the New Orleans airport by police who confused her with Shirley Jackson, who was wanted in Texas. She spent one night in jail and was detained for five days [2].

California police, relying upon information in the NCIC, twice arrested and jailed Roberto Hernandez as a suspect in a Chicago burglary case. The first time he was jailed for 12 days, while the second time he was held for a week before he was freed. They had confused him with another Roberto Hernandez, who had the same height and weight. Both Hernandezes had brown hair, brown eyes, and tattoos on their left arms. They also had the same birthday, and their Social Security numbers differed by only a single digit [3].

Someone used personal information about Michigan resident Terry Dean Rogan to obtain a California driver's license using his name. After this person was arrested for two homicides and two robberies, police entered information about these crimes into the NCIC under his false identity. Over a period of 14 months, the real Terry Dean Rogan was arrested five times by Los Angeles police, three times at gun point, even though he and Michigan police had tried to get the NCIC records corrected after his first arrest. Rogan sued the Los Angeles Police Department and was awarded $55,000 [2].

7.2.3 Analysis: Accuracy of NCIC Records

Stepping away from a requirement of the Privacy Act of 1974, the Justice Department announced in March 2003 that it would no longer require the FBI to ensure the accuracy of information about criminals and crime victims before entering it in the NCIC database [4].

Should the U.S. government take responsibility for the accuracy of the information stored in NCIC databases?

The Department of Justice argues that it is impractical for it to be responsible for the information in the NCIC database [5]: Much of the information that gets entered into the database is provided by other law enforcement and intelligence agencies. The FBI has no way of verifying that all the information is accurate, relevant, and complete. Even when the information is coming from inside the FBI, agents should be able to use their discretion to determine which information may be useful in criminal investigations. If the FBI strictly followed the provisions of the Privacy Act and verified the accuracy of every record entered into the NCIC, the amount of information in the database would be greatly curtailed. The database would be a much less useful tool for law enforcement agencies. The result could be a decrease in the number of criminals arrested by law enforcement agencies.

Privacy advocates counter that the accuracy of the NCIC databases is now more important than ever, because an increasing number of records is stored in these databases. As more erroneous records are put into the database, the probability of innocent American citizens being falsely arrested also increases.

Which argument is stronger? Let's focus on one of the oldest NCIC databases: the database of stolen vehicles. The total amount of harm caused to society by automobile theft is great. Over one million automobiles are stolen in the United States every year. Victims of car theft are subjected to emotional stress, may sustain a financial loss, and can spend a lot of time trying to recover or replace the vehicle. In addition, the prevalence of automobile theft harms everyone who owns a car by raising insurance rates. In the past, car thieves could reduce the probability that a stolen car would be recovered by transporting it across a state line. Because the NCIC database contains information about stolen vehicles throughout the United States, it enables law enforcement officials to identify cars stolen anywhere in the nation. At the present time, just over half of all stolen vehicles are recovered. If we make the conservative estimate that the NCIC has increased the percentage of recovered cars by just 10 percent, more than 50,000 additional cars are being returned to their owners each year, a significant benefit. On the other hand, if an error in the NCIC stolen car database leads to a false arrest, the harm caused to the innocent driver is great. However, there are only a few stories of false arrests stemming from errors in the NCIC stolen car database. The total amount of benefit derived from the NCIC database of stolen automobiles appears to be much greater than the total amount of harm it has caused. We conclude the creation and maintenance of this database has been the right course of action.

7.3 Software and Billing Errors

Even if the data entered into a computer are correct, the system may still produce the wrong result or collapse entirely if there are errors in the computer programs manipulating the data. Newspapers are full of stories about software bugs or "glitches." Here is a selection of stories that have appeared in print in the past few years.

7.3.1 Errors Leading to System Malfunctions

Linda Brooks of Minneapolis, Minnesota opened her mail on July 21, 2001, and found a phone bill for $57,346.20. A bug in Qwest's billing software caused it to charge some customers as much as $600 per minute for the use of their cell phones. About 1.4 percent of Qwest's customers, 14,000 in all, received incorrect bills. A Qwest spokesperson said the bug was in a newly installed billing system [6].

The U.S. Department of Agriculture implemented new livestock price-reporting guidelines after discovering that software errors had caused the USDA to understate the prices meat packers were receiving for beef. Since beef producers and packers negotiate cattle contracts based on the USDA price reports, the errors cost beef producers between $15 and $20 million [7].

In 1996 a software error at the U.S. Postal Service caused it to return to the senders two weeks' worth of mail addressed to the Patent and Trademark Office. In all, 50,000 pieces of mail were returned to sender [8].

A University of Pittsburgh study revealed that for most students, computer spelling and grammar error checkers actually increased the number of errors they made [9, 10].

Thailand's finance minister was trapped inside his BMW limousine for 10 minutes when the on-board computer crashed, locking all doors and windows and turning off the air-conditioning. Security guards had to use sledge hammers to break a window, enabling Suchart Jaovisidha and his driver to escape [11].

7.3.2 Errors Leading to System Failures

A new laboratory computer system at Los Angeles County+USC Medical Center became backlogged the day after it was turned on. For several hours on both April 16 and 17, 2003, emergency room doctors told the County of Los Angeles to stop sending ambulances, because the doctors could not get access to the laboratory results they needed. "It's almost like practicing Third World medicine," said Dr. Amanda Garner. "We rely so much on our computers and our fast-world technology that we were almost blinded" [12].

Japan's air traffic control system went down for an hour on the morning of March 1, 2003, delaying departures for hours. The backup system failed at the same time as the main system, which was out of commission for four hours. Airports kept in touch via telephone, and no passengers were put at risk. However, some flights were delayed over two hours, and 32 domestic flights had to be canceled [13].

A software error led the Chicago Board of Trade to suspend trading for an hour on January 23, 1998. Another bug caused it to suspend trading for 45 minutes on April 1, 1998. In both cases, the temporary shutdown of the trading caused some investors to lose money [14]. System errors caused trading on the London International Financial Futures and Options Exchange to be halted twice within two weeks in May 1999. The second failure idled dealers for an hour and a half [15].

Comair, a subsidiary of Delta Air Lines, canceled all 1,100 of its flights on Christmas Day, 2004, because the computer system that assigns crews to flights stopped running. Airline officials said the software could not handle the large number of flight cancellations caused by bad weather on December 23 and 24. About 30,000 travelers in 118 cities were affected by the flight cancellations [16].

7.3.3 Analysis: E-Retailer Posts Wrong Price, Refuses to Deliver

Amazon.com shut down its British Web site on March 13, 2003, after a software error led it to offer iPaq handheld computers for £7 instead of the correct price of about £275. Before Amazon.com shut down the site, electronic bargain hunters had flocked to Amazon.com's Web site, some of them ordering as many as ten iPaQs [17]. Amazon said customers who ordered at the mistaken price should not expect delivery unless they paid the difference between the advertised price and the actual price. An Amazon.com spokesperson said, "In our Pricing and Availability Policy, we state that where an item's correct price is higher than our stated price, we contact the customer before dispatching. Customers will be offered the opportunity either to cancel their order or to place new orders for the item at the correct price" [18].

Was Amazon.com wrong to refuse to fill the orders of the people who bought iPaqs for £7?

Let's analyze the problem from a rule utilitarian point of view. We can imagine a moral rule of the form: "A person or organization wishing to sell a product must always honor the advertised price." What would happen if this rule were universally followed? More time and effort would be spent proofreading advertisements, whether printed or electronic. Organizations responsible for publishing the advertisements in newspapers, magazines, and Web sites would also take more care to ensure no errors were introduced. There is a good chance companies would take out insurance policies to guard against the catastrophic losses that could result from a typo. To pay for these additional costs, the prices of the products sold by these companies would be higher. The proposed rule would harm every consumer who ended up paying more for products. The rule would benefit the few consumers who took advantage of misprints to get good deals on certain goods. We conclude the proposed moral rule has more harms than benefits, and Amazon.com did the right thing by refusing the ship the iPaqs.

We *could* argue, from a Kantian point of view, that the knowledgeable consumers who ordered the iPaqs did something wrong. The correct price was £275; the advertised price was £7. While electronic products may go on sale, retailers simply do not drop the price of their goods by 97.5 percent, even when they are being put on clearance. If consumers understood the advertised price was an error, then they were taking advantage of Amazon.com's stockholders by ordering the iPaq before the error was corrected. They were not acting in "good faith."

7.4 Notable Software System Failures

In this section we shift our focus to complicated devices or systems controlled at least in part by computers. An **embedded system** is a computer used as a component of a larger system. You can find microprocessor-based embedded systems in microwave ovens, thermostats, automobiles, traffic lights, and a myriad of other modern devices. Because computers need software to execute, every embedded system has a software component.

Software is playing an ever-larger role in system functionality [19]. There are several reasons why hardware controllers are being replaced by microprocessors controlled by software. Software controllers are faster. They can perform more sophisticated functions, taking more input data into account. They cost less, use less energy, and do not wear out. Unfortunately, while hardware controllers have a reputation for high reliability, the same cannot be said for their software replacements.

Most embedded systems are also **real-time systems:** computers that process data from sensors as events occur. The microprocessor that controls the air bags in a modern automobile is a real-time system, because it must instantly react to readings from its sensors and deploy the air bags at the time of a collision. The microprocessor in a cell phone is another example of a real-time system that converts electrical signals into radio waves, and vice versa.

This section contains five examples of computer system failures: the Patriot missile system used in the Gulf War, the Ariane 5 launch vehicle, AT&T's long-distance network, NASA's robot missions to Mars, and the automated baggage system at Denver International Airport. These are all examples of embedded, real-time systems. In every case at least part of the failure was due to errors in the software component of the system. Studying these errors provides important lessons for anyone involved in the development of an embedded system.

7.4.1 Patriot Missile

The Patriot missile system was originally designed by the U.S. Army to shoot down airplanes. In the 1991 Gulf War, the Army put the Patriot missile system to work defending against Scud missiles launched at Israel and Saudi Arabia.

At the end of the Gulf War, the Army claimed the Patriot missile defense system had been 95 percent effective at destroying incoming Scud missiles. Later analyses showed that perhaps as few as 9 percent of the Scuds were actually destroyed by Patriot missiles. As it turns out, many Scuds simply fell apart as they approached their targets—their destruction had nothing at all to do with the Patriot missiles launched at them.

The most significant failure of the Patriot missile system occurred during the night of February 25, 1991, when a Scud missile fired from Iraq hit a U.S. Army barracks in Dhahran, Saudi Arabia, killing 28 soldiers. The Patriot missile battery defending the area never even fired at the incoming Scud.

Mississippi congressman Howard Wolpe asked the General Accounting Office (GAO) to investigate this incident. The GAO report traced the failure of the Patriot system to a software error (Figure 7.1). The missile battery did detect the incoming Scud missile as it came over the horizon. However, in order to prevent the system from responding to false alarms, the computer was programmed to check multiple times for the presence of the missile. The computer predicted the flight path of the incoming missile, directed the radar to focus in on that area, and scanned a segment of the radar signal, called a range gate, for the target. In this case, the program scanned the wrong range gate. Since it did not detect the Scud, it did not fire the Patriot missile.

Why did the program scan the wrong range gate? The tracking system relied upon getting signals from the system clock. These values were stored in a floating-point variable with insufficient precision, resulting in a small mathematical error called a *truncation*. The longer the system ran, the more these truncation errors added up. The Patriot missile system was designed to operate for only a few hours at a time. However, the system at Dhahran had been in continuous operation for 100 hours. The accumulation of errors led to a difference between the actual time and the computed time of about 0.3433

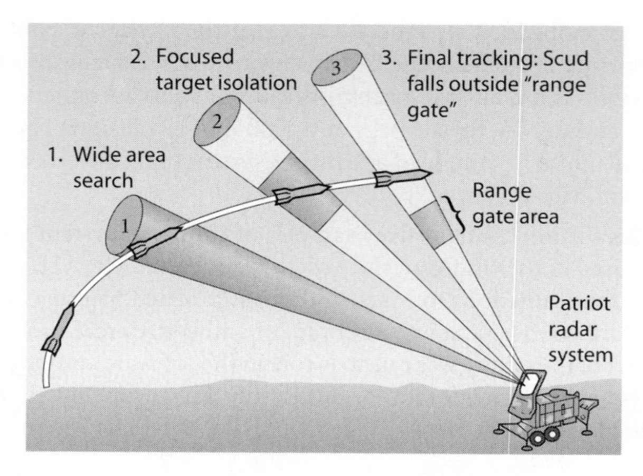

FIGURE 7.1 A software error caused the Patriot missile system to lose track of incoming Scud missiles. (1) The radar system doing a wide area search picks up the Scud missile. (2) The radar system isolates the proposed target. (3) A software error causes the system to produce a faulty range gate. The system loses track of the missile, because it does not fly through this gate. (Reprinted with permission from Marshall, SCIENCE 255:1342 (1992). Illustration: D. Defrancesco. Copyright 1992 AAAS.)

seconds. Because missiles travel at high speeds, the 0.3433-second error led to a tracking error of 687 meters (about half a mile). That was enough of an error to prevent the battery from locating the Scud in the range gate area [20].

7.4.2 Ariane 5

The Ariane 5 was a satellite launch vehicle designed by the French space agency, the Centre National d'Etudes Spatiales, and the European Space Agency. About 40 seconds into its maiden flight on June 4, 1996, a software error caused the nozzles on the solid boosters and the main rocket engine to swivel to extreme positions. As a result, the rocket veered sharply off course. When the links between the solid boosters and the core stage ruptured, the launch vehicle self-destructed. The rocket carried satellites worth $500 million, which were not insured [21].

A board of inquiry traced the software error to a piece of code that converts a 64-bit floating-point value into a 16-bit signed integer. The value to be converted exceeded the maximum value that could be stored in the integer variable, causing an exception to be raised. Unfortunately, there was no exception handling mechanism for this particular exception, so the onboard computers crashed.

The faulty piece of code had been part of the software for the Ariane 4. The 64-bit floating-point value represented the horizontal bias of the launch vehicle, which is related to its horizontal velocity. When the software module was designed, engineers determined that it would be impossible for the horizontal bias to be so large that it could not be stored in a 16-bit signed integer. There was no need for an error handler, because an error could not occur. This code was moved "as is" into the software for the Ariane 5. That proved to be an extremely costly mistake, because the Ariane 5 was faster than the Ariane 4. The original assumptions made by the designers of the software no longer held true [22].

7.4.3 AT&T Long-Distance Network

On the afternoon of January 15, 1990, AT&T's long-distance network suffered a significant disruption of service. About half of the computerized telephone-routing switches crashed, and the remainder of the switches could not handle all of the traffic. As a result of this failure, about 70 million long-distance telephone calls could not be put through, and about 60,000 people lost all telephone service. AT&T lost tens of millions of dollars of revenue. It also lost some of its credibility as a reliable provider of long-distance service.

Investigation by AT&T engineers revealed that the network crash was brought about by a single faulty line of code in an error-recovery procedure. The system was designed so that if a server discovered it was in an error state, it would reboot itself, a crude but effective way of "wiping the slate clean." After a switch rebooted itself, it would send an "OK" message to other switches, letting them know it was back on line. The software bug manifested itself when a very busy switch received an "OK" message. Under certain circumstances, handling the "OK" message would cause the busy switch to enter an error state and reboot.

On the afternoon of January 15, 1990, a System 7 switch in New York City detected an error condition and rebooted itself (Figure 7.2). When it came back on line, it broadcast an "OK" message. All of the switches receiving the "OK" messages handled them correctly, except three very busy switches in St. Louis, Detroit, and Atlanta. These switches detected an error condition and rebooted. When they came back up, all of them broadcast "OK" messages across the network, causing other switches to fail in an ever-expanding wave.

Every switch failure compounded the problem in two ways. When the switch went down, it pushed more long-distance traffic onto the other switches, making them busier. When the switch came back up, it broadcast "OK" messages to these busier switches, causing some of them to fail. Some switches rebooted repeatedly under the barrage of "OK" messages. Within 10 minutes, half of the switches in the AT&T network had failed.

The crash could have been much worse, but AT&T had converted only 80 of its network switches to the System 7 software. It had left System 6 software running on 34 of the switches, "just in case." The System 6 switches did not have the software bug and did not crash [23, 24].

7.4.4 Robot Missions to Mars

NASA designed the $125-million Mars Climate Orbiter to facilitate communications between Earth and automated probes on the surface of Mars, including the Mars Polar

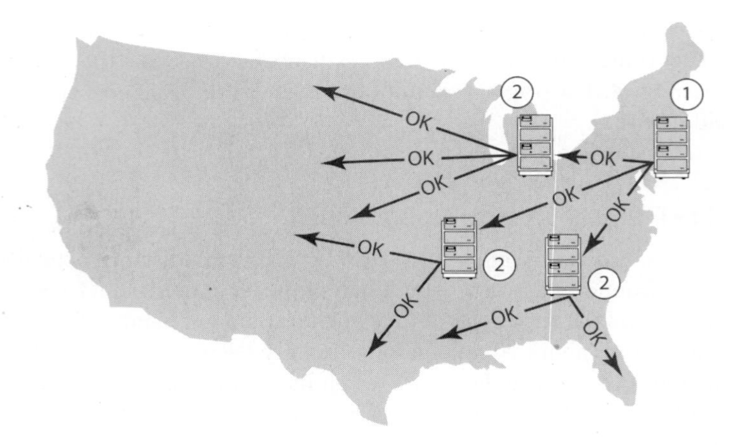

Figure 7.2 A software bug in error-recovery code made AT&T's System 7 switches crash in 1990. (1) A single switch in New York City detects an error condition and reboots. When it comes back up, it sends an "OK" message to other switches. (2) Switches in Detroit, St. Louis, and Atlanta are so busy that handling the "OK" message causes them to fail. They detect an error condition and reboot. When they come back up, they send out "OK" messages to other switches, causing some of them to fail, and so on.

Lander. Ironically, the spacecraft was lost because of a miscommunication between two support teams on Earth.

The Lockheed Martin flight operations team in Colorado designed its software to use English units. Its program output thrust in terms of foot-pounds. The navigation team at the Jet Propulsion Laboratory in California designed its software to use metric units. Its program expected thrust to be input in terms of newtons. One foot-pound equals 4.45 newtons. On September 23, 1999, the Mars Climate Orbiter neared the Red Planet. When it was time for the spacecraft to fire its engine to enter orbit, the Colorado team supplied thrust information to the California team, which relayed it to the spacecraft. Because of the units mismatch, the navigation team specified 4.45 times too much thrust. The spacecraft flew too close to the surface of Mars and burned up in its atmosphere.

A few months later NASA's Martian program suffered a second catastrophe. The Mars Polar Lander, produced at a cost of $165 million, was supposed to land on the south pole of Mars and provide data that would help scientists understand how the Martian climate has changed over time. On December 3, 1999, NASA lost contact with the Mars Polar Lander. NASA engineers suspect that the system's software got a false signal from the landing gear and shut down the engines 100 feet above the planet's surface.

Tony Spear is project manager of the Mars Pathfinder mission. He says, "It is just as hard to do Mars missions now as it was in the mid-70s. I'm a big believer that software hasn't gone anywhere. Software is the number-one problem" [25].

After Spear made this observation, NASA successfully landed two Mars Exploration Rovers on the Red Planet [26]. The rovers, named Opportunity and Spirit, were launched from Earth in June and July of 2003, successfully landing on Mars in January 2004. Mission planners had hoped that each rover would complete a three-month mission, looking for clues that the Martian surface once had enough water to sustain life. The rovers greatly exceeded this goal. Opportunity found evidence of a former saltwater lake, and after 19 months, both rovers were still operational.

7.4.5 Denver International Airport

As airline passenger traffic strained the capacity of Stapleton International Airport, the City and County of Denver planned the construction of a much larger airport. Stapleton International Airport had earned a reputation for slow baggage handling, and the project planners wanted to ensure the new airport would not suffer from the same problem. They announced an ambitious plan to create a one-of-a-kind, state-of-the-art automated baggage handling system for the Denver International Airport (DIA).

The airport authorities signed a $193-million contract with BAE Automated Systems to design and build the automated baggage-handling system, which consisted of thousands of baggage carts traveling on 21 miles of track. Each cart was essentially a tub on wheels that would follow a metal track roller-coaster style. The system would load one suitcase on each cart. Scanners would read the destination information from a label on the suitcase, and the computers would route every cart to its appropriate destination. The carts would move at 20 miles an hour. Each bag would be labeled by an agent and placed on a conveyor belt. Computers would route the bags along one or more belts until they reached a cart-loading point, where each bag would be loaded onto its own cart. Computers would then route each cart to its correct unloading point, where the bag would be unloaded from the cart onto a conveyor belt and routed to its final destination. To monitor the movement of the bags, the system used 56 bar-code scanners and 5,000 electric eyes.

There were problems from the outset of the project. The airport design was already done before the baggage handling system was chosen. As a result, the underground tunnels were small and had sharp turns, making it difficult to shoehorn in an automated baggage system. And given its ambitious goals, the project timeline was too short.

However, the most important problem with the automated baggage handler was that the complexity of the system exceeded the ability of the development team to understand it. Here are a few of the problems BAE encountered:

- Luggage carts were misrouted and failed to arrive at their destinations.
- Computers lost track of where the carts were.
- Bar-code printers didn't print tags clearly enough to be read by scanners.
- Luggage had to be properly positioned on conveyors in order to load properly.
- Bumpers on the carts interfered with the electric photocells.
- Workers painted over electric eyes or knocked photo sensors out of alignment.

- Light luggage was thrown off rapidly moving carts.
- Luggage got shredded by automated baggage handlers.
- The design did not consider the problem of fairly balancing the number of available carts among all the locations needing them.

BAE attempted to solve these problems one at a time by trial and error, but the system was too complicated to yield to this problem-solving approach. BAE should have been looking at the big picture, trying to find where the specifications for the system were wrong or unattainable.

DIA was supposed to open on October 31, 1993. The opening was delayed repeatedly because the baggage-handling system was not yet operational. Eventually, the Mayor of Denver announced the city would spend $50 million to build a conventional luggage handling system using tugs and carts. (This conventional system actually ended up costing $71 million.) On February 28, 1995, flights to and from the new airport began. However, Concourse A was not open at all. Concourse C opened with 11 airlines using a traditional baggage system. The BAE automated system, far over budget at $311 million, was used only by United Airlines in Concourse B to handle outgoing baggage originating in Denver. United used a traditional system for the rest of its baggage in Concourse B.

The failure of BAE to deliver a working system on time resulted in a 16-month delay in the opening of DIA. This delay cost Denver $1 million a *day* in interest on bonds and operating costs. As a result, DIA began charging all of the airlines a flight fee of about $20 per passenger, the highest airport fee in the nation. Airlines passed along this cost to consumers by raising ticket prices of flights going through Denver [27].

While the story of the Denver International Airport is noteworthy because of the large amount of money involved, it is not unusual for software projects to take longer than expected and to cost more than anticipated. In fact, most software projects are not completed on time and on budget. We'll explore this issue in greater detail in Section 7.7.

7.5 Therac-25

Soon after German physicist Wilhelm Roentgen discovered the x-ray in 1895, physicians began using radiation to treat cancer. Today, between 50 and 60 percent of cancer patients are treated with radiation, either to destroy cancer cells or relieve pain. Linear accelerators create high-energy electron beams to treat shallow tumors and x-ray beams to reach deeper tumors.

The Therac-25 linear accelerator was notoriously unreliable. It was not unusual for the system to malfunction 40 times a day. We devote an entire section to telling the story of the Therac-25 because it is a striking example of the harm that can be caused when the safety of a system relies solely upon the quality of its embedded software.

In a 20-month period between June 1985 and January 1987, the Therac-25 administered massive overdoses to six patients, causing the deaths of three of them. While 1987 may seem like the distant past to many of you, it does give us the advantage of 20/20 hindsight. The entire story has been thoroughly researched and documented [28]. Failures of computerized systems continue to this day, but they have not yet been fully played out and analyzed.

7.5.1 Genesis of the Therac-25

Atomic Energy of Canada Limited (AECL) and the French corporation CGR cooperated in the 1970s to build two linear accelerators: the Therac-6 and the Therac-20. Both the Therac-6 and the Therac-20 were modernizations of older CGR linear accelerators. The distinguishing feature of the Therac series was the use of a DEC PDP 11 minicomputer as a "front end." By adding the computer, the linear accelerators were easier to operate. The Therac-6 and the Therac-20 were actually capable of working independently of the PDP 11, and all of their safety features were built into the hardware.

After producing the Therac-20, AECL and CGR went their separate ways. AECL moved ahead with the development and deployment of a next-generation linear accelerator called the Therac-25. Like the Therac-6 and the Therac-20, the Therac-25 made use of a PDP 11. Unlike its predecessor machines, however, AECL designed the PDP 11 to be an integral part of the device; the linear accelerator was incapable of operating without the computer. This design decision enabled AECL to reduce costs by replacing some of the hardware safety features of the Therac-20 with software safety features in the Therac-25.

AECL also decided to reuse some of the Therac-6 and Therac-20 software in the Therac-25. Code reuse saves time and money. Theoretically, "tried and true" software is more reliable than newly written code, but as we shall see, that assumption was invalid in this case.

AECL shipped its first Therac-25 in 1983. In all, it delivered 11 systems in Canada and the United States. The Therac-25 was a large machine that was placed in its own room. Shielding in the walls, ceiling, and floor of the room prevented outsiders from being exposed to radiation. A television camera, microphone, and speaker in the room allowed the technician in an adjoining room to view and communicate with the patient undergoing treatment.

7.5.2 Chronology of Accidents and AECL Responses

MARIETTA, GEORGIA, JUNE 1985

A 61-year-old breast cancer patient was being treated at the Kennestone Regional Oncology Center. After radiation was administered to the area of her collarbone, she complained that she had been burned.

The Kennestone physicist contacted AECL and asked if it was possible that the Therac-25 had failed to diffuse the electron beam. Engineers at AECL replied that this could not happen.

The patient suffered crippling injuries as a result of the overdose, which the physicist later estimated was 75 to 100 times too large. She sued AECL and the hospital in October 1985.

HAMILTON, ONTARIO, JULY 1985

A 40-year-old woman was being treated for cervical cancer at the Ontario Cancer Foundation. When the operator tried to administer the treatment, the machine shut down after five seconds with an error message. According to the display, the linear accelerator had not yet delivered any radiation to the patient. Following standard operating procedure, the operator typed "P" for "proceed." The system shut down in the same way, indicating that the patient had not yet received a dose of radiation. (Recall it was not unusual for the machine to malfunction several dozen times a day.) The operator typed "P" three more times, always with the same result, until the system entered "treatment suspend" mode.

The operator went into the room where the patient was. The patient complained that she had been burned. The lab called in a service technician, who could find nothing wrong with the machine. The clinic reported the malfunction to AECL.

When the patient returned for further treatment three days later, she was hospitalized for a radiation overdose. It was later estimated that she had received between 65 and 85 times the normal dose of radiation. The patient died of cancer in November 1985.

FIRST AECL INVESTIGATION, JULY–SEPTEMBER 1985

After the Ontario overdose, AECL sent out an engineer to investigate. While the engineer was unable to reproduce the overdose, he did uncover design problems related to a microswitch. AECL introduced hardware and software changes to fix the microswitch problem.

YAKIMA, WASHINGTON, DECEMBER 1985

The next documented overdose accident occurred at Yakima Valley Memorial Hospital. A woman receiving a series of radiation treatments developed a strange reddening on her hip after one of the treatments. The inflammation took the form of several parallel stripes. The hospital staff tried to determine the cause of the unusual stripes. They suspected the pattern could have been caused by the slots in the accelerator's blocking trays, but these trays had already been discarded by the time the staff began their investigation. After ruling out other possible causes for the reaction, the staff suspected a radiation overdose and contacted AECL by letter and by phone.

AECL replied in a letter that neither the Therac-25 nor operator error could have produced the described damage. Two pages of the letter explained why it was technically impossible for the Therac-25 to produce an overdose. The letter also claimed that no similar accidents had been reported.

The patient survived, although the overdose scarred her and left her with a mild disability.

TYLER, TEXAS, MARCH 1986

A male patient came to the East Texas Cancer Center (ETCC) for the ninth in a series of radiation treatments for a cancerous tumor on his back. The operator entered the treatment data into the computer. She noticed that she had typed "X" (for x-ray) instead of "E" (for electron beam). This was a common mistake, because x-ray treatments are much more common. Being an experienced operator, she quickly fixed her mistake by using the up arrow key to move the cursor back to the appropriate field, changing the "X" to an "E" and moving the cursor back to the bottom of the screen. When the system displayed "beam ready," she typed "B" (for beam on). After a few seconds, the Therac-25 shut down. The console screen contained the message "Malfunction 54" and indicated a "treatment pause," a low-priority problem. The dose monitor showed that the patient had received only 6 units of treatment rather than the desired 202 units. The operator hit the "P" (proceed) key to continue the treatment.

The cancer patient and the operator were in adjoining rooms. Normally a video camera and intercom would enable the operator to monitor her patients. However, at the time of the accident neither system was operational.

The patient had received eight prior treatments, so he knew something was wrong as soon as the ninth treatment began. He was instantly aware of the overdose—he felt as if someone had poured hot coffee on his back or given him an electric shock. As he tried to get up from the table, the accelerator delivered its second dose, which hit him in the arm. The operator became aware of the problem when the patient began pounding on the door. He had received between 80 and 125 times the prescribed amount of radiation. He suffered acute pain and steadily lost bodily functions until he died from complications of the overdose five months later.

SECOND AECL INVESTIGATION, MARCH 1986

After the accident, the ETCC shut down its Therac-25 and notified AECL. AECL sent out two engineers to examine the system. Try as they may, they could not reproduce Malfunction 54. They told the physicians it was impossible for the Therac-25 to overdose a patient, and they suggested that the patient may have received an electrical shock due to a fault in the hospital's electrical system.

The ETCC checked out the electrical system and found no problems with it. After double-checking the linear accelerator's calibration, they put the Therac-25 back into service.

TYLER, TEXAS, APRIL 1986

The second Tyler, Texas accident was virtually a replay of the prior accident at ETCC. The same technician was in control of the Therac-25, and she went through the same process of entering x-ray when she meant electron beam, then going back and correcting her mistake. Once again, the machine halted with a Malfunction 54 shortly after she activated the electron beam. This time, however, the intercom was working, and she rushed to the accelerator when she heard the patient moan. There was nothing she could do to help him. The patient had received a massive dose of radiation to his brain, and he died three weeks later.

After the accident, ETCC immediately shut down the Therac-25 and contacted AECL again.

YAKIMA, WASHINGTON, JANUARY 1987

A second patient was severely burned by the Therac-25 at Yakima Valley Memorial Hospital under circumstances almost identical to those of the December 1985 accident. Four days after the treatment, the patient's skin revealed a series of parallel red stripes—the same pattern that had perplexed the radiation staff in the case of the previous patient. This time, the staff members were able to match the burns to the slots in the Therac-25's blocking tray. The patient died three months later.

THERAC-25 DECLARED DEFECTIVE, FEBRUARY 1987

On February 10, 1987, the FDA declared the Therac-25 to be defective. In order for the Therac-25 to gain back FDA approval, AECL had to demonstrate how it would make the system safe. Five months later, after five revisions, AECL produced a corrective action plan that met the approval of the FDA. This plan incorporated a variety of hardware interlocks to prevent the machine from delivering overdoses or activating the beam when the turntable was not in the correct position.

7.5.3 Software Errors

In the course of investigating the accidents, AECL discovered a variety of hardware and software problems with the Therac-25. Two of the software errors are examples of race conditions. In a **race condition,** two or more concurrent tasks share a variable, and the order in which they read or write the value of the variable can affect the behavior of the program. Race conditions are extremely difficult to identify and fix, because usually the two tasks do not interfere with each other and nothing goes wrong. Only in rare conditions will the tasks actually interfere with each other as they manipulate the variable, causing the error to occur. We describe both of these errors to give you some insight into how difficult they are to detect.

The accidents at the ETCC occurred because of a race condition associated with the command screen (Figure 7.3). One task was responsible for handling keyboard input and making changes to the command screen. A second task was responsible for monitoring the command screen for changes and moving the magnets into position. After the operator uses the first task to complete the prescription (1), the second task sees the cursor in the lower right-hand corner of the screen and begins the eight-second process of moving the magnets (2). Meanwhile, the operator sees her mistake. The first task responds to her keystrokes and lets her change the "X" to an "E" (3). She gets the cursor back to the lower right-hand corner before eight seconds are up (4). Now the second task finishes moving the magnets (5). It sees the cursor in the lower right-hand

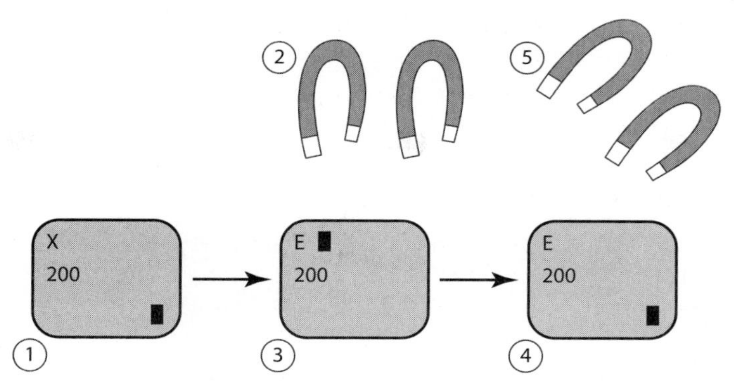

FIGURE 7.3 Illustration of a Therac-25 bug revealed by fast-typing operators. (1) The operator finishes filling in the form. The software knows the form is filled in because the cursor is in the lower right-hand corner of the screen. (2) The software instructs the magnets to move into the correct positions. While the magnets are moving, the software does not check for screen edits. (3) The operator changes the prescription from x-ray to electron beam. (4) The operator finishes the edit, returning the cursor to the lower right-hand corner of the screen. (5) The magnets finish moving. The software now checks the screen cursor. Since it is in the lower right-hand corner, the program assumes there have been no edits.

corner of the screen and incorrectly assumes the screen has not changed. The crucial substitution of electron beam for x-ray goes unnoticed.

What makes this bug particularly treacherous is that it only occurs with faster, more experienced operators. Slower operators would not be able to complete the edit and get the cursor back to the lower right-hand corner of the screen in only eight seconds. If the cursor happened to be anywhere else on the screen when the magnets stopping moving, the software would check for a screen edit and there would be no overdose. It is ironic that the safety of the system actually *decreased* as the experience of the operator *increased*.

Another race condition was responsible for the overdoses at the Yakima Valley Memorial Hospital (Figure 7.4). It occurred when the machine was putting the electron-beam gun back into position. A variable was supposed to be 0 if the gun was ready to fire. Any other value meant the gun was not ready. As long as the electron beam gun was out of position, one task kept incrementing that variable. Unfortunately, the variable could only store the values from 0 to 255. Incrementing it when it had the value 255 would result in the variable's value rolling over to 0, like a car's odometer.

Nearly every time that the operator hit the SET button when the gun was out of position, the variable was not 0 and the gun did not fire (Figure 7.4a). However, there was a very slight chance that the variable would have just rolled over when the operator hit the SET button (Figure 7.4b). In this case the accelerator would emit a charge, even though the system was not ready.

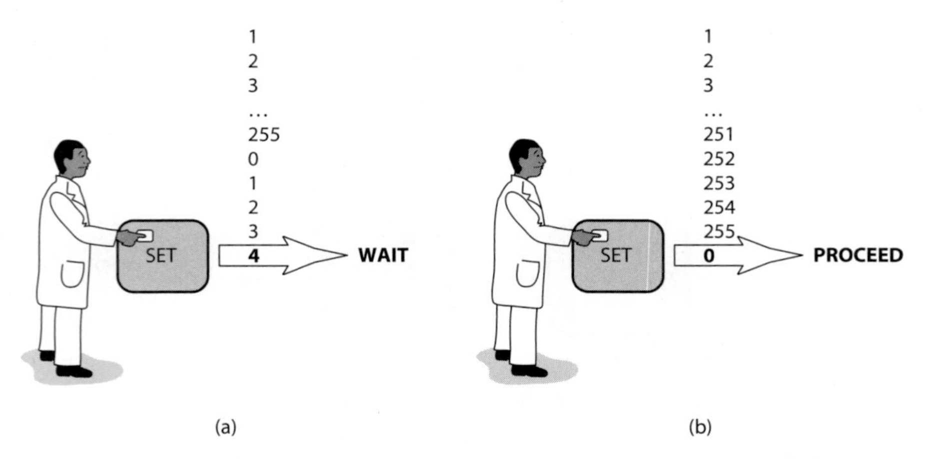

FIGURE 7.4 The Therac-25 could administer radiation too soon if the operator hit the SET button at precisely the wrong time. As long as the electron-beam gun was out of position, a software task kept incrementing an 8-bit variable. (a) Usually when the operator hit the SET button, the variable was not zero and the system would wait, just as it was supposed to. (b) If the operator hit the SET button just at the variable "rolled over" from 255 to 0, the system would administer radiation, even though the gun was out of position.

7.5.4 Post Mortem

Let's consider some of the mistakes AECL made in the design, development, and support of this system.

When accidents were reported, AECL focused on identifying and fixing particular software bugs. This approach was too narrow. As Nancy Leveson and Clark Turner point out, "most accidents are system accidents; that is, they stem from complex interactions between various components and activities" [28]. The entire system was broken, not just the software. A strategy of eliminating bugs assumes that at some point the last bug will be eradicated. But as Leveson and Turner write, "There is always another software bug" [28].

The real problem was that the system was not designed to be fail-safe. Good engineering practice dictates that a system should be designed so that no single point of failure will lead to a catastrophe. By relying completely upon software for protection against overdoses, the Therac-25 designers ignored this fundamental engineering principle.

Another flaw in the design of the Therac-25 was its lack of software or hardware devices to detect and report overdoses and shut down the accelerator immediately. Instead, the Therac-25 designers left it up to the patients to report when they had received overdoses.

There are also particular software lessons we can learn from the case of the Therac-25. First, it is very difficult to find software errors in programs where multiple tasks execute at the same time and interact through shared variables. Second, the software design needs to be as simple as possible, and design decisions must be documented to aid in the maintenance of the system. Third, the code must be reasonably documented at the time it is written.

Fourth, reusing code does not always increase the quality of the final product. AECL assumed that by reusing code from the Therac-6 and Therac-20, the software would be more reliable. After all, the code had been part of systems used by customers for years with no problems. This assumption turned out to be wrong. The earlier codes did contain errors. These errors remained undetected because the earlier machines had hardware interlocks that prevented the computer's erroneous commands from harming patients.

The tragedy was compounded because AECL did not communicate fully with its customers. For example, AECL told the physicists in Washington and Texas that an overdose was impossible, even though AECL had already been sued by the patient in Georgia.

7.5.5 Moral Responsibility of the Therac–25 Team

Should the developers and managers at AECL be held morally responsible for the deaths resulting from the use of the Therac-25 they produced?

In order for a moral agent to be responsible for a harmful event, two conditions must hold:

- *Causal condition*: the actions (or inactions) of the agent must have caused the harm.
- *Mental condition*: the actions (or inactions) must have been intended or willed by the agent.

In this case, the causal condition is easy to establish. The deaths resulted both from the action of AECL employees (creating the therapy machine that administered the overdose) and the inaction of AECL employees (failing to withdraw the machine from service or even inform other users of the machine that there had been overdoses).

What about the second condition? Surely the engineers at AECL did not intend or will to create a machine that would administer lethal overdoses of radiation. However, philosophers also extend the mental condition to include unintended harm if the moral agent's actions were the result of carelessness, recklessness, or negligence. The design team took a number of actions that fall into this category. It constructed a system without hardware interlocks to prevent overdoses or the beam from being activated when the turntable was not in a correct position. The machine had no software or hardware devices to detect an accidental overdose. Management allowed software to be developed without adequate documentation. It presumed the correctness of reused code and failed to test it thoroughly. For these reasons the mental condition holds as well, and we conclude the Therac-25 team at AECL is morally responsible for the deaths caused by the Therac-25 radiation therapy machine.

7.6 Computer Simulations

In the previous section we focused on an unreliable computer-controlled system that delivered lethal doses of radiation to cancer patients, but even systems kept behind the locked doors of a computer room can cause harm. Errors in computer simulations can result in poorly designed products, mediocre science, and bad policy decisions. In this section we review our growing reliance on computer simulations for designing products, understanding our world, and even predicting the future, and we describe ways in which computer modelers validate their simulations.

7.6.1 Uses of Simulation

Computer simulation plays a key role in contemporary science and engineering. There are many reasons why a scientist or engineer may not be able to perform a physical experiment. It may be too expensive or time-consuming, or it may be unethical or impossible to perform. Computer simulations have been used to design nuclear weapons, search for oil, create pharmaceuticals, and design safer, more fuel efficient cars. They have even been used to design consumer products such as disposable diapers [29].

Some computer simulations model past events. For example, when astrophysicists derive theories about the evolution of the universe, they can test them through computer simulations. Recently, computer simulation has demonstrated that a gas disk around a young star can fragment into giant gas planets such as Jupiter [30].

A second use of computer simulations is to understand the world around us. One of the first important uses of computer simulations was to aid in the exploration for oil. Drilling a single well costs millions of dollars, and most drillings result in "dry wells" that produce no revenue. Geologists lay out networks of microphones and set off explosive charges. Computers analyze the echoes received by the microphones to produce graphical representations of underground rock formations. Analyzing these formations helps petroleum engineers select the most promising sites to drill.

Computer simulations are also used to predict the future. Modern weather predictions are based on computer simulations. These predictions become particularly important when people are exposed to extreme weather conditions, such as floods, tornadoes, and hurricanes (Figure 7.5). Every computer simulation has an underlying mathematical model. Faster computers enable scientists and engineers to develop more sophisticated models. Over time, the quality of these models has improved.

Of course, the predictions made by computer simulations can be wrong. In 1972 the Club of Rome, an international think tank based in Germany, published a report called *The Limits to Growth*. The report predicted that a continued exponential increase in world population would lead to shortages of minerals and farm land, higher food prices, and significant increases in pollution [31]. A year after the report was published, the Arab oil embargo resulted in dramatically higher oil and gasoline prices in Western nations, giving credence to these alarming forecasts. As it turns out, the report's predictions were far too pessimistic. While the population of the earth has indeed doubled in

Figure 7.5 We rely on computer simulations to predict the path and speed of hurricanes. (Courtesy of NASA)

the past 30 years, the amount of tilled land has barely increased, food and mineral prices have dropped, and pollution is in decline in major Western cities [32].

The computer model underlying *The Limits to Growth* was flawed. It assumed all deposits of essential resources had already been discovered. In actuality, many new deposits of oil and other resources have been found in the past three decades. The model ignored the technological improvements that allow society to decrease its use of resources, such as reducing the demand for oil by improving the fuel efficiency of cars or reducing the demand for silver by replacing conventional photography with digital photography.

7.6.2 Validating Simulations

A computer simulation may produce erroneous results for two fundamentally different reasons. The program may have a bug in it, or the model upon which the program is based may be flawed. **Verification** is the process of determining if the computer program correctly implements the model. **Validation** is the process of determining if the model is an accurate representation of the real system [33]. In this section, we'll focus on the process of validation.

One way to validate a model is to make sure it duplicates the performance of the actual system. For example, automobile and truck manufacturers create computer models of their products. They use these models to see how well vehicles will perform in a variety of crash situations. Crashing an automobile on a computer is faster and much less expensive than crashing an actual car. To validate their models, manufacturers compare

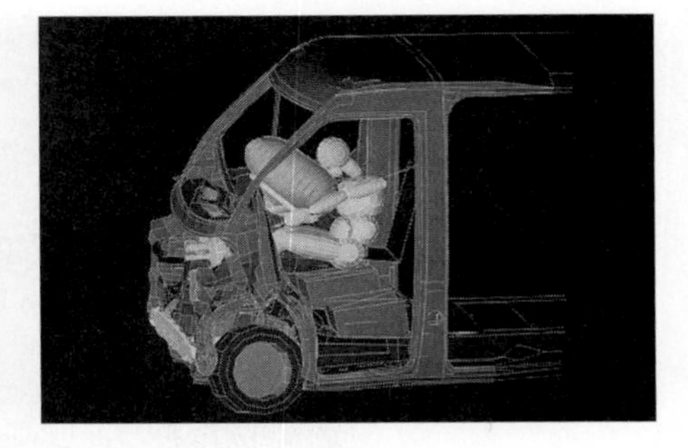

FIGURE 7.6 One way to validate a model is to compare the predicted outcome with the actual outcome in the real system. (Courtesy of Daimler-Chrysler)

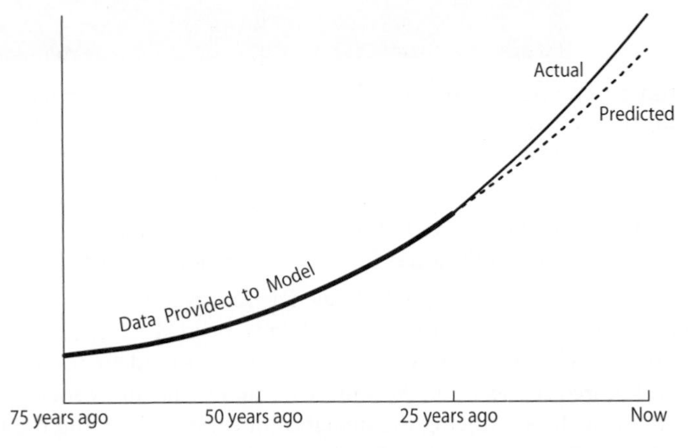

FIGURE 7.7 You can validate a model's ability to predict 25 years into the future by using it to "predict the present" with data 25 or more years old. You can then compare the model's prediction of the present with current reality.

the results of crashing an actual vehicle with the results predicted by the computer model (Figure 7.6).

Validating a model that predicts the future can introduce new difficulties. If we are predicting tomorrow's weather, it is reasonable to validate the model by waiting until tomorrow and seeing how well the prediction held up. However, suppose you are a scientist using a global warming model to estimate what the climate will be like 50 years from now. You cannot validate this model by comparing its prediction with reality, because you cannot afford to wait 50 years to see if its prediction came true. However, you can validate the model by using it to *predict the present*.

Figure 7.7 illustrates how a model can predict the present. Suppose you want to see how well your model predicts events 25 years into the future. You have access to data going back 75 years. You let the model use data at least 25 years old, but you do not let the model see any data collected in the past 25 years. The job of predicting the present, given 25-year-old data, is presumably just as hard as the job of predicting 25 years into the future, given present data. The advantage of predicting the present is that you can use current data to validate the model.

A final way to validate a computer model is to see if it has credibility with experts and decision makers. Ultimately, a model is valuable only if it is believed by those who have the power to use its results to reach a conclusion or make a decision.

7.7 Software Engineering

The field of software engineering grew out of a growing awareness of a "software crisis." In the 1960s computer architects had taken advantage of commercial integrated circuits to design much more powerful mainframe computers. These computers could execute much larger programs than their predecessors. Programmers responded by designing powerful new operating systems and applications. Unfortunately, their programming efforts were plagued by problems. The typical new software system was delivered behind schedule, cost more than expected, did not perform as specified, contained many bugs, and was too hard to modify. The informal, ad hoc methods of programming that worked fine for early software systems broke down when these systems reach a certain level of complexity.

Software engineering is an engineering discipline focused on the production of software, as well as the development of tools, methodologies, and theories supporting software production. Software engineers follow a four-step process to develop a software product [34]:

1. Specification: defining the functions to be performed by the software
2. Development: producing the software that meets the specifications
3. Validation: testing the software
4. Evolution: modifying the software to meet the changing needs of the customer

7.7.1 Specification

The process of specification focuses on determining the requirements of the system and the constraints under which it must operate. Software engineers communicate with the intended users of the system to determine what their needs are. They must decide if the software system is feasible given the budget and the schedule requirements of the customer. If a piece of software is going to replace an existing process, the software engineers study the current process to help them understand the functions the software must perform. The software engineers may develop prototypes of the user interface to confirm that the system will meet the user's needs.

The specification process results in a high-level statement of requirements and perhaps a mock-up of the user interface that the users can approve. The software engineers also produce a low-level requirements statement that provides the details needed by those who are going to actually implement the software system.

7.7.2 Development

During the development phase, the software engineers produce a working software system that matches the specifications. The first design is based on a high-level, abstract view of the system. The process of developing the high-level design reveals ambiguities, omissions, or outright errors in the specification. When these mistakes are discovered, the specification must be amended. Fixing mistakes is quicker and less expensive when the design is still at a higher, more abstract level.

Gradually the software engineers add levels of detail to the design. As this is done, the various components of the system become clear. Designers pay particular attention to ensure the interfaces between each component are clearly spelled out. They choose the algorithms to be performed and data structures to be manipulated.

Since the emergence of software engineering as a discipline, a variety of structured design methodologies have been developed. These design methodologies result in the creation of large amounts of design documentation in the form of visual diagrams. Many organizations use **computer-assisted software engineering** (CASE) **tools** to support the process of developing and documenting an ever-more-detailed design.

Another noteworthy improvement in software engineering methodologies is object-oriented design. In a traditional design, the software system is viewed as a group of functions manipulating a set of shared data structures. In an **object-oriented design,** the software system is seen as a group of objects passing each other messages. Each object has its own state and manipulates its own data based on the messages it receives.

Object-oriented systems have several advantages over systems constructed in a more traditional way:

1. *Because each object is associated with a particular component of the system, object-oriented designs can be easier to understand.*

   More easily understood designs can save time during the programming, testing, and maintenance phases of a software project.

2. *Because each object hides its state and private data from other objects, other objects cannot accidentally modify its data items.*

   The result can be fewer errors like the race conditions described in Section 7.5.

3. *Because objects are independent of each other, it is much easier to reuse components of an object-oriented system.*

   A single object definition created for one software system can be copied and inserted into a new software system without bringing along other, unnecessary objects.

When the design has reached a great enough level of detail, software engineers write the actual computer programs implementing the software system. Many different programming languages exist; each language has its strengths and weaknesses. Programmers usually implement object-oriented systems using an object-oriented programming language, such as C++, Java, or C#.

7.7.3 Validation

The purpose of validation (also called testing) is to ensure the software satisfies the specification and meets the needs of the user. In some companies, testing is an assignment given to newly hired software engineers, who soon move on to design work after proving their worth. However, good testing requires a great deal of technical skill, and some organizations promote testing as a career path.

Testing software is much harder than testing other engineered artifacts, such as bridges. We know how to construct scale models that we can use to validate our designs. To determine how much weight a model bridge can carry, we can test its response to various loads. The stresses and strains on the members and the deflection of the span change gradually as we add weight, allowing us to experiment with a manageable number of different loading scenarios. Engineers can extrapolate from the data they collect to generate predictions regarding the capabilities of a full-scale bridge. By increasing the size of various components, they can add a substantial margin of error to ensure the completed bridge will not fail.

A computer program is not at all like a bridge. Testing a program with a small problem can reveal the existence of bugs, but it cannot prove the program will work when it is fed a much larger problem. The response of a computer program to nearly identical data sets may not be continuous. Instead, programs that appear to be working just fine may fail when only a single parameter is changed by a small amount. Yet programmers cannot exhaustively test programs. Even small programs have a virtually infinite number of different inputs. Since exhaustive testing is impossible, programs can never be completely tested. Software testers strive to put together suites of test cases that exercise all the capabilities of the component or system being validated.

To reduce the complexity of validating a large software system, testing is usually performed in stages. In the first stage of testing, each individual module of the system is tested independently. It is easier to isolate and fix the causes of errors when the number of lines of code is relatively small. After each module has been debugged, modules are combined into larger subsystems for testing. Eventually, all of the subsystems are combined in the complete system. When an error is detected and a bug is fixed in a particular module, all of the test cases related to the module should be repeated to see if the change that fixed one bug accidentally introduced another bug.

EVOLUTION

Successful software systems evolve over time to meet the changing needs of their users. The evolution of a software system resembles the creation of a software system in many ways. Software engineers must understand the needs of the users, assess the strengths and weaknesses of the current system, and design modifications to the software. The same CASE tools used to create a new software system can aid in its evolution. Many of

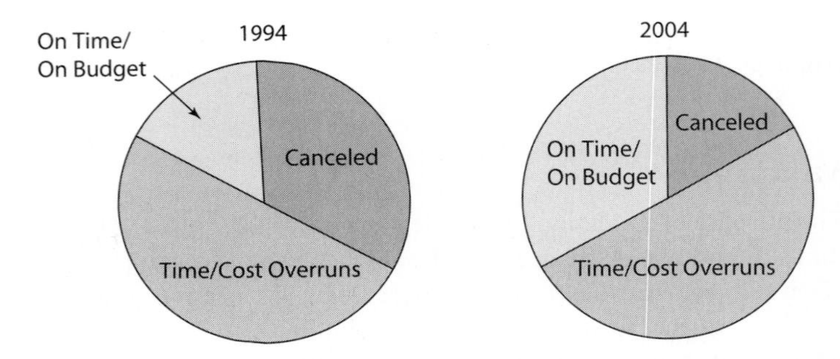

FIGURE 7.8 Research by the Standish Group reveals that the success rate of IT projects in 2004 was twice that of 1994. Today, about one-third of software projects are completed on time and on budget.

the data sets developed for the original system can be reused when validating the updated system.

7.7.4 Software Quality Is Improving

There is evidence that the field of software engineering is becoming more mature (Figure 7.8). The Standish Group regularly tracks thousands of IT projects. As recently as 1994, about one-third of all software projects were canceled before completion. About one-half of the projects were completed but had time and/or cost overruns, which were often quite large. About one-sixth of the projects were completed on time and on budget, although the completed systems often had fewer features than originally planned. Another survey by the Standish Group in 2004 showed that the probability of a software project being completed on time and on budget had doubled, to about one in three. Only about one-sixth of the software projects surveyed were canceled. As before, about half of the projects were late and/or over budget, although the time and cost overruns were not as large as in the first survey. Overall, the Standish Group reports indicate that the ability of companies to produce software on time and on budget is improving.

Still, with only about one in three software projects being completed on time and on budget, the industry has a long way to go. Rapid change is a fact of life in the software industry. In order to stay competitive, companies must release products quickly. Many organizations feel a tension between meeting tight deadlines and strictly following software engineering methodologies.

7.8 Software Warranties

As mentioned earlier, Leveson and Turner state that "there is always another software bug" [28]. If perfect software is impossible, what kind of warranty should a consumer expect to get from a software company? In this section we survey the software warranties offered by software manufacturers, how these warranties have held up in court, and the debate over new legislation spelling out conditions for licensing computer software in the United States.

7.8.1 Shrinkwrap Warranties

Consumer software is often called **shrinkwrap software** because of the plastic wrap surrounding the box containing the software and manuals. Not too many years ago, consumer software manufacturers provided no warranty for their products at all. Purchasers had to accept shrinkwrap software "as is." Today, many shrinkwrap software manufacturers, including Microsoft, provide a 90-day replacement or money-back guarantee if the program fails [35]. Here is the wording Microsoft included with its limited warranty for Microsoft Office 2000:

> LIMITED WARRANTY FOR SOFTWARE PRODUCTS ACQUIRED IN THE U.S. AND CANADA. Microsoft warrants that (a) the SOFTWARE PRODUCT will perform substantially in accordance with the accompanying written materials for a period of ninety (90) days from the date of receipt . . .
>
> CUSTOMER REMEDIES. Microsoft's and its suppliers' entire liability and your exclusive remedy shall be, at Microsoft's option, either (a) return of the price paid, if any, or (b) repair or replacement of the SOFTWARE PRODUCT that does not meet Microsoft's Limited Warranty and which is returned to Microsoft with a copy of your receipt.

At least Microsoft is willing to state that its software will actually do more or less what the documentation says it can do. The warranty for Railroad Tycoon, distributed by Gathering of Developers, promises only that you'll be able to install the software:

> LIMITED WARRANTY. Owner warrants that the original Storage Media holding the SOFTWARE is free from defects in materials and workmanship under normal use and service for a period of ninety (90) days from the date of purchase as evidenced by Your receipt. If for any reason You find defects in the Storage Media, or if you are unable to install the SOFTWARE on your home or portable computer, You may return the SOFTWARE and all ACCOMPANYING MATERIALS to the place You obtained it for a full refund. This limited warranty does not apply if You have damaged the SOFTWARE by accident or abuse.

I wonder what would happen if you actually did go back to the store with an opened game and asked for a full refund.

While vendors may be willing to give you a refund if you cannot get their software to install on your computer, they are certainly not going to accept any liability if your business is harmed because their software crashes at the wrong time. Later in the Microsoft Office 2000 Professional warranty, we find these words:

> TO THE MAXIMUM EXTENT PERMITTED BY APPLICABLE LAW, IN NO EVENT SHALL MICROSOFT OR ITS SUPPLIERS BE LIABLE FOR ANY SPECIAL, INCIDENTAL, INDIRECT, OR CONSEQUENTIAL DAMAGES WHATSOEVER (INCLUDING, WITHOUT LIMITATION, DAMAGES FOR LOSS OF

> BUSINESS PROFITS, BUSINESS INTERRUPTION, LOSS OF BUSINESS INFORMATION, OR ANY OTHER PECUNIARY LOSS) ARISING OUT OF THE USE OF OR INABILITY TO USE THE SOFTWARE PRODUCT OR THE PROVISION OF OR FAILURE TO PROVIDE SUPPORT SERVICES, EVEN IF MICROSOFT HAS BEEN ADVISED OF THE POSSIBILITY OF SUCH DAMAGES. IN ANY CASE, MICROSOFT'S ENTIRE LIABILITY UNDER ANY PROVISION OF THIS UELA SHALL BE LIMITED TO THE GREATER OF THE AMOUNT ACTUALLY PAID FOR THE SOFTWARE PRODUCT OR U.S. $5.00; PROVIDED, HOWEVER, IF YOU HAVE ENTERED INTO A MICROSOFT SUPPORT SERVICES AGREEMENT, MICROSOFT'S ENTIRE LIABILITY REGARDING SUPPORT SERVICES SHALL BE GOVERNED BY THE TERMS OF THAT AGREEMENT.

Here is even blunter language from the license agreement accompanying Harmonic Visions's Music Ace program:

> WE DO NOT WARRANT THAT THIS SOFTWARE WILL MEET YOUR REQUIREMENTS OR THAT ITS OPERATION WILL BE UNINTERRUPTED OR ERROR-FREE. WE EXCLUDE AND EXPRESSLY DISCLAIM ALL EXPRESS AND IMPLIED WARRANTIES NOT STATED HEREIN, INCLUDING THE IMPLIED WARRANTIES OF MERCHANTABILITY AND FITNESS FOR A PARTICULAR PURPOSE.

In other words, don't blame us if the program doesn't do what you hoped it would do, or if it crashes all the time, or if it is full of bugs.

7.8.2 Are Software Warranties Enforceable?

How can software manufacturers get away with disclaiming any warranties on their products? It's not clear that they can. Article 2 of the Uniform Commercial Code (UCC) governs the sale of products in the United States. In 1975 Congress passed the Magnuson-Moss Warranty Act. One goal of the act was to prevent manufacturers from putting unfair warranties on products costing more than $25. A second goal was to make it economically feasible for consumers to bring warranty suits by allowing courts to award attorneys' fees. Together, the Magnuson-Moss Warranty Act and Article 2 of the UCC protect the rights of consumers. A computer program is a product. Hence unfair warranties on shrinkwrap software could be in violation of these laws.

An early court case, *Step-Saver Data Systems v. Wyse Technology and The Software Link*, seemed to affirm the notion that software manufacturers could be held responsible for defects in their products, despite what they put in their warranties. However, two later cases seemed to indicate the opposite. In *ProCD v. Zeidenberg*, the court ruled that the customer was bound to the license agreement, even if the license agreement does not appear on the outside of the shrinkwrap box. *Mortenson v. Timberline Software* showed that a warranty disclaiming the manufacturer's liability could hold up in court.

STEP-SAVER DATA SYSTEMS V. WYSE TECHNOLOGY AND THE SOFTWARE LINK

Step-Saver Data Systems, Inc. sold timesharing computer systems consisting of an IBM PC AT server, Wyse terminals, and an operating system provided by The Software Link, Inc. (TSL). In 1986–1987 Step-Saver purchased and resold 142 copies of the Multilink Advanced operating system provided by TSL.

To purchase the software, Step-Saver called TSL and placed an order, then followed up with a purchase order. According to Step-Saver, the TSL phone sales representatives said that Multilink was compatible with most DOS applications. The box containing the Multilink software included a licensing agreement in which TSL disclaimed all express and implied warranties.

Step-Saver's timesharing systems did not work properly, and the combined efforts of Step-Saver, Wyse, and TSL could not fix the problems. Step-Saver was sued by twelve of its customers. In turn, Step-Saver sued Wyse Technology and TSL.

The Third Circuit of the U.S. Court of Appeals ruled in favor of Step-Saver [36]. It based its argument on Article 2 of the UCC. The court held that the original contract between Step-Saver and TSL consisted of the purchase order, the invoice, and the oral statements made by TSL representatives on the telephone. The license agreement had additional terms that would have materially altered the contract. However, Step-Saver never agreed to these terms.

The court wrote, "In the absence of a party's express assent to the additional or different terms of the writing, section 2-207 [of the UCC] provides a default rule that the parties intended, as the terms of their agreement, those terms to which both parties have agreed along with any terms implied by the provision of the UCC." The court noted that the president of Step-Saver had objected to the terms of the licensing agreement. He had refused to sign a document formalizing the licensing agreement. Even after this, TSL had continued to sell to Step-Saver, implying that TSL wanted the business even if the contract did not include the language in the licensing agreement. That is why the court ruled that the purchase order, the invoice, and the oral statements constituted the contract, not the license agreement.

PROCD, INC. V. ZEIDENBERG

ProCD invested more than $10 million to construct a computer database containing information from more than 3,000 telephone directories. ProCD also developed a proprietary technology to compress and encrypt the data. It created an application program enabling users to search the database for records matching criteria they specified. ProCD targeted its product, called SelectPhone, to two different markets: companies interested in generating mailing lists, and individuals interested in finding the phone numbers or addresses of particular people they wanted to call or write. Consumers who wanted SelectPhone for personal use could purchase it for $150; companies paid much more for the right to put the package to commercial use. ProCD included in the consumer version of SelectPhone a license prohibiting the commercial use of the database and program.

In addition, the license terms were displayed on the user's computer monitor every time the program was executed.

Matthew Zeidenberg purchased the consumer version of SelectPhone in 1994. He formed a company called Silken Mountain Web Services, Inc., which resold the information in the SelectPhone database. The price it charged was substantially less than the commercial price of SelectPhone. ProCD sued Matthew Zeidenberg for violating the licensing agreement.

At the trial, the defense argued that Zeidenberg could not be held to the terms of the licensing agreement, since they were not printed on the outside of the box containing the software. The U.S. Court of Appeals for the Seventh Circuit ruled in favor of ProCD. Judge Frank Easterbrook wrote, "Shrinkwrap licenses are enforceable unless their terms are objectionable on grounds applicable to contracts in general (for example, if they violate a rule of positive law, or if they are unconscionable)" [37].

MORTENSON V. TIMBERLINE SOFTWARE

M. A. Mortenson Company was a national construction contractor with a regional office in Bellevue, Washington. Timberline Software, Inc. produced software for the construction industry. Mortenson had used software from Timberline for several years. In July 1993 Mortenson purchased eight copies of a bidding package called Precision Bid Analysis.

Timberline's licensing agreement included this paragraph:

> LIMITATION OF REMEDIES AND LIABILITY.
>
> NEITHER TIMBERLINE NOR ANYONE ELSE WHO HAS BEEN INVOLVED IN THE CREATION, PRODUCTION OR DELIVERY OF THE PROGRAMS OR USER MANUALS SHALL BE LIABLE TO YOU FOR ANY DAMAGES OF ANY TIME, INCLUDING BUT NOT LIMITED TO, ANY LOST PROFITS, LOST SAVINGS, LOSS OF ANTICIPATED BENEFITS, OR OTHER INCIDENTAL, OR CONSEQUENTIAL DAMAGES, ARISING OUT OF THE USE OR INABILITY TO USE SUCH PROGRAMS, WHETHER ARISING OUT OF CONTRACT, NEGLIGENCE, STRICT TORT, OR UNDER ANY WARRANTY, OR OTHERWISE, EVEN IF TIMBERLINE HAS BEEN ADVISED OF THE POSSIBILITY OF SUCH DAMAGES OR FOR ANY OTHER CLAIM BY ANY OTHER PARTY. TIMBERLINE'S LIABILITY FOR DAMAGES IN NO EVENT SHALL EXCEED THE LICENSE FEE PAID FOR THE RIGHT TO USE THE PROGRAMS.

In December 1993 Mortenson used Precision Bid Analysis to prepare a bid for the Harborview Medical Center in Seattle. On the day the bid was due, the software malfunctioned. It printed the message "Abort: Cannot find alternate" 19 times. Mortenson continued to use the software and submitted the bid the software produced. After the firm won the contract, Mortenson discovered that its bid was $1.95 million too low.

Mortenson sued Timberline for breach of express and implied warranties. It turns out Timberline had been aware of the bug uncovered by Mortenson since May 1993. Timberline had fixed the bug and already sent a newer version of the program to some of its other customers who had encountered it. It had not sent the improved program to Mortenson. Nevertheless, Timberline argued that the lawsuit be summarily dismissed because the licensing agreement limited the consequential damages that Mortsenson could recover from Timberline. The King County Superior Court ruled in favor of Timberline. The ruling was upheld by the Washington Court of Appeals and the Supreme Court of the State of Washington [38].

7.8.3 Uniform Computer Information Transaction Act

HISTORY

The ruling against The Software Link by the Third U.S. Circuit Court of Appeals caused consternation among software developers. No large programs are error free. How can a software company stay in business if it can be held responsible for lost profits or other damages caused when one of its customers uses its product and trips over a bug? Software companies could insure themselves against these potential consequences, but the cost of the insurance would increase the price of software, perhaps substantially.

Besides, companies know that features and price are more important to consumers than 100 percent reliability. The software industry is highly competitive. Companies feel great pressure to get feature-laden products to market as quickly as possible. Putting all these factors together, companies make the deliberate decision to ship software containing bugs. They hope these bugs are minor, but sometimes they are not. They also know that consumers will discover additional bugs as they use the software. Software companies count on getting bug reports back from customers in order to improve the quality of their software. From the standpoint of a software producer, it is unrealistic to expect a program that may contain millions of lines of code to be as reliable as a spark plug.

The ruling against The Software Link, combined with the idea that software is somehow different from other products, motivated lawyers to consider an amendment to Article 2 of the UCC dealing with software.

Volunteer lawyers from the 50 states belong to the National Conference of Commissioners on Uniform State Laws (NCCUSL). The NCCUSL recommends laws that the individual states must then pass. In 1999 the NCCUSL produced the Uniform Computer Information Transaction Act (UCITA). A year later the state legislatures of Maryland and Virginia passed slightly modified versions of UCITA.

PRIMARY FEATURES OF UCITA

UCITA is a lengthy piece of legislation with many details. In a nutshell, UCITA

- affirms that manufacturers may license software to customers;
- allows manufacturers to prevent the transfer of software from one person or organization to another;
- allows software manufacturers to disclaim all liability for defects in their products; customers must accept them "as is";
- allows manufacturers to remotely disable licensed software on a customer's computer in the case of a license dispute;

- allows manufacturers to collect information about how licensees use their computers;
- applies to software in computers, but not software in embedded systems.

ARGUMENTS IN FAVOR OF UCITA

Software manufacturers and others support the proposed UCITA legislation developed by the NCCUSL. Here are the principal arguments in favor of UCITA:

1. *Article 2 of the UCC was written more than 50 years ago, long before the creation of the consumer software business.*

   Article 2 is unworkable in the realm of digital products. If we want a vital software industry, we need to recognize that software is simply not going to have the same reliability as physical products. UCITA responds to this problem. It makes absolutely clear that shrinkwrap license agreements can be enforced.

2. *UCITA recognizes that there is no such thing as a piece of perfect software.*

   All large programs contain bugs. UCITA replaces the "perfect tender" requirement of UCC's Article 2 with a "substantial conformance" requirement for commercial software.

3. *UCITA prevents fraud by allowing software manufacturers to insert electronic controls in their products to ensure they are being used according to their licenses.*

   For example, if a customer purchases a license to use a piece of software for a particular period of time, UCITA allows the manufacturer to put code in the program that makes it inoperable after the license has expired. Similarly, if the customer's license allows the program to be run on a certain number of computers, the software can include features that make it impossible for the customer to run the program on a larger number of machines.

ARGUMENTS AGAINST UCITA

The Attorneys General of 34 states have spoken out against UCITA, along with the American Law Institute, the American Intellectual Property Law Association, the Association for Computing Machinery (ACM), the Institute of Electrical and Electronics Engineers (IEEE), the Free Software Foundation, the American Library Association, Consumers Union, the National Consumer League, the United States Public Interest Research Group, and many others. These organizations believe that UCITA is bad for consumers. Here are some of the arguments they have made:

1. *Customers should be allowed to purchase software, not just license it.*
2. *If you license a piece of software and don't need it any more, you can't even give it away legally to someone who can use it.*
3. *Allowing companies to sell software "as is" removes consumer software from the protections of the Magnuson-Moss Act.*

Manufacturers do not even have to inform consumers of defects in the software that the manufacturers are aware of. Consumers can be duped into buying poor-quality software with no recourse if they want their money back.

4. *UCITA codifies the current practice of allowing consumers to see the warranty only after the package has been opened and just before the software is installed.*

   Consumers lose the right to return software as soon as they click the "I Accept" button, even before they have had the opportunity to run the software once and see if it does what they need it to do.

5. *Manufacturers are allowed to put "trap doors" into their software products, enabling them to remotely enter the computers and remove the software of customers who have violated their licensing agreements.*

   Yet manufacturers are not held responsible for damage caused to consumers by third parties using these trap doors to launch denial of service attacks.

6. *UCITA restricts free speech by allowing software manufacturers under some circumstances to forbid the publication of reviews and product comparisons mentioning their product.*

7. *The drafters of UCITA were targeting computer software. They tried to exclude embedded systems.*

   The problem is that the line between an embedded system and a computer is getting fuzzy. Consider a state-of-the-art sewing machine, the Husqvarna Viking Designer SE (Figure 7.9). It comes equipped with a touch-sensitive color LCD screen and two USB ports. Customers can download stitch and embroidery designs from a memory stick. They can even get software updates from Husqvana's Web site. The wording of UCITA will encourage manufacturers to make their systems look more like computers so that they may fall under its more generous warranty provisions.

8. *The process of creating UCITA was highly controversial, with a significant party to the discussions (the American Legal Institute) walking out.*

   That means UCITA legislation is unlikely to pass in the states without various amendments being added. This has already happened in Maryland and Virginia. Since different amendments will be added in different states, each state will end up with a different version of the law. The goal of UCITA, to have a uniform law across all 50 states, will not be met.

7.8.4 Moral Responsibility of Software Manufacturers

Should producers of shrinkwrap software be held responsible for defects in their programs?

Let's consider the consequences of holding manufacturers of shrinkwrap software liable for damages, such as lost profits, caused by errors encountered by licensees. Currently, manufacturers rely upon consumers to help them identify bugs in their products. If they must find these bugs themselves, they will need to hire many more software testers. The result will be higher prices and longer program development times.

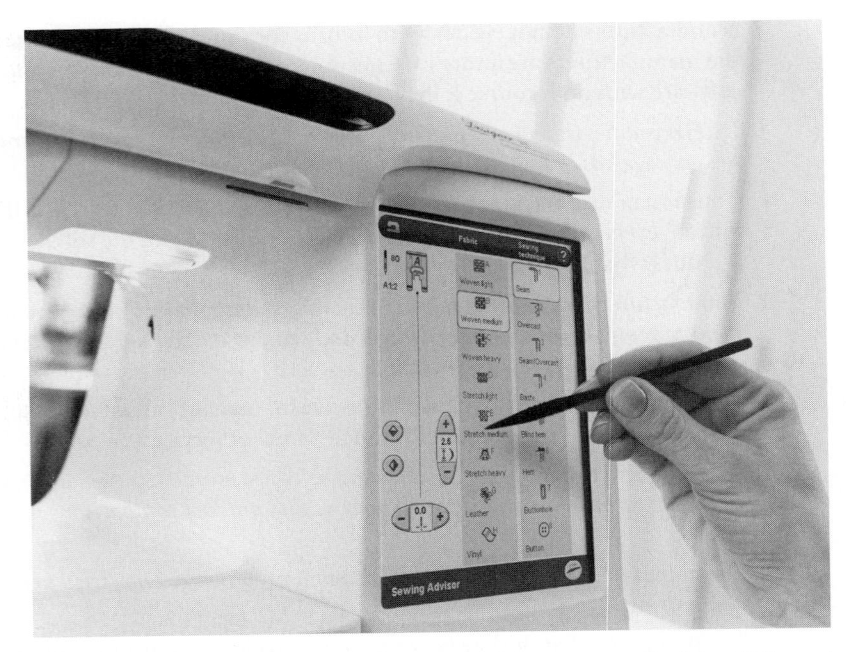

FIGURE 7.9 Is it a sewing machine or a computer? The Husqvarna Designer SE has a touch-sensitive color screen and two USB ports. (Photo courtesy of Husqvarna Viking Sewing Machine. www.husqvarnaviking.com)

Prudent companies would most likely purchase insurance to protect them from potential lawsuits. This insurance could be very expensive, depending upon the maximum liability to which a company could be held responsible. The cost of the insurance would be passed along to consumers in the form of higher prices.

These changes in the consumer software industry would affect small, start-up companies more than large, established firms. The changes would slow the entry of new companies into the field. The result would be a decrease in level of innovation and vitality in the software industry.

Consumers could license software with a higher degree of confidence, knowing that the companies stood by their products. While there might be fewer products available, and their prices would be higher, they would be more reliable.

The result of our utilitarian analysis depends upon how much weight we give to the various consequences. Let's suppose we conclude that software manufacturers should not be held liable for lost profits and other negative consequences arising from errors in their programs, because the harms are greater than the benefits. We may still ask the following question: What are the rights of consumers who license shrinkwrap software?

Consider this hypothetical scenario. A consumer goes to the store, pays $49.95, and brings home a copy of *Incredible Bulk*. The game is usable, but it contains some annoying bugs. The next year the company releases *Incredible Bulk II*. If the consumer wants all the latest bug fixes, he needs to buy the second edition. Of course, *Incredible Bulk II* has cool new features. And, as you might expect, some of these features are buggy. Never fear! The bugs will be fixed when *Incredible Bulk III* comes out in 12 months. Is this a fair arrangement?

From a social contract point of view, this arrangement is unfair. Consumers should have the right to be informed of bugs the manufacturer knows about. This knowledge allows a consumer to enter into a contract with his eyes wide open. Ideally, a manufacturer would be open about disclosing the weaknesses in its product. More realistically, consumer organizations can test software products and provide reviews for potential buyers.

If a consumer purchases the right to use product A, and the manufacturer removes defects from product A, the consumer should not have to purchase additional features in order to get access to the fixes to the original product. Manufacturers should make software patches containing bug fixes available on the Web for free downloading by their customers. Withholding these patches until the next major release of the software is wrong from a social contract point of view.

Summary

Computers are part of larger systems, and ultimately it is the reliability of the entire system that is important. A well-engineered system can tolerate the malfunction of any single component without failing. This chapter has presented many examples of how the computer turned out to be the "weak link" in the system, leading to a failure. These examples provide important lessons for computer scientists and others involved in the design, implementation, and testing of large systems.

Two sources of failure are data-entry errors and data-retrieval errors. While it's easy to focus on a particular mistake made by the person entering or retrieving the data, the system is larger than the individual person. For example, in the case of the 2000 general election in Florida, incorrect records in the computer database disqualified thousands of voters. The data-entry errors caused the *voting system* to work incorrectly. Sheila Jackson Stossier was arrested by police who confused her with Shirley Jackson. The data-retrieval error caused the *criminal justice system* to perform incorrectly.

When the topics are software and billing errors, it is easier to identify the system that is failing. For example, when Qwest sent out 14,000 incorrect bills to its cellular phone customers, it's clear that the billing system had failed.

In Sections 7.4 and 7.5, the larger systems were easy to spot. Several embedded systems were dissected to determine the causes of their failures. The program for the Patriot missile's radar tracking system had a subtle flaw: a tiny truncation error occurred every time the clock signal was stored in a floating-point variable. Over a period of 100 hours, all those tiny errors added up to a significant amount, causing the radar system to lose its target. The Ariane 5 blew up because a single assignment statement caused the onboard computers to crash. The AT&T long-distance network collapsed because of one faulty line of code.

A well-engineered system does not fail when a single component fails. In the case of hardware, this principle is easier to apply. For example, a jetliner may have three engines. It is designed to be able to fly on any two of the engines, so if a single engine fails, the plane can still fly to the nearest airport and land. When it comes to software, the goal is much harder to meet. If we have two computers in the system, that will provide redundancy in case one of the computers has a hardware failure. However, if both computers are running the same software, there is still no software redundancy. A software bug that causes one computer to fail will cause both computers to fail. The partial collapse of the AT&T long-distance network is an example of this phenomenon. All 80 switches containing the latest version of the software failed. Fortunately, 34 switches were running an older version of the software, which prevented a total collapse of AT&T's system.

Imagine what it would take to provide true redundancy in the case of software systems. Should companies maintain two entirely different billing systems so that the bills produced by one system could be double-checked by the other? Should the federal government support two completely different implementations of the National Crime Information Center? These alternatives seem unrealistic. On the other hand, redundancy seems much more feasible when we look at data-entry and data-retrieval operations. Two different data-entry operators could input records into databases, and the computer could check to make sure the records agreed. This would reduce the chance of bad data being entered into databases in the first place. Two different people could look at the results returned from a computer query, using their own common sense and understanding to see if the output makes sense.

While it may be infeasible to provide redundant software systems, safety-critical systems should never rely completely upon a single piece of software. The Therac-25 overdoses occurred because the system lacked the hardware interlocks of the earlier models.

The stories of computer system failures contain other valuable lessons. The Ariane 5 and Therac-25 failures show it can be dangerous to reuse code. Assumptions that were valid when the code was originally written may no longer be true when the code is reused. Since some of these assumptions may not be documented, the new design team may not have the opportunity to check if these assumptions still hold true in the new system.

The automated baggage system at the Denver International Airport demonstrates the difficulty of debugging a complex system. Tackling one problem at a time, solving it, and moving on to the next problem proved to be a poor approach, because the overall system design had serious flaws. For example, BAE did not even realize that simply getting luggage carts to where they were needed in a fair manner was an incredibly difficult problem. Even if BAE had solved all the low-level technical problems, this high-level problem would have prevented the system from meeting its performance goals during the busiest times.

Finally, systems can fail because of miscommunications among people. The Mars Climate Orbiter is an example of this kind of failure. The software written by the team in Colorado used English units, while the software written by the team in California used metric units. The output of one program was incompatible with the input to the other program, but a poorly specified interface allowed this error to remain undetected until after the spacecraft was destroyed.

Computer simulations are used to perform numerical experiments that lead to new scientific discoveries and help engineers create better products. For this reason, it is important that simulations provide reliable results. Simulations are validated by comparing predicted results with reality. If a simulation is designed to predict future events, it can be validated by giving it data about the past and asking it to predict the present. Finally, simulations are validated when their results are believed by domain experts and policymakers.

The discipline of software engineering emerged from a growing realization of a "software crisis." While small programs can be written in an ad hoc manner, large programs must be carefully constructed if they are to be reliable. Software engineering is the application of engineering methodologies to the creation and evolution of software artifacts. Surveys of the IT industry reveal that more projects are being completed on time and on budget, and fewer projects are being canceled. This may be evidence that software engineering is having a positive impact. However, since most projects are still not completed on time and on budget, there remains much room for improvement. For many companies, shipping a product by a particular date continues to be a higher priority than following a strict software development methodology.

Should software manufacturers be held accountable for the quality of their software, or is a program a completely different kind of product than a socket wrench? An examination of the software warranties manufacturers include in their licensing agreements reveals that while some vendors refund the purchase price of software that does not meet the needs of the purchaser, they do not want to be held liable for any damages that occur from the use of their software. These warranties seem to fly in the face of Article 2 of the Uniform Commercial Code and the Magnuson-Moss Warranty Act. Some courts have ruled that software manufacturers cannot disclaim liability for consequential damages, but other court decisions imply that software warranties disclaiming liability are enforceable. The controversy about the legality of software warranties led to the development of a model law called the Uniform Computer Information Transaction Act (UCITA). Two states—Virginia and Maryland—have passed versions of UCITA, but 34 state Attorneys General and many consumer organizations have expressed opposition to UCITA, and it seems unlikely UCITA will be adopted across the United States.

Review Questions

1. What kinds of mistakes may cause a computer to produce a faulty output?
2. What is the difference between a data-entry error and a data-retrieval error?
3. What reasons did the U.S. Department of Justice give for no longer requiring the FBI to ensure the accuracy of information kept in the NCIC databases?
4. What is an embedded system? What is a real-time system?

5. What does a linear accelerator do?

6. What was the most important difference between the Therac-20 and its successor, the Therac-25?

7. How long was the Therac-25 in operation before the first documented accident? How much longer did it take for the system to be declared unsafe?

8. What is a race condition in software? Why are race conditions difficult to debug?

9. The following reasons have been given for the failure of computerized systems:

   I. A system designed for one purpose was used for another purpose.
   II. Software was reused without adequate testing.
   III. There was an error in storing or converting a data value.
   IV. A line of code became a single point of failure.
   V. The overall system was too complicated to analyze.
   VI. There was a software race condition.
   VII. There was another software error (not listed above).

   For each of the systems listed below, select the principal reason or reasons why it failed to operate as specified.

   a. Patriot missile
   b. Ariane 5
   c. AT&T long-distance network
   d. Mars Climate Orbiter
   e. Mars Polar Lander
   f. Denver International Airport baggage system
   g. Therac-25

10. What are the advantages of allowing software users to identify and report bugs? What are the disadvantages?

11. Why are computer simulations playing an increasingly important role in science and engineering?

12. List five uses of computer simulation.

13. What is the difference between a model and a computer simulation?

14. What is the difference between verification and validation?

15. Name two different ways to validate a computer simulation.

16. What does Article 2 of the Uniform Commercial Code deal with?

17. What was the purpose of the Magnuson-Moss Warranty Act?

18. Why do some people argue that shrinkwrap software should be exempt from the Magnuson-Moss Warranty Act and Article 2 of the Uniform Commercial Code?

19. What is the significance of the court's ruling in *Step-Saver Data Systems v. Wyse Technology and The Software Link*?

20. What is the significance of the court's ruling in *ProCD, Inc. v. Zeidenberg*?

21. What is the significance of the court's ruling in *Mortenson v. Timberline Software*?

22. What is the purpose of the National Conference of Commissioners on Uniform State Laws?
23. What are the primary features of the Uniform Computer Information Transaction Act? How many states have passed UCITA legislation?
24. Why does UCITA distinguish between computer software and software inside embedded systems?
25. What are the most important arguments in favor of UCITA? What are the most important arguments against UCITA?

Discussion Questions

26. Have you ever been the victim of a software error? Whom did you blame? Now that you know more about the reliability of computer systems, do you still feel the same way?
27. Should an ecommerce site be required to honor the prices at which it offers and sells goods and services?
28. Should the FBI be responsible for the accuracy of information about criminals and crime victims it enters into the National Crime Information Center database?
29. Over a period of seven years, about 500 residents of Freeport, Texas, were overbilled for their water usage. Each resident paid on average about $170 too much, making the total amount of the overbillings about $100,000. The city council decided not to issue refunds, saying that about 300,000 bills would have had to have been examined, some residents had left town, and the individual refunds were not that large [39]. Did the city council make the right decision?
30. If a company sends a consumer an incorrect bill, should the company compensate the consumer for the time and effort the consumer takes to straighten out the mistake?
31. The chapter quotes NASA's Mars Pathfinder project manager as saying software hasn't improved in quality in the past 25 years. How could you determine whether software quality has improved in the past 25 years?
32. Perhaps programs used for business purposes ought to conform to higher standards of quality than games. With respect to software warranties, would it make sense to distinguish between software used for entertainment purposes (such as a first-person shooter game) and software used for business (such as a spreadsheet)?
33. Suppose the legislature in your state is debating the adoption of UCITA and you have been called as an expert witness. What are the three most important ideas you want your legislators to get from your testimony?
34. Read the entire end-user license agreement (EULA) from a piece of commercial software. Do any of the conditions seem shady or unreasonable? If so, which ones?
35. While waiting for an appointment with your physician, you see a brochure advertising a new surgical procedure that implants a tiny microprocessor inside your skull just behind your left ear. The purpose of the chip is to help you associate names with faces. The procedure for inserting the chip is so simple that your physician is performing it in

his office. Suppose your career takes you into sales, where such a device could help you earn higher commissions. What questions would you want to have answered before you agreed to have such a device inserted into your skull?

In-class Exercises

36. Debate the moral responsibility of three agents associated with the two Therac-25 overdoses occurring in Tyler, Texas: the radiation technician, the hospital director, and the programmer who wrote the code controlling the machine. Divide the class into six groups. Three groups (one for each of the three agents) should give reasons why their particular party should bear at least some of the moral responsibility for the deaths. The other three groups (one for each of the three agents) should give reasons why their particular party should not bear any moral responsibility.

37. California is working on an "intelligent highway" system that would allow computer-controlled automobiles to travel faster and closer together on freeways than today's human-controlled cars. What kinds of safety devices would have to be in such a system in order for you to feel comfortable using an intelligent highway? How many people in class would be comfortable being one of the first people to use the intelligent highway?

38. The New York Transit authority is transforming the L line into a partially automated subway line. A central computer system will control the speed and spacing of all trains on the line. However, each train will have an operator that starts and stops the train and has the ability to take control of it in an emergency [40]. How many people in the class would ride on a computer-controlled subway train that did not have a human operator on board?

39. Identify people in the class who have been beta testers for new software products. Ask them to tell the rest of the class about their experiences. What did acting as a beta tester teach them about software reliability?

40. A start-up company called Medick has been developing an exciting new product for handheld computers that will revolutionize the way nurses keep track of their hospitalized patients. The device will save nurses a great deal of time doing routine paperwork, reduce their stress levels, and enable them to spend more time with their patients.

    Medick's sales force has led hospital administrators to believe the product will be available next week as originally scheduled. Unfortunately, the package still contains quite a few bugs. All of the known bugs appear to be minor, but some of the planned tests have not yet been performed.

    Because of the fierce competition in the medical software industry, it is critical that this company be the first to market. It appears a well-established company will release a similar product in a few weeks. If its product appears first, Medick will probably go out of business.

    Divide the class into five groups representing the software engineers programming the device, the sales force that has been promoting the device, the managers of Medick, the venture capitalists who bankrolled Medick, and the nurses at a hospital purchasing the device. Discuss the best course of action for Medick.

Further Reading

John L. Casti. *Would-Be Worlds: How Simulation is Changing the Frontiers of Science*. John Wiley & Sons, New York, NY, 1997.

W. Wayt Gibbs. "Software's Chronic Crisis." *Scientific American*, pages 72–81, September 1994.

Stewart V. Hoover and Ronald F. Perry. *Simulation: A Problem-Solving Approach*. Addison-Wesley, Reading, MA, 1989.

Cem Kaner. "Software Engineering and UCITA." *Journal of Computer and Information Law* 18, 2 (Winter 1999/2000).

Nancy Leveson and Clark Turner. "An Investigation of the Therac-25 Accidents." *Computer*, pages 18–41, July 1993.

Hafedh Mili, Ali Mili, Sherif Yacoub, and Edward Addy. *Reuse Based Software Engineering: Techniques, Organization, and Measurement*. John Wiley & Sons, New York, NY, 2001.

Peter G. Neumann. *Computer-Related Risks*. ACM Press, New York, NY, 1995.

Proceedings of the Annual Conference on Computer Safety, Reliability, and Security (SAFECOMP), published annually by Springer-Verlag.

Ian Sommerville. *Software Engineering*. 6th ed. Addison-Wesley, Harlow, England, 2001.

Victor L. Winter and Sourav Bhattacharya, editors. *High Integrity Software*. Kluwer Academic Publishers, Boston, MA, 2001.

References

[1] Jennifer DiSabatino. "Unregulated Databases Hold Personal Data." *Computerworld*, 36(4), January 21, 2002.

[2] Peter G. Neumann. "More on False Arrests." *The Risks Digest*, 1(5), September 4, 1985.

[3] Rodney Hoffman. "NCIC Information Leads to Repeat False Arrest Suit." *The Risks Digest*, 8(71), May 17, 1989.

[4] Ted Bridis. "U.S. Lifts FBI Criminal Database Checks." *Associated Press*, March 25, 2003.

[5] Department of Justice, Federal Bureau of Investigation. "Privacy Act of 1974; Implementation." *Federal Register*, 68(56), March 24, 2003.

[6] "Computer Glitch Is to Blame for Faulty Bills, Qwest Says." *The Deseret News (Salt Lake City, Utah)*, July 24, 2001.

[7] "USDA Changes Livestock Price-Reporting Guidelines." *Amarillo (Texas) Globe-News*, July 24, 2001.

[8] "Software Error Returns Patent Office Mail." *The New York Times*, August 9, 1996.

[9] "Spelling and Grammar Checkers Add Errors." *Wired News*, March 18, 2003.

[10] D. F. Galletta, A. Durcikova, A. Everard, and B. Jones. "Does Spell-Checking Software Need a Warning Label?" *Communications of the ACM*, forthcoming.

[11] Reuters. "Official Trapped in Car After Computer Fails." *NYTimes.com*, May 12, 2003.

[12] "LA County's Main Hospital Has Computer Breakdown, Delays Ensue." *Associated Press*, April 22, 2003.

[13] "Flights at Japanese Airports Delayed." *Associated Press*, March 1, 2003.

[14] Aaron Lucchetti and Gregory Zuckerman. "Software Glitch Halts Trading on CBOT on April Fool's Day." *The Wall Street Journal*, page C19, April 2, 1998.

[15] "Liffe Glitch Halts All Electronic Trading for a Second Time." *The Wall Street Journal*, May 12, 1999.

[16] "Computer Glitches Shut Down Comair Flights." *Associated Press*, December 26, 2004.

[17] Robert Fry. "It's a Steal: Bargain-Hunting or Barefaced Robbery?" *The Times (London)*, April 8, 2003.

[18] "Amazon Pulls British Site after iPaq Fire-Sale." *NYtimes.com*, March 19, 2003.

[19] Victor L. Winter and Sourav Bhattacharya. Preface. In *High Integrity Software*, edited by Victor L. Winter and Sourav Bhattacharya. Kluwer Academic Publishers, Boston, MA, 2001.

[20] E. Marshall. "Fatal Error: How Patriot Overlooked a Scud." *Science*, 255(5050):1347, March 13, 1992.

[21] Jean-Marc Jézéquel and Bertrand Meyer. "Design by Contract: The Lessons of Ariane." *Computer*, pages 129–130, January 1997.

[22] J. L. Lions. "ARIANE 5: Flight 501 Failure, Report by the Inquiry Board." European Space Agency, July 19, 1996. ravel.esrin.esa.it/docs/esa-x-1819eng.pdf.

[23] Ivars Peterson. "Finding Fault: The Formidable Task of Eradicating Software Bugs." *Science News*, 139, February 16, 1991.

[24] Bruce Sterling. *The Hacker Crackdown: Law and Disorder on the Electronic Frontier*. Bantam Books, New York, NY, 1992.

[25] Jeff Foust. "Why Is Mars So Hard?" *The Space Review*, June 2, 2003. www.thespace review.com.

[26] Jet Propulsion Laboratory, California Institute of Technology. "NASA Facts: Mars Exploration Rover," October 2004. marsrover.jpl.nasa.gov.

[27] Richard de Neufville. "The Baggage System at Denver: Prospects and Lessons." *Journal of Air Transport Management*, 1(4):229–236, December 1994.

[28] Nancy Leveson and Clark Turner. "An Investigation of the Therac-25 Accidents." *Computer*, 26(7):18–41, 1993.

[29] William J. Kauffman III and Larry L. Smarr. *Supercomputing and the Transformation of Science*. Scientific American Library, New York, NY, 1993.

[30] Lucio Mayer, Tom Quinn, James Wadsley, and Joachim Stadel. "Forming Giant Planets via Fragmentation of Protoplanetary Disks." *Science*, November 29, 2002.

[31] Donnella H. Meadows, Dennis I. Meadows, Jorgen Randers and William W. Behrens III. *The Limits To Growth*. Universe Books, New York, NY, 1972.

[32] Bjørn Lomborg and Olivier Rubin. "Limits to Growth." *Foreign Policy*, October/November 2002.

[33] G. S. Fishman and P. J. Kiviat. "The Statistics of Discrete Event Simulation." *Simulation*, 10:185–195, 1968.

[34] Ian Sommerville. *Software Engineering*. 6th ed. Addison-Wesley, Harlow, England, 2001.

[35] Scot Petersen. "Taking the Rap for Bad Software." *PC Week*, page 29, February 28, 2000.

[36] United States Court of Appeals for the Third Circuit. *Step-Saver Data Systems, Inc. v. Wyse Technology and The Software Link, Inc.*, 1991. 939 F. 2d 91.

[37] United States Court of Appeals for the Seventh Circuit. *ProCD, Inc., v. Matthew Zeidenberg and Silken Mountain Web Services, Inc., Appeal from the United States District Court for the Western District of Wisconsin*, 1996. 96–1139.

[38] Supreme Court of the State of Washington. *M.A. Mortenson Co. v. Timberline Software Corp., et al. Opinion*, 2000.

[39] "Texans Get Soaked." *IEEE Software*, page 114, September/October 1997.

[40] Elizabeth Hays. "L Trains Do Compute." *New York Daily News*, January 18, 2004.

CHAPTER

Work and Wealth

Work keeps at bay three great evils: boredom, vice, and need.

–VOLTAIRE

8.1 Introduction

IT'S 6:30 P.M. AND A NEW SHIFT IS STARTING AT THE LIVEBRIDGE CALL CENTER. Twenty-year-old college graduate "Kristy Grover" begins phoning people to verify information they have provided on their credit card applications. She will work until 3:00 a.m. Her hours are unusual because "Kristy Grover" is actually Shilpa Thukral, and she is calling the United States from India. Companies like LiveBridge are saving billions of dollars every year by employing hundreds of thousands of Indians to staff call centers and back offices. A technical service company can hire an Indian worker for about $12,000 a year, compared to about $59,000 a year for an American employee [1].

Just a few years ago (in 1996), the Indian telecommunications infrastructure supported a mere 13,000 simultaneous overseas phone conversations. Multinational corporations invested in new, underwater fiber-optic cables, and by 2002 the overseas phone capacity had increased to more than 2.5 million simultaneous calls, making it possible for American companies to export hundreds of thousands of jobs to India [1].

The globalization of the job market is just one of many changes that information technology and automation have brought to the workplace. In this chapter we examine a variety of moral problems brought about by workplace changes. First we consider the following question: Does automation increase unemployment? Some evidence supports an affirmative answer to the question, but other evidence suggests that automation actually creates more jobs than it replaces. There is no doubt that automation has led to enormous increases in productivity. That leads us to our second question: If productivity has increased so much, why is everyone working so hard? We will examine how we have chosen to use our extra productivity.

Some futurists warn that advances in artificial intelligence and robotics will lead to massive unemployment in the not-too-distant future. We consider the morality of attempting to construct highly intelligent machines.

More mundane information technology has already led to significant changes in the way companies organize themselves. It has also led to an increase in telework (also called telecommuting), the use of temporary workers, workplace monitoring, and distributed, multinational teams. We consider how these changes have improved and harmed the lives of individual workers. Globalization is now a fact of life. Some organizations are convinced globalization benefits everyone in the world, the poor as well as the rich. Others are convinced that globalization harms everyone in the world. We'll present the evidence offered by each side to support its position. We'll also focus on the contentious issue of foreign IT workers in the United States.

Many view those without access to information technology as being severely disadvantaged. The term "digital divide" refers to the opportunity gap brought about because some people do not have access to modern information technology, particularly the Internet. We present evidence of the digital divide and study two fundamentally different models of how new technologies are diffused through a society.

Information technology has made it easier for an unequal share of benefits to accumulate in the hands of a few top performers, leading some to call this the "winner-take-all society." We explore the factors creating the winner-take-all phenomenon, the economic problems it causes, and potential remedies. Because there is a high degree of correlation between educational achievement and salaries, some people are worried about the rapid rise of the cost of a public college education. We conclude the chapter by evaluating the proposition that states ought to provide enough funding to ensure that public colleges and universities are accessible to all of their qualified high school graduates.

8.2 Automation and Unemployment

Many science fiction writers have described future worlds where machines do much of the noncreative work. Some writers paint an optimistic view of these worlds. In Isaac Asimov's short stories and novels, technology is seen as a tool for the betterment of mankind. Intelligent robots may be disliked by some people, but they are not a threat. The "Three Laws of Robotics" are etched into their positronic brains, guaranteeing that they will never turn against their creators [2]. Other writers, such as Kurt Vonnegut, Jr., describe dystopias. Vonnegut's *Player Piano* concerns a future America in which nearly all manufacturing jobs have been lost to automation. People hate machines for taking away their feelings of self-worth, yet their fascination with automation makes its triumph appear inevitable [3].

Are we about to enter an era of high unemployment caused by automation? Let's consider both sides of this question.

8.2.1 Automation and Job Destruction

LOST MANUFACTURING JOBS

There are plenty of examples of robots and other machines performing jobs that used to be done by humans. The manufacturing sector, in particular, has been affected by the introduction of automation. More than 43 million jobs disappeared in the United States between 1979 and 1994 alone. As a consequence the percentage of American workers involved in manufacturing has dropped significantly, from 35 percent in 1947 to 12 percent in 2002.

Meanwhile, manufacturing output in America continues to rise. It has doubled since 1970 [4]. In other words, productivity has increased: fewer workers are making more products. For example, in 1977 it took 35 person-hours to manufacture an automobile in the United States. By 1988 the number of person-hours had dropped to 19.1 hours [5].

LOST WHITE-COLLAR JOBS

The effects of automation are felt in the office, too. Email, voice mail, and high-speed copy machines eliminate secretarial and clerical positions. Even jobs requiring advanced degrees are vulnerable. Spreadsheets and other software packages reduce the need for accountants and bookkeepers [6]. Twenty years ago a pharmacist in a small Canadian town would fill about 8,000 prescriptions in a year. Today, Merck-Medco runs a Web-accessible pharmacy that uses robots to dispense 8,000 prescriptions an *hour* [7].

In fact, the economic recovery of 1991–1996 was notable because of the large number of white-collar, middle-management jobs that were eliminated, even as the economy grew. Unlike the recession of the early 1980s, most of the people whose jobs were eliminated in the 1990s had at least some college education. A large number of these jobs were occupied by people making more than $50,000. Only 35 percent of these higher-paid victims of downsizing were able to find jobs that paid as well. Median household income in the United States, adjusted for inflation, dropped 3 percent between 1979 and 1994 [8].

WORKING HARDER, MAKING LESS

While inflation-adjusted household incomes fell slightly between 1979 and 1994, the work week got longer. Harvard economist Juliet Schor reports that between 1970 and 1990 the average American increased the number of hours spent at work per year by 163. That's equal to an *extra month* at work every year [9].

Some believe longer work hours are a consequence of corporate downsizing, which is facilitated by the introduction of automation and information technology (Figure 8.1). When an organization sheds some of its workers, the work that needs to be done

FIGURE 8.1 When jobs are lost to automation or the introduction of information technology, the remaining workers may work harder in order to avoid being part of the next layoff.

is divided among fewer employees. Hence there is a natural tendency for the number of hours worked to increase. In addition, the fact that people have been laid off is a strong incentive for those who remain to work harder so that they won't be part of the next layoff [10].

Advances in information technology have also made it easier for people to bring work home. For example, many companies now provide their employees with laptop computers. At work, employees turn their laptop into a desktop system by plugging in a full-sized keyboard, mouse, and monitor. By bringing their laptop home, they have access to the various project files they need to continue working. Labor advocates Stanley Aronowitz, Dawn Esposito, and William DiFazio have written, "After nearly a century when homework was regarded as a wage-busting tool, computers have made it easier for employers to revive this practice. With pagers, cell phones, and laptop computers, all time becomes work time" [6].

They conclude:

> Late capitalist society is engaged in a long-term historical process of destroying job security . . . More than ever, we worry about work and are working longer hours; we are more than ever driven, nervous, seemingly trapped. At the very same time, and paradoxically, the twenty-first century bodes a time of post-work: of automation and work reorganization replacing people at faster and faster rates [6].

8.2.2 Automation and Job Creation

Other observers hold a quite different view about the effects of automation and information technology on jobs. They have concluded that while new technology may destroy certain jobs, it also creates new jobs. The net result is an increase, not a decrease in the number of available jobs. Martin Carnoy points out that the absolute number of manu-

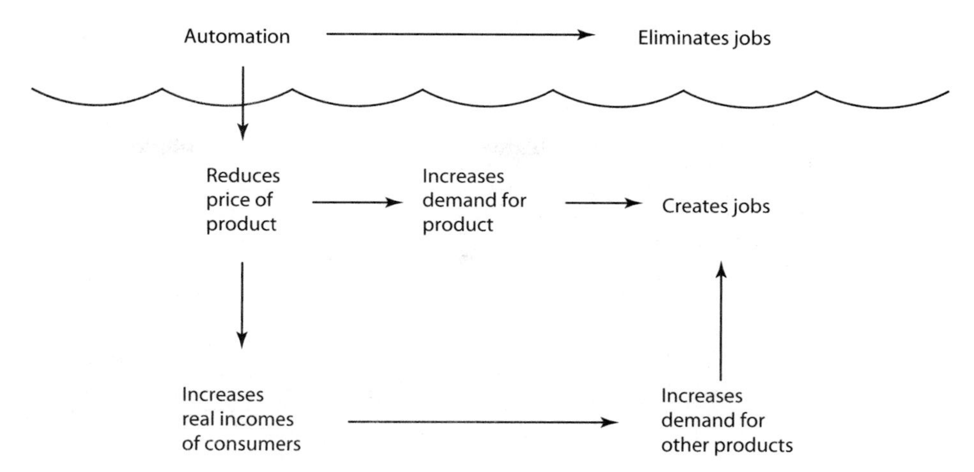

FIGURE 8.2 Superficially, automation eliminates jobs; but automation can also stimulate the creation of new jobs.

facturing jobs in the world is increasing, not decreasing. He writes, "There will be plenty of jobs in the future, and most of them will be high-paying jobs" [11].

INCREASED PURCHASING POWER

The logic of these "automation optimists" is illustrated in Figure 8.2. On the surface, it is obvious that automation eliminates certain jobs. That's what automation means. However, it's also important to look beneath the surface. Automation is introduced as a cost-saving measure: it is less expensive for a machine to perform a particular job than a human being. Because companies compete with each other, lower production costs result in lower prices for the consumer. The drop in the price of a product has two beneficial effects. First, it increases the demand for the product. In order to produce more of the product, workers must be hired. Second, people who were already purchasing the product don't have to pay as much for it. That gives them more money to spend on other things, increasing the demand for other products. This, too, results in job creation. Finally, there is an additional effect, not illustrated in the figure. Some people must be employed designing, creating, and servicing the automated devices themselves.

Two studies commissioned by the International Labor Office have reached the same conclusion as Carnoy. While automation appears to reduce employment when particular factories or companies are examined, studies of entire industries and economies do not reveal job losses [12, 13]. A study published by the U.S. National Academy Press concluded, in part, "Historically, technological change and productivity growth have been associated with expanding rather than contracting total employment and rising earnings. The future will see little change in this pattern" [14]. Larry Hirschhorn has written, "The empirical evidence suggests overall that computers have not replaced workers or destroyed jobs; if anything, they have created jobs" [15].

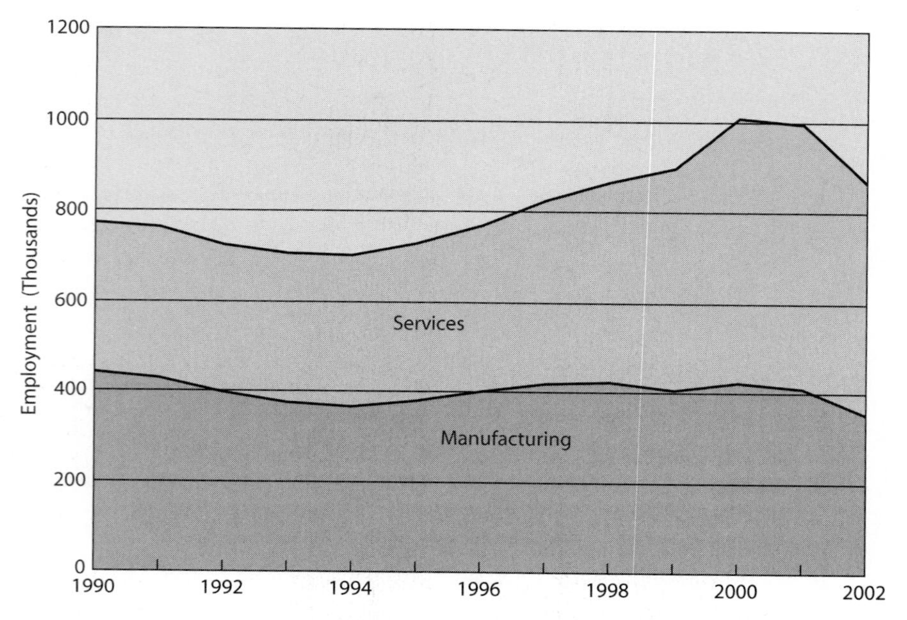

Figure 8.3 In California between 1990 and 2002, the increase in IT service jobs, such as computer systems design, Internet service providers, telecommunications, and software publishers, more than made up for the decrease in computer and electronic products manufacturing.

Figure 8.3 charts IT-related employment in California between 1990 and 2002. The number of people involved in manufacturing computer and electronic products decreased 19 percent, from 447,000 to 361,000. However, the number of people involved in providing information technology services increased during the same time period. Even though employment in IT services dropped sharply between 2001 and 2002, this sector had 54 percent more employees in 2002 than in 1990. In fact, the growth in IT services more than made up for the decline in IT manufacturing. California's total IT employment grew about 12 percent between 1990 and 2002.

WORKING LESS, MAKING MORE

Carnoy disputes the notion that layoffs are causing people who still have jobs to work longer hours now than they used to. "Workers today," he writes, "work much less than those of a century ago, produce more, earn substantially more, and have access to a greater variety of jobs. Technology displaced workers but also contributed to a much higher labor productivity and the production of new products, which helped create new jobs, economic growth, and higher incomes" [11].

8.2.3 Effects of Increase in Productivity

We have noted that productivity in the United States doubled between 1948 and 1990. Julie Schor asks us to consider what our society could have done with this dramatic increase in productivity. We could have maintained our 1948 standard of living and gone to a four-hour work day, or a six-month work year. Or every worker could be taking every other year off with pay. Instead of taking the path of working less, the average work week actually rose slightly. As a result, Americans in 1990 owned and consumed twice as much as in 1948, but had less free time in which to enjoy these things [9].

AMERICANS WORK LONG HOURS

American society is remarkable for how hard its citizens work. The number of hours worked per year in the United States is significantly higher than the number of hours worked in France or Germany. It also appears modern Americans work harder than the ancient Greeks, Romans, or Western Europeans of the Middle Ages. According to Julie Schor, "The lives of ordinary people in the Middle Ages or Ancient Greece and Rome may not have been easy, or even pleasant, but they certainly were leisurely" [9]. In the mid-fourth century the Roman Empire had 175 public festival days. In medieval England holidays added up to about four months a year; in Spain, five months; in France, six months [9].

We do not have to look back into history to find significantly shorter work weeks. Consider contemporary "stone age" societies. The Kapauku of Papua never work two days in a row. Australian aborigines and men of the Sandwich Islands work only about four hours per day. Kung Bushmen work 15 hours a week [9].

PROTESTANT WORK ETHIC

Why are Americans such hard workers? In his famous essay, *The Protestant Ethic and the Spirit of Capitalism*, Max Weber argues that the Protestant Reformation in general, and Calvinism in particular, stimulated the growth of capitalism in Western Europe. Before the Reformation, work was seen in a traditional light. Weber describes the traditional view toward labor in this way:

> A man does not "by nature" wish to earn more and more money, but simply to live as he is accustomed to live and to earn as much as is necessary for that purpose [16].

According to Weber, the Calvinist theology introduced a radically different conception of work. He writes:

> Waste of time is thus the first and in principle the deadliest of sins . . . [T]he religious valuation of restless, continuous, systematic work in a worldly calling, as the highest means to asceticism, and at the same time the surest and most evident proof of rebirth and genuine faith, must have been the most powerful conceivable lever for the expansion of that attitude toward life which we have here called the spirit of capitalism [16].

We can see an example of the "Protestant work ethic" in the early history of New England. The Puritans banished all holidays, insisting that Sunday be the sole day of rest. In 1659 the General Court of Massachusetts decreed that citizens who celebrated Christmas or other holidays by refusing to work or feasting should be fined or whipped.

TIME VERSUS POSSESSIONS

We have exchanged leisure time for material possessions. Compared to medieval Europeans or modern Bushmen, we have vastly superior health care systems, educational institutions, and transportation networks. We live in climate-controlled environments, and we have an incredible number of choices with respect to where we travel, what we wear, what we eat, and how we entertain ourselves. The cost of these freedoms and luxuries is less leisure time.

Despite our high standard of living, our expectations about what we ought to have continue to rise. In 1964, the average new American home had 1,470 square feet and one television set. Only about 20 percent of new homes had air conditioning. In 2001 the size of the average new home had risen to 2,100 square feet, and nearly 100 percent of new homes were equipped with air conditioning. The typical family home has two or three television sets. In order to maintain this lifestyle, people are working harder [10].

8.2.4 Rise of the Robots?

While automation has not yet shortened the work week of the typical American, some experts maintain that most jobs will eventually be taken over by machines. In fact, roboticist Hans Moravec predicts that by 2050 robots will have replaced human workers not just in manufacturing jobs, but in decision-making roles, too [17].

The *Encyclopedia of Computer Science* defines **artificial intelligence (AI)** as "a field of computer science and engineering concerned with the computational understanding of what is commonly called intelligent behavior, and with the creation of artifacts that exhibit such behavior" [18]. The same source defines **robots** as "programmable machines that either in performance or appearance imitate human activities" [18]. According to Moravec, developments in artificial intelligence and robotics were held back for decades by inadequate computer power. Rapid increases in microprocessor speeds over the past decade have resulted in many breakthroughs. Here are a few notable achievements in artificial intelligence and robotics over the past decade:

- A minivan equipped with a video camera and a portable workstation drove from Pittsburgh, Pennsylvania, to San Diego, California, in 1995. The computer was in control of the steering wheel 98.2% of the time [19]. (A human operator controlled the minivan's gas pedal and brakes, maintaining an average speed of about 60 miles per hour.)
- The IBM supercomputer called Deep Blue defeated world chess champion Gary Kasparov in a six-game match in 1997 [20] (Figure 8.4).

Figure 8.4 In 1997, world chess champion Gary Kasparov lost a six-game match to IBM supercomputer Deep Blue by a score of 3½ to 2½. (© Julio Donoso/CORBIS SYGMA)

- In 2000 Japanese automaker Honda created ASIMO, the first humanoid robot (android) capable of climbing and descending stairs. Two years later, engineers gave ASIMO the ability to interpret and respond to human gestures and postures [21].
- Swedish appliance giant Electrolux introduced Trilobite, the world's first domestic robotic floor vacuum cleaner, in 2001 [22].

Moravec believes these innovations are just the beginning of a new era in automation. In thirty years inexpensive desktop computers will be a million times faster than today's models, allowing them to run sophisticated AI programs. "In the [21st] century inexpensive but capable robots will displace human labor so broadly that the average workday will have to plummet to practically zero to keep everyone usefully employed" [17]. Moravec predicts humans will retire to a world of "luxurious lassitide" [17].

Perhaps Moravec has an grossly inflated view of what robots may be able to do in forty years, but what if he is right? The changes he is predicting would profoundly affect our society. For this reason, Richard Epstein suggests there is an urgent need to discuss ethical issues related to the creation of intelligent robots, before they become a reality [23]. Here are some of the questions Epstein has raised:

- Is it wrong to create machines capable of making human labor obsolete?
- Will humans become demoralized by the presence of vastly more intelligent robots? If so, is it wrong to work on the development of such robots?
- Is it morally acceptable to work on the development of an intelligent machine if it cannot be guaranteed the machine's actions will be benevolent?
- How will we ensure that intelligent robots will not be put to an evil purpose by a malevolent human?
- How will our notions of intellectual property change if computers become capable of creative work?
- How will our ideas about privacy have to change if legions of superfast computers are analyzing the electronic records of our lives?

Michael LaChat notes that "many look upon the outbreak of AI research with an uneasy amusement, an amusement masking . . . a considerable disquiet. Perhaps it is the fear that we might succeed, perhaps it is the fear that we might create a Frankenstein, or perhaps it is the fear that we might become eclipsed, in a strange oedipal drama, by our own creation" [24].

LaChat evaluates the issue in the following way. Some people would like to try to construct a **personal AI**—a machine that is conscious of its own existence. No one has proven it is impossible to create a personal AI. Therefore, assume it is theoretically possible to create one. Is it morally acceptable to attempt the construction of a personal AI?

Here is one line of reasoning: According to the second formulation of the Categorical Imperative, we should always treat other persons as ends in themselves and never treat other persons merely as means to an end. In the attempt to construct a personal AI, scientists would be treating the personal AI they created as a means to the end of increasing scientific knowledge. It is reasonable to assume that a fully conscious personal AI would be unwilling to accept its status as a piece of property. In this case, owning a personal AI would be a form of exploitation.

Are we prepared to grant a personal AI the same rights guaranteed to human persons under the United Nation's Universal Declaration of Human Rights, which (among other things) forbids slavery and servitude, and guarantees everyone freedom of movement? If we plan to treat personal AIs as property, then from a Kantian point of view any effort to bring about a personal AI would be immoral.

LaChat concedes that this line of reasoning rests on the controversial assumption that a conscious machine should be given the same moral status as a human being. The argument assumes that a personal AI would have free will and the ability to make moral choices. Perhaps any system operated by a computer program does not have free will, because it has no choice other than to execute the program's instructions as dictated by the architecture of the CPU. If a personal AI does not have free will, it cannot make moral choices, and from a Kantian point of view it should not be valued as an end in itself. Despite its intelligence, it would not have the same moral status as a human being. Creating a personal AI without free will would be morally acceptable.

We do not know whether scientists and engineers will ever be able to construct a personal AI, and we cannot say whether a personal AI would possess free will. Our predictions are uncertain because we do not understand the source of free will in humans. In fact, some philosophers, psychologists, and neuroscientists deny the existence of free will. LaChat concludes, "Though the first word of ethics is 'do no harm,' we can perhaps look forward to innovation with a thoughtful caution," knowing that we may "eclipse ourselves with our own inventions" [24].

8.3 Workplace Changes

Experts debate whether or not information technology has resulted in a net reduction in available jobs, but there is no dispute that information technology has affected *how* people work. In this section we survey a few of the ways that information technology is fundamentally changing the work experience.

8.3.1 Organizational Changes

Information technology has influenced the way manufacturing and service companies organize themselves. A typical early use of computers was to automate a back-office function, such as payroll. Using computers in this way required a company to make no changes in its organization. Later, companies began using computers inside manufacturing units. Computers enabled companies to customize products and provide better service to their customers. This use of computers delegated more responsibility to the line workers, and it encouraged a decentralization of sales and support functions, reducing a company's bureaucracy. Information technology within corporations reached a third stage with the creation of computer networks linking different parts of the business. For example, integrating cash registers with inventory systems has allowed companies to order replacements automatically.

The overall effect of the introduction of information technology is to flatten organizational structures. When the primary source of information distribution was the hand-typed, carbon-copied memorandum, most information flow followed the lines in organization charts (Figure 8.5a). Today a wide variety of technologies allow any member of an organization to contact any other member with minimal effort and cost (Figure 8.5b). As a result, new opportunities arise. Many companies assemble "tiger teams" of expert workers drawn from various parts of the organization chart. A team will work together for a short period of time to solve an urgent problem, then disband. Flexible

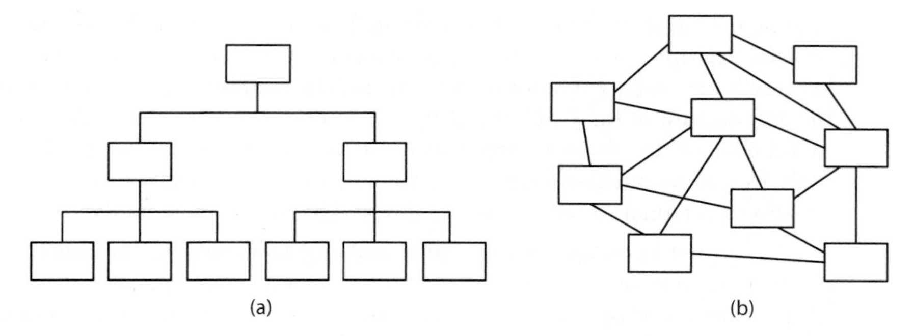

FIGURE 8.5 (a) When interactions are more expensive and time-consuming, most information flows between people and their managers. Organizations are rigid and hierarchical. (b) When interactions become inexpensive and fast, the flow of information is much more flexible. Organizations become flatter and more dynamic.

TABLE 8.1 Greater use of information technology in the workplace will increase demand for employees in certain job categories while reducing demand for employees in other categories.

Higher Demand	*Lower Demand*
Computer engineers	Bank clerks
Computer support specialists	Procurement specialists
Systems analysts	Financial records processing staff
Database administrators	Secretaries, stenographers, and typists
Desktop publishing specialists	Communications equipment operators
	Computer operators

information flow also allows companies to adopt "just-in-time" production and distribution methods, reducing inventory costs [25].

Information technology also streamlines organizations by eliminating transactional middlemen. For example, consider the automation of the supply chain. Suppose company A buys widgets from company B. In the past someone at company A called someone at company B to order the widgets. Today, many companies have adopted **supply-chain automation.** A computer at company A is linked to a computer at company B. The computers are responsible for ordering the widgets, eliminating the need for the middlemen. Automating the paperwork activities associated with purchasing supplies can reduce the number of people who produce purchase orders and invoices, pay bills, process checks, etc. The likely effect of information technology on organizations will be an increased demand in some job categories, while the demand in other categories will drop (Table 8.1) [26].

Dell Computer is a leader in supply-chain automation. Customers order computers directly from Dell using a telephone or connecting with its Web site. Seventy percent of Dell's sales are to large corporations. These companies have custom Web sites that have preconfigured systems tailored to the needs of the purchaser. Dell does not make any computers until they are ordered, allowing it to keep its inventory small—enough for only a few days' production [27].

8.3.2 Telework

Another workplace change brought about through information technology is the rise of telework. **Telework** (also called telecommuting) refers to an arrangement where employees spend a significant portion of their work day at a distance from the employer or a traditional place of work [28]. One kind of teleworking is working from a home office. Another example of teleworking is someone who commutes to a telecenter rather than the company's site. Telecenters provide employees from different firms the ability to connect to their company's computers. A third example of telework are salespersons who have no office, instead transacting all of their business from their car using cell phones and laptop computers.

According to the International Telework Association & Council (ITAC), the number of people who are teleworking is increasing rapidly. The ITAC reports that one in five working Americans engage in some form of telework [29].

ADVANTAGES OF TELEWORK

The rapid growth in the number of teleworkers is evidence there are significant benefits associated with teleworking. Here are some of the most frequently cited advantages of telework [28, 30]:

1. *Telework increases productivity.*

   A variety of studies have shown teleworkers have 10 to 43 percent more productivity than on-site workers.

2. *Telework reduces absenteeism.*

   Teleworkers are less likely to miss work than someone coming into the office.

3. *Telework improves morale.*

   Employees who are teleworking have more freedom. It is easier for them to schedule their work around their personal schedules. If they are working at home, they can dress more casually.

4. *A company can recruit and retain more top employees.*

   For example, a company that allows teleworking can recruit employees who otherwise would not be interested in the job because they are unable or unwilling to be within commuting distance of the main office. Telework allows companies to retain employees (such as mothers of young children) who would quit otherwise.

5. *Telework saves overhead.*

   With some of its workers away from the office, a company doesn't have to invest as much of its resources in office space.

6. *Telework improves the resilience of a company.*

   Because not all the employees are in one place, the company is less likely to be harmed by a natural disaster or a terrorist attack.

7. *Teleworking is good for the environment, because it reduces the amount of pollution generated by commuter traffic.*

8. *Employees may save money by teleworking.*

   They may not have to purchase as much business attire, and they may be able to avoid paying child-care expenses.

DISADVANTAGES OF TELEWORK

Telework has its detractors, too. Here are some of the reasons most frequently given why companies discourage or prohibit telework:

1. *Telework threatens the authority and control of managers.*

   When employees work at a distance from their managers, they naturally have more autonomy. How can a manager manage an employee who is not around?

2. *Telework makes it impossible for an employee to have a face-to-face interaction with customers at the company site.*

   For some jobs these interactions are crucial, meaning the job simply cannot be done from a distance.

3. *Sensitive information is less secure.*

   If a person has valuable physical or electronic files at home or in an automobile, they may be far less secure than if they were kept at the office. There is a greater chance that the information will be lost or compromised through fire or theft.

4. *When people in an organization do not keep the same hours or come into the office every day, it is more difficult to schedule team meetings.*

   Even if employees are only teleworking one or two days a week, many others in the organization can suffer significant inconvenience.

5. *Teleworkers are less visible.*

   There is a danger that teleworkers will be forgotten when it's time for raises or promotions. When somebody is "never around," others can get the idea that that person is not making a contribution to the organization.

6. *When faced with a problem or a need for information, employees at the office are less likely to contact a teleworker than another person on site.*

   Meanwhile, teleworkers are afraid to leave their telephones even for a short time, afraid that if someone from work calls them and they are not around, they will get the reputation for not being "at work."

7. *Teleworkers are isolated.*

   Some jobs require people to bounce ideas off co-workers. What are people working at home supposed to do?

8. *Teleworkers end up working longer hours for the same pay.*

   When everything a person needs to do his job is right there at home, he is more likely to keep coming back to it. How does someone leave her work at the office when her home *is* her office? Critics of telework say that overwork is the reason why teleworkers exhibit higher productivity.

8.3.3 Temporary Work

The modern business environment is highly competitive and rapidly fluctuating. As a result, the level of commitment companies are willing to make to their employees is dropping. Some companies once boasted that they took care of their employees and did not engage in layoffs during business downturns. Those days are gone. The dot-com bust led to massive layoffs in the information technology industry.

Companies are giving themselves more flexibility and saving money on benefits by hiring more subcontractors and temporary employees. Workers cannot count on long-term employment with a single firm. Instead they must rely on their "knowledge portfolios," which they carry from job to job [11].

8.3.4 Monitoring

Information technology has given companies many new tools to monitor the activities of their employees. An American Management Association survey of 1,627 large companies in 2001 revealed that 82 percent of them monitored the activities of their workers in some way. Examples of employee monitoring include tracking Internet usage, monitoring phone calls, monitoring emails, reviewing computer files, and videotaping. Some software products allow managers to observe every keystroke made by employees at their computers and even take snapshots of employees' computer monitors. Other software produces summary statistics, such as the amount of time the employee spent surfing the Web versus the time spent using an accounting program. Monitoring can serve many purposes.

The principal purpose of monitoring is to identify inappropriate use of company resources [31]. A quarter of companies in the United Kingdom have fired employees for improper use of the Internet. In the majority of these cases, the employee was surfing the Web for pornography. Another study of employee emails concluded that eliminating email containing gossip and jokes would cut the time staff spend reading email by 30 percent [32]. A study conducted by IDC concluded that between 30 and 40 percent of Internet use by employees was not work related [33].

Monitoring can help detect illegal activities of employees, as well. By monitoring instant messaging conversations, employers have caught employees who had performed various misdeeds, including an employee who hacked into a company computer after being denied a promotion [34].

Monitoring is also used to ensure that customers are getting the products and services they need. Reviewing customer phone calls to help desks can reveal if the company ought to be providing its customers with better documentation or training [35].

About 10 percent of companies use monitoring to gauge the productivity of workers [31]. Telemarketing firms keep track of how many calls their employees make per hour. Sometimes monitoring can help an organization assess the quality of the work done by its employees. Major League Baseball has introduced QuesTec's Umpire Information System to evaluate how well umpires are calling balls and strikes [36].

Companies are beginning to investigate the use of wireless networks to track the locations of their employees. Knowing the location of service technicians would enable an automated system to respond to a breakdown by alerting the technician closest to the malfunctioning piece of equipment. A system that tracked the locations of hospital physicians could upload a patient's file into the wireless laptop held by a doctor approaching a hospital bed.

More schools are using video cameras to increase security [37]. The school district in Biloxi, Mississippi, used gambling-generated tax receipts to install digital cameras in all 500 of its classrooms. An elementary school principal gushes, "It's like truth serum. When we have a he-said, she-said situation, nine times out of 10, all we have to do is ask children if they want us to go back and look at the camera, and they fess up" [38].

It's an open question whether monitoring is ultimately beneficial to an organization. Obviously, organizations institute monitoring because they have reason to believe it will improve the quantity and/or quality of the work performed by its employees. There is evidence that employee monitoring makes employees more focused on their tasks, but also reduces job satisfaction [39].

8.3.5 Multinational Teams

In the 1980s General Electric and Citibank set up software development teams in India. Since then, many corporations have established field offices in India, including Analog Devices, Cadence Design Systems, Cisco, Intel, Microsoft, and Sun Microsystems. Bangalore, in particular, has made an effort to become the Silicon Valley of India. Companies use Indian companies to write software, process credit card applications, and do billing. Texas Instrument's chip-design team in Bangalore has 200 patents to its name. Hewlett Packard and Oracle both have thousands of employees in India. SAP has 500 engineers in Bangalore.

Multinational teams allow a company to have people at work more hours during the day. It becomes easier to have a call support center open 24 hours a day. It is even possible for projects to be shuttled between multiple sites, allowing around-the-clock progress to be made on time-sensitive products. For example, a team in Palo Alto can spend its day finding bugs in a piece of software, then hand the bug reports over to a team in Bangalore that spends *its* day fixing the bugs [40].

However, the main attraction of India is cost savings. Wages in India are substantially lower than in the United States or Western Europe. The total cost of an Indian computer programmer is about $20,000 a year. Companies say they need to lower their expenses in order to stay in business. If they go out of business, their U.S. employees will lose their jobs. Hence creating multinational teams is a way for companies to stay in business and preserve jobs in the United States [41].

Creating multinational teams has disadvantages, too. The principal disadvantage is that the infrastructure in less developed countries can make business more difficult. For example, because India has only two international airports—one in New Delhi and the other in Mumbai—it is hard to travel to and from Bangalore. The highway system in India is primitive, and electrical power is unreliable.

Despite the difficulties, corporations are increasingly making use of multinational teams. In the year 2000 about 27,000 computer jobs moved overseas from the United States. Forrester Research estimates that the total number of foreign IT employees will grow to half a million by 2005, and more than three million by 2015. Of these jobs, about 70 percent will be in India, 20 percent in the Philippines, and 10 percent in China. *The Economist* magazine notes that while three million IT-related jobs moving overseas sounds like a huge number, it represents only about two percent of total American employment [42]. On the other hand, three million IT-related jobs represents about three times the total IT employment in California.

8.4 Globalization

Globalization refers to the process of creating a worldwide network of businesses and markets. Globalization results in a greater mobility of goods, services, and capital around the world. Investments are made across national boundaries. Products manufactured in one country are sold in another. Consumers calling a telephone help center get connected with support technicians located on the other side of the world.

The rapidly decreasing cost of information technology has made globalization possible (Figure 8.6). The cost of computing dropped by 99.99 percent between 1975 and 1995. The cost of an international telephone call from New York to London dropped by 99 percent between 1930 and 1996 [43]. Companies have made extensive use of low-cost information technology to coordinate operations distributed around the planet.

8.4.1 Arguments for Globalization

Those who favor globalization seek the removal of trade barriers between nations. The North American Free Trade Agreement (NAFTA) between Canada, the United States, and Mexico is a step toward globalization. The World Trade Organization (WTO) is an international body that devises rules for international trade. It promotes the goal of free trade among nations.

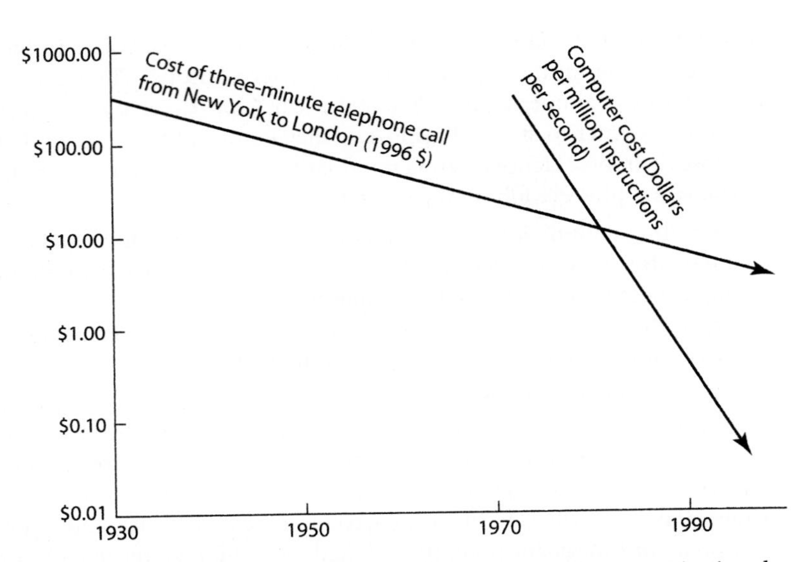

FIGURE 8.6 The dramatic declines in the cost of computing and communications have made global enterprises feasible.

The WTO and other proponents of globalization support free trade with these arguments:

1. *Globalization increases competition among multiple possible providers of the same product.*

   Competition ensures that higher-quality products are sold at the best possible prices. Consumers get better prices when each area produces the goods or services it does best: corn in Kansas, automobiles in Ontario, semiconductors in Singapore, and so on. When prices are lower, the real purchasing power of consumers is higher. Hence globalization increases everyone's standard of living.

2. *People in poorer countries deserve jobs, too.*

   When they gain employment, their prosperity increases.

3. *Every example in the past century of a poor country becoming more prosperous has been the result of that country producing goods for the world market rather than trying for self-sufficiency* [44].

4. *Creating jobs around the world reduces unrest and leads to more stability.*

   Countries with interdependent economies are less likely to go to war with each other.

8.4.2 Arguments against Globalization

Ralph Nader, American trade unions, the European farm lobby, and organizations such as Friends of the Earth, Greenpeace, and Oxfam oppose globalization. They give these reasons why globalization is a bad trend:

1. *The United States and other governments should not be subordinate to the WTO.*

   The WTO makes the rules for globalization, but nobody elected it. It makes its decisions behind closed doors. Every member country, from the United States to the tiniest dictatorship, has one vote in the WTO.

2. *American workers should not be forced to compete with foreign workers who do not receive decent pay and working conditions.*

   The WTO does not require member countries to protect the rights of their workers. It has not banned child labor. Dictatorships such as the People's Republic of China are allowed to participate in the WTO even though they do not let their workers organize into labor unions.

3. *Globalization has accelerated the loss of both manufacturing jobs and white-collar jobs overseas.*

4. *The removal of trade barriers hurts workers in foreign countries, too.*

   For example, NAFTA removed tariffs between Canada, Mexico, and the United States. Because they receive agricultural subsidies from the U.S. government, large American agribusinesses grow corn and wheat for less than its true cost of production and sell the grain in Mexico. Mexican farmers who cannot compete with these prices are driven out of business. Most of them cannot find jobs in Mexico and end up immigrating to the United States [45].

Even if globalization is a good idea, there are reasons why a company may not choose to move its facilities to the place where labor is the least expensive. Interestingly, these arguments are more relevant to "blue collar" jobs such as manufacturing than they are to "white collar" jobs such as computer programming. With automation, the cost of labor becomes a smaller percentage of the total cost of a product. Once the labor cost is reduced to a small enough fraction, it makes little difference whether the factory is located in China or the United States. Meanwhile, there are definite additional costs associated with foreign factories. If you include products in transit, foreign factories carry more inventory than identical factories in the United States. There are also more worries about security when the product is being made in a foreign country. For these reasons, moving a factory to a less-developed country is not always in the best interest of a company [4].

8.4.3 Dot-Com Bust Increases IT Sector Unemployment

In the 1990s Intel's stock rose 3,900 percent, Microsoft's stock increased in value 7,500 percent, and Cisco System's stock soared an incredible 66,000 percent. That means $1,000 of Cisco stock purchased in 1990 was worth $661,000 at the end of 1999. Investors looking for new opportunities for high returns focused on **dot-coms,** Internet-related start-up companies. Speculators pushed up the values of many companies that had never earned a profit. Early in 2000 the total valuation of 370 Internet start-ups was $1.5 trillion, even though they had only $40 billion in sales (that's *sales*, not profits) [46].

In early 2000 the speculative bubble burst, and the prices of dot-com stocks fell rapidly. The resulting "dot-com bust" resulting in 862 high-tech start-ups going out of business between January 2000 and June 2002. Across the United States, the high-tech industry shed half a million jobs [47]. In San Francisco and Silicon Valley, the dot-com bust resulted in the loss of 13 percent of nonagricultural jobs, the worst downturn since the Great Depression [48].

8.4.4 Foreign Workers in the American IT Industry

Even while hundreds of thousands of information technology workers were losing their jobs, U.S. companies hired tens of thousands of foreigners to work in the United States. The U.S. government grants these workers visas allowing them to work in America. The two most common visas are called the H-1B and the L-1.

An H-1B visa allows a foreigner to work in the United States for up to six years. In order for a company to get an H-1B visa for a foreign employee, the company must demonstrate that there are no Americans qualified to do the job. The company must also pay the foreign worker the prevailing wage for the job. Information technology companies have made extensive use of H-1B visas to bring in skilled foreign workers and to hire foreign students graduating from U.S. universities.

In the midst of the high-tech downturn, the U.S. government continued to issue tens of thousands of H-1B visas: 163,600 in 2000–2001 and 79,100 in 2001–2002. Meanwhile, the unemployment rate among American computer science professionals was about 5.1 percent. Many of the 100,000 unemployed computer scientists complained to Congress about the large number of H-1B visas being issued. Congress decided to drop the H-1B quota to 65,000 for the fiscal year beginning October 1, 2003, and it initially set a quota of 65,000 for the following fiscal year. However, the 65,000 H-1B visas approved for 2004–2005 were filled in a single day; representatives of universities and technology companies said the quota was set too low [49]. Bill Gates said, "Anyone who's got the education and the experience, they're not out there unemployed" [50]. Congress responded in May 2005 by allowing an additional 20,000 foreigners to get H-1B visas. Some professional organizations argue against giving out any H-1B visas at all [51].

The other important work visa is called the L-1. American companies use L-1 visas to move workers from overseas facilities to the United States. For example, Intel employees in Bangalore, India, could be transferred to Hillsboro, Oregon, if they held an L-1 visa. Employees brought in to the United States under an L-1 visa do not need to be paid the prevailing wage. That saves employers money. Critics of L-1 visas claim lower-paid foreign workers are replacing higher-paid American workers within the walls of high-tech facilities located in the United States. The number of new L-1 visas issued has increased from 41,739 in 1999 to 57,700 in 2002 [52].

8.4.5 Foreign Competition

The debate over the number of visas to grant foreign workers seeking employment in the United States should not mask another trend: the increasing capabilities of IT companies within developing nations, particularly China and India.

In 2004 IBM agreed to sell its PC division to Chinese computer manufacturer Lenovo for $1.75 billion, making Lenovo the number three manufacturer of PCs in the world [53]. A few months later, Chinese Premier Wen Jiabao visited India to encourage new collaborations between Chinese hardware companies and Indian software companies [54].

India's IT outsourcing industry is growing rapidly; Indian companies now employ more than a million people and have annual sales exceeding $17 billion. About 70 percent of these sales are in software engineering work, such as designing, programming, and maintaining computer programs. The other 30 percent of these sales are in IT-related services, such as call centers, medical transcription, and X-ray interpretation [55].

The number of college students in China is increasing rapidly, from 11 million in 2000 to 16 million in 2005 [56]. Some Chinese universities are becoming recognized for their research expertise. For example, Intel's new Pentium Extreme Edition chip makes use of a compiler developed at China's Tsinghua University [57].

More evidence of global competition comes from the annual Association for Computing Machinery Collegiate Programming Contest. When the contest began 29 years ago, only schools from North America and Europe competed. In 2005, 4109 teams from 71 countries entered the contest. The top 78 teams did well enough in regional competitions to attend the world finals, held in Shanghai, China. The winning team was from Shanghai Jiao Tong University, and the second and third place teams were from Russian universities. The top American team came from the University of Illinois, which tied for 17th place [58, 59].

8.5 The Digital Divide

The **digital divide** refers to the situation where some people have access to modern information technology while others do not. The underlying assumption motivating the term is that people who use telephones, computers, and the Internet have opportunities denied to people without access to these devices. The idea of a digital divide became popular in the mid-1990s with the rapid growth in popularity of the World Wide Web.

According to Pippa Norris, the digital divide has two fundamentally different dimensions. The **global divide** refers to the disparity in Internet access between more industrialized and less industrialized nations. The **social divide** refers to the difference in access between the rich and poor within a particular country [60].

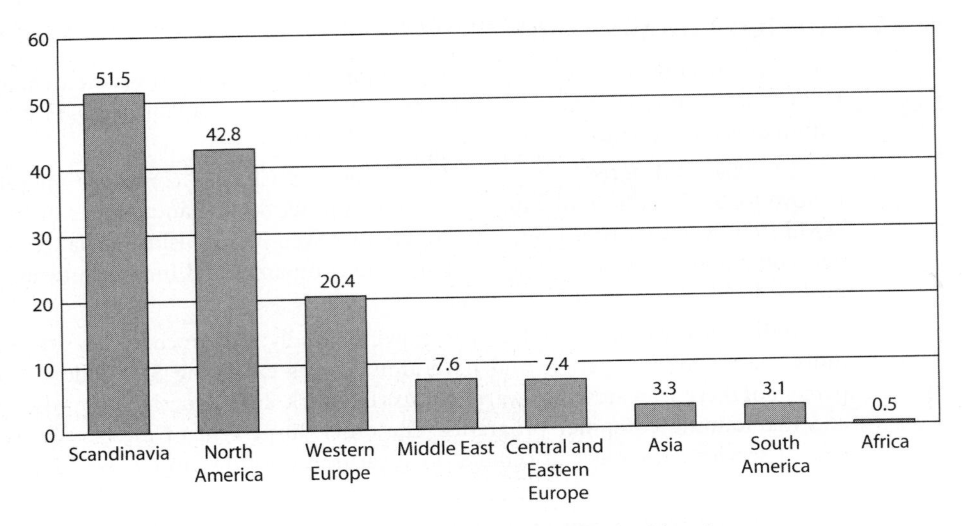

FIGURE 8.7 Percentage of people with Internet access, by world region.

8.5.1 Evidence of the Digital Divide

GLOBAL DIVIDE

There is plenty of evidence of what Norris calls the global divide. One piece of evidence is the percentage of people with Internet access (Figure 8.7). In August 2001 about half a billion people, representing about 8.4 percent of the world's population, had access to the Internet [27]. Access to the Internet in Scandinavia, North America, and Western Europe was significantly above this average, while the rest of the world's regions were below this average. Only 0.5 percent of the population—1 out of every 200 persons—had Internet access in Africa in 2000.

What is hampering Internet development in less technologically developed countries?

1. *Often there is little wealth.*

   In many of these countries there is not enough money to provide everyone in the country with the necessities of life, much less pay for Internet connections.

2. *Many of these countries have an inadequate telecommunications infrastructure.*

   Less than 2 percent of the people in less-developed countries have access to a telephone. Many poor people have no access to newspapers, radio, or television [60].

3. *The primary language is not English.*

   English is the dominant language for business and scientific development.

4. *Literacy is low, and education is inadequate.*

Half of the population in poorer countries has no opportunity to attend secondary schools. There is a strong correlation between literacy and wealth, both for individuals and for societies [27].

5. *The country's culture may not make participating in the Information Age a priority* [61].

SOCIAL DIVIDE

Even within wealthy countries such as the United States, the extent to which people use the Internet varies widely according to age and educational achievement. Pew Surveys polled Americans to find out how many made use of the Internet in the year 2000. Online access varied from 66 percent of 18–29 year olds to 13 percent for those 65 and over. While 74 percent of those with a college degree used the Internet, only 18 percent of those who dropped out of high school went online. The variance in Internet access was not as dramatic across racial groups. For example, Pew Surveys reported that 50 percent of whites, 46 percent of Hispanics, and 35 percent of blacks went online in 2000 [60].

8.5.2 Models of Technological Diffusion

New technologies are usually expensive. Hence the first people to adopt new technologies are those who are better off. As the technology matures, its price drops dramatically, enabling more people to acquire it. Eventually the price of the technology becomes low enough that it becomes available to nearly everyone.

The history of the consumer VCR illustrates this phenomenon. The first VHS VCR, introduced by RCA in 1977, retailed for $1,000. That's $3,255 in 2003 dollars. In late 2003 you could buy a VHS VCR from a mass-marketer for about $50. Between 1976 and 2003 the price of a VCR in constant dollars fell by 98.5 percent. As the price declined, more people could afford to purchase a VCR and sales increased rapidly. The VCR progressed from a luxury that only the rich could afford into a consumer product found in nearly every American household.

Technological diffusion refers to the rate at which a new technology is assimilated into a society. Two different theories predict how a new technology is acquired by people in a society, based on their socioeconomic status (Figure 8.8). We divide society into three groups. People with the highest socioeconomic status are in group A, people with the lowest socioeconomic status are in group C, and group B consists of those people in the middle.

In the **normalization model** (Figure 8.8a), group A begins to adopt the technology first, followed by group B, and finally group C. However, at some point nearly everyone in all three groups is using the new technology.

In the **stratification model** (Figure 8.8b), the order of adoption is the same. However, in this model the eventual number of people in group C who adopt the technology is lower than the number of adoptees in group A. The percentage of people in group B who adopt the technology is somewhere between the levels of the other two groups.

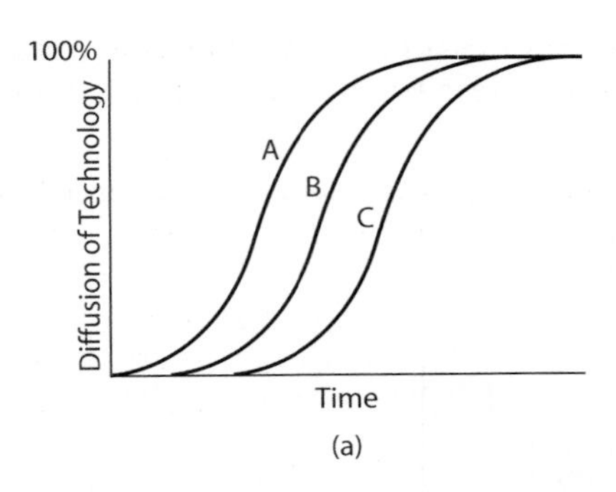

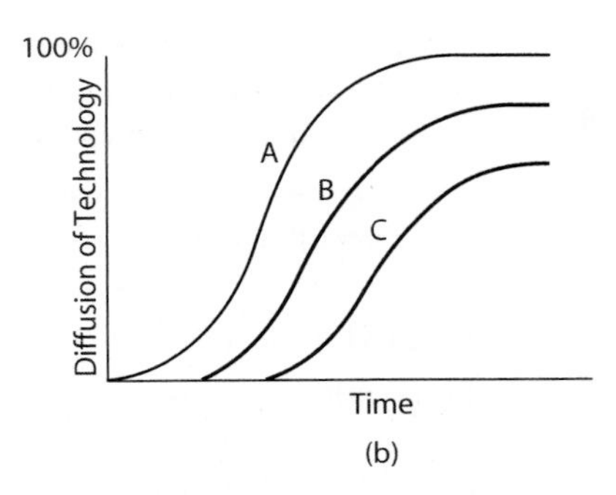

FIGURE 8.8 Two models for technological diffusion. In both models the most advantaged group A is the first to adopt a new technology, while the least advantaged group C is the last to adopt it. (a) In the normalization model, the technology is eventually embraced by nearly everyone in all groups. (b) In the stratification model, the eventual adoption rate of the technology is lower for less-advantaged groups.

Technological optimists believe the global adoption of information technology will follow the normalization model. Information technology will make the world a better place by reducing poverty in developing countries. Creating opportunities elsewhere will reduce the number of people trying to immigrate into the United States.

Technological pessimists believe information technology adoption will follow the stratification model, leading to a permanent condition of "haves" and "have nots." Information technology will only exacerbate existing inequalities between rich and poor nations and between rich and poor people within each nation [60].

Technological pessimists point out that the gap between the richest 20 countries and the poorest 20 countries grew between 1960 and 1995. In 1960 the average gross domestic product (GDP) of the richest countries was 18 times larger than the average GDP of the poorest countries. By 1995 the gap had grown to become 37 times greater. Some of the poorest countries grew even poorer during the last third of the twentieth century [27].

8.5.3 Critiques of the Digital Divide

Mark Warschauer has suggested three reasons why the term "digital divide" is not helpful. First, it tends to promote the idea that the difference between the "haves" and the "have nots" is simply a question of access. Some politicians have jumped to the conclusion that providing technology will close the divide. Warschauer says this approach will not work. To back his claim, he gives as an example the story of a small town in Ireland.

While many factories in Ireland produce IT products, there is not a lot of use of IT among Irish citizens. Ireland's telecommunications company held a contest in 1997 to select and fund an "Information Age Town." The winner was Ennis, a town of 15,000 in western Ireland. The $22 million of prize money represented $1,200 per resident, a large sum for a poor community. Every business was equipped with an Integrated Services Digital Network (ISDN) line, a Web site, and a smart-card reader. Every family received a smart card and a personal computer.

Three years later, there was little evidence of people using the new technology. Devices had been introduced without adequately explaining to the people why they might want to use them. The benefits were not obvious. Sometimes the technology competed with social systems that were working just fine. For example, before the introduction of the new technology, unemployed workers visited the social welfare office three times a week to sign in and get an unemployment payment. These visits served an important social function for the unemployed people. It gave them an opportunity to visit with other people and keep their spirits up. Once the PCs were introduced, the workers were supposed to "sign in" and receive their payments over the Internet. Many of the workers did not like the new system. It appears that many of the PCs were sold on the black market. The unemployed workers simply went back to reporting in person to the social welfare office.

For IT to make a difference, social systems must change as well. The introduction of information technology must take into account local culture, which includes language, literacy, and community values.

Warschauer's second criticism of the term "digital divide" is that it implies everyone is on one side or another of a huge canyon. Everybody is put into one of two categories: "haves" and "have nots." In reality access is a continuum, and each individual occupies a particular place on it. For example, how do you categorize someone who has a 56K modem connecting his PC to the Internet? Certainly that person has online access, but he is not able to retrieve the same wealth of material as someone with a broadband connection.

Thirdly, Warschauer says that the term "digital divide" implies that a lack of access will lead to a less advantaged position in society. Is that the proper causality? Models of technological diffusion show that those with a less advantaged position in society tend to adopt new technologies at a later time, which is an argument that the causality goes the other way. In reality, there is no simple causality. Each factor affects the other [27].

Rob Kling has put it this way:

> [The] big problem with "the digital divide" framing is that it tends to connote "digital solutions," i.e., computers and telecommunications, without engaging the important set of complementary resources and complex interventions to support social inclusion, of which informational technology applications may be enabling elements, but are certainly insufficient when simply added to the status quo mix of resources and relationships" [27].

Finally, Warschauer points out that the Internet does not represent the pinnacle of information technology. In the next few decades dramatic, new technologies will be created. We will see these new technologies being adopted at different speeds, too.

8.6 The "Winner–Take–All Society"

8.6.1 The Winner–Take–All Phenomenon

The Declaration of Independence states that "all men are created equal," but we live in a society in which some people have far more wealth and power than others. What if everyone were guaranteed roughly the same amount of income? The traditional answer to this question is that there would be little motivation for people to exert themselves, either mentally or physically. If everyone were paid the same, there would be no point in getting an education, taking risks, or working hard. Productivity would be low, and the overall standard of living would be poor. For this reason, many people believe a superior alternative is a market economy that rewards innovation, hard work, and risk-taking by compensating people according to the value of the goods they produce.

In their book *The Winner-Take-All Society*, economists Robert Frank and Philip Cook explore the growth of markets where a few top performers receive a disproportionate share of the rewards. Their book is the primary source for this section [62].

Frank and Cook observe that the winner-take-all phenomenon has existed for quite awhile in the realms of sports, entertainment, and the arts. A few "superstar" athletes, actors, and novelists earn millions from their work and garner lucrative endorsements, while those who perform at a slightly lower level make far less. However, the winner-take-all phenomenon has now spread throughout our global economy. Sometimes the qualitative difference between the top product and the second-best product is very slight, yet that can be the difference between success and failure. Hence corporations compete for the top executive talent that can give them the edge over their competition. The compensation of CEOs at America's largest corporations has risen much faster than the wages of production workers (Figure 8.9) [63].

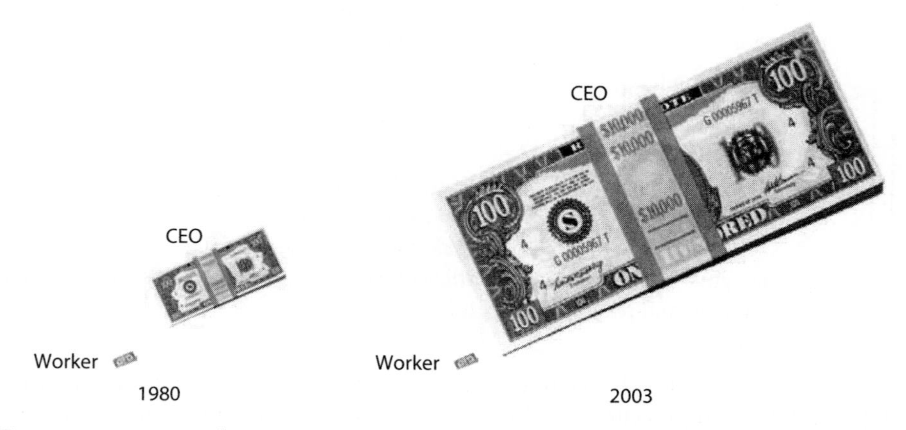

FIGURE 8.9 In 1980 the average pay for a CEO at a large American company was about 40 times the pay of a production worker. By 2003 the ratio had risen to about 400 to 1.

Several factors have led toward winner-take-all phenomena in our economy:

1. *Information technology and efficient transportation systems make it easier for a leading product to dominate the worldwide market.*

   For example, consider a music studio that has a digital recording of the world's best orchestra playing Beethoven's Fifth Symphony. The studio can produce millions of perfect copies of this recording, enough for every classical music lover on the planet. Why would anyone want to listen to the second-best orchestra when a CD of the best orchestra is available for virtually the same price?

2. *Network economies encourage people to flock to the same product.*

   If by chance you should need to use someone else's computer, it is far more likely that person will own a Windows PC than a Macintosh. In this respect knowing how to use a Windows computer has greater utility than knowing how to use a Macintosh. If a person cannot decide which computer to purchase, this factor alone may encourage someone to buy a Windows PC.

3. *English has become the de facto language of international business.*

   English is the native language in 12 countries, including the United States, which is the dominant economic power on the planet. Another 56 countries teach English in their schools. The dominance of English makes it easier for products to find a worldwide market.

4. *Business norms have changed.*

   In the past, large businesses promoted from within and would not recruit executives from other firms. Today, firms vigorously compete with each other for top executive talent.

8.6.2 Harmful Effects of Winner-Take-All

Frank and Cook argue that winner-take-all effects are bad for the economy, for a variety of reasons. First, winner-take-all markets increase the gap between the rich and the poor. Between 1979 and 1989 the inflation-adjusted incomes of the top one percent of U.S. wage-earners doubled, while the median income was flat, and the average income of the bottom 20 percent actually declined.

Winner-take-all effects draw some of the most talented people into socially unproductive work. The problem with winner-take-all contests is that they attract too many contestants. For every comedian who hosts a late-night talk show, tens of thousands of comedians struggle in nightclubs, hoping for their big break. The multimillion-dollar incomes of a relatively few high-profile attorneys help attract many of the brightest college students toward law school. We end up with a glut of lawyers. Meanwhile, there is a shortage of nurses and nuclear engineers.

Winner-take-all markets create wasteful investment and consumption. For example, there is fierce competition among candidates for slots in the top business and law schools. No one wants to go for an interview looking less than his or her best. For this reason, male interviewees are reluctant to show up for an interview wearing a suit that costs less than $600. But if everyone is wearing a $600 suit, no one has an advantage over the others due to his attire. If they had all spent $300 on their suits, there would have been the same relative equity. The behavior of business school applicants is similar to an arms race. The desire to seek an advantage leads to an escalation of consumption, even if the eventual result is simply parity.

A disproportionate share of the best and brightest college students become concentrated in a few elite institutions. "The day has already arrived," write Frank and Cook, "when failure to have an elite undergraduate degree closes certain doors completely, no matter what other stellar credentials a student might possess" [62]. Many Wall Street firms will not even interview candidates who did not graduate from one of a very small number of top law schools. These law schools show a preference for graduates of elite undergraduate programs. Hence high school students interested in reaching the top of the legal profession know their best chance is to do their undergraduate work at an elite school. The result is a tremendous competition for a relatively small number of openings at these colleges, while in truth there are hundreds of top-quality public and private colleges and universities in the United States.

Winner-take-all is not fair, because it gives much greater rewards to the top performers than those whose performance is only slightly inferior. Here is an example from the world of professional sports, where winnings and performance data are objective and publicly available. Paul Azinger and Brad Faxon both play on the PGA Tour. Their skill levels are very close (see Table 8.2). Brad Faxon is certainly not ten times as good a golfer as Paul Azinger, but near the end of the 2003 season, Faxon had won 10 times as much in prize money as Azinger.

Winner-take-all markets harm our culture. Here's why. People are social; they like to read the same books and see the same movies as their friends. It gives them something to talk about. Suppose two books have about the same appeal to a consumer, but one of them is on a best-seller list. The consumer is more likely to select the book on the best-seller list, because it increases the probability she will encounter a friend who has read it. But that means it's really important for a book publisher to get its books on the best-seller list. Publishers know that books written by "name" authors have a greater

TABLE 8.2 Comparison of personal statistics of PGA Tour professionals Paul Azinger and Brad Faxon near the end of the 2003 season.

Metric	*Paul Azinger*	*Brad Faxon*
Driving distance (yards)	284.1	276.5
Driving accuracy (%)	59.7	58.9
Greens in regulation (%)	63.8	62.5
Putting average	1.794	1.761
Scoring average	71.88	70.18
Tournaments entered	23	24
Winnings	$200,004	$2,004,445

chance of making the best-seller list than books written by new authors. This knowledge can lead a publisher to give a big advance to a well-known author to produce a second-rate work, rather than invest the same resources in developing an unknown, but more talented, author. The same effect happens with movie producers. Hoping for the largest possible sales on the first weekend, they bankroll second-rate sequels to big hits rather than original stories filmed by lesser-known directors.

8.6.3 Reducing Winner-Take-All Effects

If winner-take-all markets have harmful consequences on our economy and society, what can be done? Frank and Cook suggest four ways to reduce winner-take-all effects. First, societies can enact laws limiting the number of hours that stores remain open for business. These laws ensure parity among competing businesses and prevent them from engaging in positional arms races. Without these laws, one business may extend its hours in order to gain an advantage over its competitors. Soon all of its competitors follow suit. Parity is restored, but now all the employees must bear the burden of the longer hours. Regulations on business hours are often called "blue laws."

Second, in the absence of laws, businesses can form cooperative agreements to reduce positional arms races. An example is when a group of professional sport team owners agree to establish a cap on team salaries.

Third, more progressive tax structures reduce excess competition for the few handsomely rewarded positions. Back in 1961, the marginal tax rate on income in the highest tax bracket was 91 percent. By 1989 the highest marginal income tax rate had been lowered to 28 percent. Consumption taxes and luxury taxes are other ways of targeting the wealthiest people. Heavily taxing those with the highest incomes makes a higher income less attractive and dissuades some people from competing for the highest-paying jobs. Society benefits when these people engage in more productive work.

Finally, campaign finance reform can reduce the political power of the wealthiest one percent of the population, who control 37 percent of the wealth. Reducing the political power of the very wealthy is another way to reduce the attraction of competing for the highest-paying positions.

8.7 Access to Public Colleges

Many people attend college because they believe a college degree will help them get a better-paying job. These beliefs are well founded; the average income of Americans aged 25–64 rises rapidly with the highest level of education achieved (Figure 8.10).

8.7.1 Effects of Tuition Increases

In the past 20 years tuition at public colleges and universities has risen much faster than inflation. The share of state revenues directed toward higher education has shrunk significantly, from 9.8 percent in 1980 to 6.9 percent in 2000. Meanwhile, the share of colleges' annual budgets derived from tuition rose from 12.9 percent to 18.5 percent [64].

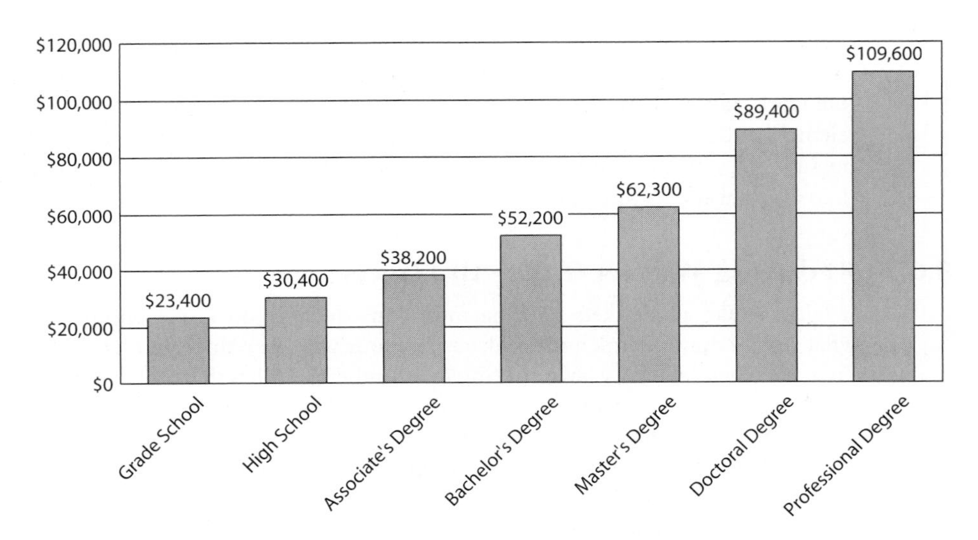

FIGURE 8.10 U.S. Census Bureau statistics reveal that people with more education have higher incomes on average. This chart documents the average income of Americans aged 25–64 according to the highest diploma or degree earned. Statistics were collected in 1997–1999 from people with full-time, year-round employment.

As states found themselves in budget trouble in 2001–2003, they cut university budgets even further. Universities made up for the cuts by raising tuition significantly [65].

Meanwhile, family incomes have risen much more slowly. This means that a public college education is relatively more expensive now than it was 20 years ago. In 1980 tuition at public colleges represented 13 percent of the earnings of a low-income family; by 2000 the cost of tuition had risen to 25 percent of a family's earnings. Not surprisingly, a smaller percentage of low-income families are sending their children on to four-year colleges [66].

America is moving away from the notion that states ought to make public colleges and universities accessible to all of their citizens. A survey revealed that 63 percent of Americans believe the majority of the cost of a public college education ought to be paid by students and their families [67]. Is this right?

8.7.2 Ethical Analysis

As we have done in previous chapters, we will use our toolbox of workable ethical theories to evaluate the problem. We'll state the question this way: "Should a state make a public college education available to all of its qualified high school graduates by funding the difference between the cost of the education and the financial resources of that person and his or her family?"

UTILITARIAN ANALYSIS

First we analyze the question using a utilitarian methodology. The goal of the state is to guarantee access. Hence it only has to make up the difference between the true cost of the education and what students and their family are able to pay. Most families are capable of financing a portion of the cost of a student's college education. Some families have the means to pay the entire cost. If the true cost of a public college education is $15,000 per year, and students earn a degree in four years, the maximum cost to the state per student is $60,000. After taking into account each family's ability to pay, the cost to the state will be less, perhaps amounting to an average of $40,000 per student.

Now let's consider the benefits. The student is the primary beneficiary of the education. The average college graduate makes about $20,000 more per year than the average high school graduate. Over a 35-year career, the difference in incomes between a college graduate and a high school graduate would be roughly $700,000. State and local taxes typically run about 12 percent of total income. That means in the course of a career a college graduate will pay about $84,000 more in taxes than a high school graduate. The extra amount the government receives in taxes from college graduates is much more than the amount it originally invested in their education.

The state receives other benefits from increasing the number of college graduates. College graduates are less likely to be unemployed, reducing the potential of future expenses to the state budget. They are less likely to be imprisoned, reducing the cost to the state of incarcerating inmates. College graduates are more likely to be involved in civic activities, and they are more likely to fulfill the duties of citizens, such as voting.

However, other factors weigh against a state providing its qualified citizens with a public college education. Establishing such a policy would undoubtedly increase the number of people pursuing an advanced education. As more and more people received college degrees, the value of the degree would decline and the difference in wages between high school and college graduates would shrink, reducing the future economic benefits to the state. It might even turn the net gain in tax revenue into a net loss.

Even if this program pencils out as a long-term money-maker for the state (and hence a good action), it must compete with other good actions the state can take with its tax dollars available this year. States cannot raise taxes indiscriminately. If taxes get too high, companies or individuals might leave the state, reducing the overall tax receipts. For this reason, it is more realistic to assume that the state has a fixed amount of money to spend on various activities. Besides funding public colleges and universities, the state also provides unemployment checks to people out of work, funds health clinics for uninsured citizens, staffs prisons, and provides money for K–12 education, to name just a few of its programs.

Higher education may simply be a lower priority than other state programs. First, the state cannot force prisoners, kindergarten students, or many beneficiaries of state services to pay for the services they receive. In contrast, public university students can be asked to pay tuition and fees to help pay for their education. Students also have access to loans to help finance their education. The state police cannot take out a loan from a bank to keep its patrol cars in operation. The second reason why higher education may not be a priority is because it is not a matter of life or death, or public safety. A college degree may be a ticket to a better future, but many successful people do not have college degrees. For these reasons a state may determine that the net benefits of maintaining social programs may exceed the net benefits of ensuring all qualified students have access to an education at a public college.

KANTIAN ANALYSIS

Now let's turn to a Kantian analysis. Kant distinguished between perfect duties and imperfect duties. A **perfect duty** is a duty you are obliged to fulfill in each instance. If you are a parent, you have a perfect duty to care for your children. An **imperfect duty** is a duty you ought to fulfill in general, but not in every instance. You have an imperfect duty to develop your talents. If you happen to have musical talents, you ought to find a way to develop them, but you do not have to take up every instrument in the orchestra. According to Kant, the duty to help others is an imperfect duty. To fulfill this duty, you need to help others, but you do not have to help everybody on every possible occasion. You do not have the time, the physical stamina, or the economic resources to save the entire world. Besides, sometimes it is inappropriate to lend a helping hand. You don't want to help people if that would destroy their self-esteem, for instance.

There are many good things people can do with their financial resources. Similarly, there are many good things the people of a state can do with tax revenues. While it is important that a state find ways to help those in need, it is not wrong for a state to decide which needs are most urgent and put its resources there. From a Kantian point of view, providing access to a public college education would be a good thing for a state to do, but it would not be mandatory.

ANALYSIS USING RAWLS'S PRINCIPLES OF JUSTICE

Finally, let's evaluate the proposal from the point of view of social contract theory, specifically focusing on Rawls's principles of justice. According to the second principle of justice, while the distribution of income and wealth need not be equal among all citizens, the inequalities that do exist must be to everyone's advantage, and all qualified people should have equal access to positions of responsibility. We have seen that there is a high degree of correlation between the level of a person's education and the annual salary that a person can expect to earn. Higher-paying jobs tend to be associated with positions of greater responsibility. Hence public higher education promotes inequality in the sense that those who get a higher education will derive socioeconomic benefits not acquired by those who do not go to college. Statistics show that the percentage of high school graduates from poor families going on to college has declined as states have withdrawn subsidies for higher education. Therefore, children from poor families do not have a fair and equal opportunity to achieve positions of responsibility in society. To satisfy Rawl's second principle of justice, society ought to ensure that its less well-off members have access to a public higher education.

Summary

This chapter has explored a variety of ways in which information technology and automation have affected the workplace. It began by asking the question: Does automation increase unemployment? On the surface, the answer to this question seems obvious: Of course automation increases unemployment. That is what automation means: replacing human labor with machine labor. Industrial robots, voice mail systems, and a myriad of other devices have displaced millions of workers in the past fifty years. However, a deeper look reveals how automation can create jobs, too. When products are less expensive, more people want to buy them, increasing the number that must be made. If products are less expensive, consumers have more money left to spend, which increases demand for other products. Finally, some people are involved in creating and maintaining the machines themselves. For these reasons, the rapid introduction of automation has not yet led to widespread unemployment in the countries where automation is used the most. In fact, the total number of manufacturing jobs worldwide continues to increase.

Thanks to automation, productivity has more than doubled since World War II. However, the length of the work week in the most highly industrialized nations has not decreased by half. Instead, productivity has been used to increase the standard of living. This choice is understandable, since our society defines success in terms of wealth and material possessions. However, not all cultures have the same values. People in some "primitive" cultures choose to work much shorter hours. Even in Western Europe, the amount of time spent at work in ancient and medieval societies was less than it is today.

Intelligent robots have been a fixture of science fiction novels for more than sixty years. In the past decade, however, faster microprocessors have enabled AI researchers to create systems capable of amazing feats, such as steering a minivan driven across America. A few ethicists have suggested that we temper our efforts to create ever-more-intelligent computers with some reflection about how highly intelligent computers would affect society.

Information technology has transformed the way businesses organize themselves. Rapid and inexpensive communications allow many more information channels to open up within organizations, which can speed processes and eliminate middlemen. Evidence of more flexible organizational structures include the rise of telework and multinational teams. Improvements in information technology have also given management unprecedented access to the moment-by-moment activities of employees. Workplace monitoring has become the rule, rather than the exception, in large corporations.

As modern information technology has spread around the world, corporations form tightly connected networks and sell their products and services in many markets. This process is called globalization. Advocates of globalization claim it creates jobs for people in poorer countries and increases competition, resulting in lower prices and a higher standard of living for everyone. Critics of globalization say it forces workers in highly developed countries to compete with people willing to work for a fraction of the pay.

The notion that only manufacturing jobs could be lost to overseas competition has been disproved by recent events. While the dot-com bust has put hundreds of thousands of IT professionals out of work in the United States, American companies have shipped hundreds of thousands of jobs to India and other countries where well-educated people will work for a fraction of what an American earns. Unemployed American high-tech workers have criticized companies for hiring large numbers of foreigners to work in the United States under H-1B or L-1 visas. Companies respond that reducing labor costs is a necessity in a competitive marketplace. In order to survive and thrive, companies must keep prices down and profits up.

The "digital divide" is a way of dividing people into two groups: those with access to information technology and those who do not have access. The term is based on the premise that access to information technology is a prerequisite for success in the information age. Some also assume that simply giving people access to the technology will solve the problem. Pippa Norris points out that there are several fundamentally different dimensions to the digital divide. One dimension separates the more industrialized nations from the less industrialized nations. Another dimension separates rich and poor within a particular country. Mark Warschauer says the notion of a digital divide is too simplistic for three reasons. First, people have widely varying access to information technology. Access should be seen as continuum, not a division into "haves" and "have nots." Second, simply giving people information technology devices, such as computers, cell phones, and Internet accounts, does not guarantee they will take full advantage of the opportunities they make available. For IT to make a difference, social systems must be taken into account. The use of information technology "is a social practice, involving access to physical artifacts, content, skills, and social support" [27]. Third, it's too simplistic to say that a lack of access causes someone to have lower socioeconomic status. You could just as easily say that people with lower socioeconomic status adopt new technologies later. In reality, each factor influences the other.

Frank and Cook invented the term "winner-take-all society" to refer to the way that information technology, the spread of English, network effects, and other factors are creating marketplaces where a few top performers gain a disproportionate share of the rewards. They present evidence that winner-take-all effects harm our economy and our culture, and they suggest actions that can be taken to reduce the winner-take-all phenomenon.

Should a state ensure all of its qualified high school graduates have the means to attend college? This is an important issue in the Information Age, because many of the most lucrative jobs are unavailable to people without a college education. While a utilitarian analysis reveals many financial benefits to such a policy, it also reveals why higher education may be a lower priority than other services provided by the state. A Kantian analysis concluded that providing access to colleges and universities was an imperfect duty. By contrast, the analysis based on Rawls's principles of justice resulted in the conclusion that society has a moral obligation to ensure all its qualified high-school graduates have an equal opportunity to attend college.

Review Questions

1. Explain these terms in your own words:
   a. digital divide, global divide, and social divide
   b. dot-com
   c. globalization
   d. imperfect duty
   e. perfect duty
   f. Protestant work ethic
   g. supply-chain automation
   h. telework
   i. winner-take-all society
2. What are some benefits brought about by automation? What are some harms brought about by automation?
3. What evidence has been given to show that automation eliminates jobs? What evidence has been given to show that automation creates more jobs than it destroys?
4. If automation has doubled productivity since World War II, why hasn't the work week gotten shorter?
5. How can information technology lead to changes in the structure of an organization?
6. How can telework improve the environment?
7. Why do teleworkers fret about being less visible?
8. Proponents of globalization claim that it helps workers in developing countries. Opponents of globalization claim the opposite. Summarize the arguments pro and con.
9. How does Norris categorize the digital divide?
10. Why does Warschauer say the notion of the digital divide is too simplistic and perhaps harmful?
11. Why has the percentage of low-income U.S. families sending children to four-year public colleges dropped in the past three decades?

Discussion Questions

12. Do you agree with Voltaire that a lack of work results in boredom and vice?
13. Would you accept a salaried position (paying a certain amount each month) if you knew it would require you to work at least 50 hours per week in order to complete the required work?
14. If automation leads to chronic and widespread unemployment, should the government provide long-term unemployed adult citizens with the opportunity to do meaningful work at a wage that will keep them out of poverty? Why or why not?
15. Labor advocates Stanley Aronowitz, Dawn Esposito, and William DiFazio say the United States should institute a guaranteed income that would provide each adult with enough

money for food, housing, clothing, health care, and recreation. What are the merits and demerits of this proposal?

16. Is it wrong to create machines capable of making human labor obsolete?

17. Will humans become demoralized by the presence of vastly more intelligent robots? If so, is it wrong to work on the development of such robots?

18. Is it morally acceptable to work on the development of an intelligent machine if it cannot be guaranteed the machine's actions will be benevolent?

19. How will our notions of intellectual property change if computers become capable of creative work?

20. How will our ideas about privacy have to change if legions of superfast computers are analyzing the electronic records of our lives?

21. Kant says that humans should always be treated as ends in themselves, never merely as means to an end. Are there any circumstances under which an intelligent computer should be given the same consideration?

22. A multinational corporation has an office in Palo Alto, California and an office in Bangalore, India. A 21-year-old American computer science graduate works as a software tester at the Palo Alto office. A 21-year-old Indian computer science graduate has an identical position at the Bangalore office. The American earns $65,000 per year in salary and benefits; the Indian earns $15,000 per year in salary and benefits. Is this arrangement moral? Should the company give equal pay and benefits for equal work?

23. Would the music industry be healthier if winner-take-all effects were reduced? If so, which of the proposed solutions in Section 8.6.3 would make the most sense for the music industry?

24. Should the federal government discourage companies from taking advantage of their salaried employees by requiring firms to pay overtime to *any* employee who worked more than 40 hours in one week?

In-class Exercises

25. A multinational corporation transfers a foreign employee to the United States on an L-1 visa. The foreign employee is a computer programmer, working alongside an American computer programmer doing the same work. Both programmers joined the company five years ago, after graduating from college. Their training, skills, and experience are virtually identical.

    Divide the class into two groups, pro and con, to debate the following proposition: "The salaries and benefits of the two computer programmers should be roughly equivalent."

26. You lead a group of five software engineers involved in the testing of a new product. Your manager tells you that because of a company-wide layoff, you will need to give notice to one member of your team. From your interactions with the team members, you can easily identify the two members who are least productive, but you are not sure which of them you should lay off. You know that the company keeps track of all Internet traffic

to each person's computer, although you have never shared this information with your team. You could use this information to determine how much time, if any, these two employees are spending surfing the Web. Is it wrong to access these records?

27. A company runs a large technical support office. At any time, about 50 technical support specialists are on duty, answering phone calls from customers. The company is considering paying the technical support specialists based on two criteria: the average number of phone calls they answer per hour and the results of occasional customer satisfaction surveys. Debate the pros and cons of the proposed method of determining wages.

28. In this role-playing exercise students weigh the pros and cons of working for companies with different philosophies about work.

Company A is a large, established hardware and software company. Employees have a reasonable level of job security, although there have been layoffs in the past few years. Salaries are highly competitive. The company offers stock options, but the stock price is not rising rapidly, and employees know they are not going to get rich from selling their options. The typical programmer works about 45 hours a week.

Company B is a medium-sized, mature software company that plays a dominant role in a specialized market. The company has never had to lay off employees. Salaries are a little low by industry standards, but programmers get paid overtime when they work more than 40 hours a week. The company discourages managers from resorting to overtime work on projects. Many employees are involved in community activities, such as coaching their kids' sports teams.

Company C is a small start-up company trying to be the first to bring a new kind of shopping experience to the Web. Salaries are not high, but all of the employees have a lot of stock options. If the product is successful, everyone expects to become a multimillionaire when the company goes public in a couple of years. In return for the stock options, the founders expect a total commitment from all the employees until the product is released. Every programmer in the company is working 10 hours a day, 7 days a week.

Divide the class into four groups: three groups of recruiters and one group of students about to graduate from college. Each group of recruiters, representing one of the three companies, should make a "pitch" that highlights the reasons why their company represents the best opportunity. The graduates should raise possible negative aspects of working for each company.

29. Debate the following proposition: "It is immoral for a corporation to pay its chief executive officer (CEO) four-hundred times as much as a production worker."

30. State colleges and universities serve the interests of the people of their respective states. Debate the following proposition: "All financial aid given out by a state college or university should be based on financial need. In other words, there should be no merit-based scholarships."

31. It costs more money for a college to offer engineering and computer science courses than courses in the liberal arts and business. The average starting salary of a student graduating with an engineering or computer science degree is much higher than the average salary of a classmate graduating with a degree in the liberal arts.

Debate the following proposition: "Engineering and computer science students should pay higher tuition than students majoring in the liberal arts."

Further Reading

The Downsizing of America. Times Books (Random House), New York, NY, 1996.

Stanley Aronowitz and Jonathan Cutler, editors. *Post-Work: The Wages of Cybernation*. Routledge, New York, NY, 1998.

Martin Carnoy. *Sustaining the New Economy: Work, Family, and Community in the Information Age*. Russell Sage Foundation (Harvard University Press), New York, NY (Cambridge, MA), 2000.

Philip K. Dick. *Do Androids Dream of Electric Sheep?* Del Ray, 1996.

Robert H. Frank and Philip J. Cook. *The Winner-Take-All Society*. The Free Press, New York, NY, 1995.

Pippa Norris. *Digital Divide: Civic Engagement, Information Poverty, and the Internet Worldwide*. Cambridge University Press, Cambridge, England, 2001.

Juliet B. Schor. *The Overworked American: The Unexpected Decline of Leisure*. BasicBooks, 1991.

Kurt Vonnegut, Jr. *Player Piano*. Delacourte Press, New York, NY, 1952. (A paperback edition published by Dell in 1999 is still in print.)

Mark Warschauer. *Technology and Social Inclusion: Rethinking the Digital Divide*. The MIT Press, Cambridge, MA, 2003.

References

[1] Jeffrey Kosseff. "U.S. Calls, India Answers." *The Sunday Oregonian (Portland, Oregon)*, October 5, 2003.

[2] Isaad Asimov. "Runaround." *Amazing Science Fiction*, March 1942.

[3] Kurt Vonnegut, Jr. *Player Piano*. Delacourte Press, New York, NY, 1952.

[4] "The Misery of Manufacturing." *The Economist*, pages 61–62, September 27, 2003.

[5] R. Reich. *The Work of Nations: Preparing Ourselves for 21st Century Capitalism*. Knopf, New York, NY, 1991.

[6] Stanley Aronowitz, Dawn Esposito, William DiFazio, and Margaret Yard. "The Post-Work Manifesto." In *Post-Work: The Wages of Cybernation*, edited by Stanley Aronowitz and Jonathan Cutler, pages 31–80. Routledge, New York, NY, 1998.

[7] Michael Gurstein. "Perspectives on Urban and Rural Community Informatics: Theory and Performance, Community Informatics and Strategies for Flexible Networking." In *Closing the Digital Divide: Transforming Regional Economies and Communities with Information Technology*, edited by Stewart Marshall, Wallace Taylor, and Xinghuo Yu, pages 1–11. Praeger, Westport, CT, 2003.

[8] Louis Uchitelle and N. R. Kleinfield. "The Price of Jobs Lost." In *The Downsizing of America*. Times Books / Random House, New York, NY, 1996.

[9] Julie B. Schor. *The Overworked American: The Unexpected Decline of Leisure*. Basic Books, New York, NY, 1991.

[10] Melissa Will. "Hyper Business or Just . . . Hyperbusy." *Women in Business*, 53(3), May/June 2001.

[11] Martin Carnoy. *Sustaining the New Economy: Work, Family, and the Community in the Information Age*. Russel Sage Foundation / Harvard University Press, New York, NY / Cambridge, MA, 2000.

[12] Raphael Kaplinsky. *Microelectronics and Work: A Review*. International Labour Office, Geneva, Switzerland, 1987.

[13] John Bessant. *Microelectronics and Change at Work*. International Labour Office, Geneva, Switzerland, 1989.

[14] Richard M. Cyert and David C. Mowery, editors. *Technology and Employment: Innovatin and Growth in the U.S. Economy*. National Academy Press, 1987. Panel of Technology and Employment, Committee on Science, Engineering, and Public Policy, National Academy of Sciences.

[15] Larry Hirschhorn. "Computers and Jobs: Services and the New Mode of Production." In *The Impact of Technological Change on Employment and Economic Growth*, pages 377–415. Ballinger Publishing Company, Cambridge, MA, 1988.

[16] Max Weber. *The Protestant Ethic and the Spirit of Capitalism*. Charles Scribner's Sons, New York, NY, 1958. Translated by Talcott Parsons, with a foreword by R. H. Tawney.

[17] Hans Moravec. *Robot: Mere Machine to Transcendent Mind*. Oxford University Press, Oxford, England, 1999.

[18] Anthony Ralton, Edwin D. Reilly, and David Hemmendinger, editors. *Encyclopedia of Computer Science*. Groves Dictionaries, New York, NY, fourth edition, 2000.

[19] Steven Ashley. "Driving Between the Lines." *Mechanical Engineering*, 117(11), November 1995.

[20] Monty Newborn. *Deep Blue: An Artificial Intelligence Milestone*. Springer, 2002.

[21] "Humanoids on the March." *The Economist*, March 12, 2005.

[22] Electrolux. "The Trilobite 2.0," August 3, 2005. www.electrolux.com/node613.asp.

[23] Richard G. Epstein. Review article. "Ethics and Information Technology," 1:227–236, 1999.

[24] Michael R. LaChat. "Artificial Intelligence and Ethics: An Exercise in the Moral Imagination." *The AI Magazine*, pages 70–79, Summer 1986.

[25] M. Castells. "The Informational Economy and the New International Division of Labor." In *The New Global Economy in the Information Age: Reflections on Our Changing World*, edited by M. Carnoy, M. Castells, S. S. Cohen, and F. H. Cardoso, pages 15–43. Pennsylvania State University Press, University Park, PA, 1993.

[26] Charles Babcock, Doug Brown, and Louis Trager. "Do You Live in the Internet's Rust Belt?" *Interactive Week*, September 4, 2000.

[27] Mark Warschauer. *Technology and Social Inclusion: Rethinking the Digital Divide*. The MIT Press, Cambridge, MA, 2003.

[28] Mike Gray, Noel Hodson, and Gil Gordon. *Teleworking Explained*. John Wiley & Sons, Chichester, England, 1993.

[29] International Telework Association & Council. "Number of Teleworkers Increases by 17 Percent," October 3, 2000. www.workingfromanywhere.org.

[30] Joel Kugelmass. *Telecommuting: A Manager's Guide to Flexible Work Arrangements*. Lexington Books, New York, NY, 1995.

[31] "Employers Take a Closer Look." *informationweek.com*, pages 40–41, July 15, 2002.

[32] Rachel Fielding. "Management Week: Web Misuse Rife in UK Firms." *VNU NET*, July 15, 2002.

[33] "Stopping Workplace Internet Abuse—First Step Is Identifying Scope of the Problem." *PR Newswire*, October 7, 2002.

[34] Carl Weinschenk. "Prying Eyes." *Information Security*, August 2002.

[35] Melissa Solomon. "Watching Workers; the Dos and Don'ts of Monitoring Employee Productivity." *Computerworld*, July 8, 2002.

[36] Murray Chass. "Umpires Renew Objections to Computer System." *NYTimes.com*, March 4, 2003.

[37] Katie Hafner. "Where the Hall Monitor Is a Webcam." *NYTimes.com*, February 27, 2003.

[38] Sam Dillon. "Classroom Cameras Catch Every Move." *The Sunday Oregonian (Portland, Oregon)*, September 28, 2003.

[39] Andrew Urbaczewski and Leonard M. Jessup. "Does Electronic Monitoring of Employee Internet Usage Work?" *Communications of the ACM*, 45(1):80–83, January 2002.

[40] Robert X. Cringely. "Holy Cow! What Are All These Programmers Doing in India?" *i, cringely*, July 10, 1997. www.pbs.org/cringely.

[41] Cindy Easton. "Offshore Software Development: Is It Helping or Hurting Our Economy?" *The Cursor (Software Association of Oregon)*, February 2003.

[42] "The New Geography of the IT Industry." *The Economist*, pages 47–49, July 19, 2003.

[43] "One World?" *The Economist*, October 16, 1997.

[44] Paul Krugman. "Enemies of the WTO; Bogus Arguments against the World Trade Organization." *Slate*, November 24, 1999. slate.msn.com.

[45] Kristi Disney. *Globalization: The Migration of Work and the Workers*. Oxfam America, Fall 2001. www.oxfamamerica.org/publications/art917.html.

[46] Anthony Perkins. "Investors: Brace Yourselves for the Next Bubble Bath." *Red Herring*, pages 21–22, November 13, 2000.

[47] Reuters. "Technology Sector Lost 560,000 Jobs in Two Years." *NYTimes.com*, March 19, 2003.

[48] Joseph Menn. "Data Reveals Severity of Tech's Pain." *The Los Angeles Times*, March 7, 2003.

[49] Patrick Thibodeau. "Feds to Research 20,000 H-1B Visas Next Week." *Computerworld*, May 4, 2005.

[50] Eric Chabrow. "Opposing Views: The Debate over the H-1B Visa Program." *Information Week*, May 9, 2005.

[51] Patrick Thibodeau. "H-1B Visa Count Down, Anger Up." *Computerworld*, February 3, 2003.

[52] Art Jahnke. "Should We Put a Cap on the Number of L-1 Visas?" *CIO*, June 5, 2003.

[53] "Chinese Firm Buys IBM PC Business." *BBC News*, December 8, 2004.

[54] S. Srinivasan. "Chinese PM Seeks Indian Tech Cooperation." *Associated Press*, April 10, 2005.

[55] Stella M. Hopkins. "Offshoring to India Taking Off." *The Charlotte (NC) Observer*, June 28, 2005.

[56] Sam Dillon. "U.S. Slips in Attracting World's Best Students." *NYTimes.com*, December 21, 2004.

[57] The Deal. "Setting the Stage of China's Tech Future." *CNet News.com*, May 30, 2005.

[58] Birgitta Forsberg. "American Universities Fall Way Behind in Programming." *San Francisco Chronicle*, April 9, 2005.

[59] Jen Lin-Liu. "Battle of the Coders." *IEEE Spectrum Online*, May 23, 2005.

[60] Pippa Norris. *Digital Divide: Civic Engagement, Information Poverty, and the Internet Worldwide*. Cambridge University Press, Cambridge, England, 2001.

[61] Elena Murelli. *Breaking the Digital Divide: Implications for Developing Countries*. SFI Publishing, 2002. Edited and with a foreword by Rogers W'o Okot-Uma.

[62] Robert H. Frank and Philip J. Cook. *The Winner-Take-All Society*. The Free Press, New York, NY, 1995.

[63] "Where's the Stick?" *The Economist*, page 13, October 11, 2003.

[64] Jeffrey Selingo. "The Disappearing State in Public Higher Education." *The Chronicle of Higher Education*, February 28, 2003.

[65] Michael Arnone. "Students Face Another Year of Big Tuition Increases in Many States." *The Chronicle of Higher Education*, August 15, 2003.

[66] Sara Hebel. "Report Urges Disciplined Spending by States to Make College More Affordable." *The Chronicle of Higher Education*, May 10, 2002.

[67] Jeffrey Selingo. "What Americans Think about Higher Education." *The Chronicle of Higher Education*, May 2, 2003.

AN INTERVIEW WITH

Jerry Berman

Jerry Berman is the founder and President of the Center for Democracy and Technology (CDT). CDT is a Washington, D.C.–based Internet public policy organization founded in December of 1994. CDT plays a leading role in free speech, privacy, Internet Governance, and architecture issues affecting democracy and civil liberties on the global Internet. Mr. Berman has written widely on Internet and civil liberties issues, and often appears in print and television media. He testifies regularly before the Congress on Internet policy and civil liberties issues.

Prior to founding the Center for Democracy and Technology, Mr. Berman was Director of the Electronic Frontier Foundation. From 1978–1988, Mr. Berman was Chief Legislative Counsel at the ACLU and founder and director of ACLU Projects on Privacy and Information Technology. Mr. Berman received his BA, MA, and LLB from the University of California, Berkeley.

How did you get involved in Internet law?

When I worked on civil liberties and privacy at the ACLU in the early 1980s, the prevailing view was computer databases and the rise of the computer state posed a major threat to privacy. This is true. But at the same time there was the beginning of the use of the computer as a communications device, and the start of data networks for communications purposes—the beginnings of the Internet. While recognizing the threat to privacy, I saw that the Internet had the potential to facilitate and broaden First Amendment speech.

In many ways, my colleagues and I have been involved in trying to frame the law and to define privacy, free speech, and how the Internet is governed. We're trying to sort out the "Constitution" for this new social space. By analogy, the Internet business community wants to make sure there's a "commerce clause" to encourage robust commercial transactions over the Internet. We agree with that, but also see the need for a "Bill of Rights" to protect speech, privacy, and other democratic values. We have had some successes, but the work is very much in progress.

The Internet is a more powerful communication medium than newspapers or television, because it allows everyone with an Internet connection to express their views. How can the Internet be anything but democratic?

Like any other technology, the Internet can be regulated. Other countries are exercising considerable control over what ISPs can connect to and what can reside on a server. Even a well-intentioned Congress attempting to protect intellectual property to reduce theft of music and movies could mandate technological changes to computers that make it difficult to use the computer in an open, interconnected way. So one way the Internet can be less than democratic is from bad laws and bad policy.

Another threat is from bad actors provoking bad law. Hackers, people stealing music, using spyware, and engaging in online fraud can provoke policy responses that may have the unintended consequence of undermining the openness of the Internet. We're seeing this now in legitimate efforts to combat spam, spyware, and piracy. We need appropriate laws which combat these harms without harming the openness of the Net. Finding the right solutions is what CDT is about.

One of the great challenges is that given the freedom to connect and communicate that everyone has on the Internet, there is a corollary concept of responsibility. Unless there are shared ethics that respect property, privacy, pluralism, diversity, and the rule of law, the Internet will never realize its potential.

Responding to public pressure, the U.S. Congress passed the Communications Decency Act to restrict access of children to sexually explicit materials on the Web. Why did you organize a legal challenge to the CDA?

In enacting the CDA over our objections, Congress attempted to treat the Internet the same way as other broadcast mass media (TV, radio). The first filed challenge to the CDA, *ACLU v Reno*, was designed to persuade the courts that if you restrict speech for children, you also necessarily restrict adults' free-speech rights, because the definition of indecency covers Constitutionally protected speech for adults. If ISPs had to block all indecent content for children, that content would not reach adults who are entitled to it, because adults and children are all on the same Internet network.

We filed a second challenge to the CDA, and eventually the ACLU suit and the CDT suit were joined and argued together. CDT brought together a broad coalition of Internet technology companies, news organizations, and librarians to educate the courts that the Internet was architecturally different from broadcast media. Traditional media is a one-to-many [communication] and the Internet is a many-to-many communication, much like print. It was also critical to explain that the Internet is a global medium: it isn't effective to censor speech in the US if it's also available on the Internet from outside the US. It is impossible for ISPs to prevent content flowing from sources they do not control, and any ISP censorship would violate Constitutional rights. The architecture of the Internet leads to different analysis and different policy solutions to both protect free speech and protect children from inappropriate content. The lawyers for our coalition argued the case in the Supreme Court on behalf of all the plaintiffs and made the case for user control and user empowerment. The only effective way to deal with unwanted content is for parents and other users (rather than the government) to voluntarily employ available filtering tools and parental controls offered by ISPs and other vendors.

In issues of Constitutionally protected speech, the courts seek to determine if Congress has chosen the least restrictive means for achieving their public purpose. We were able to show that blocking content at the provider end is neither effective nor the least restrictive means for protecting children from inappropriate content. Voluntary filtering is a less restrictive means because it allows users to decide what comes into their homes and, given the global nature of the Internet, gives them the most effective means to do that.

Should an ordinary American citizen's Web site enjoy the same Constitutional protections as The New York Times?

On the Internet everyone can be publishers. And if they're holding themselves out as publishers, they have the same credentials as *The New York Times*, since no one's handing out credentials on the Internet. The Supreme Court heard the Communications Decency Act (CDA) challenge and ruled that the Internet communicator enjoys the maximum protection afforded under the First Amendment. Like the print media, the Internet is not subject to equal time, to the fairness doctrine, or various spectrum allocations. The whole technology of the Internet and the ability of anyone to be a publisher suggests the Internet publisher should, if anything, enjoy greater protection than *The New York Times*. For example, if a newspaper libels someone with false charges, it may require a lawsuit to restore a reputation. On the Internet, anyone can answer back in the blogosphere and reputations are often quickly restored. Thus Courts may narrow the scope of libel suits when the Internet is concerned in favor of more robust debate and "give and take" on the Internet.

Why should a person who has committed no crime be concerned about electronic information gathering and data mining by government agencies?

These databases contain vast amounts of information on all of us, including very personal information—our medical histories, financial transactions, what we purchase, and what we read. Under our concept of privacy, people who have done nothing wrong should have every expectation that the government is not viewing, collecting, or analyzing information about them. So asking "why should I worry if I have nothing to hide" is the wrong formulation. The question should be, "since I have done nothing wrong, why should the government be investigating me?"

The government can look at records that pertain to a suspected terrorist. Yet with data mining, the government may have no articulable suspicion pointing at anyone, but simply mines personal data from airlines, banks, and commercial entities to look at patterns of behavior that might indicate someone may be a terrorist, is associated with a terrorist, knows a terrorist, or is engaged in a behavior that may fit a pattern that the government thinks applies to terrorists. These types of data mining and data analysis can result in significant false positives—innocent people get caught up in investigations—and this can have consequences. First, just being investigated can be an intrusion into privacy. Second, consequences flow from fitting a pattern—you may be denied the right to get on a plane, or be passed over for employment because you lived in an apartment building at the same time as a tenant with the same name as a terrorist's.

Privacy advocates argue that the government needs to have an articulable reason to collect or analyze personal information: the government should need a court order from a judge and should show why they believe a data mining project is likely to result in identifying suspected or potential terrorists. We do need to realize the government has almost carte blanche to conduct these investigations because they have significant authority under current law to engage in data mining exercises. There are very few privacy protections under the Constitution or statute pertaining to these vast databases of personal information. We need stronger privacy laws to deal with data mining.

CHAPTER

9 Professional Ethics

We have come through a strange cycle in programming, starting with the creation of programming itself as a human activity. Executives with the tiniest smattering of knowledge assume that anyone can write a program, and only now are programmers beginning to win their battle for recognition as true professionals.

—GERALD WEINBERG, *The Psychology of Computer Programming*, 1971

9.1 Introduction

INFORMALLY, A **PROFESSION** IS A VOCATION that requires a high level of education and practical experience in the field. Medicine and law are two well-known professions. We pay doctors and lawyers well, trusting that they will correctly ascertain and treat our medical and legal problems, respectively. While knowledge and skill are certainly necessary, they are not sufficient. In return for the trust they are given, professionals have a special obligation to ensure their actions are for the good of those who depend on them. Moral choices made by professionals can have a much wider impact than the choices made by those holding less responsible positions in society.

In this chapter we focus on moral problems faced by software engineers. We begin by considering the extent to which software engineering is a profession along the lines of medicine or law. Next, we present and analyze the Software Engineering Code of Ethics and Professional Practice, jointly developed by the two largest computing societies. Our analysis leads us into a discussion of virtue ethics, an ethical theory based on the idea that good character is the source of correct moral decisions. Three case studies give us the opportunity to use the Software Engineering Code of Ethics and Professional Practice as a tool for ethical analysis.

Finally, we discuss whistleblowing: a situation in which a member of an organization breaks ranks to reveal actual or potential harm to the public. Whistleblowing raises important moral questions about loyalty, trust, and responsibility. Two accounts of whistleblowing illuminate these moral questions and give us the opportunity to ponder the personal sacrifices some have made for the greater good of society. We consider the important role management plays in creating an organizational atmosphere that either allows or suppresses internal dissent.

9.2 Is Software Engineering a Profession?

The term **software engineer** refers to someone engaged in the development or maintenance of software, or someone who teaches in this area. Is software engineering a profession like medicine or law? In this section we examine characteristics of well-developed professions. We will see that in most respects software engineering falls short of the mark.

9.2.1 Characteristics of a Profession

A fully developed profession has a well-organized infrastructure for certifying new members and supporting those who already belong to the profession. Ford and Gibbs have identified eight components of a mature professional infrastructure [1]:

- *Initial professional education*—formal course work completed by candidates before they begin practicing the profession
- *Accreditation*—assures that the formal course work meets the standards of the profession
- *Skills development*—activities that provide candidates with the opportunity to gain practical skills needed to practice the profession
- *Certification*—process by which candidates are evaluated to determine their readiness to enter the profession
- *Licensing*—the process giving candidates the legal right to practice the profession
- *Professional development*—formal course work completed by professionals in order to maintain and develop their knowledge and skills
- *Code of ethics*—mechanism by which a profession ensures that its members will use their knowledge and skills for the benefit of society
- *Professional society*—organization promoting the welfare of the profession, typically consisting of most if not all of the members of the profession

Figure 9.1 illustrates how these components work together to support the profession. A person desiring to join the profession undertakes some initial professional ed-

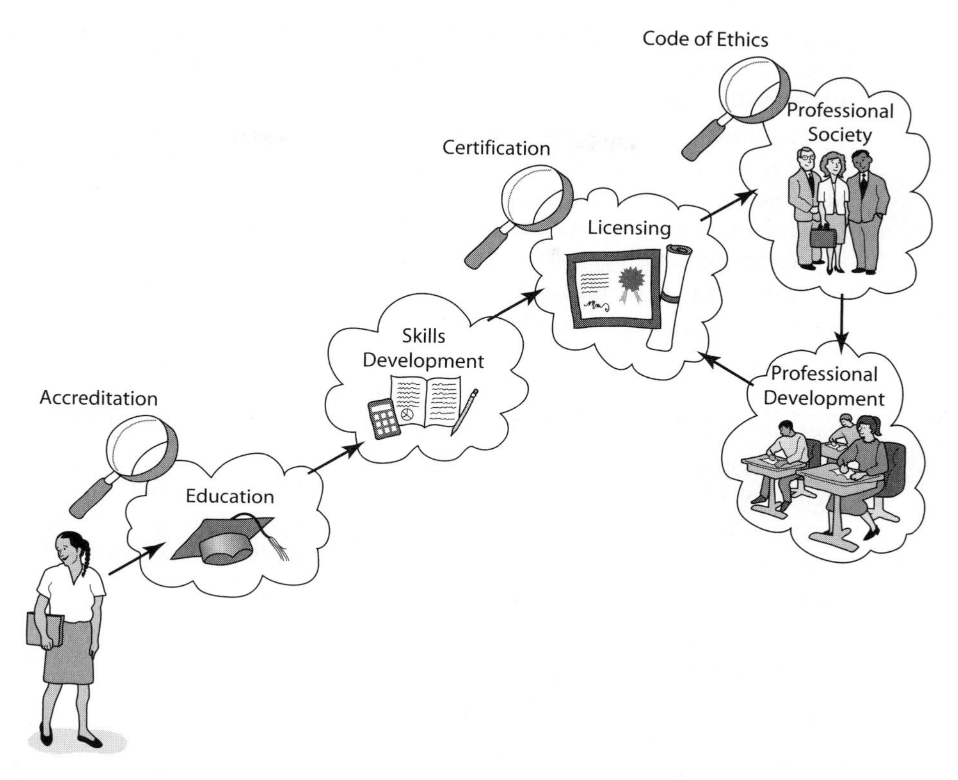

Figure 9.1 A mature profession has eight attributes that enable it to certify new members and support existing members [1].

ucation. A process of accreditation assures that the educational process is sound. After completing their formal education, candidates gain skills through practical experience working in the field. Another check determines if the candidate is ready to be certified. Successful candidates are licensed to practice the profession.

When the public can trust the competence and integrity of the members of a profession, every one of its members benefits. For this reason professionals have a stake in ensuring that fellow members of the profession are capable and act appropriately. For mature professions, professional societies establish codes of ethics and require their members to keep their knowledge current through continuing education and training. Professionals who do not follow the code of ethics or fail to keep up with changes in the field can lose their licenses.

9.2.2 Certified Public Accountants

To illustrate these steps, let's consider how a person becomes a Certified Public Accountant (CPA). We choose accounting because it is a fully developed profession that does not require graduate study for membership. In this respect it is more similar to software engineering than the medical or legal professions, which require their members to earn advanced degrees.

The first step for someone wishing to become a CPA is to graduate with 150 semester credit hours and at least a bachelor's degree from an accredited college or university. Many people pursuing a CPA choose to major in accounting, although it is not strictly necessary. However, the candidate must have completed at least 24 semester credit hours in accounting, auditing, business law, finance, and tax subjects.

After graduation, the candidate gets practical training in the profession by finding employment as an accountant working under the supervision of a CPA.

Finally, candidates must sit for the CPA exam, which has four sections. Candidates who do not pass at least two parts must re-take the entire exam. Candidates who pass at least two parts of the exam must pass the remaining parts within five years.

Completion of the necessary formal education, plus satisfactory scores on every section of the CPA exam, plus two years' work experience enable an accountant to become a Certified Public Accountant. In order to retain certification, CPAs must fulfill continuing education requirements and abide by the profession's code of ethics.

9.2.3 Software Engineers

IS SOFTWARE ENGINEERING A PROFESSION?

It is easy to find a crucial difference between software engineering and mature professions. At the heart of every mature profession is certification and licensing. Certification and licensing allow a profession to determine who will be allowed to practice the profession. For example, a person may not practice law in a state without passing the bar exam and being granted a license. In contrast, people may develop computer programs, either as consultants, sole proprietors, or members of larger firms, without being certified or having been granted a license.

Without certification and licensing, the rest of the characteristics of a mature profession become irrelevant. A person does not have to complete college or serve an apprenticeship under the guidance of an experienced mentor in order to gain employment as a programmer. The vast majority of software engineers do not belong to either of computing's professional societies. It is up to particular employers to monitor the behavior of their employees and guide their continuing education—no professional organization has the authority to forbid someone from practicing software engineering.

In another important respect software engineers differ from most professionals, such as dentists and ministers. In most cases, professionals work directly with individual clients. A dentist treats one patient at a time. An accountant audits one business at a time. Most software engineers work inside a company as part of a team that includes many other software engineers as well as managers. In this environment the responsibility of an individual software engineer is more difficult to discern. Low-level technical decisions are made by groups, and final authority rests with management.

STATUS OF CERTIFICATION AND LICENSING

The two largest organizations supporting the computing field are the IEEE Computer Society (IEEE-CS), with about 100,000 members, and the Association for Computing Machinery (ACM), with about 75,000 members. Like organizations supporting mature professions, the IEEE-CS and the ACM strive to advance the discipline and support their members through publications, conferences, local chapters, student chapters, technical committees, and the development of standards.

In 1993 the IEEE-CS and ACM set up a joint steering committee to explore the establishment of software engineering as a profession. The joint steering committee created several task forces to address particular issues. One task force conducted a survey of practitioners with the goal of understanding the knowledge and skills required by software engineers. Another task force developed accreditation criteria for undergraduate programs in software engineering. A third task force developed a code of ethics for software engineers.

In May 1999 the ACM Council passed a resolution that stated, in part, "ACM is opposed to the licensing of software engineers at this time because ACM believes that it is premature and would not be effective in addressing the problems of software quality and reliability" [2].

ABILITY TO HARM PUBLIC

In one key respect—the ability to harm members of the public—software engineering is like other professions. The Therac-25 killed or gravely injured at least six people, in part because of defective software. While most software engineers do not write code for safety-critical systems such as linear accelerators, society does depend on the quality of their work. People make important business decisions based on the results they get from their spreadsheet programs. Millions rely upon commercial software to help them produce their income tax returns. Errors in programs can result in such harms as lost time, incorrect businesses decisions, and fines.

The ability to cause harm to members of the public is a powerful reason why software engineers should follow a code of ethics, even if they are not professionals in the same sense as physicians, lawyers, and CPAs. Despite the ACM's decision not to move forward with professional licensing, both the ACM and IEEE-CS have endorsed the Software Engineering Code of Ethics and Professional Practice, presented in the next section.

9.3 Software Engineering Code of Ethics

The Software Engineering Code of Ethics and Professional Practice is a practical framework for moral decision making related to problems that software engineers may encounter.

9.3.1 Preamble

Computers have a central and growing role in commerce, industry, government, medicine, education, entertainment and society at large. Software engineers are those who contribute by direct participation or by teaching, to the analysis, specification, design, development, certification, maintenance and testing of software systems. Because of their roles in developing software systems, software engineers have significant opportunities to do good or cause harm, to enable others to do good or cause harm, or to influence others to do good or cause harm. To ensure, as much as possible, that their efforts will be used for good, software engineers must commit themselves to making software engineering a beneficial and respected profession. In accordance with that commitment, software engineers shall adhere to the following Code of Ethics and Professional Practice.

The Code contains eight Principles related to the behavior of and decisions made by professional software engineers, including practitioners, educators, managers, supervisors and policymakers, as well as trainees and students of the profession. The Principles identify the ethically responsible relationships in which individuals, groups, and organizations participate and the primary obligations within these relationships. The Clauses of each Principle are illustrations of some of the obligations included in these relationships. These obligations are founded in the software engineer's humanity, in special care owed to people affected by the work of software engineers, and the unique elements of the practice of software engineering. The Code prescribes these as obligations of anyone claiming to be or aspiring to be a software engineer.

It is not intended that the individual parts of the Code be used in isolation to justify errors of omission or commission. The list of Principles and Clauses is not exhaustive. The Clauses should not be read as separating the acceptable from the unacceptable in professional conduct in all practical situations. The Code is not a simple ethical algorithm that generates ethical decisions. In some situations standards may be in tension with each other or with standards from other sources. These situations require the software engineer to use ethical judgment to act in a manner which is most consistent with the spirit of the Code of Ethics and Professional Practice, given the circumstances.

Ethical tensions can best be addressed by thoughtful consideration of fundamental principles, rather than blind reliance on detailed regulations. These Principles should influence software engineers to consider broadly who is affected by their work; to examine if they and their colleagues are treating other human beings with due respect; to consider how the public, if reasonably well informed, would view their decisions; to analyze how the least empowered will be affected by their decisions; and to consider whether their acts would be judged worthy of the ideal professional working as a software engineer. In all these judgments concern for the health, safety and welfare of the public is primary; that is, the "Public Interest" is central to this Code.

The dynamic and demanding context of software engineering requires a code that is adaptable and relevant to new situations as they occur. However, even in this generality, the Code provides support for software engineers and managers of software engineers who need to take positive action in a specific case by documenting the ethical stance of the profession. The Code provides an ethical foundation to which individuals within teams and the team as a whole can appeal. The Code helps to define those actions that are ethically improper to request of a software engineer or teams of software engineers.

The Code is not simply for adjudicating the nature of questionable acts; it also has an important educational function. As this Code expresses the consensus of the profession on ethical issues, it is a means to educate both the public and aspiring professionals about the ethical obligations of all software engineers.

9.3.2 Principles

PRINCIPLE 1: PUBLIC

Software engineers shall act consistently with the public interest. In particular, software engineers shall, as appropriate:

1.01 Accept full responsibility for their own work.

1.02 Moderate the interests of the software engineer, the employer, the client and the users with the public good.

1.03 Approve software only if they have a well-founded belief that it is safe, meets specifications, passes appropriate tests, and does not diminish quality of life, diminish

Figure 9.2 Software engineers shall approve software only if they have a well-founded belief that it is safe, meets specifications, passes appropriate tests, and does not diminish quality of life, diminish privacy, or harm the environment. The ultimate effect of the work should be to the public good (Clause 1.03).

privacy or harm the environment. The ultimate effect of the work should be to the public good.

1.04 Disclose to appropriate persons or authorities any actual or potential danger to the user, the public, or the environment, that they reasonably believe to be associated with software or related documents.

1.05 Cooperate in efforts to address matters of grave public concern caused by software, its installation, maintenance, support or documentation.

1.06 Be fair and avoid deception in all statements, particularly public ones, concerning software or related documents, methods and tools.

1.07 Consider issues of physical disabilities, allocation of resources, economic disadvantage and other factors that can diminish access to the benefits of software.

1.08 Be encouraged to volunteer professional skills to good causes and contribute to public education concerning the discipline.

PRINCIPLE 2: CLIENT AND EMPLOYER

Software engineers shall act in a manner that is in the best interests of their client and employer, consistent with the public interest. In particular, software engineers shall, as appropriate:

2.01 Provide service in their areas of competence, being honest and forthright about any limitations of their experience and education.

2.02 Not knowingly use software that is obtained or retained either illegally or unethically.

FIGURE 9.3 Software engineers shall not knowingly use software that is obtained or retained either illegally or unethically (Clause 2.02).

2.03 Use the property of a client or employer only in ways properly authorized, and with the client's or employer's knowledge and consent.

2.04 Ensure that any document upon which they rely has been approved, when required, by someone authorized to approve it.

2.05 Keep private any confidential information gained in their professional work, where such confidentiality is consistent with the public interest and consistent with the law.

2.06 Identify, document, collect evidence and report to the client or the employer promptly if, in their opinion, a project is likely to fail, to prove too expensive, to violate intellectual property law, or otherwise to be problematic.

2.07 Identify, document, and report significant issues of social concern, of which they are aware, in software or related documents, to the employer or the client.

2.08 Accept no outside work detrimental to the work they perform for their primary employer.

2.09 Promote no interest adverse to their employer or client, unless a higher ethical concern is being compromised; in that case, inform the employer or another appropriate authority of the ethical concern.

PRINCIPLE 3: PRODUCT

Software engineers shall ensure that their products and related modifications meet the highest professional standards possible. In particular, software engineers shall, as appropriate:

3.01 Strive for high quality, acceptable cost and a reasonable schedule, ensuring significant tradeoffs are clear to and accepted by the employer and the client, and are available for consideration by the user and the public.

3.02 Ensure proper and achievable goals and objectives for any project on which they work or propose.

3.03 Identify, define and address ethical, economic, cultural, legal and environmental issues related to work projects.

3.04 Ensure that they are qualified for any project on which they work or propose to work by an appropriate combination of education and training, and experience.

3.05 Ensure an appropriate method is used for any project on which they work or propose to work.

3.06 Work to follow professional standards, when available, that are most appropriate for the task at hand, departing from these only when ethically or technically justified.

3.07 Strive to fully understand the specifications for software on which they work.

3.08 Ensure that specifications for software on which they work have been well documented, satisfy the users' requirements and have the appropriate approvals.

FIGURE 9.4 Software engineers shall ensure proper and achievable goals and objectives for any project on which they work or propose (Clause 3.02).

3.09 Ensure realistic quantitative estimates of cost, scheduling, personnel, quality and outcomes on any project on which they work or propose to work and provide an uncertainty assessment of these estimates.

3.10 Ensure adequate testing, debugging, and review of software and related documents on which they work.

3.11 Ensure adequate documentation, including significant problems discovered and solutions adopted, for any project on which they work.

3.12 Work to develop software and related documents that respect the privacy of those who will be affected by that software.

3.13 Be careful to use only accurate data derived by ethical and lawful means, and use it only in ways properly authorized.

3.14 Maintain the integrity of data, being sensitive to outdated or flawed occurrences.

3.15 Treat all forms of software maintenance with the same professionalism as new development.

PRINCIPLE 4: JUDGMENT

Software engineers shall maintain integrity and independence in their professional judgment. In particular, software engineers shall, as appropriate:

4.01 Temper all technical judgments by the need to support and maintain human values.

4.02 Only endorse documents either prepared under their supervision or within their areas of competence and with which they are in agreement.

4.03 Maintain professional objectivity with respect to any software or related documents they are asked to evaluate.

4.04 Not engage in deceptive financial practices such as bribery, double billing, or other improper financial practices.

4.05 Disclose to all concerned parties those conflicts of interest that cannot reasonably be avoided or escaped.

4.06 Refuse to participate, as members or advisors, in a private, governmental or professional body concerned with software related issues, in which they, their employers or their clients have undisclosed potential conflicts of interest.

PRINCIPLE 5: MANAGEMENT

Software engineering managers and leaders shall subscribe to and promote an ethical approach to the management of software development and maintenance. In particular, those managing or leading software engineers shall, as appropriate:

5.01 Ensure good management for any project on which they work, including effective procedures for promotion of quality and reduction of risk.

5.02 Ensure that software engineers are informed of standards before being held to them.

5.03 Ensure that software engineers know the employer's policies and procedures for protecting passwords, files and information that is confidential to the employer or confidential to others.

5.04 Assign work only after taking into account appropriate contributions of education and experience tempered with a desire to further that education and experience.

5.05 Ensure realistic quantitative estimates of cost, scheduling, personnel, quality and outcomes on any project on which they work or propose to work, and provide an uncertainty assessment of these estimates.

5.06 Attract potential software engineers only by a full and accurate description of the conditions of employment.

5.07 Offer fair and just remuneration.

5.08 Not unjustly prevent someone from taking a position for which that person is suitably qualified.

5.09 Ensure that there is a fair agreement concerning ownership of any software, processes, research, writing, or other intellectual property to which a software engineer has contributed.

5.10 Provide for due process in hearing charges of violation of an employer's policy or of this Code.

5.11 Not ask a software engineer to do anything inconsistent with this Code.

5.12 Not punish anyone for expressing ethical concerns about a project.

FIGURE 9.5 Software engineers shall help develop an organizational environment favorable to acting ethically (Clause 6.01).

PRINCIPLE 6: PROFESSION

Software engineers shall advance the integrity and reputation of the profession consistent with the public interest. In particular, software engineers shall, as appropriate:

6.01 Help develop an organizational environment favorable to acting ethically.

6.02 Promote public knowledge of software engineering.

6.03 Extend software engineering knowledge by appropriate participation in professional organizations, meetings and publications.

6.04 Support, as members of a profession, other software engineers striving to follow this Code.

6.05 Not promote their own interest at the expense of the profession, client or employer.

6.06 Obey all laws governing their work, unless, in exceptional circumstances, such compliance is inconsistent with the public interest.

6.07 Be accurate in stating the characteristics of software on which they work, avoiding not only false claims but also claims that might reasonably be supposed to be speculative, vacuous, deceptive, misleading, or doubtful.

6.08 Take responsibility for detecting, correcting, and reporting errors in software and associated documents on which they work.

6.09 Ensure that clients, employers, and supervisors know of the software engineer's commitment to this Code of ethics, and the subsequent ramifications of such commitment.

6.10 Avoid associations with businesses and organizations which are in conflict with this code.

6.11 Recognize that violations of this Code are inconsistent with being a professional software engineer.

6.12 Express concerns to the people involved when significant violations of this Code are detected unless this is impossible, counter-productive, or dangerous.

6.13 Report significant violations of this Code to appropriate authorities when it is clear that consultation with people involved in these significant violations is impossible, counter-productive or dangerous.

PRINCIPLE 7: COLLEAGUES

Software engineers shall be fair to and supportive of their colleagues. In particular, software engineers shall, as appropriate:

7.01 Encourage colleagues to adhere to this Code.

7.02 Assist colleagues in professional development.

7.03 Credit fully the work of others and refrain from taking undue credit.

7.04 Review the work of others in an objective, candid, and properly documented way.

7.05 Give a fair hearing to the opinions, concerns, or complaints of a colleague.

7.06 Assist colleagues in being fully aware of current standard work practices including policies and procedures for protecting passwords, files and other confidential information, and security measures in general.

7.07 Not unfairly intervene in the career of any colleague; however, concern for the employer, the client or public interest may compel software engineers, in good faith, to question the competence of a colleague.

7.08 In situations outside of their own areas of competence, call upon the opinions of other professionals who have competence in that area.

PRINCIPLE 8: SELF

Software engineers shall participate in lifelong learning regarding the practice of their profession and shall promote an ethical approach to the practice of the profession. In particular, software engineers shall continually endeavor to:

8.01 Further their knowledge of developments in the analysis, specification, design, development, maintenance and testing of software and related documents, together with the management of the development process.

8.02 Improve their ability to create safe, reliable, and useful quality software at reasonable cost and within a reasonable time.

8.03 Improve their ability to produce accurate, informative, and well-written documentation.

8.04 Improve their understanding of the software and related documents on which they work and of the environment in which they will be used.

8.05 Improve their knowledge of relevant standards and the law governing the software and related documents on which they work.

FIGURE 9.6 Software engineers shall continually endeavor to improve their ability to create safe, reliable, and useful quality software at reasonable cost and within a reasonable time (Clause 8.02).

8.06 Improve their knowledge of this Code, its interpretation, and its application to their work.

8.07 Not give unfair treatment to anyone because of any irrelevant prejudices.

8.08 Not influence others to undertake any action that involves a breach of this Code.

8.09 Recognize that personal violations of this Code are inconsistent with being a professional software engineer.

9.4 Analysis of the Code

In this section we analyze the Code and derive an alternate set of underlying principles upon which it rests.

9.4.1 Preamble

The preamble to the Code points out that there is no mechanical process for determining the correct actions to take when faced with a moral problem. Our experience evaluating moral problems related to the introduction and use of information technology confirms this statement. Even two people with similar philosophies may reach different conclusions when confronted with a moral problem. Two Kantians may agree on the basic facts of a moral problem, but disagree on how to characterize the will of the moral agent. Two utilitarians may agree on the benefits and harms resulting from a proposed action, but assign different weights to the outcomes, causing them to reach opposite conclusions.

The preamble also warns against taking an overly legalistic view of the Code. Simply because an action is not expressly forbidden by the Code does not mean it is morally acceptable. Instead, judgment is needed to detect when a moral problem has arisen and to determine the right thing to do in a particular situation.

While the Code is expressed as a collection of rules, these rules are based on principles grounded in different ethical theories. This is not surprising, considering that the Code was drafted by a committee. When we encounter a situation where two rules conflict, the preamble urges us to ask questions that will help us consider the principles underlying the rules. These questions demonstrate the multifaceted grounding of the Code:

1. *Who is affected?*

   Utilitarians focus on determining how an action benefits or harms other people.
2. *Am I treating other human beings with respect?*

   Kant's Categorical Imperative tells us to treat others as ends in themselves, rather than simply as a means to an end.
3. *Would my decision hold up to public scrutiny?*

   A social relativist is concerned about whether an action conforms with the mores of society.
4. *How will those who are least empowered be affected?*

   Rawls's second principle of justice requires us to consider whether inequalities are to the greatest benefit of the least-advantaged members of society.
5. *Are my acts worthy of the ideal professional?*

   The ethics of virtue is based on imitation of morally superior role models. Since we did not discuss virtue ethics in Chapter 2, let's examine it now.

9.4.2 Virtue Ethics

ORIGIN OF VIRTUE ETHICS

In *The Nicomachean Ethics*, Aristotle expresses the opinion that happiness results from living a life of virtue [3]. He distinguishes between *intellectual virtue*, which is developed through education, and *moral virtue*, which comes about through repetition of the appropriate acts (Figure 9.7). You can acquire the virtue of honesty, for example, by habitually telling the truth. According to Aristotle, deriving pleasure from a virtuous act is a sign that you have acquired that virtue.

There is a wealth of virtues, of course. Here is a brief list of two dozen virtues given by James Rachels: benevolence, civility, compassion, conscientiousness, cooperativeness, courage, courteousness, dependability, fairness, friendliness, generosity, honesty, industriousness, justice, loyalty, moderation, patience, prudence, reasonableness, self-discipline, self-reliance, tactfulness, thoughtfulness, and tolerance [4].

A person who possesses many moral virtues has a strong moral character. According to Aristotle, when people with strong character face a moral problem, they know the

Figure 9.7 According to Aristotle, happiness derives from living a life of virtue. You acquire moral virtues by repeating the appropriate acts.

right thing to do, because the action will be consistent with their character. As Justin Oakley and Dean Cocking put it, "An action is right if and only if it is what an agent with a virtuous character would do in the circumstances" [5].

STRENGTHS OF VIRTUE ETHICS

Virtue ethics has two advantages over the ethical theories we considered in Chapter 2. First, it provides a motivation for good behavior. The calculus of utility and the categorical imperative say nothing about motivation. A utilitarian or a Kantian may do the right thing, but the reasoning behind the action is cold and analytical. Virtue ethics, on the other hand, stresses the importance of loyalty, thoughtfulness, courteousness, dependability, and other characteristics of healthy social interactions.

A second advantage of virtue ethics is that it provides a solution to the problem of impartiality. Recall that utilitarianism, Kantianism, and social contract theory require us to be completely impartial and treat all human beings as equals. This assumption leads to moral evaluations that are hard for most people to accept. For example, when a couple is faced with the choice between using $4,000 to take their children to Disneyland for a week or feeding 1,000 starving Africans for a month, the calculus of utility would conclude saving 1,000 lives was the right thing to do. However, most of us expect that good parents will show more kindness to their children than to people living on the other side of the world.

Virtue ethics avoids the pitfall of impartiality by rejecting the notion that every action must be designed to produce the maximum benefit for people overall [5]. Instead, some virtues are partial toward certain people, while others are impartial and treat everyone equally. Love, friendship, and loyalty are examples of virtues that allow a person to be partial toward friends and family members. Honesty, civility, and courteousness are examples of virtues that a person would extend equally to all human beings.

WEAKNESS OF VIRTUE ETHICS

However, virtue ethics has a significant liability. Using virtue ethics alone, it is often difficult to determine what to do in a particular situation. Suppose you are in charge of dispatching crews to fight brush fires in southern California. Three fires erupt at the same time: one near a mountain resort frequented by the wealthy; another near a town of 10,000 people with a high rate of poverty; and a third close to a middle-class suburb. You only have the resources to fight two of the fires. Which fires do you attack? Looking back on the list of virtues, which ones come into play? Compassion? Fairness? Justice? Prudence?

Suppose we decide the most relevant virtue is prudence. What is the prudent thing to do? Perhaps prudence dictates that you allocate fire crews to minimize property damage, but making your decision based on the total value of the property that each fire is threatening is an example of the utilitarian approach to moral problem solving.

Perhaps the most relevant virtue is justice. Suppose only two of the fires are inside the fire-control district that funds the fire-fighting brigade. Using justice as your virtue, you may decide to abide by the fire district policies. Following written policies looks suspiciously like a Kantian approach to decision making.

You may argue that the desire to be prudent came before the decision to take a utilitarian approach, or the will to be just was an antecedent to a Kantian analysis. Even if this were so, a fundamental problem remains. If the desire to exercise different virtues compels you toward different actions, which action should you take? Put another way, what is the methodology for answering the question, "What would a person with strong moral character do in these circumstances?"

VIRTUE ETHICS COMPLEMENTS OTHER THEORIES

Rather than treating virtue ethics as a stand-alone theory, some ethicists believe it makes more sense to see virtue ethics as a complement to one of the other theories, such as utilitarianism. Adding virtue ethics allows ethical decision makers to consider their rationale for taking the action as well as the beneficial or harmful effects of the action.

Remember the problem of moral luck, one of the major criticisms of act utilitarianism? Since an action is judged right or wrong based solely on its consequences, an unlucky, unintended consequence can result in an action being considered wrong. Suppose your mother-in-law is in the hospital and you send her an expensive and beautiful bouquet of flowers. Unfortunately, she gets an allergic reaction to one of the flowers in the bouquet. As a result, she must spend an additional four days in the hospital. From a purely act utilitarian point of view, you did the wrong thing when you sent your mother-in-law the flowers. In a mixed act utilitarian/virtue ethics theory, we would also take into account that you were acting out of thoughtfulness, a virtue. If nothing else, introducing the virtue ethics component makes it easier for us to think about some of the other consequences of the action. Despite the allergic reaction, your mother-in-law appreciated your kind gesture, a benefit. In addition, you strengthened your habit of thoughtfulness by practicing it on your mother-in-law, another benefit.

9.4.3 Alternative List of Fundamental Principles

The start of each section of the Code begins with the statement of a fundamental principle. For example, the first section begins with the fundamental principle, "Software engineers shall act consistently with the public interest." All of these statements of fundamental principles are expressed from the point of view of what software engineers ought to do.

Another way to devise a list of fundamental principles is to consider those virtues we would like to instill among all the members of a profession. We end up with a set of general, discipline-independent rules that cut across the eight categories of the Code. Here is an alternative list of fundamental principles derived using that approach:

1. *Be impartial.*

   The good of the general public is equally important to the good of your organization or company. The good of your profession and your company are equally important to your personal good. It is wrong to promote your agenda at the expense of your firm, and it is wrong to promote the interests of your firm at the expense of society. (Supports Clauses 1.02, 1.03, 1.05, 1.07, 3.03, 3.12, 4.01, and 6.05.)

2. *Disclose information that others ought to know.*

   Do not let others come to harm by concealing information from them. Do not make misleading or deceptive statements. Disclose potential conflicts of interest. (Supports Clauses 1.04, 1.06, 2.06, 2.07, 3.01, 4.05, 4.06, 5.05, 5.06, 6.07, 6.08, 6.09, 6.12, and 6.13.)

3. *Respect the rights of others.*

   Do not infringe on the privacy rights, property rights, or intellectual property rights of others. (Supports Clauses 2.02, 2.03, 2.05, and 3.13.)

4. *Treat others justly.*

   Everyone deserves fair wages and appropriate credit for work performed. Do not discriminate against others for attributes unrelated to the job they must do. Do not penalize others for following the Code. (Supports Clauses 5.06, 5.07, 5.08, 5.09, 5.10, 5.11, 5.12, 7.03, 7.04, 7.05, 7.07, and 8.07.)

5. *Take responsibility for your actions and inactions.*

   As a moral agent, you are responsible for the things you do, both good and bad. You may also be responsible for bad things that you allow to happen through your inaction. (Supports Clauses 1.01, 3.04, 3.05, 3.06, 3.07, 3.08, 3.10, 3.11, 3.14, 3.15, 4.02, and 7.08.)

6. *Take responsibility for the actions of those you supervise.*

   Managers are responsible for setting up work assignments and training opportunities to promote quality and reduce risk. They should create effective communication channels with subordinates so that they can monitor the work being done and be aware of any quality or risk issues that arise. (Supports Clauses 5.01, 5.02, 5.03, and 5.04.)

7. *Maintain your integrity.*

   Deliver on your commitments and be loyal to your employer, while obeying the law. Do not ask someone else to do something you would not be willing to do yourself. (Supports Clauses 2.01, 2.04, 2.08, 2.09, 3.01, 3.02, 3.09, 4.03, 4.04, 6.06, 6.10, 6.11, 8.08, and 8.09.)

8. *Continually improve your abilities.*

   Take advantage of opportunities to improve your software engineering skills and your ability to put the Code to use. (Supports Clauses 8.01, 8.02, 8.03, 8.04, 8.05, and 8.06.)

9. *Share your knowledge, expertise, and values.*

   Volunteer your time and skills to worthy causes. Help bring others to your level of knowledge about software engineering and professional ethics. (Supports Clauses 1.08, 6.01, 6.02, 6.03, 6.04, 7.01, 7.02, and 7.06.)

In the following section we will use these fundamental principles to guide our analysis in three case studies.

9.5 Case Studies

Throughout this text we have evaluated a wide range of moral problems. Our methodology has been to evaluate the moral problem from the point of view of Kantianism, act utilitarianism, rule utilitarianism, and social contract theory.

Another way to evaluate information technology–related moral problems is to make use of the Software Code of Ethics and Professional Practice. We follow a three-step process:

1. Consult the list of fundamental principles and identify those that are relevant to the moral problem.
2. Search the list of clauses accompanying each of the relevant fundamental principles to see which speak most directly to the issue.
3. Determine whether the contemplated action aligns with or contradicts the statements in the clauses. If the action is in agreement with all of the clauses, that provides strong evidence the action is moral. If the action is in disagreement with all of the clauses, it is safe to say the action is immoral.

Usually, the contemplated action will be supported by some clauses and opposed by others. When this happens, we must use our judgement to determine which of the clauses are most important before we can reach a conclusion about the morality of the contemplated action.

In the remainder of this section we will apply this methodology to three case studies.

9.5.1 Software Recommendation

SCENARIO

Sam Shaw calls the Department of Computer Science at East Dakota State University seeking advice on how to improve the security of his business's local area network. A secretary in the department routes Mr. Shaw's call to Professor Jane Smith, an internationally recognized expert in the field. Professor Smith answers several questions posed by Mr. Shaw regarding network security. When Mr. Shaw asks Professor Smith to recommend a software package to identify security problems, Professor Smith tells him that NetCheks got the personal computer magazine's top rating. She does not mention that the same magazine gave a "best buy" rating to another product with fewer features but a much lower price. She also fails to mention that NetCheks is a product of a spin-off company started by one of her former students and that she owns 10 percent of the company.

Analysis

From our list of nine fundamental principles, three are most relevant here:

- Be impartial.
- Disclose information that others ought to know.
- Share your knowledge, expertise, and values.

Searching the list of clauses identified with these fundamental principles, the following ones seem to fit the case study most closely:

- *1.06. Be fair and avoid deception in all statements, particularly public ones, concerning software or related documents, methods and tools.*
  Professor Smith was deceptive when she mentioned the most highly rated software package but not the one rated to be a "best buy."
- *1.08. Be encouraged to volunteer professional skills to good causes and contribute to public education concerning the discipline.*
- *6.02. Promote public knowledge of software engineering.*
  Professor Smith freely provided Sam Shaw with valuable information about network security.
- *4.05. Disclose to all concerned parties those conflicts of interest that cannot reasonably be avoided or escaped.*
- *6.05. Not promote their own interest at the expense of the profession, client or employer.*

Professor Smith did not tell Sam Shaw that she had a personal stake in the success of the NetCheks software. She did not tell him about the "best buy" package that may have provided him every feature he needed at a much lower price.

Mr. Shaw was asking Professor Smith for free advice, and she provided it. When she freely shared her knowledge about network security, she was acting in the spirit of Clauses 1.08 and 6.02, and doing a good thing.

However, Professor Smith appears to have violated the other three clauses to at least some degree. Most importantly, she did not reveal her personal interest in NetCheks, which could lead her to be biased. The fact that the she did not mention the "best buy" package is evidence that she was neither evenhanded nor completely forthcoming when she answered Mr. Shaw's question about software packages.

Perhaps Mr. Shaw should have heeded the admonition, "Free advice is worth what you pay for it." Nevertheless, the ignorance or foolishness of one person does not excuse the bad behavior of another. Professor Smith should have revealed her conflict of interest. At that point, Mr. Shaw could have chosen to get another opinion, if he so desired.

9.5.2 Child Pornography

Scenario

Joe Green, a system administrator for a large corporation, is installing a new software package on the PC used by employee Chuck Dennis. The company has not authorized Joe to read other people's emails, Web logs, or personal files. However, in the course of installing the software he accidentally comes across directories containing files with suspicious-looking names. He opens a few of the files and discovers they contain child pornography. Joe believes possessing such images is against federal law. What should he do?

Analysis

Looking over the list of nine fundamental principles, we find these to be most relevant to our scenario:

- Be impartial
- Respect the rights of others.
- Treat others justly.
- Maintain your integrity.

We examine the lists of clauses associated with these four fundamental principles and identify those which are most relevant:

- *2.03. Use the property of a client or employer only in ways properly authorized, and with the client's or employer's knowledge and consent.*

Somebody has misused the company's PC by using it to store images of child pornography. By this principle Joe has an obligation to report what he discovered.

- *2.09. Promote no interest adverse to their employer or client, unless a higher ethical concern is being compromised; in that case, inform the employer or another appropriate authority of the ethical concern.*
  While revealing the existence of the child pornography may harm the employee, possessing child pornography is illegal. Applying this principle would lead Joe to disclose what he discovered.
- *3.13. Be careful to use only accurate data derived by ethical and lawful means, and use it only in ways properly authorized.*
  Joe discovered the child pornography by violating the company's policy against examining files on personal computers used by employees.
- *5.10. Provide for due process in hearing charges of violation of an employer's policy or of this Code.*
  Simply because Chuck had these files on his computer does not necessarily mean he is guilty. Perhaps someone else broke into Chuck's computer and stored the images there.

Our analysis is more complicated because Joe violated company policy to uncover the child pornography on Chuck's PC. Once he has this knowledge, however, the remaining principles guide Joe to reveal what he has discovered to the relevant authorities within the corporation, even though management may punish Joe for breaking the privacy policy. There is the possibility that Chuck is a victim. Someone else may be trying to frame Chuck or use his computer as a safe stash for their collection of images. Joe should be discreet until a complete investigation is completed and Chuck has had the opportunity to defend himself.

9.5.3 Anti-Worm

SCENARIO

The Internet is plagued by a new worm that infects PCs by exploiting a security hole in a popular operating system. Tim Smart creates an anti-worm that exploits the same security hole to spread from PC to PC. When Tim's anti-worm gets into a PC, it automatically downloads a software patch that plugs the security hole. In other words, it fixes the PC so that it is no longer vulnerable to attacks via that security hole [6].

Tim releases the anti-worm, taking precautions to ensure that it cannot be traced back to him. The anti-worm quickly spreads throughout the Internet, consuming large amounts of network bandwidth and entering millions of computers. To system administrators, it looks just like another worm, and they battle its spread the same way they fight all other worms [7].

Analysis

These fundamental principles are most relevant to the anti-worm scenario:

- Continually improve your abilities.
- Share your knowledge, expertise, and values.
- Respect the rights of others.
- Take responsibility for your actions and inactions.

Examining the list of clauses associated with each of these fundamental principles reveals those that are most relevant to our case study:

- *1.01. Accept full responsibility for their own work.*
  Tim tried to prevent others from discovering that he was the author of the anti-worm. He did not accept responsibility for what he had done.
- *1.08. Be encouraged to volunteer professional skills to good causes and contribute to public education concerning the discipline.*
  The anti-worm did something good by patching security holes in PCs. Tim provided the anti-worm to the Internet community without charge. However, system administrators spent a lot of time trying to halt the spread of the anti-worm, a harmful effect.
- *2.03. Use the property of a client or employer only in ways properly authorized, and with the client's or the employer's knowledge and consent.*
  Tim's "client" is the community of Internet PC owners who happen to use the operating system with the security hole. While his anti-worm was designed to benefit them, it entered their systems without their knowledge or consent. The anti-worm also consumed a great deal of network bandwidth without the consent of the relevant telecommunications companies.
- *8.01. Further their knowledge of developments in the analysis, specification, design, development, maintenance, and testing of software and related documents, together with the management of the development process.*
- *8.02. Improve their ability to create safe, reliable, and useful quality software at reasonable cost and within a reasonable time.*
- *8.06. Improve their knowledge of this Code, its interpretations, and its application to their work.*
  Tim followed the letter of these three clauses when he acquired a copy of the worm, figured out how it worked, and created a reliable anti-worm in a short period of time. The experience improved his knowledge and skills. Perhaps he should invest some time improving his ability to interpret and use the Code of Ethics!

According to some of these principles, Tim did the right thing. According to others, Tim was wrong to release the anti-worm. How do we resolve this dilemma? We can simplify our analysis by deciding that Tim's welfare is less important than the public good. Using this logic, we will no longer consider the fact that Tim improved his technical knowledge and skills by developing and releasing the anti-worm.

That leaves us with three clauses remaining (1.01, 1.08, and 2.03). From the point of view of Clause 1.01, what Tim did was wrong. By attempting to hide his identity, Tim refused to accept responsibility for launching the anti-worm. He has clearly violated the Code of Ethics in this regard.

When we evaluate Tim's action from the point of view of Clause 1.08, we must determine whether his efforts were directed to a "good cause." Certainly Tim's anti-worm benefited the PCs it infected by removing a security vulnerability. However, it harmed the Internet by consuming large amounts of bandwidth, and it harmed system administrators, who spent time battling it. Because there were harmful as well as beneficial consequences, we cannot say that Tim's efforts were directed to a completely good cause.

Finally, let's evaluate Tim's action from the point of view of Clause 2.03. Even though the anti-worm was completely benevolent, Tim violated the property rights of the PC owners, because the anti-worm infected their PCs without authorization. Hence Tim's release of the anti-worm was wrong from the point of view of this Clause.

To summarize our analysis, Tim's release of the anti-worm is clearly wrong from the point of view of Clauses 1.01 and 2.03. It is also hard to argue that he satisfied the spirit of Clause 1.08. We conclude that Tim's action violated the Software Engineering Code of Ethics and Professional Practice.

9.6 Whistleblowing

A **whistleblower** is someone who breaks ranks with an organization in order to make an unauthorized disclosure of information about a harmful situation after attempts to report the concerns through authorized organizational channels have been ignored or rebuffed [8]. Sometimes employees become whistleblowers out of fear that actions taken by their employer may harm the public; other times they have identified fraudulent use of tax dollars [9].

Whistleblowing is alluded to in Clauses 1.02, 1.03, 1.04, 1.05, 2.05, 2.09, 3.01, 6.06, and 6.13 of the Software Engineering Code of Ethics and Professional Practice. These clauses provide a justification for whistleblowing in a variety of circumstances.

As you might expect, whistleblowers are usually punished for disclosing information that organizations have tried to keep under wraps. If they do not lose their job outright, they have probably lost all chances for future advancement within the organization. Whistleblowers and their families typically suffer emotional distress and economic hardship.

Nevertheless, whistleblowers often serve the public good. For this reason the U.S. government has passed two pieces of legislation to encourage whistleblowing: the False Claims Act and the Whistleblower Protection Act of 1989.

The False Claims Act was first enacted by Congress in 1863 in response to massive fraud perpetrated by companies providing supplies to the Union Army during the Civil War. The law allowed a whistleblower to sue, on behalf of the government, a person or company that was submitting falsified claims to the government. If the organization was found guilty and forced to pay a settlement to the government, the whistleblower received half of the settlement.

In 1943 Congress amended the False Claims Act, drastically reducing the share of the settlement a whistleblower would receive and limiting the evidence or information a whistleblower could use in the lawsuit. As a result, the law fell into disuse.

In the mid-1980s the media carried numerous stories about defense contractors perpetrating fraud against the government. Congress responded by amending the False Claims Act once again, making it easier for people to put together a successful lawsuit and allowing whistleblowers to receive between 15 and 30 percent of settlements. The False Claims Act also provides certain protections to whistleblowers against retaliation by their employers.

The Whistleblower Protection Act of 1989 establishes certain safeguards for federal employees and former employees who claim negative personnel actions have been taken against them for whistleblowing. Whistleblowers can appeal to the U.S. Merit Systems Protection Board.

In this section we study two famous whistleblowing cases. The first case relates to events leading to the loss of the space shuttle *Challenger*. The second case focuses on fraudulent activities of the defense contractor Hughes Aircraft. We then survey ethical responses to the act of whistleblowing.

9.6.1 Morton Thiokol/NASA

On January 28, 1986, the space shuttle *Challenger* lifted off from Cape Canaveral. On board were seven astronauts, including schoolteacher Christa McAuliffe, the first civilian to fly into space. Just 73 seconds after lift-off, hot gases leaking from one of the booster rockets led to an explosion that destroyed the *Challenger* and killed everyone on board (Figure 9.8).

Engineer Roger Boisjoly was in charge of inspecting the O-rings on the boosters recovered after launches of the space shuttle. The O-rings were supposed to seal connections between sections of the booster rockets. On two occasions in 1985 he had seen evidence that a primary O-ring seal had failed. Boisjoly presented a report on his findings to NASA officials at the Marshall Space Flight Center. Frustrated that NASA officials were not giving sufficient attention to the problem, he wrote a memo to Vice President for Engineering Robert Lund stating that an O-ring failure could lead to the loss of a shuttle flight and the launch pad. Despite Boisjoly's persistent efforts to get the seals redesigned, the problem was not fixed.

On January 27, 1986, Boisjoly and a group of Morton Thiokol engineers met to discuss the proposed launch for the following day. Florida was in the middle of an unusual cold snap; the weather forecast for northern Florida called for an overnight low

FIGURE 9.8 The explosion of the *Challenger* killed seven astronauts, including the first schoolteacher in space, Christa McAuliffe. (Courtesy of NASA)

of 18 degrees Fahrenheit. The engineers knew that frigid temperatures greatly increased the probability that an O-ring would fail, allowing hot gases to escape from a booster rocket. They prepared a set of 14 slides that documented their concern about a low-temperature launch.

The evening of January 27, Morton Thiokol had a teleconference with the Marshall Space Flight Center and the Kennedy Space Center. Morton Thiokol's presentation ended with the engineers' recommendation that NASA not launch the *Challenger* if the temperature was below 53 degrees. NASA asked Morton Thiokol Vice President Joe Kilminster for a go/no-go decision. Kilminster said his recommendation was not to launch.

NASA officials were displeased to get this recommendation from Morton Thiokol. The launch had already been delayed several times. They were eager to launch the space shuttle before the President's State of the Union address the following evening, so that the President could include the mission in his speech. After NASA officials expressed their dismay with the recommendation, Kilminster asked for a five-minute break in the proceedings.

During the recess, Morton Thiokol's four top managers huddled away from the engineers. Senior Vice President Jerald Mason and Vice President Calvin Wiggins supported the launch, while Vice Presidents Joseph Kilminster and Robert Lund were opposed. However, Lund changed his mind after Mason "told him to take off his engineering hat and put on his management hat" [10]. (It is worth noting that more than half of Morton-Thiokol's profits came from its work for NASA.)

When Morton Thiokol rejoined the teleconference, Kilminster told NASA officials that Morton Thiokol recommended the launch go ahead. NASA officials at the Marshall Flight Center prevented the engineers' negative recommendation from being communicated to the NASA officials with final authority to approve or delay the launch.

A month after the loss of the *Challenger*, Boisjoly testified before a Presidential commission appointed to investigate the disaster. Morton Thiokol lawyers had advised Boisjoly to reply to every question with a simple "yes" or "no." Instead, Boisjoly shared with the commission his hypothesis about how the cold temperature had caused the failure of an O-ring. In later meetings with commission members, he presented documents that supported his hypothesis, including his 1985 memo. Boisjoly's testimony and documents contradicted the testimony of Morton Thiokol management. The company responded by isolating Boisjoly from NASA personnel and the O-ring redesign effort [10, 11].

Distressed by the hostile environment, Boisjoly stopped working for Morton Thiokol in July 1986. Two years later, he found work as a forensic engineer.

9.6.2 Hughes Aircraft

In the 1980s Hughes Aircraft manufactured military-grade hybrid computer chips at its Micro-electronic Circuit Division in Newport Beach, California. (A *hybrid computer chip* contains both digital and analog circuits.) The division produced about 100,000 hybrid chips per year. The military put these chips in a variety of sophisticated weapons systems, such as fighter planes and air-to-air missiles. Manufacturing these chips was a lucrative business for Hughes Aircraft; the government paid between $300 and $5,000 for each chip.

In return for paying these high prices, the government insisted that the chips pass stringent quality assurance tests. Hughes Aircraft technicians made two kinds of tests. First, they ensured the chips functioned correctly. Second, they checked the chips for resistance to shocks, high temperatures, and moisture. About 10 percent of the chips failed at least one of these tests. A common problem was that a chip would have a defective seal, which let moisture in. These chips were called "leakers."

Margaret Goodearl and Donald LaRue supervised the testing area. The company hired Ruth Ibarra to be an independent quality control agent.

In August 1986 floor worker Lisa Lightner found a leaker. Donald LaRue ordered her to pass the chip. Lightner told Goodearl, and Goodearl reported the incident to upper management. Hughes Aircraft management threatened to fire Goodearl if she didn't reveal the identity of the worker who had complained.

Two months later, LaRue ordered Shirley Reddick, another floor worker, to reseal lids on some hybrid chips, in violation of the required process for handling leakers. Reddick reported the incident to Goodearl, who relayed the report to upper management. Again, Goodearl was told she might be fired if she kept up this pattern of behavior.

In the same month, LaRue asked tester Rachel Janesch to certify that a defective hybrid chip had passed the leak test. Goodearl played a role in reporting the incident to Hughes Aircraft management. In this case, the chips were retested.

Goodearl and Ibarra found a box of hybrid chips with blank paperwork, meaning the necessary tests had not been performed. When Goodearl reported this discovery to her superiors, they told her she was no longer part of the team. Goodearl filed a formal harassment complaint. A mid-level manager in Personnel called her into his office, tore up her complaint, threw his glasses at her, and said, "If you ever do anything like that again, I will fire your ass" [9].

Goodearl's performance evaluations, which had been excellent, dropped sharply as soon as she began complaining about irregularities in the chip testing facility. In late 1986 Goodearl and Ibarra contacted the Office of the Inspector General, part of the U.S. Department of Justice. A joint decision was made for Goodearl and Ibarra to find a clear-cut case of fraud.

One day LaRue put two leaky hybrid chips on his desk, planning to approve them after Goodearl had gone home. Goodearl and Ibarra made photocopies of the documentation showing the chips had failed the leak test. After the chips were shipped from Hughes Aircraft, the Department of Defense tested them and found them to be leakers. As a result of this incident, the Office of the Inspector General began a formal investigation of fraud at Hughes Aircraft.

Hughes Aircraft fired Goodearl in 1989. Ibarra had left Hughes Aircraft in 1988, "after being relieved of all meaningful responsibilities and put in a cubicle with nothing to do" [12]. In 1990, Goodearl and Ruth Ibarra (now known under her married name, Ruth Aldred) filed a civil suit against Hughes Aircraft, claiming that Hughes Aircraft had violated the False Claims Act by falsifying records in order to defraud the government. This civil suit was put on hold until the end of the criminal trial.

The Inspector General's criminal investigation led to a trial in 1992. The jury found Hughes Aircraft guilty of conspiring to defraud the government. Hughes Aircraft appealed the verdict, but the verdict was upheld. Since a criminal conviction can be used as evidence in a civil trial, the verdict nearly assured that Goodearl and Alred would prevail in their civil suit. Hughes Aircraft began negotiating a settlement in the civil suit.

Four years later, Hughes Aircraft was ordered to pay \$4.05 million in damages. Goodearl and Aldred received 22 percent of the settlement, or \$891,000. In addition, Hughes Aircraft was required to pay their legal fees, which amounted to \$450,000 [9, 13].

Goodearl and Aldred paid a high price for whistleblowing. Both were unemployed for an extended period of time. Aldred and her husband went on welfare until they could find work. Goodearl and her husband had to file for bankruptcy, and they eventually divorced. Despite these hardships, both whistleblowers said they "would do it all again" [14].

9.6.3 Morality of Whistleblowing

Are whistleblowers heroes or traitors? Marcia Miceli and Janet Near point out that people become whistleblowers for different reasons. They suggest we ought to consider their motives before we decide if they were acting morally [15]. While it is fair to say that all whistleblowers are trying to bring an end to wrongdoing, they may well have other reasons for publicizing a problem. We can evaluate the morality of whistleblowing by considering whether the whistleblower is motivated by a desire to help others or harm others.

Consider a person who has known about a dangerous product for years, but only becomes a whistleblower after he has been turned down for a raise or promotion. If the disgruntled employee whistleblows in order to exact revenge on an organization that has let him down, the primary motivation is to hurt the company, not help the public.

Another example of questionable whistleblowing is the case of employees who have been involved in a cover-up for some period of time, realize that they are about to be caught, and then cooperate with the authorities to identify other guilty parties in order to avoid punishment.

But suppose a person doesn't have ulterior motives for whistleblowing and is doing it simply to inform the public of a dangerous situation or a misappropriation of funds. There are three general reactions to altruistic whistleblowing [11].

WHISTLEBLOWERS CAUSE HARM

The typical corporate response to whistleblowing is to condemn it. Whistleblowers are disloyal to their companies. Through their actions they generate bad publicity, disrupt the social fabric of an organization, and make it more difficult for everyone to work as part of a team. In other words, their betrayal causes short-term and long-term damage to the company. While it is the responsibility of engineers to point out technical problems, the management of a company is ultimately responsible for the decisions being made, both good and bad. If management makes a mistake, the public has recourse through the legal system to seek damages from the company, and the Board of Directors or CEO can replace the managers who have used bad judgment.

The weakness with this response is its cavalier and overly legalistic attitude toward public harm. If people are hurt or killed, they or their heirs can always sue for damages. Yet surely society is better off if people are not harmed in the first place. A monetary settlement is a poor replacement for a human life.

WHISTLEBLOWING IS A SIGN OF ORGANIZATIONAL FAILURE

A second response to whistleblowing is to view it as a symptom of an organizational failure that results in harm all around [16]. The company suffers from bad publicity. The careers of accused managers can be ruined. It makes people suspicious of one another, eroding team spirit. Whistleblowers typically suffer retaliation and become estranged from their coworkers. Labeled as troublemakers, their long-term prospects with the company are dim.

Since whistleblowing is a sign of failure, organizations need to find a way to prevent it from happening in the first place. Some suggest that organizations can eliminate the need for whistleblowing by creating management structures and communication processes that allow concerns to be raised, discussed, and resolved.

This may be easier said than done. Robert Spitzer observes that organizations have shifted away from principle-based decision making to utilitarian decision making. A characteristic of rule-oriented ethical decision making is its absolute nature. According to Kantianism or social contract theory, the end never justifies the means. If an action violates a moral rule, it shouldn't be done, period. In contrast, a utilitarian process weighs expected benefits and harms. Once an organization begins using utilitarian thinking, the question is no longer, "Should we do it?" but, "How much of it can we do without harm?" Spitzer writes, "One can see situations in which it would be permissible to use an evil means to achieve a good so long as enough benefit can be actualized." He suggests that organizations should return to using principle-based ethics in their decision making [17].

WHISTLEBLOWING AS A MORAL DUTY

A third response is to assert that under certain circumstances people have a moral duty to whistleblow. Richard De George believes whistleblowers should ask themselves five questions:

1. Do you believe the problem may result in "serious and considerable harm to the public"?
2. Have you told your manager your concerns about the potential harm?
3. Have you tried every possible channel within the organization to resolve the problem?
4. Have you documented evidence that would persuade a neutral outsider that your view is correct?
5. Are you reasonably sure that if you do bring this matter to public attention, something can be done to prevent the anticipated harm?

According to De George, you have a right to whistleblow if you answer "yes" to the first three questions; if you answer "yes" to all five questions, you have a duty to whistleblow [18].

De George's five requirements are controversial. Some would say whistleblowing is justified even when fewer requirements are met. For example, what if the potential whistleblower knows about a problem that could result in the death or injury of millions of people, such as a meltdown inside a nuclear power plant? The whistleblower has communicated his concerns to his manager, but there is not time to lobby every potential decision maker in the company. He is reasonably sure that if he contacted a television station, something could be done to prevent the meltdown. At the very least the media could alert people so they could get out of harm's way. Shouldn't that person be obliged to whistleblow, even though the answer to the third question is no?

To others, insisting that the whistleblower have convincing documentation is too strict a condition to be met in order for whistleblowing to be a moral imperative. After all, once the whistleblower has revealed the wrong to another organization, that organization may be in a better position to gather supporting evidence than the whistleblower [19].

Along the same line, some argue that whistleblowing should be considered an obligation even when only the first three requirements are met. They hold that people should be willing to sacrifice their good and the good of their family for the greater good of society.

Others believe De George goes too far when he gives conditions under which people are morally *required* to whistleblow. These commentators suggest that a person's obligation to whistleblow must be weighed against that person's other obligations, such as the duty to take care of one's family. Whistleblowing often results in significant emotional stress and the loss of employment. If it results in a person being labeled a troublemaker, whistleblowing can end a career. Hence there are serious emotional and financial consequences to whistleblowing that affect not only whistleblowers but also their spouses and children [11].

Put another way, it is reasonable to take a strictly utilitarian approach to whistleblowing? Should we expect potential whistleblowers to weigh the benefits to a large number of people against the harm to themselves and their family, and decide to go public? After all, the whistleblower has already gone out on a limb to inform management of the dangerous situation. It is the managers who made the immoral decision to cover up the problem, not the whistleblower. We are asking a lot when we ask innocent people to sacrifice their career and the welfare of their family for the benefit of strangers. We shouldn't be surprised to learn that when whistleblower Al Ripskis was asked what advice he would give potential whistleblowers, his immediate reply was "Forget it!" [20].

On the other hand, whistleblower Carlos G. Bell, Jr., chastises fellow engineers for the way they duck responsibility:

> We engineers are almost without exception only too willing to assign moral responsibility to any administrator or executive or politician under whom we can place ourselves. Our reward for living in such ways is a part of the American dream: we are involved in very few arguments and year-by-year, we build up sizable pensions for our old age [21].

Moral responsibility is different from other kinds of responsibility. First of all, moral responsibility must be borne by people. While the Fourteenth Amendment to the Constitution may make a corporation a person in the legal sense of the word, a corporation is not a moral agent. We cannot assign moral responsibility to a corporation or any other organization [22].

Second, moral responsibility is different from role responsibility, causal responsibility, and legal responsibility in that it is not exclusive [22]. **Role responsibility** is responsibility borne because of a person's assigned duties. A company may hire a bookkeeper to send out invoices and pay the bills. It is the bookkeeper's responsibility to get the bills paid on time. **Causal responsibility** is responsibility assigned to people because they did something (or failed to do something) that caused something to happen. "Joe is responsible for the network being down, because he released the virus onto our local computer network." **Legal responsibility** is responsibility assigned by law. Homeowners are responsible for the medical bills of a postal carrier who slips and falls on their driveway. Role responsibility, causal responsibility, and legal responsibility can be exclusive. For example, if one person is responsible for paying the bills, the other employees are not. Moral responsibility is not exclusive. For example, if an infant is brought into a home, both the mother and the father are responsible for the baby's well-being.

Because moral responsibility is not exclusive, people cannot pass the buck by saying, "My boss made the final decision, not me," or by saying, "I just wrote the software; I wasn't responsible for testing it." When people abdicate their moral responsibility, great harms can be done. In the 1970s executives at Ford Motor Company were anxious to begin selling a 2,000 pound, $2,000 alternative to Japanese imports. Unfortunately, prototypes of the Ford Pinto could not pass the mandatory collision test, because the windshield kept popping out. Forbidden from making design changes that would increase the weight of the car or delay its introduction, engineers solved the problem by redirecting the energy of the collision down the drive train to the gas tank. They knew this change would make the gas tank more likely to rupture, but the car did not have to pass a fuel tank integrity test. Covering up design problems allowed Ford to get its subcompact car to market. However, Ford eventually paid millions of dollars to settle dozens of lawsuits resulting from fiery crashes involving Pintos. Moreover, unfavorable media attention harmed Ford's reputation for years [20].

Michael McFarland argues that a team of engineers should be held to a higher level of moral responsibility than any of its individual members. There may well be situations where a person has a duty to speak the truth. To this duty, McFarland adds another duty held by moral agents: the duty to help others in need. If whistleblowing should be done, and no individual has the strength to do it, then it must be done by the group acting collectively [23].

Summary

Software engineering is not a full-fledged profession like medicine or law, because you do not need to be certified and licensed in order to work as a software engineer. Nevertheless, software engineers can, through inadequate education, insufficient practical training, or bad choices, cause a great deal of harm to members of the public. In this respect the responsibility of software engineers is similar to that held by members of fully developed professions. For these reasons the two largest computing societies have worked together to develop a code of ethics to guide the behavior of software engineers.

The Software Engineering Code of Ethics and Professional Practice is based upon eight general principles related to the following subjects: the public, client and employer, product, judgment, management, profession, colleagues, and self. Each of these general principles contains a list of clauses related to specific areas of potential moral concern for the practicing software engineer. You can use the Code of Ethics as a practical guide to help you weigh moral choices that you may face as a practicing software engineer. Good judgment is still needed, however. In many cases there is a conflict between two or more of the relevant clauses. At these times you must determine which of the clauses is most relevant and/or most important.

The Code of Ethics asks software engineers to ponder if their actions are worthy of the ideal professional. The ethics of virtue, or virtue ethics, is based on the imitation of morally superior role models. Virtue ethics arises from Aristotle's belief that happiness is the result of living a virtuous life. One of the strengths of virtue ethics is that it makes clear how good deeds are motivated by friendship, loyalty, dependability, and other praiseworthy attributes of a good person. Another strength of virtue ethics—at least according to its supporters—is that it does not demand that every action produce the maximum benefit, solving the problem of impartiality that plagues Kantianism, utilitarianism, and social contract theory. On the other hand, virtue ethics does not provide a formal process for moral decision making: using virtue ethics alone, it is not always clear what a person is supposed to do in a particular situation. For this reason some philosophers argue that virtue ethics should be used as a complement to another theory, such as utilitarianism, rather than as a stand-alone ethical theory.

To many, whistleblowing is a heroic act requiring great moral courage. A whistleblower brings to light a real or potential harm to the public, such as an abuse of taxpayer's money or a defective product, after trying and failing to get the problem resolved within the organization. Inevitably, whistleblowers and their families suffer emotionally and economically. It may take a decade for a whistleblower to be vindicated in court.

Different commentators have taken widely different views about whistleblowing. Some say whistleblowing does so much harm to the whistleblower and the organization that it is never the right thing to do. At the other extreme are those who say whistleblowing is a moral imperative when several conditions are met. They go so far to say that any harm done to whistleblowers and their families is outweighed by the benefit done to society. In the middle are those who argue that any decision for or against whistleblowing must be made on a case-by-case basis.

If whistleblowing is ever called for, it is only as a last resort. Everyone agrees that people who discover real or potential harms to the public should first attempt to get the problem fixed within the organization. It would be better if there were never a need for whistleblowing. Organizations ought to have communication and decision-making structures that make it easier to identify and deal with financial irregularities or product defects.

The predominant American corporate mindset does not align well with this ideal. Managers focused on maximizing "the bottom line" may well make decisions on utilitarian grounds, weighing the costs and benefits of each alternative. Utilitarian thinking allows an organization to do something that is slightly bad in order to reap a greater good. Undisclosed bad deeds are less harmful than those brought to the light. Hence utilitarian thinking can create an atmosphere in which the free communication of organizational actions is suppressed. In this environment, those who wish to report financial irregularities or product defects are ignored or silenced. The financial scandals at Enron, Tyco International, WorldCom, Adelphia Communications, and other corporations that cost investors billions of dollars have prompted some ethicists to call for a return to principle-based decision making.

"I'm making this decision on principle, just to see how it feels."

Review Questions

1. What is a profession? In which respects is software engineering like a fully developed profession such as medicine? In which respects is software engineering not like a fully developed profession?
2. Why did the ACM pass a resolution opposed to the licensing of software engineers?
3. Identify as many clauses as you can in the Software Engineering Code of Ethics and Professional Practice that refer to issues related to privacy.
4. Identify as many clauses as you can in the Software Engineering Code of Ethics and Professional Practice that refer to issues related to intellectual property.
5. Identify five clauses in the Software Engineering Code of Ethics and Professional Practice that reflect a utilitarian ethical viewpoint. Identify five clauses in the Code that reflect a Kantian viewpoint.
6. Describe virtue ethics in your own words. What are the advantages and disadvantages of this ethical theory?
7. The text gives James Rachels's short list of 24 virtues. Come up with a list of 5 additional virtues.
8. What is whistleblowing? What harms does it cause? What benefits may it provide?

9. Which clauses in the Software Engineering Code of Ethics and Professional Practice support the legitimacy of whistleblowing? Which clauses in the Code may be violated by a whistleblower (assuming the whistleblower is telling the truth)?

Discussion Questions

10. The *Challenger* disaster led to the deaths of seven astronauts and the loss of millions of dollars worth of equipment. How much moral responsibility should each of the following groups hold for this tragedy: Morton-Thiokol engineers, Morton-Thiokol senior management, NASA management?

11. In the criminal proceedings resulting from the government's investigation of fraud at the Micro-electronic Circuit Division, the jury found Hughes Aircraft guilty, but it found supervisor Donald LaRue not guilty. The jury felt LaRue was simply following orders from management. Was the jury's decision a just one?

12. Do you agree with Michael McFarland that a team of engineers has greater moral responsibility than any individual engineer on the team?

13. You are a manager in charge of a section of 30 employees in a large corporation. This morning one of your employees—Jane Lee—enters your office and tells you she thinks two members of your staff are having an affair. These employees are married—but not to each other. Jane is afraid that if it is true, others in the office will inevitably find out about it and morale will suffer. She suggest that you discreetly monitor their emails to see if it provides evidence of an affair. If you find evidence, you can nip the problem in the bud. If there is no problem, you do not have to embarrass yourself by talking with the employees. What should you do? [24]

14. According to virtue ethics, the right action to take in a particular situation is the action that a person with strong moral character would take. If you decide to practice virtue ethics, you need to find a moral role model. How would you choose a role model?

15. Two weeks ago you started a new job as system administrator for a computer lab at a small college. Wanting to make a good impression, you immediately set out to learn more about the various applications provided to the users of the lab. One of the packages, an engineering design tool, seemed way out of date. You looked through the lab's file of licensing agreements to see how much it would cost to get an upgrade. To your horror, you discovered that the college never purchased a license for the software—it is running a bootlegged copy!

    When you bring this to the attention of your boss, the college's Director of Information Technology, he says, "The license for this software would cost us $10,000, which we don't have in our budget right now. This software is absolutely needed for our engineering students, though. Maybe we can get the license next year. For the time being, just keep the current version running."

    How would you respond to your manager?

16. You are a junior computer science major. You sent your resume to a half-dozen companies hoping to get a summer internship. Two weeks ago XYZ Corporation contacted you and offered you a paid summer internship. One week ago you accepted their offer.

Today you received a much better internship offer from ABC Corporation. What should you do?

17. You are the manager of a software development group within a large corporation. Your group would be more productive if the PCs were upgraded, but you do not have any money left in your annual equipment budget. Because of employee turnover, you do have plenty of money left in your personnel budget, but corporate rules do not allow you to spend personnel funds on equipment.

    If you overspend your equipment budget, you will receive a negative performance review. You also know that whatever money is left over in your budget at the end of the fiscal year is "swept up" by the corporation. In other words, you cannot carry over a surplus from one year to the next—your group loses the money.

    You complain about your situation to the manager of another group, who has the opposite problem. She has plenty of money left in her equipment budget, but her personnel expenses are going to exceed her labor budget unless she does something. She offers to buy you the $50,000 of equipment you need out of her budget, if you pick up $50,000 of her personnel expenses out of your budget. If you take this action, both groups will get what they need, and neither group will exceed any of its budgets.

    Discuss the morality of the proposed course of action.

In-class Exercises

18. A college equips its large lecture halls with wireless networks, and it requires all of its students to purchase a laptop computer when they enroll. A computer science professor plans to streamline how quizzes are administered in his introductory programming class. Students will take the quizzes online as they sit in the classroom. A computer will grade the quizzes instantly, providing the students with instant feedback. The computer will also provide the professor with information about how well the students did on each question, which will enable him to spend more of his lecture time focusing on those topics that the students are having the hardest time understanding. Discuss the benefits and risks associated with implementing the proposed system.

19. Company X wants to open a dating service Web site. It hires Company Y to develop the software. Company Y hires Gina as a private contractor to provide a piece of instant messaging software for the package. Gina's contract says she is not responsible for the security of the site. Company Y is supposed to perform that bit of programming. However, software development runs behind schedule, and Company Y implements a simplistic security scheme that allows all messages to be sent in plain text, which is clearly insecure.

    Gina brings her concerns to the management of Company Y. Company Y thanks her for her concern, but indicates it still plans to deliver the software without telling Company X. Company Y reminds Gina that she has signed a confidentiality agreement that forbids her from talking about the software to anyone, including Company X.

    What should Gina do?

20. You are members of the information services team at a large corporation. The President has asked for a confidential meeting with your group to talk about ways to improve

productivity. The President wants to ensure that people are not sending personal emails or surfing the Web for entertainment while they are supposed to be working. The Chief Information Officer suggests that employees be informed that their emails and Web surfing will be monitored. In truth, the company does not have the resources to do this and does not plan to implement any monitoring. The CIO strictly forbids anyone in the information services team from revealing this fact. Debate the morality of management making such an announcement.

21. The members of the class are the employees of a small, privately held company that produces computer games. Everyone shares in the profits of the company. The company has been making electronic versions of popular board games for established game companies. Business is steady, but profits have not been large. The marketing team says that a first-person shooter game based on the Gulf War conflict would generate a huge amount of publicity for the company and could be highly profitable. Debate the morality of producing such a game.

Further Reading

Association for Computing Machinery Web site. www.acm.org.

Margaret Coady and Sidney Bloch, editors. *Codes of Ethics and the Professions*. Melbourne University Press, Melbourne, Australia, 1996.

ComputingCases.org (Web site).

Myron Peretz Glazer and Penina Migdal Glazer. *The Whistleblowers: Exposing Corruption in Government and Industry*. Basic Books, New York, NY, 1989.

IEEE Computer Society Web site. www.computer.org.

Deborah G. Johnson. *Ethical Issues in Engineering*. Prentice Hall, Englewood Cliffs, NJ, 1991.

Alasdair MacIntyre. *After Virtue*. 2nd ed. University of Notre Dame Press, Notre Dame, IN, 1984.

Mike W. Martin. *Meaningful Work: Rethinking Professional Ethics*. Oxford University Press, New York, NY, 2000.

Justin Oakley and Dean Cocking. *Virtue Ethics and Professional Roles*. Cambridge University Press, Cambridge, England, 2001.

References

[1] Gary Ford and Norman E. Gibbs. "A Mature Profession of Software Engineering." Technical report, Carnegie-Mellon University, January 1996. CMU/ SEI-96-TR-004, ESC-TR-96-004.

[2] Fran Allen (co chair), Barry Boehm, Fred Brooks, Jim Browne, Dave Farber, Sue Graham, Jim Gray, Paula Hawthorn (co chair), Ken Kennedy, Nancy Leveson, Dave Nagel, Peter Neumann, Dave Parnas, and Bill Wulf. "ACM Panel and Professional Licensing in Software Engineering Report to Council." May 15, 1999. www.acm.org/serving/se_policy.

[3] Aristotle. *The Nicomachean Ethics*. Oxford University Press, Oxford, England, 1998. Translated by F. H. Peters and M. Ostwald.

[4] James Rachels. *The Elements of Moral Philosophy*. 4th ed. McGraw-Hill, Boston, MA, 2003.

[5] Justin Oakley and Dean Cocking. *Virtue Ethics and Professional Roles*. Cambridge University Press, Cambridge, England, 2001.

[6] J. Eric Smith. "Anti-worm Worm Makes Rounds, Cleanses Systems of Infection." *GEEK.com*, August 20, 2003.

[7] Florence Olsen. "Attacks Threaten Computer Networks as Students Arrive for the Fall Semester." *The Chronicle of Higher Education*, September 5, 2003.

[8] Irena Blonder. "Blowing the Whistle." In *Codes of Ethics and the Professions*, pages 166–190. Melbourne University Press, Melbourne, Australia, 1996.

[9] Kevin W. Bowyer. "Goodearl and Aldred versus Hughes Aircraft: A Whistle-Blowing Case Study." In *Frontiers in Education*, pages S2F2–S2F7. October 2000.

[10] Roger M. Boisjoly. "The Challenger Disaster: Moral Responsibility and the Working Engineer." In *Ethical Issues in Engineering*, pages 6–14. Edited by Deborah G. Johnson. Prentice Hall, Englewood Cliffs, NJ, 1991.

[11] Mike W. Martin. *Meaningful Work: Rethinking Professional Ethics*. Oxford University Press, New York, NY, 2000.

[12] Taxpayers Against Fraud. *U.S. Department of Justice Joins Whistle-blowers in Lawsuit Against Hughes Aircraft Seeking Several Hundred Million Dollars*. December 15, 1992. Press release.

[13] "The Hughes Whistleblowing Case." *ComputingCases.org*.

[14] Andre Mouchard. "Whistle-Blowers Set to Use Their Reward." *The Orange County Register (California)*, September 11, 1996.

[15] Marcia P. Miceli and Janet P. Near. "Whistle-Blowing as Antisocial Behavior." In *Antisocial Behavior in Organizations*. SAGE Publications, Thousand Oaks, CA, 1997.

[16] Michael Davis. "Avoiding the Tragedy of Whistleblowing." *Business and Professional Ethics Journal*, 8(4):3–19, Winter 1989.

[17] Robert J. Spitzer, S.J. "For Good Reason, 'Organizational Ethics' a Hot Topic Nowadays." *Gonzaga (Gonzaga University newsletter)*, 5(2):2, Fall 2003.

[18] Richard T. DeGeorge. *Business Ethics*. 3rd ed. Macmillan, New York, NY, 1990.

[19] Gene G. James. "Whistle Blowing: Its Moral Justification." In *Business Ethics*, 2nd ed., pages 332–344. McGraw-Hill, New York, NY, 1990.

[20] Myron Peretz Glazer and Penina Migdal Glazer. *The Whistleblowers: Exposing Corruption in Government and Industry*. Basic Books, New York, NY, 1989.

[21] Bell, Carlos G., Jr. "One Ethical Problem Faced by the Atomic Energy Commission and Its Contractors." In *Beyond Whistleblowing: Defining Engineers' Responsibilities, Proceedings of the Second National Conference on Ethics in Engineering*, pages 250–258. Illinois Institute of Technology, Chicago, IL, 1983.

[22] John Ladd. "Collective and Individual Moral Responsibility in Engineering: Some Questions." In *Beyond Whistleblowing: Defining Engineers' Responsibilities, Proceedings*

of the Second National Conference on Ethics in Engineering, pages 90–113. Illinois Institute of Technology, Chicago, IL, 1983.

[23] Michael McFarland. "The Public Health, Safety, and Welfare: An Analysis of the Social Responsibility of Engineers." In *Ethical Issues in Engineering*, edited by D. G. Johnson, pages 159–174. Prentice Hall, Englewood Cliffs, NJ, 1991.

[24] Herbert W. Lovelace. "When Affairs of the Heart Raise IT Privacy Issues." *informationweek.com*, December 10, 2001.

AN INTERVIEW WITH

Paul Axtell

Paul Axtell is a corporate communications consultant who has helped clients such as John Deere, American Express, Hewlett-Packard, Kodak, Monsanto, Oregon State University, Ohio State University, and a number of K–12 school districts to enhance individual and group performance and organizational effectiveness. Paul has worked in Brazil and Canada, and has done training for young African political leaders.

Some of Paul's current projects include coaching manufacturing teams, supporting cultural change within a university, managing a year-long training program for organizational consultants, changing working relationships within management groups, and developing training programs for elementary and middle school reading coaches.

Paul has three areas of focus. The first is contextual education. That is, how do you give people the ability to shift their thinking, broaden their point of view, and change the way they fundamentally relate to current issues and each other. The second area of focus is process skills. How do you set up and manage conversations in a way that works for everyone involved? The third area is creating an awareness for the best practices of working together in group situations.

Paul received his B.S. in Chemical Engineering from South Dakota School of Mines and Technology and an MBA from Washington University in St. Louis.

Some commentators have suggested that whistleblowing is a sign of organizational failure. They suggest that organizations can eliminate the need for whistleblowing by creating management structures and communication processes that allow concerns to be raised, discussed, and resolved. Do you agree with this assessment?

Whistleblowing is a check and balance that is needed in certain circumstances. It is certainly alarming when a situation gets so out of hand that an employee must go to outside authorities to get someone to pay attention. When this happens, it not only reveals a set of circumstances that are not working, but it also adds to the distrust of the people who lead corporate organizations.

It should not be surprising, however, that we encounter such situations. Almost all of our relationships, both personal and organizational, have problems that begin with not being able to talk. We are raised in a culture that says it's not safe to share our thinking, voice our concerns, or push back in conversations. We debate about bringing things up at home with loved ones and at work in meetings. Given this fundamental approach to relationship and conversation, we should expect problems.

So, whatever attention we can give to create structures and processes and permission and safety is well worth it. And the need for it goes well beyond uncovering misdeeds and poor practices. The real benefit would be in a sense of belonging and caring that goes with an open and honest relationship.

In addition to setting up structures that protect people, we need to embrace a far larger goal. We need to set out to establish a cultural norm of freedom, permission, and safety. It will be very difficult to obtain because we are not raised or trained to create such a culture. Still, how can it not be the right path to be on?

Training will probably be required on both sides. We all need training on how to raise questions and concerns with a bit more set-up and graciousness. We need to be clear that we are on the same side and speak consistent with that context. And we all need training on how to hear and respond to questions and complaints, especially when they are not presented in the best way.

If a corporation wants to change the culture, it needs to pay attention to the people it is grooming to be supervisors, managers, and directors. As columnist Dave Berry says, If your date is rude to the waiter, you are dating a rude person. That makes a lot of sense as we promote people. We do need to do the training, but most behavior starts with perception.

From your experience, what are the principal barriers to improved communication inside a large corporation?

I would say that there are three principal barriers. First, people are raised to be careful and not speak up. Second, most supervisors and managers do not have the skill set to thoughtfully deal with questions and complaints. Third, our lack of follow-up and follow-through makes things worse when we invite feedback and then don't deliver.

How can a corporation remove these barriers?

Constantly invite people to raise issues and concerns. However, don't ask if you don't intend to follow through. Have a great response to missteps by employees. It's management's reaction to problems that determine whether employees feel safe. Lastly, get to know people. It's very difficult to speak up when I don't know you.

Can you give examples of how email can be harmful to communication within an organization?

Email has a couple of potential pitfalls. The most common is a lack of context for the message. Context is usually communicated by either tone of voice or set-up. Obviously tone of voice is missing on email, and people usually are very brief, which means they don't do adequate set-up. On the receiving end of the message sits a human being who by default tends to take things personally. That creates a defensive response.

How can these problems be avoided?

It's simple to avoid these problems. If people are aware of the issues discussed above, they do them less.

What is the most challenging part of your job?

The first challenge is to re-teach people how to learn. We were all wonderful learners when we were 2- and 3-years old. We observed. We mimicked. We paid attention to the people around us. We practiced until we could do things. We had little or no concern with looking foolish or not knowing how to do something. Then later we came to value knowing and information as having more relevance than tacit knowledge. Amazingly, it's the really good people who still want to learn, who still want feedback. Most of us are highly selective about who can give us feedback about what. We are not wide open to feedback. We are not even looking for it, for the most part.

The second challenge is to get people to acknowledge the impact of conversation in their lives—even to the point of arguing that they don't really have much else to work with. After their technical competence, it is the quality of their conversations that determines how things turn out. Conversation is the basis for their relationships. Conversation is the basis for having influence in an organization. Conversation determines the culture. Conversation determines how they are viewed.

The third challenge is in working to change the perceptions or views of individuals who have somehow gotten to a place in life where they are not responsible for what happens. As soon as you and I say to ourselves or others that "it's not our fault" or "it's not our job," we essentially are at the mercy of the circumstances. Certainly the things that happen in our lives often control the outcomes, but truly effective people don't give in completely to the circumstances. They maintain the view that they can make a difference in how things turn out. Interestingly, these people rarely give excuses.

What surprises you the most about communications in the corporate world?

I'm not so sure I'm surprised, but I think there are a number of things that seem to be missing that would make a big difference if they were present:

1. If managers wrote more of their own communication pieces and signed them, they would come across as more authentic. Employees are also highly skeptical about positive spin writing and admire a more direct, what-is-so, way of writing.
2. Written notes of acknowledgment and appreciation mean so much to people. We keep them for years. Yet handwritten notes to individuals and groups are a lost art.
3. An essential part of being effective is having the ability to set up a conversation, keep it on track, and wrap it up, not only in meetings but also in hallway conversations. These process skills are often missing at all levels of the organization.
4. Making specific commitments with clear due dates would reduce the amount of upsets that occur with unfulfilled expectations and lack of progress.
5. Checking in with people about their families, projects, weekends, and then engaging and enjoying in the conversation that follows is another piece that technical folks tend not to do.

APPENDIX

A Plagiarism

Consequences of Plagiarism

According to the Council of Writing Program Administrators (WPA), "plagiarism occurs when a writer deliberately uses someone else's language, ideas, or other original (not common-knowledge) material without acknowledging its source" [1]. The consequences of plagiarism can be severe. Newspaper reporters and college professors have lost their jobs because they plagiarized the work of others [2, 3]. Colleges and universities view plagiarism as a form of cheating. A few years ago at the University of Virginia, 48 students either quit or were expelled for plagiarism [4].

The vast amount of information freely available on the Internet, the power of search engines, and the cut-and-paste capability of contemporary computer programs have made it easier than ever to commit plagiarism. Of course, Web search engines can also make it easy for teachers to detect plagiarism [5].

Types of Plagiarism

You are plagiarizing if you deliberately do any of the following:

- Copy the words of another without both (1) putting the copied text in quotation marks and (2) citing the source
- Paraphrase the words of another without citing the source

- Incorporate the figures or drawings of another person without crediting the source
- Include facts that are not common knowledge without citing the source
- Use another person's ideas or theories without giving that person credit

Guidelines for Citing Sources

Common knowledge means information that is available in many places and known to a large number of people. For example, it is common knowledge that Delaware was the first state to ratify the United States Constitution. You do not have to cite a source when presenting common knowledge.

However, you *should* cite a source when you present facts that are not common knowledge. For example, it is not common knowledge that the percentage of college freshmen in the United States interested in majoring in computer science dropped by more than 60 percent between 2000 and 2004 [6].

You must cite a source if you present another person's interpretation of the facts, whether or not you acknowledge the person by name. For example, Cass Sustein argues that information technology may weaken democracy by allowing people to filter out news that contradicts their view of the world [7]. If you repeat someone else's idea, you must cite where you found it.

How to Avoid Plagiarism

Always put quotation marks around text you have obtained from another source, and write down enough information about the source that you can cite it properly. Do this when you are collecting your notes, so that when you are writing your paper, you will not forget that the words are a direct quotation or whom you are quoting.

When you are paraphrasing the work of another, read over the material, then put it aside before you begin writing. That will help ensure you are using your own words to express the ideas. Check your paraphrase against the source document. Make sure you have not distorted the original meaning. Whenever you have used a phrase from the another person's work, you must put the phrase in quotation marks. Always cite the source of the ideas you are paraphrasing, even if there are no direct quotes.

Finally, remember to cite the sources of illustrations and figures that you reproduce.

Misuse of Sources

The WPA definition of plagiarism emphasizes that it is the *deliberate* attempt to conceal the source of the words or ideas. This aligns with our definition of ethics as being focused on the *voluntary* moral choices people make. If a person has no intention of deceiving, but fails to cite sources or use quotation marks correctly, that person's actions constitute **misuse of sources.**

Additional Information

For more information, read "Defining and Avoiding Plagiarism: The WPA Statement on Best Practices," which is the principal source document for this appendix [1].

References

[1] Council of Writing Program Administrators. *Defining and Avoiding Plagiarism: The WPA Statement on Best Practices*, January, 2003. www.wpacouncil.org.

[2] "Corrections." *The New York Times*, May 2, 2003.

[3] Scott Smallwood. "Arts Professor at New School U. Resigns after Admitting Plagiarism." *The Chronicle of Higher Education*, September 20, 2004.

[4] Brian Hansen. "Combating Plagiarism: Is the Internet Causing More Students to Copy?" *The CQ Researcher*, 13(32), 2003.

[5] Katie Hafner. "Lessons in Internet Plagiarism." *The New York Times*, June 28, 2001.

[6] Jay Vegso. "Interest in CS as a Major Drops among Incoming Freshmen." *Computing Research News*, 17(3), May 2005.

[7] Cass Sunstein. *republic.com*. Princeton University Press, Princeton, NJ, 2001.

Index

A

Abelson, Hal, 197
Abernathy, Ralph, 238
Aborigines, 5, 375
Abraham, 68
Absolute rights, 84
Accolade company, 186
Accreditation, 416–417
ACK messages, 305
ACLU (American Civil Liberties Union)
  CIPA, 133
  Internet law, 411
  Patriot Act, 241
  Real ID Act, 254
  spam, 116
ACLU v. Reno, 412
ACM (Association for Computing Machinery)
  Collegiate Programming Contest, 389
  software engineering licensing, 419
  TIA program, 247
  UCITA, 356
Act utilitarianism
  case against, 77–78
  case for, 76–77
  CIPA evaluation in, 134
  law breaking in, 93
  Principle of Utility in, 74–75
  scenario evaluation for, 75–76
  spam evaluation in, 113–114
Ad-blocking software, 147
Adams, Douglas, 325
Add-ons, security as, 321
Addiction
  Internet
    contributing factors to, 142
    ethical evaluation of, 143
    reality of, 140–142
  online games, 122
Adding machines, 6–9
Addresses
  email, 109
  URLs, 121
Adleman, Leonard, 256
ADV subject line, 117
Advanced Research Projects Agency (ARPA), 28–30
Advertising restrictions, 131
AECL (Atomic Energy of Canada Limited), 337–343
Aftab, Parry, 137
AI (artificial intelligence), 105, 376–379
Aiken, Howard, 12
Air Force Office of Special Investigations, 297
Aldred, Ruth, 441–442
Algebra, Boolean, 10–11
Allen, Paul, 20–21
Altair 8800 computer, 20
AltaVista search engine, 38
Alto computer, 34–35
Amazon.com
  customer profiles, 221
  DoS attacks against, 304
  wrong prices on, 329–330
Amber, 22
America Online (AOL)
  chat rooms, 137
  identity theft, 136
  music service, 184
  spam filter, 112
  spyware protection, 228
  Web filter, 132
American Civil Liberties Union (ACLU)
  CIPA, 133
  Internet law, 411
  Patriot Act, 241
  Real ID Act, 254
  spam, 116
American Intellectual Property Law Association, 356
American Law Institute, 356
American Library Association, 133, 242, 356
American Management Association survey, 383
American Society of Composers, Authors, and Publishers (ASCAP), 166–167
Amish, 4
Amtrak, 285
Analog Devices company, 384
Analytical Engine, 9–10
AND operators in Boolean algebra, 10–11
Anonymous electronic money systems, 260
Anti-worm scenario, 436–438
Antisocial online community members, 154
AOL (America Online)
  chat rooms, 137
  identity theft, 136
  music service, 184

AOL (America Online) *(continued)*
spam filter, 112
spyware protection, 228
Web filter, 132
AOL Guardian program, 132
Apache web server, 191
Appalachian Spring, 196
Apple Computer, Inc. v. Franklin Computer Corp., 186
Apple computers
development of, 35–36
Franklin Computer lawsuit, 186
hypertext systems, 36–37
Linux pressure on, 192
music service, 184
personal computer, 21
ApplyYourself software, 303–304
Ariane satellite launch device, 332–333
Aristotle, 429–430
Arithmetic Machine, 6
Armstrong, Neil, 27
Aronowitz, Stanley, 372
ARPA (Advanced Research Projects Agency), 28–30
ARPANET, 28–30, 321
Arrests, false, 327, 413
Articles of Confederation, 130
Artificial intelligence, 105, 376–379
Artificial Passenger software, 47
Aryan Nations site, 123
"As We May Think" (Bush), 33, 51
ASCAP (American Society of Composers, Authors, and Publishers), 166–167
ASIMO robot, 377
Asimov, Isaac, 370
Association for Computing Machinery (ACM)
Collegiate Programming Contest, 389
software engineering licensing, 419
TIA program, 247
UCITA, 356
Asymmetric attacks, 304
Asymmetric encryption, 255
AT&T
long distance network collapse, 299, 333
semiconductor development, 17
Surity 3600 device, 258
system hacked, 293
Atanasoff, John, 13
Atanasoff-Berry Computer, 13
Atomic Energy of Canada Limited (AECL), 337–343
Attachments, viruses in, 280–283
Audio Home Recording Act, 210
Audits, IRS, 246
Augmentation Research Center, 34, 51
Australian railroads, 286
Authentication, 291
Authorization, 291
Automation
jobs gained from, 372–374
jobs lost from, 370–371
productivity changes from, 375–376
robots for, 376–379
supply-chain, 380–381
Automobiles, 4
automated, 376
black boxes, 225
thefts of, 328
Axtell, Paul, interview with, 455–457
Azinger, Paul, 396

B

Baase, Sara, 259
Babbage, Charles, 8–10
Back Orifice RAT, 291
Back pain, 3
Bad patents, 187
BAE Automated Systems, 335–336
Baez, Joan, 238
Baggage handling systems, 335–336
Ballistic Research Laboratory (BRL), 13
Ballots, butterfly, 309
Bamford, James, 259
Band-Aid brand, 164
Bangalore, 385
Banking, digital cash for, 259–261
Baran, Paul, 29
Bardeen, John, 17
Bardo, Robert, 212
Basic Books v. Kinko's Graphics Corporation, 167
BASIC interpreter, 21
Batteries, invention of, 22–23
BBC News, 123
BBN, 29
BBSs (bulletin board systems)
E911 posted on, 298–300
pirate boards, 296
BearShare software, 182
Beings in act utilitarianism, 75
Bell, Alexander Graham, 24–25
Bell, Carlos G., Jr., 445
Bell Labs, 17–18
Bell Labs Technical Journal, 297
BellSouth, 298–300
Benn, Stanley, 215, 219–220
Bentham, Jeremy, 74–76, 134
Berman, Jerry, interview with, 411–413
Berners-Lee, Tim, 36–37, 120–121
Bernoulli numbers, 10
Berry, Clifford, 13
Best Buy, 136
Bible
in divine command theory, 66–68
Gutenberg, 32–33
Bill of Rights
First Amendment
history of, 129–130
and Internet, 411
limitations on, 130–132
Fourth Amendment
and electronic surveillance, 212, 234–235
and Patriot Act, 241–242

Third Amendment, 217
Billing errors, 328–329
Biloxi school district, 384
BIND program, 191
Biometrics, 291
BIOS, 187
Birsch, Douglas, 73
Bishop, Matt, interview with, 321–323
BitLaw Web site, 184
BitTorrent networks
file downloads from, 179–180
shutting down of, 210
Black boxes, automobile, 225
Blacklists
for spam, 115–117
for Web filters, 132–133
Blankenship, Loyd, 300
Blaster worm, 79, 285
Blind signatures for digital cash, 260
Blogs, 123
Bloustein, Edward, 213
Blue boxes, 35, 296
Blue laws, 397
BMI, 166
Bob's ACL Knee Community, 153
Body scanners, 224–225
Boisjoly, Roger, 439–441
Bombing, email, 306–307
Boole, George, 10
Boolean algebra, 10–11
Bootstrap Institute, 51
Bork, Robert, 229
Bowen, Debra, 120
Boyle, James, 197
Brain virus, 282
Brand names, 164
Brandeis, Louis, 217–218, 235
Brattain, Walter, 17
Breaking laws, 91
in act utilitarianism, 93
conclusion, 93
in Kantianism, 91–92
in rule utilitarianism, 92–93
in social contract theory, 91
Bricklin, Dan, 21
British coast guard, 286
British Open University, 153
BRL (Ballistic Research Laboratory), 13
Broadband connections, 31
Brooks, Linda, 328
Brooks Brothers, 224
Brown, Peter, 36
Browsers, 37
Brunner, John, 283
Buchanan, Pat, 308–309
Buchwald, Art, 163
Buckman, John, 198
Buffer overflows, 287–288, 322
Bugs, 237
Bulletin board systems (BBSs)
E911 posted on, 298–300
pirate boards, 296
Burma, Internet access in, 123
Bush, George W., 27–28, 308–309
Bush, Vannevar, 33, 51
Bush differential analyzer, 14
Busicom company, 20
Business Week, 303–304
Businesses, online, 121
Butterfly ballots, 309
Bynum, Terrell, 103
Byrne, Edmund, 213
Byron, Augusta Ada, 10
Byron, Lord, 10

C

Cactus Data Shield, 177
Cadence Design Systems, 384
CALEA (Communications Assistance for Law Enforcement Act), 239–240
California IT employment, 374
Calvinism, 375–376
CAMI (Columbia Artists Management, Inc.), 166–167
Campaign finance reform, 397
CAN SPAM Act, 117–120
Canada, privacy practices in, 277
Carlin, George, 131
Carnegie Mellon University, 304
Carnoy, Martin, 46, 372–374
CASE (Computer-assisted software engineering) tools, 348–349
Cash, digital, 259–261
Categorical Imperative, 69–72
Cathode ray tubes (CRTs), 14–15
Causal responsibility, 343, 445–446
Cavoukian, Ann, interview with, 275–277
CBS election predictions, 16–17
CDA (Communications Decency Act), 124, 412
CDs
burners, 173
copying, 210
copyproof, 177
CDT (Center for Democracy and Technology), 411–412
Cell phones, 3–4
Censorship, 126
direct, 126–127
ethical perspectives on, 128–129
on Internet, 127–128
self-censorship, 127
Census Bureau, U.S.
early computer use by, 15–16
records kept by, 230–231
Center for Democracy and Technology (CDT), 411–412
CEO pay compared to production workers, 394
Cerf, Vinton, 29
CERN, 36–37, 121
CERT (Computer Emergency Response Team), 308
Certification, 416, 418–419
Certified Public Accountants (CPAs), 417–418
CGR corporation, 337
Challenger disaster, 439–441
Charles Katz v. United States, 237
Charlton, John, 142
Chase Manhattan, 261
Chat room predators, 137–139

Checks, electronic, 260
Chess program, 376
Chicago Board of Trade, 329
Child Internet Protection Act (CIPA), 124
  ethical evaluation of, 133–135
  passage of, 133
Child Online Protection Act, 124
Children, 132
  and chat room predators, 137–139
  and pornography, 124–125, 435–436
  Web filters for, 132–133
Children's Online Privacy Protection Act (COPPA), 229
Chilling Effects Clearinghouse, 209–210
China
  dictatorships in, 387
  IBM assets acquired by, 389
  Internet access in, 123–124
  IT development in, 389
  job movement to, 385
Chips, 258
  hybrid, 441–442
Choice as privacy factor, 275
ChoicePoint, 250
Choke worm, 286
Christianity, 66
Christmas holiday, banishment of, 376
Church, Christian
  and divine command theory, 66
  and privacy, 221
  Protestant work ethic, 375–376
CIA, 297
Cigarette advertising, 131
CIPA (Child Internet Protection Act), 124
  ethical evaluation of, 133–135
  passage of, 133
Circuit-switched networking, 29–30
Cisco Systems
  hacker attacks on, 303
  India offices for, 384
  stock prices, 387–388
Citibank
  digital cash project, 261
  India offices for, 384
  phished site, 136
Civil War, 231, 439
Clark, Melville, 168
Clay, Henry, 24
Clayton F. Summy Corporation, 162
Clean rooms, 187
Cleaver, Eldrige, 238
Client principles in Software Engineering Code of Ethics, 422–423
Clinton, William, 258
Clipper chips, 258
Club of Rome, 344
CNN
  blocked by China, 123
  DoS attacks against, 304
  self-censorship by, 127
Coca-Cola, 163
Cocking, Dean, 430
Code of Fair Information Practices, 233
Code Red worm, 285
Code reuse dangers, 343
Codes of ethics. *See* Software Engineering Code of Ethics and Professional Practice
Codex, 31–32
Coercive Acts, 217
Cohen, Bram, 179
Coins, electronic, 260
COIs (Communities of Interest), 153
Colgate University, 112
Collaborative filtering, 244
Colleague principles in Software Engineering Code of Ethics, 427
Collective IQ, 52
College access, 397
  ethical analysis, 398–400
  tuition increases, 397–398
Collegiate Programming Contest, 389
Colossus computer, 13
Columbia Artists Management, Inc. (CAMI), 166–167
Comair, 329
Coming to America, 163
Commercial computers, first, 15–17
Common Law Courts, 130
Commons, 196–199
Communication in corporations, 456–457
Communications Assistance for Law Enforcement Act (CALEA), 239–240
Communications Decency Act (CDA), 124, 412
Communities
  costs and benefits of, 53
  online, 25, 153–154
Communities of Interest (COIs), 153
Communities of Practice (COPs), 153
Competition, DRM effect on, 178
Compilers
  developed in China, 389
  GNU, 191
Complex Number Calculator, 28
Compression, multimedia, 174
Compulsion to program, 140–141
Computer-assisted software engineering (CASE) tools, 348–349
Computer crime
  chat room predators, 137–138
  hackers. *See* Hackers
  identity theft. *See* Identity theft
Computer Emergency Response Team (CERT), 308
Computer Fraud and Abuse Act, 289, 302–303
Computer simulations, 344–347

Computing milestones
adding machines, 6–9
Analytical Engine, 9–10
Atanasoff-Berry Computer, 13
Boolean algebra, 10–11
Colossus, 13
commercial computers, 15–17
ENIAC, 13–14
Harvard Mark 1, 12
IBM System/360 computers, 19
integrated circuits, 18
microprocessors, 20–22
Small-Scale Experimental Machine, 14–15
transistors, 17–18
Z series computers, 11–12
Conditions for moral responsibility, 343
Confidentiality, 160–161
Conflict of interest scenario, 434–435
Consequentialist theories, 75, 104–105
Conspiracy Theory, 140
Constitution of the United States
First Amendment
history of, 129–130
and Internet, 411
limitations on, 130–132
Fourteenth Amendment, 445
Fourth Amendment
and electronic surveillance, 212, 234–235
and Patriot Act, 241–242
Third Amendment, 217
Consumers Union, 356
Consumption taxes, 397
Content Scramble System (CSS), 176
Contexts in cultural relativism, 62–66
Convenience vs. privacy, 276
Cook, Philip, 394–397
Cookies, 227–228
Cooperative agreements for winner-take-all effects, 397
Copland, Aaron, 196
COPPA (Children's Online Privacy Protection Act), 229
COPs (Communities of Practice), 153
copyproof CDs, 177
Copyright Act of 1909, 169
Copyright Act of 1976, 185
Copyright Office, 185
Copyrights
extension of rights, 168–169
litigation, 166–167
principle rights of, 165–166
software, 185–186
Core values
acting on, 54
shared, 65
Corporations, communication in, 456–457
Counterfeit music and software, 155–156, 173
Court of Star Chamber, 129–130
Covert surveillance, 234–238
CPAs (Certified Public Accountants), 417–418
Crawler-based search engines, 38
Creative Commons, 196–199
Credentials, 221
Credit bureaus, 228
Credit card fraud, 249
Crime, computer
chat room predators, 137–138
hackers. *See* Hackers
identity theft. *See* Identity theft
Crito, 59
Cronkite, Walter, 17
CRTs (Cathode ray tubes), 14–15
CSA code, 275
CSS (Content Scramble System), 176
Cuba
Internet access in, 123
and watch lists, 238
Cuckoo's Egg, The, 297
Cultural relativism, 62–66
Customer service, monitoring, 384
Customs, 62
Cyber Angels, 137
Cyber Sentinel program, 132
Cybercafes, 122
Cyberpunk: Outlaws and Hackers on the Computer Frontier, 286, 292
Cyberterrorism, 285

D

Daoud, Ouhnane, 243
DARPA (Defense Advanced Research Projects Agency), 247
Data-entry errors, 326–328
Data mining, 244–246
household data, 246
IRS audits, 246
opt-in and opt-out, 248–249
P3P, 249
privacy violations from, 413
Syndrome Surveillance System, 246
TOA, 247–248
transaction information, 248
Data-retrieval errors, 326–328
Databases
Hippocratic, 249
mining. *See* Data mining
Davey Jones Locker, 167
Davies, Donald, 29
Davies, Glyn, 259
DDoS (distributed denial-of-service) attacks, 307–308, 311–312
De Forest, Lee, 17
De George, Richard, 444–445
De Scandalis Magnatum law, 129–130
Declaration of Independence, 394
DeCSS program, 176
Deep Blue program, 376
Defense Advanced Research Projects Agency (DARPA), 247
Defense Intelligence Agency, 297

Dell Computer
DoS attacks against, 304
supply-chain automation at, 381
Delta Airlines, 286
Democracy, World Wide Web for, 40
Denial-of-service (DoS) attacks, 304
defensive measures, 307
distributed, 307–308, 311–312
resource consumption by, 304–307
SATAN for, 308
Denning, Dorothy, 249, 293
Denver International Airport, 335–336
Department of Agriculture, 328
Department of Health, Education, and Welfare, 232–233
Department of Justice
Clipper chip, 258
NCIC databases, 327
whistleblowing, 442
Depew, Chauncey, 25
Development
professional and skills, 416
software, 348–349
Diagnostic and Statistical Manual of Mental Disorders, 141–142
Diamond Multimedia Systems, Inc., 171–172
Diamond v. Diehr, 186
Dictionary attacks, 112, 291
Diebold, 209
DIF (Discriminant function), 246
DiFazio, William, 372
Difference Engine, 8–9
Difference principle, 85–86
Differences, method of, 8–9
Diffie, Whitfield, 255
Digital Angel Corporation, 227
Digital cash, 259–261
Digital divide, 389
critiques of, 392–393
evidence of, 390–391
models of technological diffusion, 391–392
Digital Millennium Copyright Act (DMCA), 174–175, 180–181
Digital rights management (DRM), 175, 177–178
Digital video recorders (DVRs), 224–225
Direct mail advertising, 111, 245
Direct Marketing Association, 120
Disclosures, information, 221–223
Discriminant function (DIF), 246
Disfranchised voters, 326
Distributed denial-of-service (DDoS) attacks, 307–308, 311–312
Divine command theory, 66–69
DMCA (Digital Millennium Copyright Act), 174–175, 180–181
DNS (domain name service), 191
Do Not Call Registry, 220
Do not email lists, 146
Domain, public, 162
Domain name service (DNS), 191
Domestic Security Enhancement Act, 242
Donne, John, 53
DoS (denial-of-service) attacks, 304
defensive measures, 307
distributed, 307–308, 311–312
resource consumption by, 304–307
SATAN for, 308
DOS operating system, 22
Dot-com bust, 387–388
Downsizing, corporate, 371–372
Driver's Privacy Protection Act, 212
DRM (digital rights management), 175, 177–178
Dumpster diving, 250, 295–296
Dutifulness, 69
Duty, imperfect and perfect, 400
DVDs
copying, 210
encrypting, 176
DVRs (digital video recorders), 224–225

E

eBay
as community, 153
DoS attacks against, 304
E911 Document, 298–301
EarthLink, 136
East Texas Cancer Center, 339–340
Easterbrook, Frank, 354
"Eavesdropper, The", 298
eBlaster program, 132
Eckert, J. Presper, 13–14, 17
Eckert-Mauchley Computer Corporation, 15–16
Economic equality, 85–86
Economist, The, 385
ECPA (Electronic Communications Privacy Act), 239, 300–301, 303
Education
college access, 397
ethical analysis, 398–400
tuition increases, 397–398
in professional process, 417
EDVAC (Electronic Discrete Variable Automatic Computer) computer, 14
EFF (Electronic Frontier Foundation)
activities of, 209
and copyright laws, 155–156
creation of, 300–301
Eisenhower, Dwight, 16–17
Eldred, Eric, 197
Elections, U.S. Presidential
1896, 25
1952, 16–17
1984, 46
2000, 27–28, 308–309, 326

Electricity, 22–23
Electrolux, 377
Electromagnetism, 22–23
Electromagnets, 23
Electronic checks, 260
Electronic coins, 260
Electronic Communications Privacy Act (ECPA), 239, 300–301, 303
Electronic Discrete Variable Automatic Computer (EDVAC) computer, 14
Electronic Frontier Foundation (EFF)
  activities of, 209
  and copyright laws, 155–156
  creation of, 300–301
Electronic money, 259–261
Electronic Numerical Integrator and Calculator (ENIAC) computer, 13–14
Electronic voting machines, 209
Electrostatic memory, 15
Elitism, harmful effects of, 396
Ellsberg, Daniel, 300
Email
  ARPAnet, 29
  benefits of, 39
  bombing, 306–307
  growth of, 110
  operation of, 109–110
  pitfalls in, 456
  popularity of, 2
  spam. *See* Spam
Embedded computers, 39–40, 330–331
Employee monitoring, 383–384
Employer principles in Software Engineering Code of Ethics, 422–423
Employment. *See* Jobs
Encryption, 254
  asymmetric, 255
  Clipper chips, 258
  for digital cash, 259–261
  DVDs, 176
  export restrictions on, 258–259
  Pretty Good Privacy, 255, 258–259
  public key, 255
  symmetric, 254–255
Encyclopedias, online, 122
Engelbart, Douglas
  interview with, 51–52
  mother of all demos by, 34–35
Engineering, software, 347–350
English language dominance, 390
Enhanced 911 service, 225–226
ENIAC (Electronic Numerical Integrator and Calculator) computer, 13–14
Enigma cipher machine, 13
Enlightenment, 128
Ennis, Ireland, 392–393
Enquire program, 36
Epstein, Richard, 378
Equality, economic and social, 85–86
Equifax credit reporting agency, 228
Equivalence fallacy, 68
Errors
  data-entry and data-retrieval, 326–328
  software and billing, 328–330
Esposito, Dawn, 372
Ethernet, creation of, 35
Ethics
  definitions in, 54–55
  discussion questions, 97–98
  further readings, 99
  in-class exercises, 98–99
  introduction, 53–54
  law breaking, 91–93
  professional. *See* Professional ethics
  references, 99–101
  review questions, 95–97
  scenarios, 55–59
  summary, 94–95
  theories, 59–60
    act utilitarianism, 74–78
    comparing, 90–91
    cultural relativism, 62–66
    divine command theory, 66–69
    Kantianism, 69–75
    rule utilitarianism, 79–82
    social contract theory, 82–90
    subjective relativism, 60–62
    virtue ethics, 429–432
European Commission, 286
European farm lobby, 387
Euthyphro, 68
Evolution of software systems, 349–350
Exit polling, 28
Experian credit reporting agency, 228
Export restrictions on encryption, 258–259

F

Faggin, Federico, 20
Fair Credit Reporting Act, 228
Fair use, 169
  and DRM, 177
  examples, 169–170
  space shifting, 171–172
  time shifting, 170–171
Fairchild Semiconductor, 18
Fall Joint Computer Conference, 34–35
False arrests, 327, 413
False Claims Act, 438–439, 442
False information in networking, 139–140
False values in Boolean algebra, 10–11
Family Education Rights and Privacy Act (FERPA), 229
Farm lobby, 387
Farmer, Dan, 308
Farnsworth, Philo, 27
FastTrack network, 179–180
Faxes, unsolicited, 118
Faxon, Brad, 396
FBI. *See* Federal Bureau of Investigation (FBI)

FCC (Federal Communications Commission)
CALEA guidelines, 238
Enhanced 911 service, 225
and First Amendment, 131–132
FCC v. Pacifica Foundation et al., 131–132
FDA (Food and Drug Administration), 340
Federal Bureau of Investigation (FBI)
CALEA, 239
chat room investigations by, 137
Clipper chip, 258
Cuckoo's egg, 297
Mayfield detention by, 243
NCIC, 231–232, 327–328
Patriot Act, 241–244
secret wiretapping, 236–237
Federal Communications Act, 236, 238
Federal Communications Commission (FCC)
CALEA guidelines, 238
Enhanced 911 service, 225
and First Amendment, 131–132
Federal Trade Commission (FTC), 220
Felten, Edward, 175–176
FERPA (Family Education Rights and Privacy Act), 229
Ferranti Mark 1 computer, 15–16
Filters
collaborative, 244
packet, 292
spam, 112
web, 132–133
"Filthy Words" monologue, 131
Final Rip Off, The, 111
Financial Services Modernization Act, 229
finger program, 292
fingerd program, 287, 289
Fingerprint authentication, 291
Fingerprint misidentification, 243
Fire alarm telegraphs, 24
Firewalls, 291–292
First Amendment to the U.S. Constitution
history of, 129–130
and Internet, 411
limitations on, 130–132
Fischer, Constance, 215
Flood attacks, 305–306
Folkways, 62
Fonda, Jane, 238
Food and Drug Administration (FDA), 340
Ford, Gary, 416
Ford Motor Company, 446
Foreign competition, 389
Foreign workers, 388
Forrester Research, 385
Fourteenth Amendment to the U.S. Constitution, 445
Fourth Amendment to the U.S. Constitution
and electronic surveillance, 212, 234–235
and Patriot Act, 241–242
Frank, Robert, 394–397
Franklin, Benjamin, 161
Franklin ACE computer, 186
Frankston, Bob, 21
Fraudulent messages, 135–136
Free File service, 231
Free press, 32
Free Software Foundation, 356
Freedom of choice in privacy, 275
Freedom of expression, 129
FCC v. Pacifica Foundation et al., 131–132
history of, 129–130
limitations on, 130–131
Mill on, 128
spam as, 116
Fried, Charles, 215
Friends of the Earth, 387
Friis, Janus, 179
Fry Guy (alias), 298–299
FTC (Federal Trade Commission), 220
ftp application, 287

G

Galilei, Galileo, 126
Galileo space probe, 284–285
Gambling, online, 123
Games, on-line, 122
GAO (General Accounting Office), 331
Garner, Amanda, 329
Gates, Bill, 20–21, 388
Gathering of Developers, 351
Genealogy, online, 122
General Accounting Office (GAO), 331
General Electric, 384
Generic trademarks, 164
Genesis, 68
Gentul, Jack, 182
Georgetown University, 211–212
Germany, Web censorship in, 123
Gershwin Publishing Corporation v. Columbia Artists Management, Inc., 166–167
Gert, Bernard, 81
Gibbs, Norman, 416
Glaucon, 98, 103
Glidden, Carlos, 25
Global divide, 389–391
Global warming, 52
Globalization, 369–370, 385
arguments against, 387
arguments for, 386–387
foreign competition, 389
foreign workers, 388
multinational teams, 384–385
GNU Project, 191–192
GNU Public License, 191
Golden Rule, 97
Good vs. God, 68
Good will in Kantianism, 69–72
Goodearl, Margaret, 441–442
Google search engine
pages indexed by, 38

phone number service, 211
rankings in, 140
Gore, Al, 27–28, 308–309
Government monopolization, censorship through, 126
Gramm-Leach-Bliley Act, 229
Graphical user interfaces, 35–36
Greatest Happiness Principle, 74–75
Green Stamps, 223
Greenpeace, 387
Greensboro, N.C. sit-ins, 88
Griffiths, Mark, 142
Grokster network, 179–180, 183–184
Guide software, 36
Gutenberg, Johannes, 32
Gutenberg Bible, 32–33
Gyges, legend of, 100, 103

H

H&R Block, 231
H1-B visas, 388
Hack SDMI challenge, 175–176
Hacker Crackdown, The, 292, 301–302
Hackers, 279–280
dumpster diving and social engineering by, 295–296
Fry Guy, 298–299
Legion of Doom, 297–298
malicious, 296
original definition of, 292–293
on PDP-1, 293–295
penalties against, 302–303
recent incidents, 303–304
retrospective, 301–302
Riggs, 299–300
Steve Jackson Games, 300–301
Hackers: Heroes of the Computer Revolution, 292
Hafner, Katie, 286, 292
Hamlet scenario, 159–161
Hampden-Turner, Charles, 62
Handshakes, three-way, 305
Hanger, Charles, 232
"Happy Birthday to You", 162
Harmonic telegraph, 24
Harmonic Vision warranties, 352
Harrison, George, 187
Harvard Mark 1 computer, 12
Harvard University, admissions software for, 304
Hayakawa, S. I., 51
Health care, 5
Health Insurance Portability and Accountability Act, 230
Hellman, Martin, 255
Hello worm, 286
Henry, Joseph, 23
Henry, Patrick, 130
Hernandez, Roberto, 327
Hertz, Heinrich, 26
He's So Fine, 187
Hewlett Packard, 384
Hippocratic databases, 249
Hirschhorn, Larry, 373
History
computing. *See* Computing milestones
information storage and retrieval. *See* Information storage and retrieval milestones
networking. *See* Networking
Hitler, Adolf, 61, 123
Hobbes, Thomas, 82–84
Hoff, Ted, 20
Holidays, 375
Hollaar, Lee, 184
Holocaust, 139
Homebrew Computer Club, 35
Honda Motor Company, 46, 377
Hoover, J. Edgar, 231, 236–237
Hopper, Grace Murray, 12
Hosts for viruses, 280
Hughes Aircraft case, 441–442
Husqvarna Viking Designer, 357–358
Hussein, Saddam, overthrow of
looting after, 82
self-censorship after, 127
Hybrid computer chips, 441–442
HyperCard system, 36–37
Hypertext systems
evolution of, 33–34
networked. *See* World Wide Web (WWW)
single-computer, 36–37

I

Ibarra, Ruth, 441–442
IBM
Artificial Passenger software, 47
BIOS, 187
Chinese acquisition of assets, 389
IBM PC computers, 21–22, 36
System/360 computers, 19
Identification cards, national, 252–254
Identified electronic money systems, 260
Identity theft
background, 249–251
phishing, 135–136
Social Security numbers in, 251–252
and TIA, 247
Identity Theft and Assumption Act, 251
Identity Theft and Assumption Deterrence Act, 303
Identity Theft Penalty Enhancement Act, 251
IEEE (Institute of Electrical and Electronic Engineers), 356
IEEE Computer Society, 419
Illicit transactions, 261
Illuminati BBS, 300–301
IMP (Interface Message Processor), 29
Impartiality problem, 430–431
Imperfect duty, 400
Implanted chips, 227
Inanimate objects, 105
Income tax, 231
India
globalization of jobs to, 369
IT development in, 389

India *(continued)*
for multinational teams, 384–385
Information Age, catalysts for, 2
Information storage and retrieval milestones
codex, 31–32
hypertext, 33–34, 36–37
newspapers, 32
personal computers, 34–36
printing press, 32
search engines, 38–39
World Wide Web, 36–38
Information technology (IT)
data mining, 244–249
declining costs of, 385
digital divide, 389–393
dot-com bust, 387–388
employment in, 374
foreign workers in, 388
history
computing. *See* Computing milestones
information storage and retrieval. *See* Information storage and retrieval milestones
networking. *See* Networking
issues in, 39–40
networking. *See* Networking
supply-chain automation, 380–381
Inktomi search engine, 39
Innovation commons, 197
Inquisition, 126
Instant messaging
definition, 137
worms in, 286
Instant photography patents, 165
Institute of Electrical and Electronic Engineers (IEEE), 356
Integrated circuits, 18
Intel Corporation, 389
4004 microprocessor, 20
8008 microprocessor, 20
8080 microprocessor, 20
founding of, 20
India offices for, 384
Lintel computers, 192
software growth encouraged by, 195
stock price, 387
Intellectual property
CDs, 177
Creative Commons, 196–199
definition of, 157
digital rights management, 175, 177–178
discussion questions, 203–204
DMCA, 174–175
DVDs, 176
fair use, 169–174
further reading, 204–205
in-class exercises, 204
introduction, 155–157
open-source movement, 188–193
peer-to-peer networks, 178–184
protection of, 39
benefits of, 161
copyrights, 165–169
limits to, 161–162
patents, 165
trade secrets, 163–164
trademarks and service marks, 164
references, 205–208
review questions, 202–203
rights to, 157–162
SDMI, 175–176
Seltzer interview, 209–210
software protection, 184–185
copyrights, 185–186
legitimacy of, 193–196
patents, 186–187
safe development, 187
summary, 198–202
Intellectual virtues, 429
Intelligent machines, 105
Interface Message Processor (IMP), 29
Internal Revenue Service (IRS)
audits by, 246
Operation Shamrock, 238
Privacy Act of 1974, 233
records disclosed by, 231
Social Security number requirement, 251
International Telework Association & Council (ITAC), 383
Internet, 29
addiction to
contributing factors to, 142
ethical evaluation of, 143
reality of, 140–142
censorship on, 127–128
computational power of, 107–108
global divide, 390–391
high-speed connections and piracy, 174
law for, 411
music services, 184
networking. *See* Networking
Internet Explorer browser, 37
Internet Protocol (IP), 29
Internet radio, 174
Internet worm, 286–290
IP (Internet Protocol), 29
IP spoofing, 306–307
iPaq handheld computer, 329–330
iPod portable music player, 184
IQ, collective, 52
Iraq War
looting in, 82
self-censorship in, 127
Ireland, 392–393
IRS (Internal Revenue Service)
audits by, 246
Operation Shamrock, 238
Privacy Act of 1974, 233
records disclosed by, 231
Social Security number requirement, 251
Isaac, 68
Islam, 66
IT. *See* Information technology (IT)
ITAC (International Telework Association & Council), 383

ITT Communications, 237
iTunes Music Store, 184

J

Jackson, Shirley, 327
Janesch, Rachel, 442
Jaovisidha, Suchart, 329
Japanese-Americans, privacy violations of, 231
Jaschan, Sven, 286
Jet Propulsion Laboratory, 334
Jobs
  automation effects on
    jobs gained from, 372–374
    jobs lost from, 370–371
    productivity changes from, 375–376
    robots for, 376–379
    supply-chain, 380–381
  college access for, 397–400
  discussion questions, 403–404
  further reading, 406
  global movement of, 4–5, 40, 369–370, 385–389
  in-class exercises, 404–406
  monitoring, 383–384
  multinational teams, 384–385
  organization changes in, 379–381
  references, 406–409
  review questions, 403
  summary, 401–402
  telework, 381–383
  temporary work, 383
Jobs, Steve, 35
Johansen, Jon, 176
Johnson, Lyndon, 238
Johnson & Johnson, 164
Jonson, Ben, in *Hamlet* scenario, 159–161
Jordon, Eason, 127
Judaism, 66
Judgment principles in Software Engineering Code of Ethics, 424–425
Junk faxes, 118
Just consequentialism, 104–105
Just deserts argument, 194
Juvenal, 126

K

Kahn, Robert, 29
Kant, Immanuel, 69
  on addiction, 143
  on censorship, 128
  on duties, 84
  on pornography, 124
Kantianism, 69
  case against, 73
  case for, 72
  ethical evaluations in
    CIPA, 134
    Internet addiction, 143
    Internet worm, 289
    MAPS, 117
    online voting, 312
    PDP-1 unauthorized update, 294
    public college access, 400
    scenario for, 71–74
    spam, 113
    sting operation, 138–139
  good will and Categorical Imperative in, 69–72
  law breaking in, 91–92
Kapauku, 375
Kasparov, Gary, 376–377
Katz, Charles, 237
Kay, Alan, 34–35
KaZaA network, 179–181, 280
Keegan, Joseph, 215
Kelvir worm, 286
Kenadek, Richard, 167
Kennedy, Robert, 238
Kennestone Regional Oncology Center, 337–338
Kenswil, Larry, 174
key2audio protection system, 177
KGB, 297
Kilburn, Tom, 15
Kilby, Jack, 18
Kilminster, Joe, 440–441
Kim, Selena, 177–178
Kimberly-Clark, 164
King, Martin Luther, Jr., 232, 238
King for a Day, 163
Kinko's Graphics Corporation, 167, 169
Kleinrock, Leonard, 29
Kling, Rob, 393
Knowledge portfolios, 383
Knuth, Donald, 191
Kodak, 165
Kolata, Gina, 286
Krum, Charles and Howard, 26
Ku Klux Klan, 231
Kufrovich, Francis, 137
Kung Bushmen, 375

L

L-1 visas, 388
LaChat, Michael, 378–379
Ladder of privacy, 216
LaMacchia, David, 167–168
Land, Edwin, 165
LANs (local area networks), 110
LaRue, Donald, 441–442
Lawrence Berkeley Laboratory, 297
Laws
  breaking, 91
    in act utilitarianism, 93
    conclusion, 93
    in Kantianism, 91–92
    in rule utilitarianism, 92–93
    in social contract theory, 91
  for Internet, 411
  of Robotics, 370
  for winner-take-all effects, 397
Leach, Edmund, 214
Learning, online, 122
Legal responsibility, 446
Legion of Doom, 297–298, 300
Legion of Doom Technical Journal, The, 297, 299
Leonard, Andrew, 192
Lessig, Lawrence, 197
Leveson, Nancy, 342, 350
Leviathan, 82
Levin, Carl, 247
Levine, Morton, 215, 217
Leviticus, 66
Levy, Steven, 292

Libel laws, 130
Licensing, 416
  direct censorship through, 126–127
  inventions, 165
  for open-source software, 189
  software engineers, 418–419
Licensing Act, 32
Licklider, J.C.R., 28
Lightner, Lisa, 441
Limited rights, 84
Limits to Growth, The, 344–345
Lindh, John Walker, 243
Lineage game, 122
Linear accelerators. *See* Therac-25 linear accelerator
Linford, Steve, 304
Linkedin software, 154
Lintel servers, 192
Linux
  DVD support for, 176
  in open-source movement, 191–193
Lisa computer, 36
LiveBridge company, 369
Loan applications, 39
Local area networks (LANs), 110
Locke, John, 84, 157–159, 194
Lockheed Martin, 334
Logarithmic functions, 8
Logic functions in Boolean algebra, 10–11
Login names in hacking, 295
London International Financial Futures and Options Exchange, 329
Long-distance telephone systems
  collapse of, 299, 333
  phone phreaking, 296–297
LookSmart search engine, 39
Los Angeles County-USC Medical Center, 329
Lotus Development Corporation, 246
Love Bug virus, 282–283
Lovelace, Countess of, 10
Loyalty programs, 223–224
Lund, Robert, 439, 441
Luther, Martin, 32
Luthor, Lex (alias), 297
Luxury taxes, 397

M

M. A. Mortenson Company, 354–355
Macintosh computers, 36
Macrovision software, 177
Madison, James, 130
Mafiaboy (alias), 304
Magellan shopping carts, 224
Magnatune record label, 198
Magnuson-Moss Warranty Act, 352, 356
Mail Abuse Prevention System (MAPS), 115–117
Mainframe computers, 19
Major League Baseball, 384
Management principles in Software Engineering Code of Ethics, 425
Mankind's collective IQ, 52
Manufacturing jobs lost, 371
Many/any fallacy, 65
MAPS (Mail Abuse Prevention System), 115–117
Marconi, Guglielmo, 26
Mark-8 personal computer, 20
Marketplace: Households CD, 246
Markoff, John, 286, 289, 292
Mars missions, 333–335
Mason, George, 130
Mason, Jerald, 441
Massachusetts Institute of Technology
  admissions software for, 304
  hacking at, 292–293
Mauchley, John, 13–14
Maxwell, James Clerk, 26
May, Christopher, 175
Mayfield, Brandon, 243
Mazor, Stanley, 20
McAuliffe, Christa, 439–440
McDonald's, 299
McFarland, Michael, 446
McNealy, P.J., 177
McVeigh, Timothy, 232
Mechanical adding machines, 6–9
Media Center PC, 174
Media Desktop, 179
MediaCloq software, 177
Medicine, 5
Melissa virus, 282
Memex system, 33–34
Memory
  in EDVAC, 14
  in radar systems, 14–15
Menebrea, 10
Mental condition for moral responsibility, 343
Merchandise 7X formula, 163
Merck-Medco pharmacy, 371
Mercury Theater, 26–27
Method of differences, 8–9
MGM, 183–184
Miceli, Marcia, 443
Michelangelo virus, 282
Michigan Technical University, 182
Mickelson, Sig, 16
Microprocessors
  invention of, 20–22
  uses for, 3
Microsoft Corporation
  BASIC interpreter, 21
  DOS, 22
  India offices for, 384
  Linux pressure on, 192
  and security, 285–286
  and software piracy, 155–156, 195
  stock prices of, 387
  warranties of, 351–352
  Windows systems, 36, 285–286
Midbar Tech, 177
Middlemen, eliminating, 380
Milestones
  computing. *See* Computing milestones
  information storage and retrieval. *See* Information storage and retrieval milestones
  networking. *See* Networking

Mill, John Stuart
and act utilitarianism, 74
on censorship, 128
on pleasures, 143
on Principle of Harm, 129
and rule utilitarianism, 79
Mining, data, 244–246
household data, 246
IRS audits, 246
opt-in and opt-out, 248–249
P3P, 249
privacy violations from, 413
Syndrome Surveillance System, 246
TOA, 247–248
transaction information, 248
Missile system failures, 331–332
Mitnick, Kevin, 296
Models
computer, 344–347
technological diffusion, 391–392
Mondale, Walter, 46
Money, electronic, 259–261
Monitoring in workplace, 383–384
Monks, 32
Monty Python's Flying Circus, 111
Moon landings
faked, 139–140
transmission of, 27
Moor, James
core values, 54
interview with, 103–105
Moore, Gordon, 18, 20
Moore School, 13–14
Moral luck problem, 78
Moral responsibilities
characteristics of, 445–446
conditions for, 343
of software manufacturers, 357–359
Moral virtues, 429
Morality vs. ethics, 55–56
Moravec, Hans, 376–378
Morpheus technology, 179
Morris, Robert Tappan, Jr., 286–290
Morse, Samuel, 22–24
Morse code, 24
Mortenson v. Timberline Software, 352
Morton Thiokol, 439–441
Mosaic browser, 37
Mostly Harmless, 325
Mouse, invention of, 34
Mozilla Firefox browser, 37
MP3 spoofing, 181–182
MSN Search program, 39
MTS company, 20–21
Mueller, Scott, 119
Multinational teams, 384–385
Muris, Timothy, 220
Murphy, Eddie, 163
Murrah Federal Building bombing, 232
Music
legal Internet services, 184
piracy, 155–156, 173–174
Music Ace program, 352
Musical telegraph, 24
My Sweet Lord, 187

N

Nachi worm, 79–80
Nader, Ralph, 387
NAFTA (North American Free Trade Agreement), 385, 387
Napster network, 179, 184
Nardone v. United States, 236
NASA
Mars missions, 333–335
Space Shuttle disaster, 439–441
and WANK Worm, 284–285
National Archives, 122
National Bureau of Standards, 15
National Center for Supercomputer Applications, 37
National Conference of Commissioners on Uniform State Laws (NCCUSL), 355
National Consumer League, 356
National Crime Information Center (NCIC), 231–232, 327–328
National Do Not Call Registry, 220
National ID cards, 252–254
National Organization of Women, 133
National Security Agency (NSA)
and Clipper chip, 258
computers at, 259
Cuckoo's egg, 297
watch lists, 238
National Stolen Property Act, 303
Natural rights
privacy, 217–220
property, 157–158
software protection, 194
Navigation tables, 7–8
Navigator browser, 37
NCCUSL (National Conference of Commissioners on Uniform State Laws), 355
NCIC (National Crime Information Center), 231–232, 327–328
NCsoft company, 122
Near, Janet, 443
Negative rights, 84
Neidorf, Craig, 298–300
Nelson, Stewart, 292–295
Nelson, Ted, 34
Neonet technology, 179
Netscape Navigator browser, 37
Networked hypertext systems. *See* World Wide Web (WWW)
Networking
addiction to, 140–143
censorship in, 126–129
chat room predators, 137–139
and children, 132–135
discussion questions, 146–147
email, 109–115
false information in, 139–140
freedom of expression in, 129–132

Networking *(continued)*
further reading, 148
identity theft from, 135–136
in-class exercises, 147–148
introduction, 107–109
milestones, 22
ARPAnet, 28–30
broadband, 31
electricity and electromagnetism, 22–23
email, 29
Internet, 29
NSFNet, 31
radio, 26–27
remote computing, 28
telegraph, 23–24
telephone, 24–25
television, 27–28
typewriters and teletypes, 25–26
peer-to-peer, 178–184
pornography ethical perspectives, 124–126
references, 148–152
review questions, 145–146
spam fighting on, 115–120
summary, 144–145
World Wide Web, 120–124
Neville, Robert, 215
New Jersey Institute of Technology, 182
New York Times, The, 300
News agencies, self-censorship by, 127
Newspapers, 32
NeXT computer, 36–37
Nicomachean Ethics, The, 429
911 service, 225–226
Nipkow, Paul, 27
Nixon, Richard, 238
NLS system, 34
No Electronic Theft Act, 167–168
Nokia 6800, 3
Nonpolitical speech, freedom of, 131
Normalization model, 391–392
Norms, 61
Norris, Pippa, 389–390
North American Free Trade Agreement (NAFTA), 385, 387
North Korea, Internet access in, 123
NOT operators in Boolean algebra, 10–11
Noyce, Robert, 18, 20
Nozck, Robert, 194
NSA (National Security Agency)
and Clipper chip, 258
computers at, 259
Cuckoo's egg, 297
watch lists, 238
NSFNet, 31
Nuclear power plants, 5

O

Oakley, Justin, 430
Oberlin College, 283
Object-oriented design, 349
Objectivism, 90
Oersted, Christian, 23
Off-line digital cash, 260
Office of the Inspector General, 442
Office Workstations, Ltd., 36
Oil exploration, 344–345
Oklahoma City bombing, 232
Olmstead, Roy, 235
Olmstead v. United States, 235
Olson, Theodore, 133
On-line communities, 25, 153–154
On-line digital cash, 260
On-line voting
ethical evaluations for, 310–313
motivation for, 308–309
proposals for, 309–310
Online Policy Group v. Diebold, 209
Online System, 34
Ontario Cancer Foundation, 338
Open Directory search engine, 38
Open Source Initiative, 189
Open-source software, 188
benefits of, 189–190
criticisms of, 192–193
definition of, 189
examples of, 190–191
GNU Project and Linux, 191–192
impact of, 192
vs. proprietary, 188–189
Operation Shamrock, 237–238, 241
Operation Sundevil, 302
Opt-in lists, 111, 117
Opt-in policies, 248–249, 276
Opt-out policies, 248–249, 276
Optimists, technological, 392
OR operators in Boolean algebra, 10–11
Oracle, 384
Ordeal, reputation through, 221
Organization changes in workplace, 379–381
Orkut software, 154
Orzack, Maressa, 141
Overflow, buffer, 287–288, 322
Overpeer company, 182
Ovid, 126
Oxfam, 387

P

P3P (Platform for Privacy Preferences), 249
Pacifica Foundation, 131–132
Packet-switched networking, 29–30
Packets
email, 109–110
filtering, 292
Palm Beach County "butterfly ballots", 309
Palm Beach County Probation Department, 298
Palo Alto Research Center (PARC), 34–35
Paper, invention of, 32
Papert, Seymour, 42
Papua, 375
Papyrus scrolls, 31

Paradoxes, intellectual property, 159–161
Paramount Pictures, 163
PARC (Palo Alto Research Center), 34–35
Parchment, 31
Parker, John, 22
Pascal, Blaise, 6
Passwords
  for authentication, 291
  hacking, 295
Patches, security, 292, 322–323
Patent and Trademark Office, 186
Patents
  purpose of, 165
  software, 186–187
Patriot Act, 212, 240
  follow-on legislation, 242
  provisions of, 240–241
  renewal of, 243–244
  responses to, 241–242
  successes and failures, 243
Patriot Act II, 242
Patriot missile system failures, 331–332
Pay ratios, CEOs to production worker, 394
PayPal service, 135–136
PC bangs, 122
PDP-1 computer, 293–295
Pedophilia, 137–139
Peele, Stanton, 142–143
Peeping Toms, high-tech, 223
Peer-to-peer networks, 178–184
Pen registers, 239–240, 242
Pennsylvania State University, 182–183
Pentagon, terrorist attack on, 240
Pentagon Papers, 300
People's Republic of China
  dictatorships in, 387
  IBM assets acquired by, 389
  Internet access in, 123–124
  IT development in, 389
  job movement to, 385
Perfect duty, 400
Perl language, 191
Persistent online games, 122
Personal AI machines, 378
Personal computers, invention of, 21–22, 34–36
Personal information, 222
Personal video recorders (PVRs), 174
Pessimists, technological, 392
Petersen, Joe, 137
Pew Surveys, 391
PGA Tour, 396
PGP (Pretty Good Privacy), 255, 258–259
Pharmacies, online, 371
Philippines, job movement to, 385
Phishing, 135, 250
Phoenix Project, 300
Phone phreaking, 296–297
Phone-sex hotline, 298–299
Phonebook services
  Google, 211
  ProCD, 352–354
PHP language, 191
Phrack, 298, 300
Physical effects, 3
Ping messages, 306
Pinto, Ford, 446
Piracy, 155–156, 173–174
Pirate boards, 296
Platform for Privacy Preferences (P3P), 249
Plato
  Crito, 59
  Euthyphro, 68
  Republic, The, 100, 103
Player Piano, 370–371
Player piano rolls, 168–169
Plovernet BBS, 297
Podcasting, 154
Polaroid Corporation, 165
Police sting operations, 138–139
Poor and rich, growing gap between, 392
Pornography
  child pornography scenario, 435–436
  ethical perspectives on, 124–126
  in spam, 112
  Web restrictions on, 123–124, 412
Port Washington Community Concert Association, 167
Positive rights, 84
Possessions vs. time, 376
Postal Service, 329
Potter, Stewart, 237
Precision Bid Analysis package, 354–355
Predators, chat room, 137–139
Predictions, simulations for, 344–346
Preece, Jennifer, interview with, 153–154
Prepublication reviews, 126
Presidential elections, U.S.
  1896, 25
  1952, 16–17
  1984, 46
  2000, 27–28, 308–309
Pretty Good Privacy (PGP), 255, 258–259
Prime numbers, 108
Princeton University, 182
Principle of Harm, 129
Principle of Utility, 74–75, 79
Principles of justice, 85–86
Printing press, 32, 126
Prior restraints on expression, 130
Privacy, 211–213
  for autonomous moral agents, 219–220
  benefits of, 215–216
  Cavoukian interview, 275–277
  covert government surveillance, 234–238
  data mining, 244–249
  defining, 213–214
  disclosing information, 221–223
  discussion questions, 266–268
  DRM effect on, 178
  encryption. *See* Encryption
  further reading, 269
  harms of, 214

Privacy *(continued)*
historical evolution of, 220–221
in-class exercises, 268–269
ladder of, 216
legislation on, 228–230, 238–244
natural right to, 217–220
Patriot Act, 240–244
as prudential right, 220
public information, 223–228
public records, 230–234
references, 269–274
review questions, 264–266
rights associated with, 218–219
summary, 262–264
and telemarketing, 220
Privacy Act of 1974, 232–234, 327
Privacy Rights Clearinghouse, 250
Problem of moral luck, 78
ProCD, Inc. v. Zeidenberg, 352–354
Proctor & Gamble, 222–223
Product principles in Software Engineering Code of Ethics, 423–424
Production workers, pay ratios to CEOs, 394
Productivity
from automation, 375–376
changes in, 371
monitoring for, 384
Profession principles in Software Engineering Code of Ethics, 425–427
Professional development, 416
Professional ethics, 415–416
Axtell interview, 455–457
case studies, 433–434
anti-worm, 436–438
child pornography, 435–436
software recommendation, 434–435
discussion questions, 449–450
further reading, 451
in-class exercises, 450–451
references, 451–453
review questions, 448–449
software engineering. *See* Software Engineering Code of Ethics and Professional Practice
summary, 446–447
virtue ethics, 429–432
Whistleblowing, 438–439
Hughes Aircraft case, 441–442
morality of, 443–446
Morton Thiokol case, 439–441
and organizational failure, 455–456
Professional societies, 416
Professions
characteristics of, 416–417
CPAs, 417–418
software engineering as, 418–419
Professor Publishing business, 167
Programming, addiction to, 140–141
Programming languages for security, 322
Property
intellectual. *See* Intellectual property
natural right to, 157–158
Proprietary software, 188–189
Protestant Ethic and the Spirit of Capitalism, The, 375
Protestant Reformation, 32, 375
Protestant work ethic, 375–376
Prudence virtue, 431
Prudential rights, 220
Psychology of Computer Programming, The, 415
Public college access, 397
ethical analysis, 398–400
tuition increases, 397–398
Public domain, 162
Public information, 222
911 service, 225–226
automobile black boxes, 225
body scanners, 224–225
cookies, 227–228
digital video recorders, 224–225
implanted chips, 227
rewards programs, 223–224
RFIDs, 226–227
spyware, 228
Public key cryptography, 255
Public life vs. private life, 214
Public principles in Software Engineering Code of Ethics, 421–422
Public records, 222, 230
census records, 230–231
IRS, 231
NCIC, 231–232
Privacy Act of 1974, 232–234
Purchasing power, 373–374
Puritans, 376
PVRs (personal video recorders), 174
Pythagoras, 155
Python language, 191

Q

Quality, software, 350
Quartering Act, 217
Queries, Boolean, 11
Qwest company, 328

R

Race conditions, 340–342
Rachels, James
on core values, 65
on privacy, 215
on social contract theory, 83
on virtues, 429–430
Radar systems, 14–15
Radiation treatments. *See* Therac-25 linear accelerator
Radio
Internet, 174
invention of, 26–27
Radio frequency ID (RFID) tags, 226–227, 254
Radio Shack computers, 21
Railroad Tycoon software, 351

Ratings systems, voluntary, 127
RATs (remote access Trojans), 291, 311
Rawls, John
on freedom of expression, 135
principles of justice, 85–86, 400
Ray, James Earl, 232
Raymond, Eric, 190
RBL (Realtime Blackhole List), 115–117
RCA Communications, 237
Reagan, Ronald, 46, 229
Real ID Act, 253–254
Real-time systems, 331
Realtime Blackhole List (RBL), 115–117
Recording Industry Association of America (RIAA)
Diamond Multimedia Systems lawsuit, 171–172
Felton letter, 175
lawsuits against individuals, 155, 179–180, 182, 302
Napster lawsuit, 179
Verizon subpoena, 180
Red Book standard, 177
Reddick, Shirley, 442
Reformation, Protestant, 32, 375
Refrigerators, 4
Registration, censorship through, 126–127
Regulation of Internet, 411
Rehnquist, William, 133
Reid, Richard, 243
Reid, Samuel, 22
Reiman, Jeffrey
on privacy, 215–216, 219–220
on vices, 143
Reinertson, Mark, 118
Relativism
cultural, 62–66
subjective, 60–62
Reliability, 325–326
data-entry and data-retrieval errors, 326–328
discussion questions, 363–364
further reading, 365
in-class exercises, 364
references, 365–367
review questions, 361–363
simulations, 344–347
software and billing errors, 328–330
software engineering, 347–350
software system failures, 330–336
software warranties, 350–351
enforceability of, 352–355
moral responsibilities, 357–359
shrinkwrap software, 351–352
UCITA effect on, 355–357
summary, 359–361
Therac-25. *See* Therac-25 linear accelerator
Religion in divine command theory, 66–69
Remington & Company, 25
Remington Rand Corporation, 16
Remote access Trojans (RATs), 291, 311
Remote computing, 28
Rensselaer Polytechnic Institute, 182
Reprogramming ENIAC, 14
Republic, The, 100, 103
Responsibility
characteristics of, 445–446
conditions for, 343
of software manufacturers, 357–359
Retinal scans, 291
Return to Castle Wolfenstein, 122
Reuse of code, 343
Reverse engineering, 163
Rewards programs, 223–224
RFID (radio frequency ID) tags, 226–227, 254
Rhapsody music service, 184
RIAA. *See* Recording Industry Association of America (RIAA)
RIAA v. Diamond Multimedia Systems, Inc., 171–172
Rich and poor, growing gap between, 392
Richardson, William, 232–233
Ridge, Tom, 243
Ries, Richard, 142
Riggs, Robert, 298–299
"Right to Privacy, The", 217
Rights
intellectual property, 159–162
privacy, 217–220
property, 157–158
in social contract theory, 84
and software protection, 193–194
Ring of Gyges, 100, 103
Ringo software, 154
Rio portable music player, 171–172
Ripskis, Al, 445
Rivest, Ron, 256
Roberts, Lawrence, 29
Robotics, Three Laws of, 370
Robots, 376–379
Roentgen, Wilhelm, 336
Rogan, Terry Dean, 327
Role playing games, 122
Role responsibility, 445
Roosevelt, Franklin, 236, 251
Rosenberg, Alexander, 220
Ross, W. D., 78
Rousseau, Jean-Jacques, 83–84
Routers, email, 109–110
Roving surveillance, 240–241
Roving wiretaps, 239
Royal Astronomical Society, 8–9
RSA encryption, 255–257
Ruby language, 191
Rule utilitarianism
basis of, 79
case against, 81–82
case for, 80–81
ethical evaluations in
PDP-1 unauthorized update, 294–295
scenario for, 79–80
spam, 114
law breaking in, 92–93
Russell, Steve, 293

S

S&H Green Stamps, 223
SafeAudio program, 177
Safeway stores, 224
Saffo, Paul, 34
Saltzman, Eric, 197
Sandwich Islands, 375
Sapphire (Slammer) worm, 285
Sarnoff, David, 26
Sasser worm, 286
SATAN (Security Administrator Tool for Analyzing Networks), 308
Saturday Evening Gazette, 217–218
Saudi Arabia, Internet access in, 123
Scanlan, Michael, 159
Scanners, body, 224–225
Scantlebury, Roger, 29
Schaeffer, Rebecca, 212
Scheutz, Georg and Edvard, 9
Schneier, Bruce, 260–261, 312
Schoeman, Ferdinand, 214
Schools, monitoring in, 384
Schor, Juliet, 371, 375
Science Museum of London, 9
Scott, Gini Graham, 215
Scrolls, papyrus, 31
Scud missiles, 331–332
SDMI (Secure Digital Music Initiative), 175–176
Search engines, 38–39
SEC (Securities and Exchange Commission), 238
Second Treatise of Government, The, 157
Secondary use of data, 244
Secret Service, 299–302
Secrets, trade, 163–164
Secure Digital Music Initiative (SDMI), 175–176
Secure Electronic Registration and Voting Experiment, 310
Securepoint company, 286
Securities and Exchange Commission (SEC), 238
Security, 279–280
  Bishop interview, 321–323
  defensive measures, 291–292
  denial-of-service attacks, 304–308
  discussion questions, 315–316
  further reading, 317
  hackers. *See* Hackers
  in-class exercises, 316–317
  monitoring for, 384
  online voting, 308–312
  phone phreaking, 296–297
  vs. privacy, 276
  references, 318–320
  review questions, 314
  summary, 313–314
  Trojan horses, 290–291
  viruses, 280–283
  worms. *See* Worms
Security Administrator Tool for Analyzing Networks (SATAN), 308
Sega Genesis game console, 186
Sega v. Accolade, 186
Segregation protests, 88
SelectPhone product, 353–354
Self-censorship, 127
Self principles in Software Engineering Code of Ethics, 427–428
Seltzer, Wendy, interview with, 209–210
Semaphore telegraph, 22–23
sendmail application, 191, 287, 289
Servers, email, 109
Service marks, 164
SESAC, 166
SETI@home project, 108
Shadow length computations, 6–7
Shakespeare, William, in *Hamlet* scenario, 159–161
Shamir, Adi, 256
Shamrock, Operation, 237–238, 241
Shanghai Jiao Tong University, 389
Shannon, Claude, 10, 28
Shedd, John, 279
Sherman, Cary, 302
Shockley, Bill, 17–18
Shockley Semiconductor, 18
Shockwave Rider, The, 283
Sholes, Christopher, 25
Shopping, online, 121
Shoulder surfing, 250, 296
Shrinkwrap software, 351–352
Signal Security Agency, 237
Signatures for digital cash, 260
Silken Mountain Web Services, Inc., 354
"Simple Gifts", 196
Simulations, 344–347
Single-computer hypertext systems, 36–37
Sit-ins, 88
60 Minutes, 140
Skills development, 416
Skimmers, 250
Slammer (Sapphire) worm, 285
Slavery, 378
Sleep deprivation, 141
Small-Scale Experimental Machine, 14–15
Smallpox cure, 108
Smith, David L., 282
Smith, Paul, 133
Smurf attacks, 306
Social Contract, The, 83
Social contract theory, 82
  case against, 89–90
  case for, 87–88
  ethical evaluations in
    CIPA, 134–135
    Internet addiction, 143
    Internet worm, 289
    MAPS, 116
    PDP-1 unauthorized update, 294
    public college access, 400
    scenario for, 87
    spam, 114
    sting operation, 139
  law breaking in, 91
  social contract in, 82–84
  theory of justice, 84–86
Social divide, 389, 391

Social engineering, 295–296
Social equality, 84–86
Social Security Act, 251
Social Security numbers, 251–252
Societies
in cultural relativism, 62–66
definition of, 54–55
professional, 416
winner-take-all, 394–397
Socrates, 59, 68
Software
errors in, 328–330
open-source. *See* Open-source software
piracy, 155–156, 173–174
protections for, 184–185
copyrights, 185–186
legitimacy of, 193–196
patents, 186–187
safe development, 187
quality of, 350
shrinkwrap, 351–352
software system failures, 330–331
Ariane, 332–333
AT&T long distance network collapse, 333
baggage handling systems, 335–336
Mars missions, 333–335
Patriot missile system, 331–332
warranties, 350–351
enforceability of, 352–355
moral responsibilities, 357–359
shrinkwrap software, 351–352
UCITA effect on, 355–357
Software Engineering Code of Ethics and Professional Practice, 419–420
analysis of, 428–433
case studies, 433–434
anti-worm, 436–438
child pornography, 435–436
software recommendation, 434–435
discipline-independent rules in, 432–433
preamble, 420–421, 428–429
principles, 421–428
Software engineers, 347–350
definition of, 416
as professionals, 416–419
Software Link, The, 353, 355
Software Patent Institute, 187
Software recommendation scenario, 434–435
Sony, 170–171
Sony v. Universal City Studios, 170–171, 183–184
Soo, Kim Hyun, 122
Soule, Samual, 25
South Korea
gaming in, 122
home shopping in, 121
Southbound on the Freeway, 1–2
Soviet Union, censorship in, 126
Space shifting, 171–172
Space Shuttle
Challenger disaster, 439–441
and WANK Worm, 284–285
Spacewar game, 293
Spain, terrorist attack in, 243
Spam, 39
CAN SPAM Act, 118–120
ethical evaluations of, 112–115
growth of, 110–111
MAPs for, 115–117
popular subject lines for, 110
proposed solutions to, 117–118
and spim, 120
tactics, 111–112
Spamhaus company, 304
Spann, Bill, 182
Sparapani, Timothy, 254
SPARC architecture, 165
Spear, Tony, 334–335
Spears, Britney, 155
Specialization, 53
Specification process for software, 347–348
Spector PRO program, 132
Speculation, 387–388
Spider programs, 38
Spim, 120
Spitzer, Robert, 444
Spock, Benjamin, 238
Spoofing
IP, 306–307
MP3, 181–182
spam, 112–113
Spreadsheet programs, 21
Spyware, 228
SQL Server, 285
Stack attacks, 288
Stacks, HyperCard, 36–37
Stallman, Richard, 188, 192, 195
Standard of living, 376
Star Chamber, Court of, 129–130
State of nature, 83
Step Reckoner, 6–7
Step-Saver Data Systems v. Wyse Technology and The Software Link, 352–353
Sterling, Bruce, 292, 301–302
Steve Jackson Games, 300–301
Stevenson, Adlai, 16–17
Stibitz, George, 28
Sting operations, 138–139
Stoll, Clifford, 297
Stored-program concept, 14
Stossier, Sheila Jackson, 327
Stratification model, 391–392
StreamCast, 180, 183–184
Sturgeon, William, 23
Subjective relativism, 60–62
SubSeven RAT, 291, 311
Sudduth, Andy, 287, 289
Sumner, William Graham, 62
Sun Microsystems
India offices for, 384
SPARC architecture, 165
Supply-chain automation, 380–381
Supreme Court, United States
Bork nomination, 229
Diamond v. Diehr, 186
FCC v. Pacifica Foundation et al., 132

Supreme Court, United States *(continued)*
Katz v. United States, 237
Nardone v. United States, 236
Ohmstead v. United States, 235
Sony v. Universal City Studios, 170–171, 183–184
United States v. American Library Association, 133
Weiss v. United States, 236
White-Smith Music Publishing Co. v. Apollo Co., 168–169
wiretapping, 239
Surity 3600 device, 258
Surveillance, covert government, 234–238
Sweeney, Harold, 17
Swenson, May, 1–2
Sykes, Charles, 215, 221
Symmetric encryption, 254–255
SYN flood attacks, 305–306
Syndromic Surveillance System, 246
Syrus, Publilius, 211
System/360, IBM, 19
System administrator responsibilities, 291–292, 307
System malfunctions, 328–329

T

T-Mobile, 251
Tables, navigation, 7–8
Tandy computers, 21
Tarbox, Katie, 137
Task Force on Heroin Suppression, 238
Taxes
online payments, 122
for winner-take-all effect reduction, 397
TCL/TK language, 191
TCP (Transmission Control Protocol), 29
TCP/IP protocol, 29, 37
Tech Model Railroad Club, 292–293
Technological diffusion, 391–392
Technology, dependence on, 3–4
Telecommuting, 40
Telegraph
harmonic, 24
invention of, 23–24
semaphore, 22–23
Telemarketing, 220
Telephone Consumer Protection Act, 118
Telephone systems, 24–25
Clipper chips for, 258
hacking, 293
phone phreaking, 296–297
Teletype machines, 25–26
Television
closed circuit, 223
invention of, 27–28
Telework, 381
advantages of, 381–382
disadvantages of, 382–383
Templeton, Brad, 111
Temporary work, 383
TeraGrid network, 303
Teresa, Mother, 61
Terrorism
and Patriot Act, 240–244
and TIA project, 247–248
TEX typesetting system, 191
Texas Instruments
India offices for, 384
integrated circuits developed by, 18
Theory of justice, 84–86, 400
Therac-25 linear accelerator, 336–337
accidents, 337–340
analysis, 342–343
harm from, 419
moral responsibility for, 343
origin of, 337
software errors in, 340–342
Third Amendment, 217
Thomson, Judith Jarvis, 218–219
Three Laws of Robotics, 370
Three-way handshakes, 305
Thukral, Shilpa, 369
TIA (Total Information Awareness) project, 212, 247–248
Tiger teams, 379
Timberline Software, Inc., 354–355
Time shifting, 170–171
Time vs. possessions, 376
Title III of the Omnibus Crime Control and Safe Streets Act, 239
Titus, Jon, 20
TiVo, 224–225
Tolerance, 61
Tollbooth information, 245–246
Tomlinson, Ray, 29
Tootill, Geoff, 15
Torah, 66
Torvalds, Linus, 192
Total Information Awareness (TIA) project, 212, 247–248
Trade secrets, 163–164
Trade unions, 387
Trademarks, 164
Trans Union credit reporting agency, 228
Transactions
illicit, 261
privacy of, 247
Transistors, 17–18
Transmission Control Protocol (TCP), 29
Trap-and-trace device, 239–240
Trigonometric functions, 6–7
Trilobite robot, 377
Trojan horses
examples, 290–291
minimizing, 322
Trompenaars, Fons, 62
TRS-80 computers, 21
True values in Boolean algebra, 10–11
Truncation errors, 331–332
Tsinghua University, 389
Tuition increases, 397–398
Turner, Clark, 342, 350
Twain, Mark, 26
2600 Magazine, 176
Typewriters, 25–26
Tysver, Daniel A., 184

U

UCC (Uniform Commercial Code), 352, 355
UCITA (Uniform Computer Information Transaction Act), 355–357
Umpire Information System, 384
Unemployment from automation, 370–379
Uniform Commercial Code (UCC), 352, 355
Uniform Computer Information Transaction Act (UCITA), 355–357
Uniform Resource Locators (URLs), 121
United Devices, 108
United Nations, 378
United States Naval Academy, 182
United States Public Interest Research Group, 356
United States v. American Library Association, 133
UNIVAC computer, 15–17
Universal City Studios, 170–171, 183–184
Universal Declaration of Human Rights, 378
Universal law in Categorical Imperative, 70
Universities, music sharing at, 182–183
University of California at Berkeley, 108
University of Kansas, 303
University of Missouri, 298
University of Nevada, 303
University of Phoenix, 122
Unix operating system
  GNU project, 191
  security holes in, 286–289
URLs (Uniform Resource Locators), 121
USA PATRIOT Act, 240
  follow-on legislation, 242
  provisions of, 240–241
  renewal of, 243–244
  responses to, 241–242
  successes and failures, 243
USA Today, 296
User identifiers, 291
Utilitarianism
  act, 74–78
  ethical evaluations in
    CIPA, 134
    Internet addiction, 143
    Internet worm, 289–290
    MAPS, 116–117
    on-line voting, 310–313
    PDP-1 unauthorized update, 294–295
    public college access, 399–400
    spam, 113–114
    sting operation, 138
  law breaking in, 92–93
  rule, 79–82
  on software protection, 194
Utility, Principle of, 74–75, 79

V

Vacuum tubes for signal amplifiers, 17
Vaidhyanathan, Siva, 168
Validating
  simulations, 345–347
  software, 349–350
Variable attacks, 288
Vellum, 31
Verifying simulations, 345
Verizon, 181
Video games, first, 293
Video Privacy Protection Act, 229
Video recorders, digital, 224–225
Vietnam war
  Pentagon Papers, 300
  watch lists during, 238
Virginia Declaration of Rights, 130
Virtual worlds, 122
Virtue ethics, 429–430
  complement to other theories, 431–432
  strengths of, 430–431
  weaknesses of, 431
Viruses
  current, 283
  minimizing, 322
  spread of, 280–283
  well-known, 282–283
Visas, work, 388
VisiCalc program, 21
VisiCorp Corporation, 36
Vivendi Universal, 312
Volta, Alessandro, 22
Voltaire, 369
Voluntary information disclosures, 222
von Leibniz, Gottfried, 6–7
von Neumann, John, 14
Vonnegut, Kurt, Jr., 370–371
Voter News Service, 28
Voters, disfranchised, 326
Voting, on-line, 308–312
Voting machines, 209

W

Wages
  CEOs vs. production worker, 394
  and globalization of jobs, 369
Walt Disney Productions, 171
WANK worm, 284–285
War in Iraq
  looting in, 82
  self-censorship in, 127
War of the Worlds, 26–27
WarGames, 295
Warner, John, 247
Warranties, software, 350–351
  enforceability of, 352–355
  moral responsibilities, 357–359
  shrinkwrap software, 351–352
  UCITA effect on, 355–357
Warren, Samuel, 217–218
Warschauer, Mark, 392–393
Washington Post, 300
Watch lists for crime figures, 238
Watson, Thomas A., 25
WBAI radio station, 131
Wealth
  college access, 397–400
  digital divide, 389–393

Wealth *(continued)*
discussion questions, 403–404
dot-com bust, 387–388
further reading, 406
in-class exercises, 404–406
references, 406–409
review questions, 403
summary, 401–402
winner-take-all society, 394–397
Web. *See* World Wide Web (WWW)
Weber, Max, 375
Wedges, 250
Weinberg, Gerald, 415
Weiss v. United States, 236
Weizenbaum, Joseph, 140–141
Welles, Orson, 26–27
Western Union, 237
Whistleblower Protection Act, 438–439
Whistleblowing, 438–439
Hughes Aircraft case, 441–442
morality of, 443–446
Morton Thiokol case, 439–441
and organizational failure, 455–456
White-collar jobs lost, 371
White-Smith Music Company, 168
Whitman, Walt, 107
Wiggins, Calvin, 441
Wikipedia, 122
Williams, F. C., 15
Williams, Samuel, 28
Williams Tube, 15
Windows Media Player, 178
Windows systems, 36, 285–286
Winner-take-all society
development of, 394–395
harmful effects of, 395–397
reducing harm from, 397
Winner-Take-All Society, The, 394
Wire Fraud Act, 303
Wireless networks, 384
Wiretapping
history of, 235–237
legislation authorizing, 238–244
roving, 239
Wolpe, Howard, 331
Wood engraving, 32
Woolworth's store, 88
Word documents, 282
Work ethic, Protestant, 375–376
Workplace. *See also* Jobs
monitoring in, 383–384
multinational teams, 384–385
organization changes in, 379–381
telework, 381–383
temporary work, 383
World Trade Center, 240
World Trade Organization (WTO), 385–387
World War II, 12–15, 236–237
World Wide Web (WWW)
attributes of, 120–121
children on, 132–135
controlling, 123–124
as democracy tool, 40
development of, 36–38
false information in, 139–140
filters for, 132–133
open-source servers, 191
popularity of, 3
search engines, 38–39
uses of, 121–123
Worms
anti-worm scenario, 436–438
Blaster, 285
Code Red, 285
definition of, 283–284
instant messaging, 286
Internet, 286–290
minimizing, 322
Sapphire, 285
Sasser, 286
WANK, 284–285
Wozniak, Steve, 35
Writs of assistance, 234
WTO (World Trade Organization), 385–387
WWW. *See* World Wide Web (WWW)
Wyden, Ron, 119
Wyse Technology, 353

X

X-rays. *See* Therac-25 linear accelerator
X Window System, 191
Xanadu network, 34
Xerox Corporation, 35–36, 164
Xerox PARC, 34–35

Y

Yahoo, DoS attacks against, 304
Yakima Valley Memorial Hospital, 338, 340–341
Yellow Book standard, 177
Young, Kimberly, 141–142

Z

Z series computers, 11–12
Zeidenberg, Matthew, 353–354
Zennistrom, Niklas, 179
Zilog Z80 microprocessors, 21
Zimmerman, Phil, 255, 258
Zoning laws, 131
Zope language, 191
Zuse, Konrad, 11–12